Clinical Anatomy of the Knee

Murat Bozkurt • Halil İbrahim Açar
Editors

Clinical Anatomy of the Knee

An Atlas

Springer

Editors
Murat Bozkurt
Department of Orthopaedics and Traumatology
Faculty of Medicine
Ankara Yildirim Beyazit University
Ankara, Turkey

Halil İbrahim Açar
Department of Anatomy
Faculty of Medicine
Ankara University
Ankara, Turkey

ISBN 978-3-030-57577-9 ISBN 978-3-030-57578-6 (eBook)
https://doi.org/10.1007/978-3-030-57578-6

This Springer imprint is published by the registered company Springer Nature Switzerland AG
The registered company address is: Gewerbestrasse 11, 6330 Cham, Switzerland

Preface

Anatomical treatment methods for the knee joint have recently become more preferred. Our primary goal in editing *Clinical Anatomy of the Knee* is to create a valuable resource that includes a rich visual content for those physicians, residents, fellows, or students practicing or interested in knee problems. With this book, we have combined the detailed anatomy of the knee joint with the biomechanics and radiology of the knee joint, and we have correlated all of this basic information together with some of the treatment methods that we have applied. In particular, we think that this transfer of basic knowledge to clinical applications will be used effectively in both diagnosis and appropriate treatment practices.

We would like to express our sincere gratitude to all of the authors who contributed to this book.

Ankara, Turkey

Murat Bozkurt, MD, PhD
Halil İbrahim Açar, MD

Contents

Functional Anatomy of Knee 1

Halil İbrahim Açar, Yiğit Güngör,
and Murat Bozkurt

1.1 Introduction

The knee joint is the largest joint of the body. It includes many important structures, specific to the knee such as the menisci and cruciate ligaments. Another important feature is the joint surfaces that are not highly compatible to bring together bones. To increase compatibility and provide stability, there are several certain structures in the joint.

The knee joint is basically formed between the tibia and the femur. The patellofemoral joint, which is made of the femur and patella, is a part of the knee joint with very important properties. Although the fibula is not a direct part of the knee joint, it constitutes a significant area holding important ligaments and muscles related to the joint [1, 2].

In this section, different aspects of the knee are considered. First, the properties of the bony structures in the knee joint are defined. Then, the anatomic structures are evaluated layer by layer from the perspective of the dissector, and the relationships between them are emphasized in integrity.

H. İ. Açar (✉) · Y. Güngör
Department of Anatomy, Faculty of Medicine, Ankara University, Ankara, Turkey

M. Bozkurt
Department of Orthopaedics and Traumatology, Faculty of Medicine, Ankara Yildirim Beyazit University, Ankara, Turkey

M. Bozkurt, H. İ. Açar (eds.), *Clinical Anatomy of the Knee*,
https://doi.org/10.1007/978-3-030-57578-6_1

1.2 Bones

1.2.1 Distal End of the Femur

Femur is the longest and largest bone of the body. It extends from superior to inferior, from lateral to medial, slightly oblique. The anatomical axis of femur passes between the shaft of femur and intercondylar notch. It extends slightly medially, to 9° angle between the vertical axis. The mechanical axis passes between the center of the head of femur and the intercondylar notch. There is a 3° angle between the mechanical axis and the vertical axis (Fig. 1.1) [1–4].

Femur articulates with the tibia via its condyles and with the patella via the patellar surface.

The lateral and medial femoral condyles are the most significant structures observed in the distal femur. Compared to the lateral condyle, the medial condyle extends further distally. However, in anatomical position, as the femur shaft lies obliquely from lateral to medial, both the condyles end at the same horizontal level (Fig. 1.1) [1]. The femoral condyles are not symmetrical. The sagittal axis of the lateral condyle is longer than the medial. The lateral condyle axis is located in the sagittal plane. However, there is an angle of approximately 22° between the medial condyle axis and the sagittal plane (Fig. 1.2).

The most prominent point on the outer surface of the lateral condyle is the lateral epicondyle

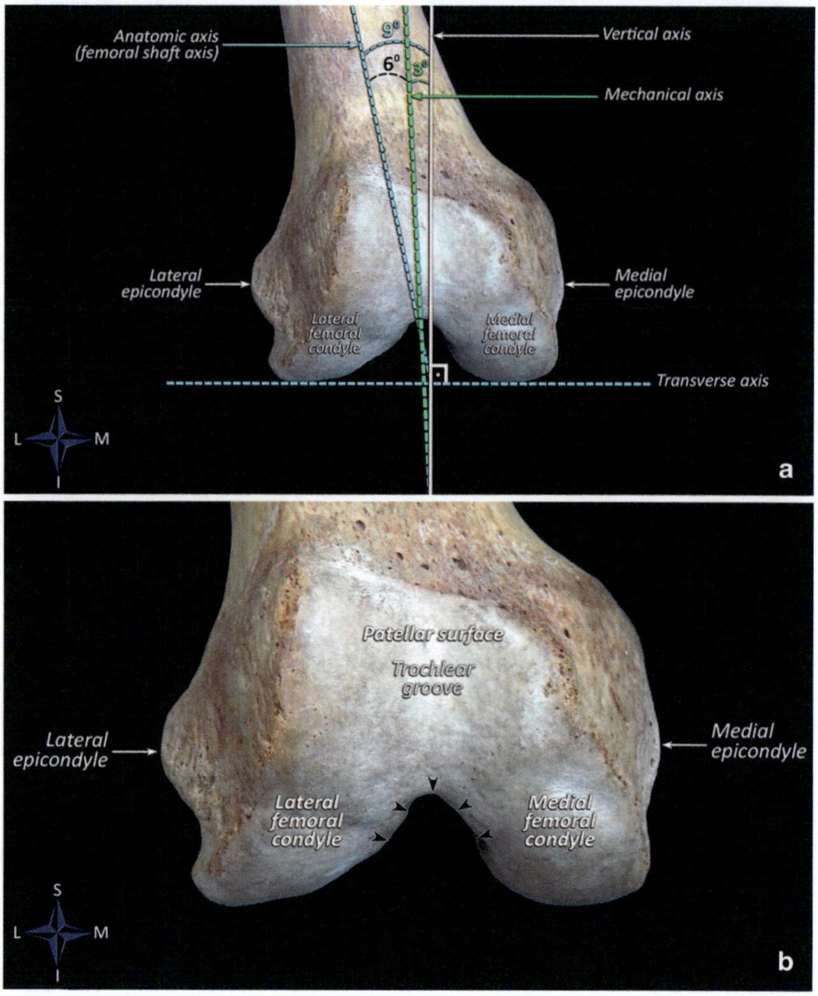

Fig. 1.1 Anterior view of the right distal femur. (**a**) The axes of the femur. (**b**) Close-up view of the distal femur. Black arrowheads indicate anterior border of the intercondylar notch. *S* superior, *I* inferior, *L* lateral, *M* medial, on the star showing directions

(Fig. 1.2). The lateral collateral ligament (LCL) attaches to just proximal and posterior to the lateral epicondyle of the femur (Fig. 1.3) [2, 5, 6]. Immediately below the lateral epicondyle, a shallow groove is observed, in which the tendon of popliteus passes. The popliteus tendon inserts to the outer surface of the lateral condyle on the portion immediately anterior-inferior to the lateral epicondyle (Fig. 1.3) [2, 6, 7]. Another important structure attached to the outer surface of the lateral femoral condyle is the lateral head of the gastrocnemius muscle [6]. This tendon originates posterior and superior to the attachment site of the lateral collateral ligament (Fig. 1.3). The medial surface of the lateral femoral condyle forms the lateral wall of the intercondylar notch.

The medial epicondyle, which is the attachment area of the medial collateral ligament, is located on the medial surface of the medial femoral condyle (Fig. 1.4) [2, 8, 9]. The adductor tubercle where the adductor magnus tendon inserts is located superior and posterior to the medial epicondyle (Fig. 1.4). As on the lateral side, the origin of the medial head of the gastrocnemius muscle is in the posterior-superior part of the medial condyle (Fig. 1.4) [2, 8, 9]. The lateral surface of the medial condyle forms the medial wall of the intercondylar notch.

Intercondylar notch is located between the condyles (Fig. 1.5). This notch contains the attachment areas of the anterior and posterior cruciate ligaments [1, 2]. The cartilage covering the trochlear groove forms the anterior border of

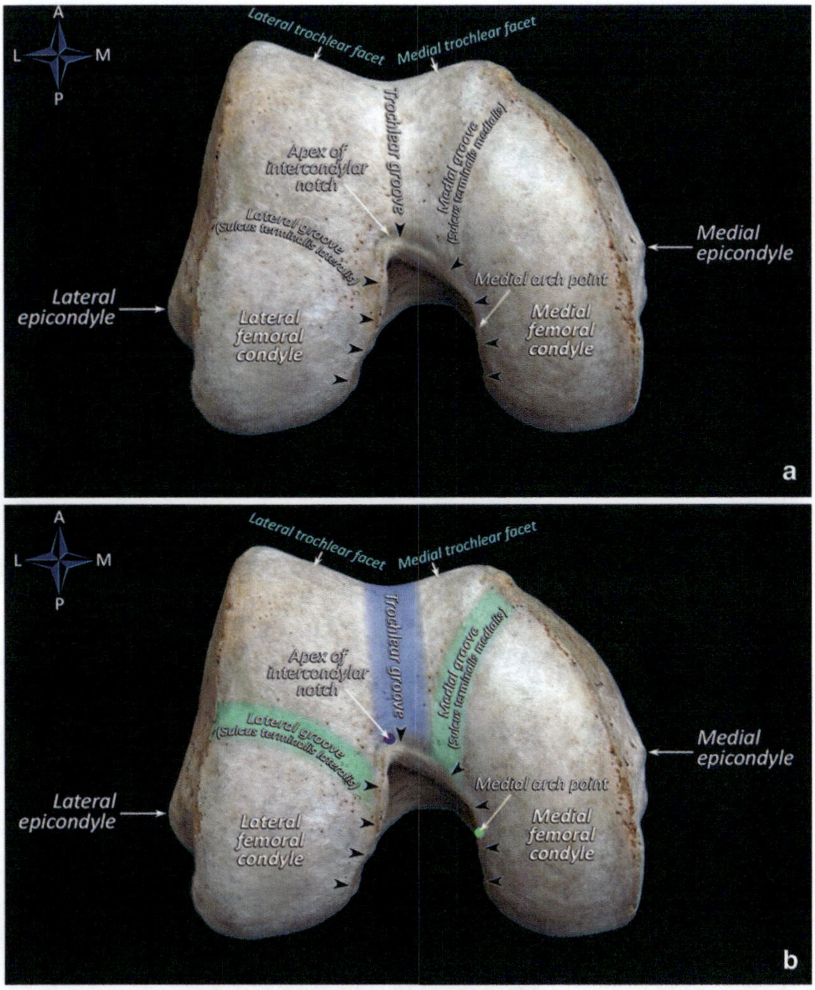

Fig. 1.2 Inferior view of the right distal femur. (**a**) Native view. (**b**) Colored view. *A* anterior, *P* posterior, *L* lateral, *M* medial, on the star showing directions

the intercondylar notch. The notch is separated from the popliteal surface by the intercondylar line posteriorly. The attachment area of the anterior cruciate ligament (ACL) is on the lateral wall of the notch, in other words, the posterior and superior parts of the medial surface of the lateral femoral condyle. This area can be observed as a slight depression (Fig. 1.5). Similarly, the attachment area of the posterior cruciate ligament (PCL) places on the medial wall of the notch, in other words, the anterior and superior parts of the lateral surface of the medial femoral condyle (Fig. 1.5).

The condyles join anteriorly to form a joint surface for the patella (Fig. 1.1). This surface, known as the patellar surface, extends further proximally on the anterior aspect of the lateral condyle (Figs. 1.1 and 1.2). The lateral facet of the patellar surface is separated from the outer surface of the lateral condyle with a more vertical and more prominent edge (Fig. 1.2). The medial facet of the patellar surface is flatter (Fig. 1.2) [2]. The trochlear groove is a significant structure for the stability of the patella. The decrease in the slope of the groove, especially the lateral side, may lead to dislocations of the patella.

1.2.2 Proximal End of the Tibia

Just as at the distal end of the femur, the most significant structures at the proximal end of the tibia are the lateral and medial condyles (Fig. 1.6). The lateral and medial joint facets covered with

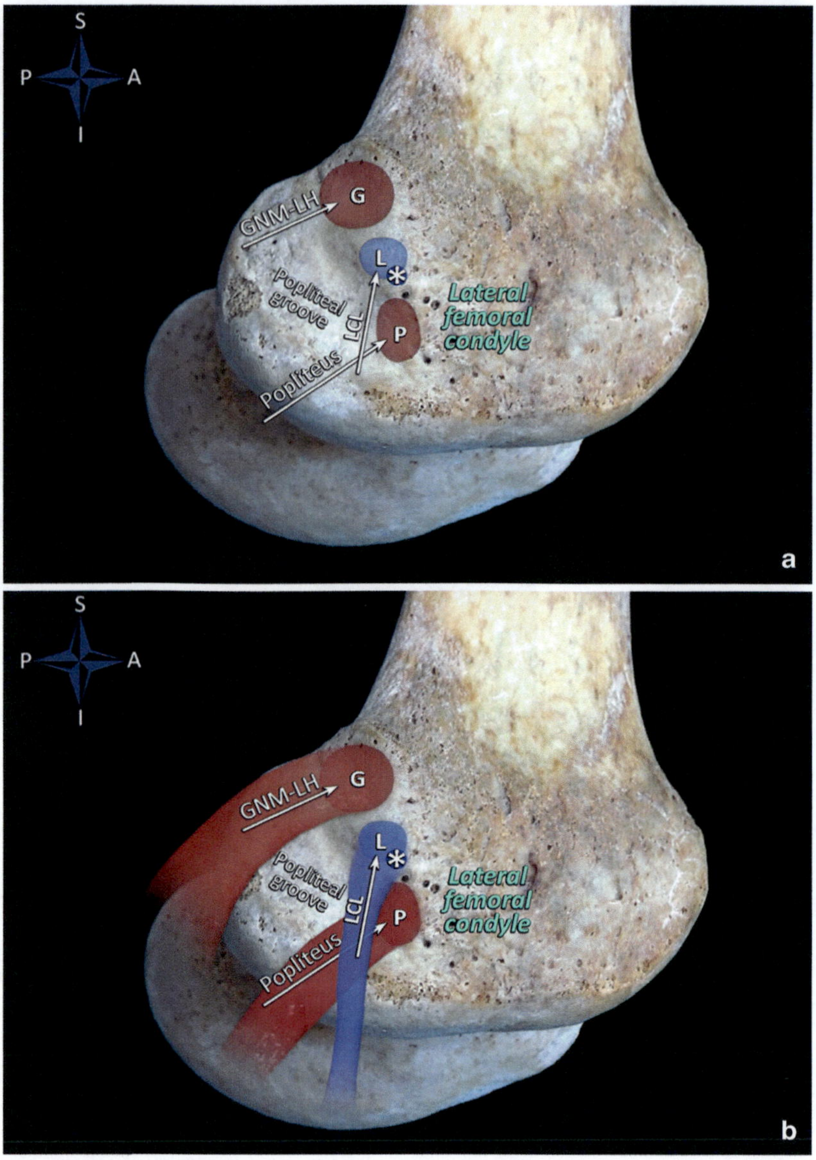

Fig. 1.3 Lateral view of the right distal femur. Structures that attach to the lateral side of the lateral femoral condyle. (**a**) Placement of footprints. (**b**) Extension of the attached structures. Asterisk is on the lateral epicondyle. *G* lateral head of the gastrocnemius (GNM-LH), *L* lateral collateral ligament (LCL), *P* popliteus. *S* superior, *I* inferior, *P* posterior, *A* anterior, on the star showing directions

cartilage are located on the superior articular surface, known clinically as the tibial plateau (Fig. 1.7). The menisci are located on the condyles. The central parts of the medial and lateral facets are in contact with the femur and the peripheral parts with the menisci. The joint surfaces are not completely in the horizontal plane. They are slightly inclined posteriorly and inferiorly according to tibia shaft (Fig. 1.8). Moreover, this inclination differs between the lateral and medial condyles. The intercondylar area is located between the condyles (Fig. 1.6). The cruciate ligaments are attached to this area with the anterior and posterior roots of the menisci (Fig. 1.7). The proximal tibia slopes posteriorly in the sagittal plane (Fig. 1.8). Because of this slope, the centers of the condyles (centers of the joint surfaces) come over the posterior part of the tibia shaft [2].

The joint surface of the medial condyle (*medial articular facet*) is oval shaped with its long axis in the anteroposterior direction

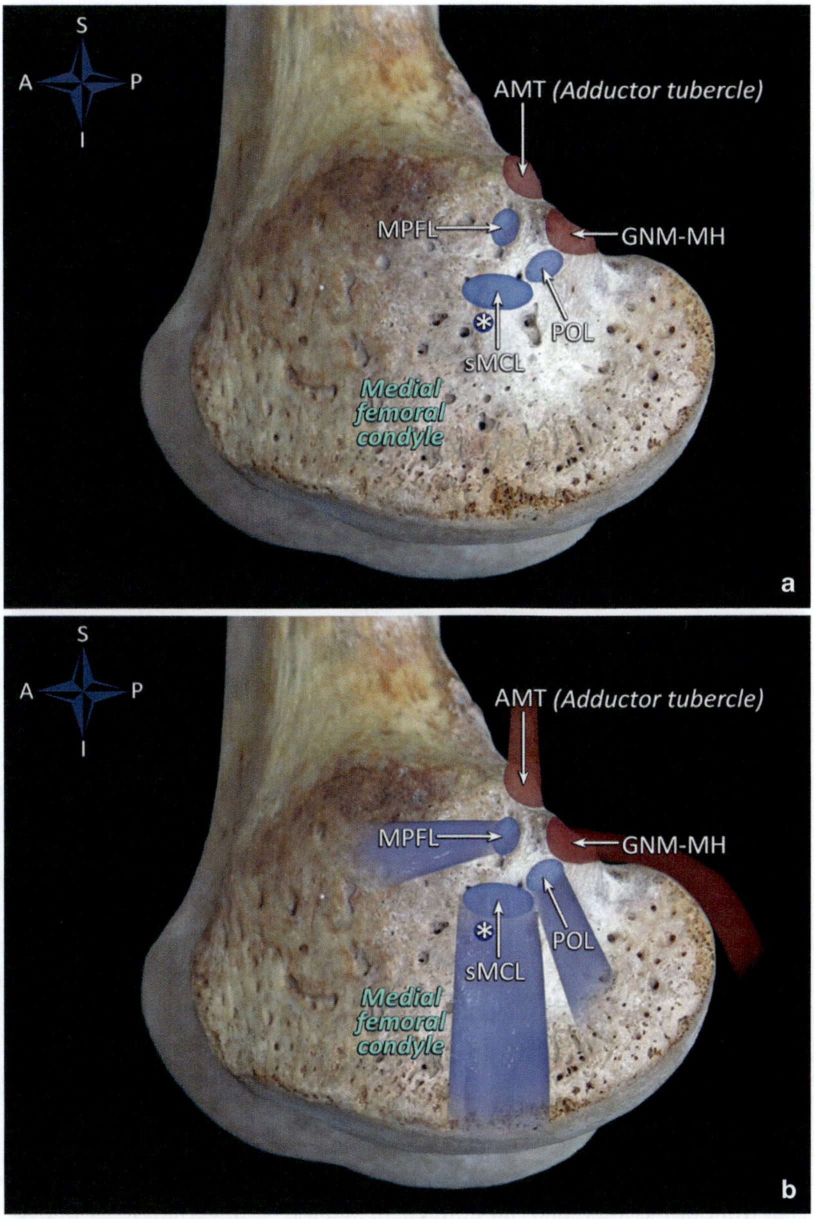

Fig. 1.4 Medial view of the right distal femur. Structures that attach to the medial side of the medial femoral condyle. (**a**) Placement of footprints. (**b**) Extension of the attached structures. Asterisk is on the medial epicondyle. *AMT* adductor magnus tendon, *MPFL* medial patellofemoral ligament, *sMCL* superficial medial collateral ligament, *POL* posterior oblique ligament, *GNM-MH* medial head of the gastrocnemius. *S* superior, *I* inferior, *A* *a*nterior, *P* posterior, on the star showing directions

(Fig. 1.7). The trace of the medial meniscus is narrower at the anterior and wider at the posterior direction (Fig. 1.7). The meniscus covers more space at the posterior part of the facet, and the anterior part has a mild slope (approximately 10°) to superior for providing a concavity. The joint surface of the lateral condyle (*lateral articular facet*) is smaller and rounder than the medial (Fig. 1.7). It is slightly concave in the transverse axis and slightly convex in the sagittal axis. Medial and lateral intercondylar tubercles are observed on the close sides of both the facets (Fig. 1.6).

An intercondylar area with an irregular surface is seen between the medial and lateral facets (Fig. 1.6). The middle region of the intercondylar area formed by the medial and lateral intercondylar tubercles is named as the intercondylar emi-

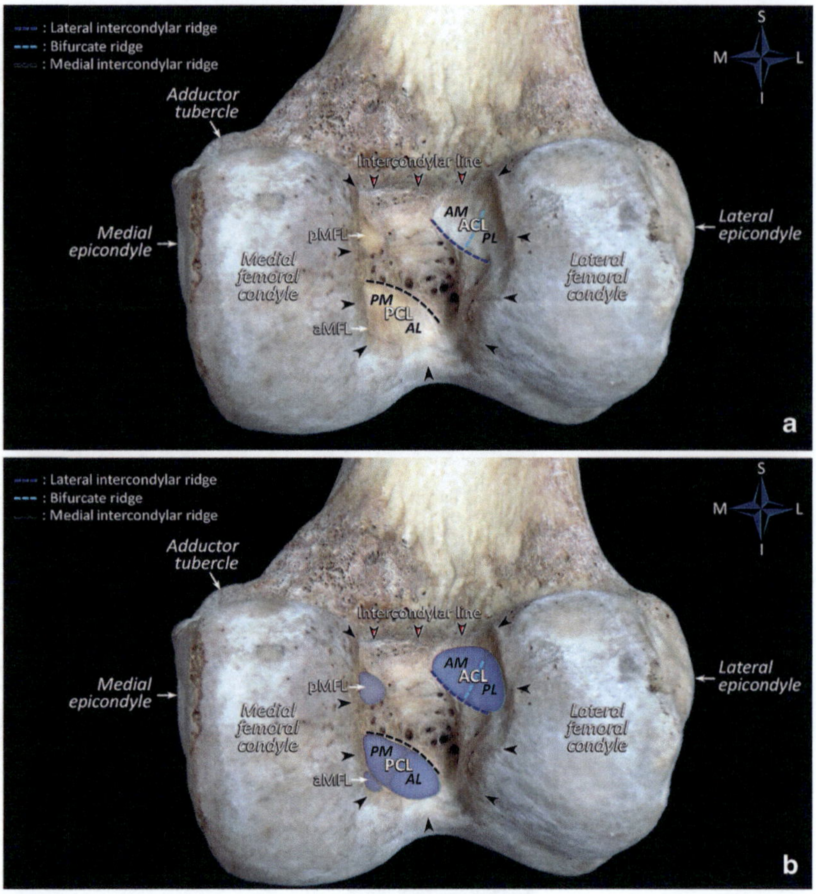

Fig. 1.5 Posterior view of the right distal femur. Structures that attach to the intercondylar notch. (**a**) Native view. (**b**) Colored view. Black arrowheads indicate border of the intercondylar notch. *ACL* anterior cruciate ligament, *AM* anteromedial bundle of ACL, *PL* posterolateral bundle of ACL, *PCL* posterior cruciate ligament, *AL* anterolateral bundle of PCL, *PM* posteromedial bundle of PCL, *aMFL* anterior meniscofemoral ligament, *pMFL* posterior meniscofemoral ligament. *S* superior, *I* inferior, *M* medial, *L* lateral, on the star showing directions

nence. The eminence is more prominent and narrower region of the intercondylar area. The ACL and the anterior roots of the menisci attach to the anterior intercondylar area in front of the intercondylar eminence (Fig. 1.7) [10, 11]. The footprint of the anterior root of the medial meniscus is seen in the anteromedial of the anterior intercondylar area [12]. The footprint of the ACL is in front of the intercondylar eminence, and the footprint of the anterior root of the lateral meniscus is immediately posterolateral to it (Fig. 1.7) [13]. The PCL and the posterior roots of the menisci are attached to the posterior intercondylar area which is posterior to the intercondylar eminence (Fig. 1.7) [14–16]. In the posterior intercondylar area, the posterior root of the lateral meniscus is attached to the flat area posterior to the lateral intercondylar tubercle. The posterior root of the medial meniscus is attached to the depressed area posterior to the medial intercondylar tubercle just anterior to PCL (Fig. 1.7) [12]. The footprint of the PCL extends more posteriorly and slightly inferiorly from the tibial plateau (Fig. 1.7) [14].

A triangular area is seen on the anterior surface of the proximal end of the tibia. The base of this triangle is above, and it is formed by the line joining the anterior edges of the condyles. The top of the triangle is marked by the tibial tuberosity (Fig. 1.9) [1]. The tibial tuberosity is formed of two areas which are flatter at the superior and rougher at the inferior. The patellar tendon is attached to the inferior part and the infrapatellar bursa is located beneath this tendon in the superior part [2]. The lateral edge of this triangle is more evident than the medial.

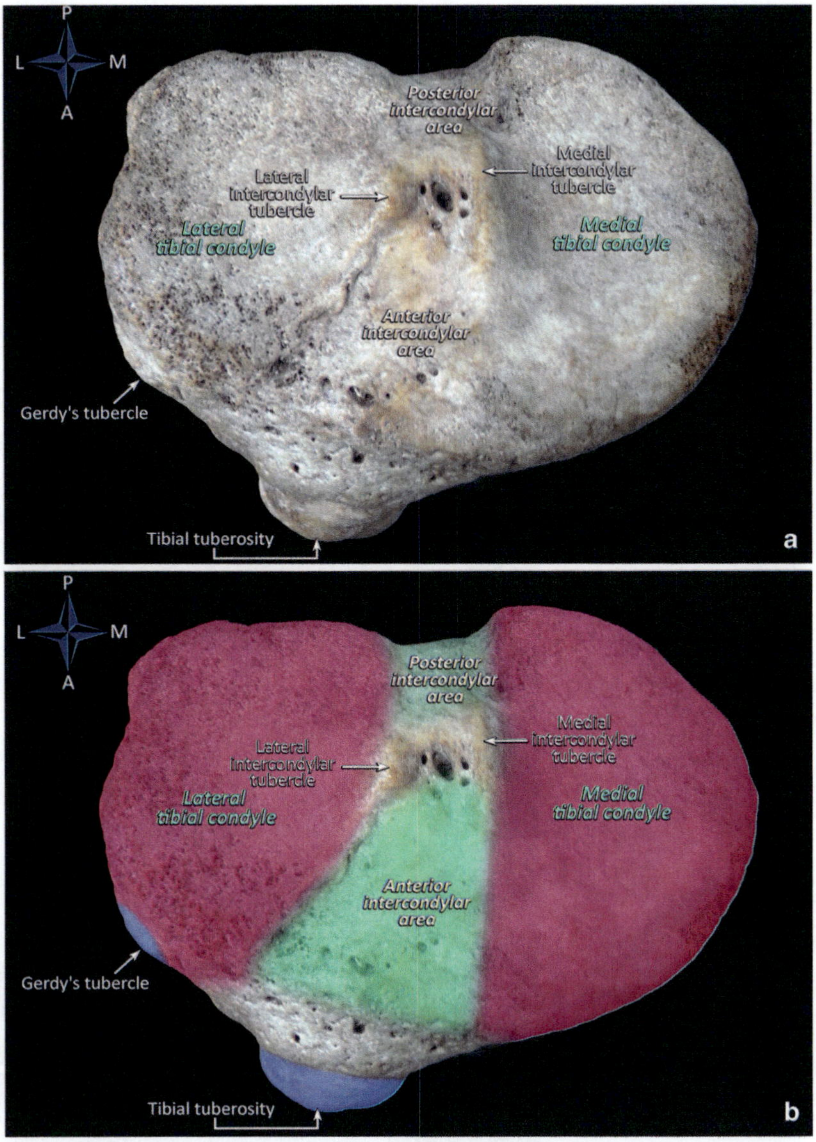

Fig. 1.6 Superior view of the right proximal tibia. (**a**) Native view. (**b**) Colored view. *P* posterior, *A* anterior, *L* lateral, *M* medial, on the star showing directions

Majority of the iliotibial tract fibers are attached to the most prominent point on this edge. This protuberance is known as Gerdy's tubercle (Fig. 1.9) [6].

The posterior and inferior surfaces of the lateral condyle of tibia make a joint with the fibula head. The fibular articular facet is smooth and oval shaped. The slope of the facet varies considerably between individuals (Fig. 1.10). A shallow groove where the popliteus tendon is located is observed at the medial side of the facet. Semimembranosus inserts are on the posterior side of the medial condyle. A groove is observed for semimembranosus tendon, above the insertion of this muscle. The upper part of this groove appears vertical, and semimembranosus tendon is located on it. The lower part of this groove appears transverse, and the anterior arm of semimembranosus tendon is attached to it. The attachment area of the posterior oblique ligament (POL) is observed medial to semimembranosus. This area is between the attachments of semi-

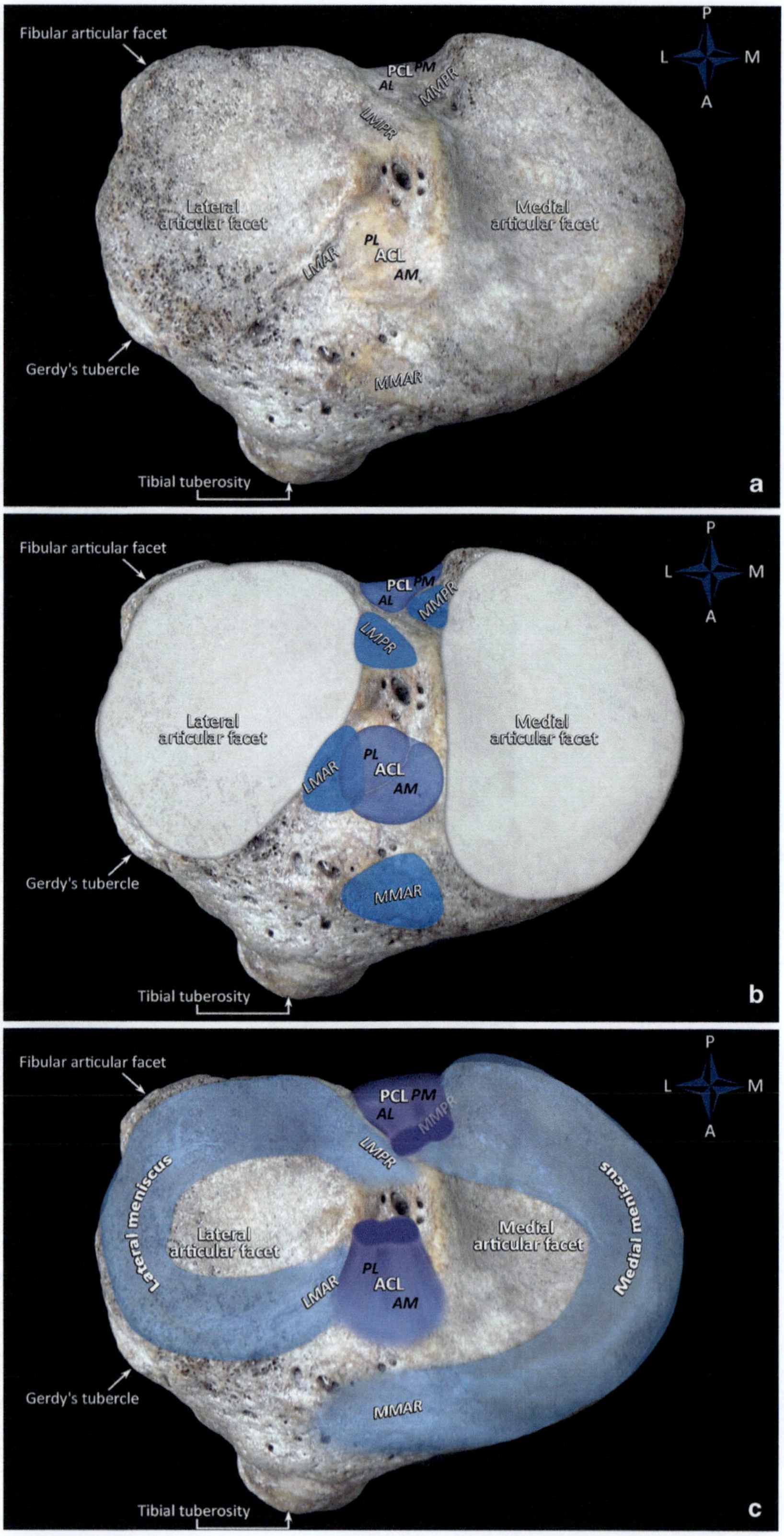

Fig. 1.7 Superior view of the right proximal tibia. Structures that attach to the intercondylar areas. (**a**) Native view. (**b**) Colored view. (**c**) Extension of the attached structures. *ACL* anterior cruciate ligament, *AM* anteromedial bundle of ACL, *PL* posterolateral bundle of ACL, *PCL* posterior cruciate ligament, *AL* anterolateral bundle of PCL, *PM* posteromedial bundle of PCL, *LMAR* lateral meniscus anterior root, *LMPR* lateral meniscus posterior root, *MMAR* medial meniscus anterior root, *MMPR* medial meniscus posterior root. *P* posterior, *A* anterior, *L* lateral, *M* medial, on the star showing directions

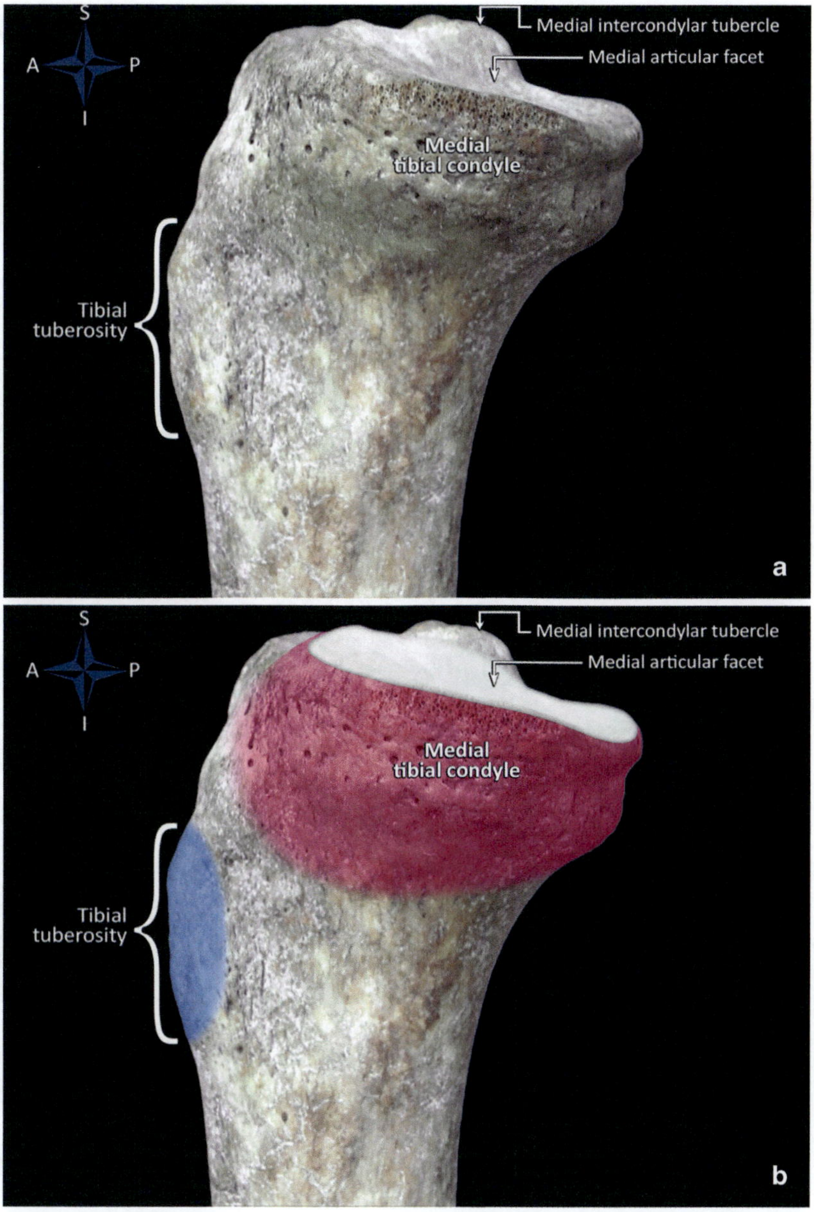

Fig. 1.8 Medial view of the right proximal tibia. (**a**) Native view. (**b**) Colored view. *S* superior, *I* inferior, *A* anterior, *P* posterior, on the star showing directions

membranosus and PCL. The posterior intercondylar area extends a few centimeters distal to the tibial plateau level between the two condyles in the form of a groove (Fig. 1.10).

Soleus muscle is attached to the soleal line on the posterior side of the tibia at proximal. Popliteus attaches to the triangular area which is supero-medial to soleal line. The tibialis posterior attaches to the area which is inferolateral to the soleal line (Fig. 1.10).

1.2.3 Proximal End of the Fibula

The proximal end of the fibula consists mainly of the head of the fibula. The neck of the fibula is located just distal to the fibular head (Fig. 1.11). The facet of the fibular head articulates with the posteroinferior of the lateral condyle of the tibia (Figs. 1.9 and 1.11). The inclination of the articular facet varies considerably between individuals. It can be closer to the horizontal plane or have an

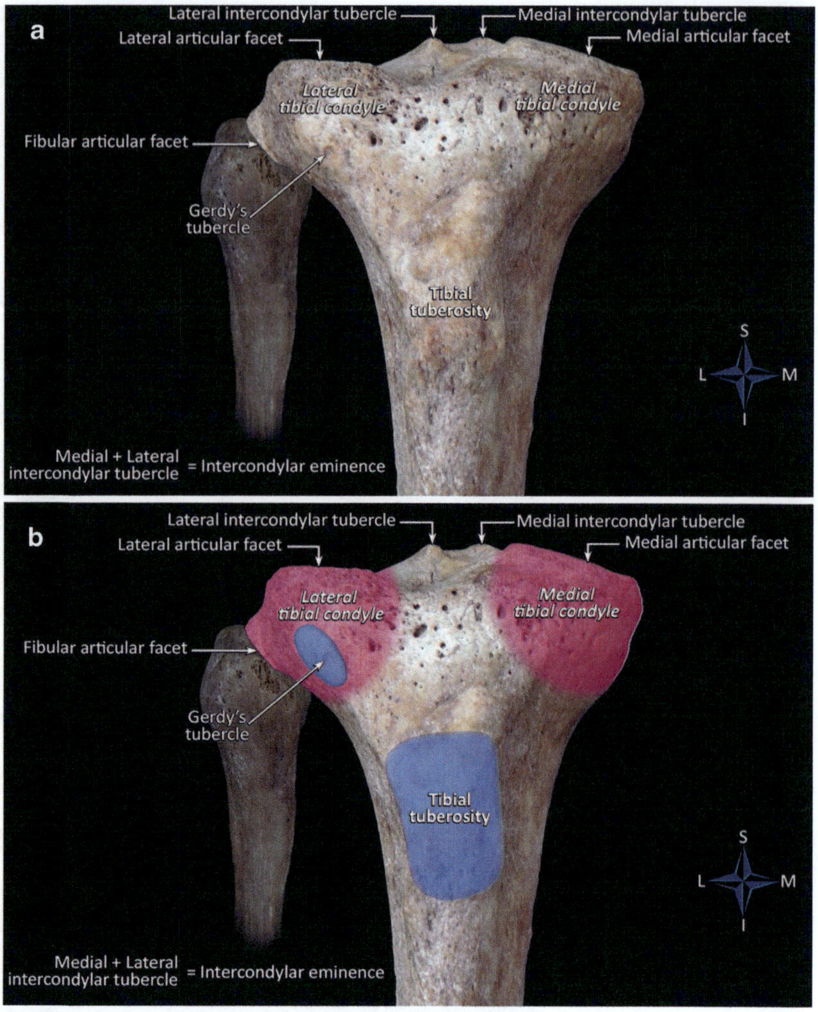

Fig. 1.9 Anterior view of the right proximal tibia. (**a**) Native view. (**b**) Colored view. *S* superior, *I* inferior, *L* lateral, *M* medial, on the star showing directions

oblique course up to 45° [2]. A prominence is observed over the head, which is named the apex of the head or styloid process. The LCL and biceps femoris are attached to the lateral of the fibular head. The popliteofibular ligament is more posteriorly attached the styloid process (Fig. 1.12).

1.2.4 Patella

The patella is the largest sesamoid bone in the body [1]. It is located inside the tendon of the quadriceps femoris. It is a triangular bone. The apex of the patella is at the inferior and base of that is at the superior (Fig. 1.13). There are two flat joint surfaces divided as lateral and medial patellar facets by a vertical ridge (Fig. 1.13) [17]. These surfaces provide fitness with the trochlear groove and facets on the joint surface facing the femur. The lateral joint surface is larger in order to fit with the longer and wider lateral trochlear facet of the patellar surface of the femur. The proximal part of anterior surface slopes slightly from superior to inferior and from posterior to anterior (Fig. 1.13). The rectus femoris is attached to the anterior and inferior of this surface, which is separated with a blunt edge from the middle part of anterior surface. The

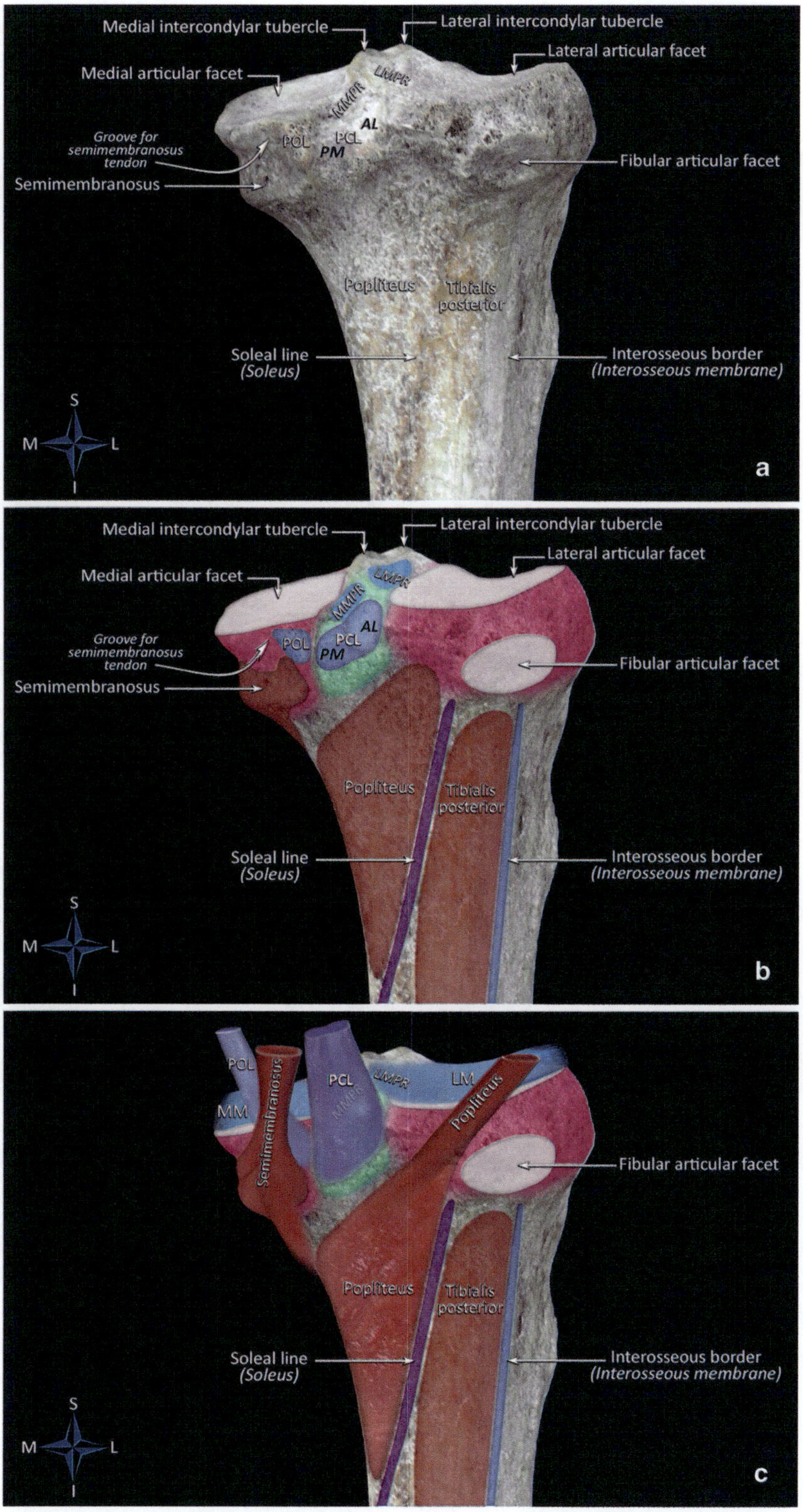

Fig. 1.10 Posterior view of the right proximal tibia. (**a**) Native view. (**b**) Colored view. (**c**) Extension of the attached structures. *PCL* posterior cruciate ligament, *AL* anterolateral bundle of PCL, *PM* posteromedial bundle of PCL, *LMPR* lateral meniscus posterior root, *MMPR* medial meniscus posterior root, *POL* posterior oblique ligament. *S* superior, *I* inferior, *L* lateral, *M* medial, on the star showing directions

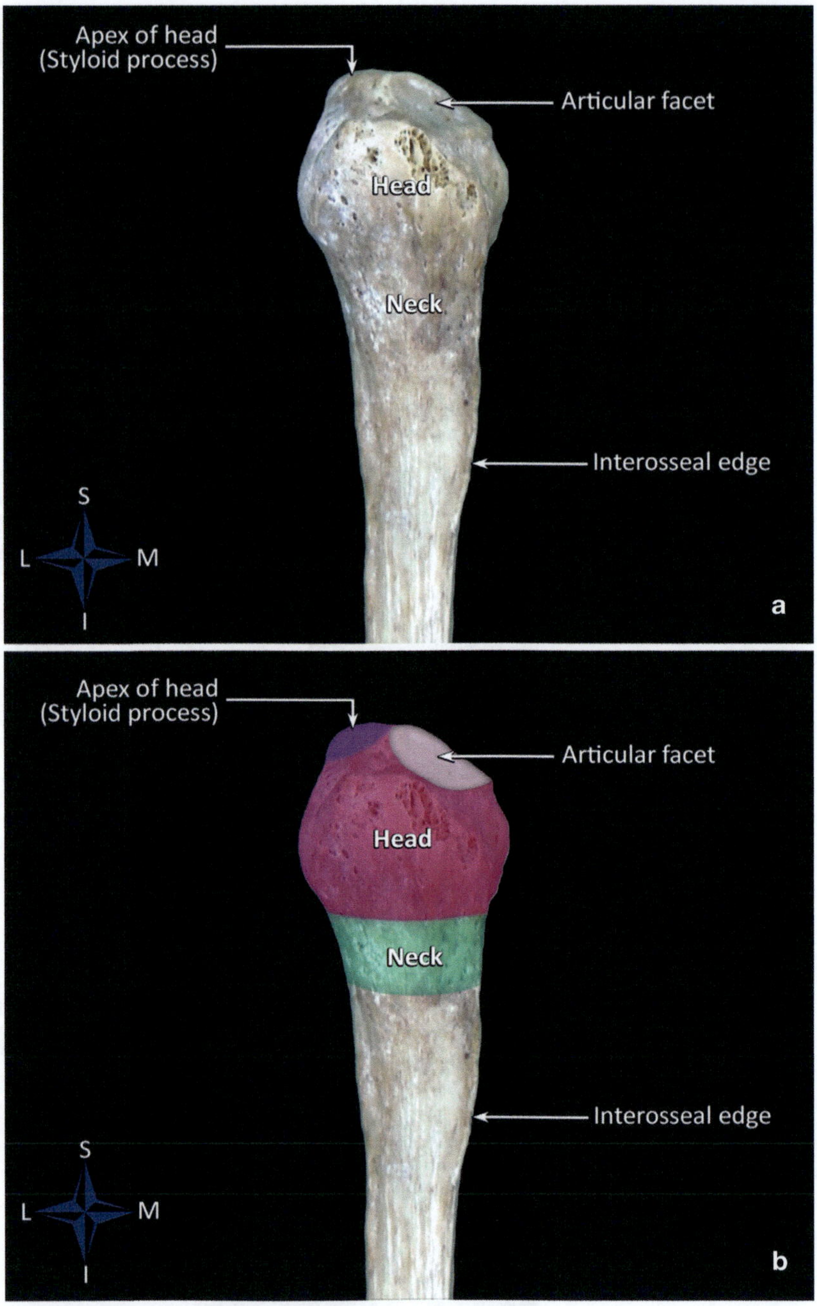

Fig. 1.11 Anterior view of the right proximal fibula. (**a**) Native view. (**b**) Colored view. *S* superior, *I* inferior, *L* lateral, *M* medial, on the star showing directions

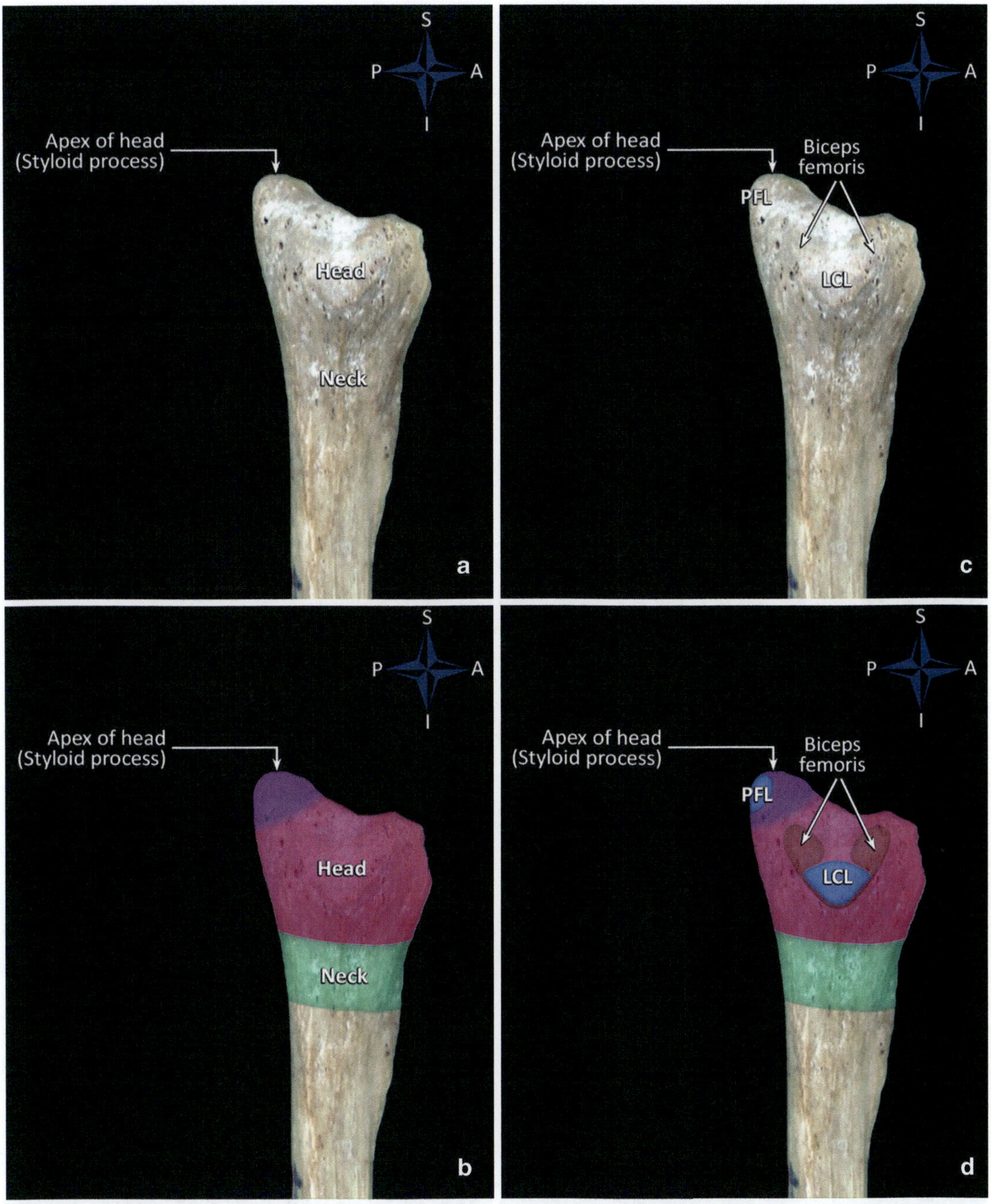

Fig. 1.12 Lateral view of the right proximal fibula. (**a**, **b**) Parts of the proximal fibula. (**c, d**) Structures that attach to the proximal fibula. (**a**, **c**) Native view. (**b**, **d**) Colored view. *LCL* lateral collateral ligament, *PFL* popliteofibular ligament. *S* superior, *I* inferior, *P* posterior, *A* anterior, on the star showing directions

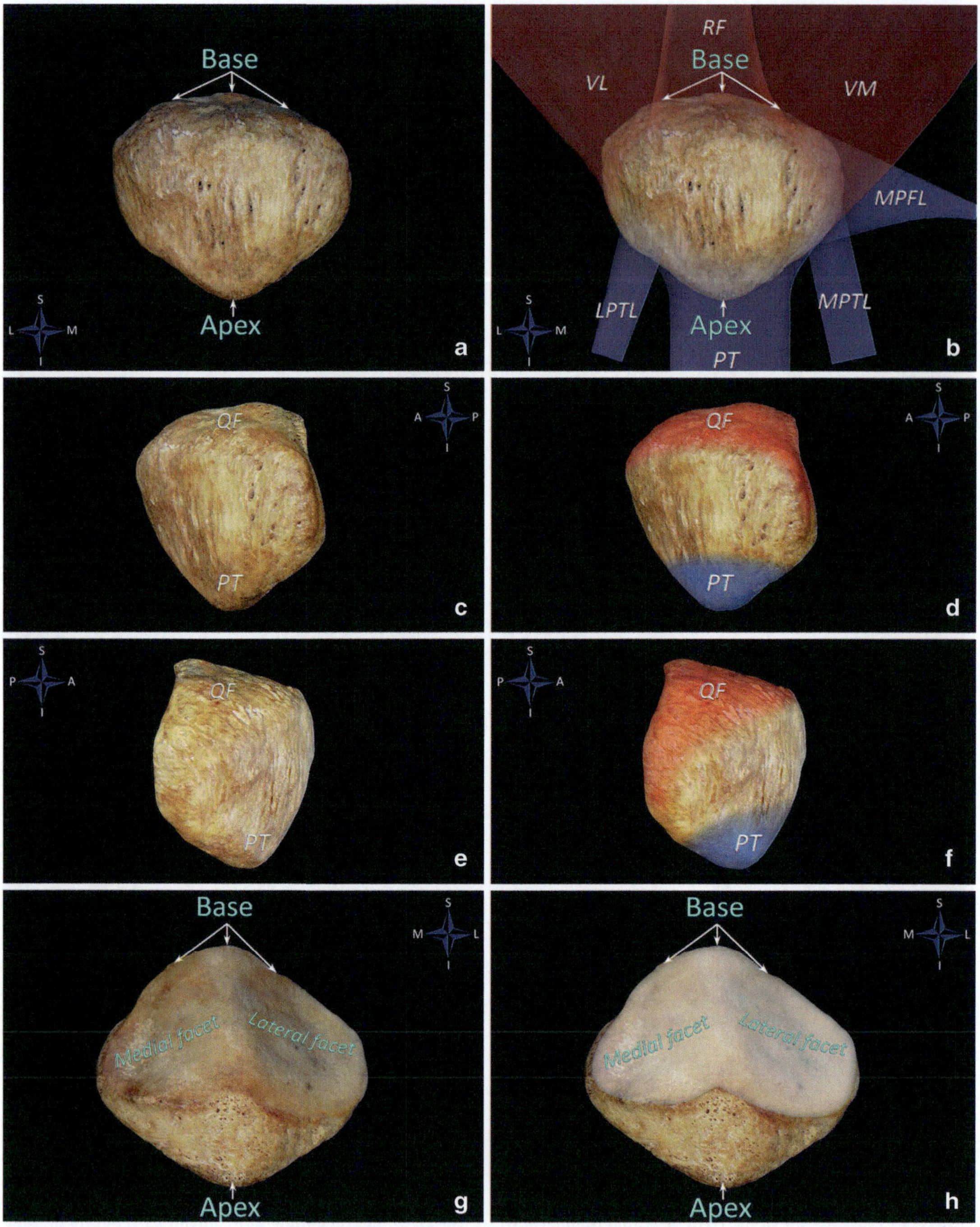

Fig. 1.13 Right patella. (**a**, **b**) Anterior views. (**c**, **d**) Anteromedial views. (**e**, **f**) Anterolateral views. (**g**, **h**) Posterior views. (**a**, **c**, **e**, **g**) Native views. (**b**, **d**, **f**, **h**) Colored views. Extension of the structures attached on patella are shown on **b**. *RF* rectus femoris, *VL* vastus lateralis, *VM* vastus medialis, *MPFL* medial patellofemoral ligament, *PT* patellar tendon, *MPTL* medial patellotibial ligament, *LPTL* lateral patellotibial ligament, *QF* quadriceps femoris. *S* superior, *I* inferior, *L* lateral, *M* medial, *P* posterior, *A* anterior, on the star showing directions

vastus intermedius is attached to the center of the remaining posterior and superior, while the vastus lateralis and medialis are attached to each side of this surface. The distal parts of the tendons of vastus lateralis and medialis are attached to the upper halves of the lateral and medial edges of the patella (Fig. 1.13) [2, 17]. In particular, the inferior part of the vastus medialis extends more distally and courses more obliquely (named *vastus medialis obliquus*) [1, 2].

1.3 Lateral and Medial Sides of the Knee

The structures on the lateral and medial sides of the knee are similarly organized in layers. The differences between references are observed in the definitions of the structures in these layers. Nevertheless, these definitions provide a great convenience for the safe operation of lateral and medial knee surgery.

1.3.1 The Medial Side of the Knee

Structures in the medial side of the knee can be examined in three layers [2, 18]. Medial support and stability of the knee is provided by these anatomic structures located from superficial to deep. The different layers have important roles and functions in the mechanics of the knee joint.

Medial subcutaneous tissue: Significant neurovascular structures are found in the subcutaneous tissue over the important medial stabilizers. The great saphenous vein and saphenous nerve must be considered in this region (Fig. 1.14).

Great saphenous vein (long saphenous vein): This vein starts from the medial of the foot and extends superiorly from immediately anterior of the medial malleolus. It extends from the medial of the leg to the posteromedial of the knee. It is located posterior to patella as far as approximately a palm-size from the medial edge of the patella (Fig. 1.15) [1, 2].

The saphenous nerve enters the adductor canal together with femoral vessels. It separates from the vessels close to the lower end of the canal. It penetrates the anteromedial intermuscular septum (subsartorial fascia), which forms the anteromedial wall of the canal, and passes beneath the sartorius. It becomes superficial by penetrating the fascia lata between the sartorius and gracilis tendon, together with the saphenous branch of the descending genicular artery (Fig. 1.16) [1, 2, 19]. From here, it subcutaneously accompanies the long saphenous vein in the medial of the leg. It gives branches to the medial of the leg (medial crural cutaneous nerve) and extends to the medial of the foot with the vein. The infrapatellar branch of the saphenous nerve often separates from saphenous nerve immediately at the posterior edge of the sartorius and then curves laterally for distributing to the infrapatellar area (Figs. 1.15 and 1.16). However, variations are frequent [20]. It can also pass in front of or through sartorius to the infrapatellar region. The nerve is observed more than one branch in approximately three fourth of the cases [20–22]. These branches may appear in different courses in the same case. The infrapatellar branch can be transected in a medial parapatellar incision or during the opening anteromedial arthroscopy portals. The course and distribution of the nerve explain the sensory loss lateral to the incision site.

Layer 1: Layer 1 is the most superficial layer underneath the subcutaneous tissue. Basically, it is formed by the insertion of the sartorius muscle which is in aponeurotic structure. The medial patellar retinaculum is observed anterior to the sartorial fascia (Figs. 1.15 and 1.16).

Sartorial fascia: Since the ending of sartorius is observed as a fascia rather than a tendon, it is called "sartorial fascia" in many references. Sartorius fascia covers the last part of the gracilis and semitendinosus tendons on the medial side of the knee (Figs. 1.15 and 1.17). Most of the fibers attach to the anterolateral side of the tibia along a

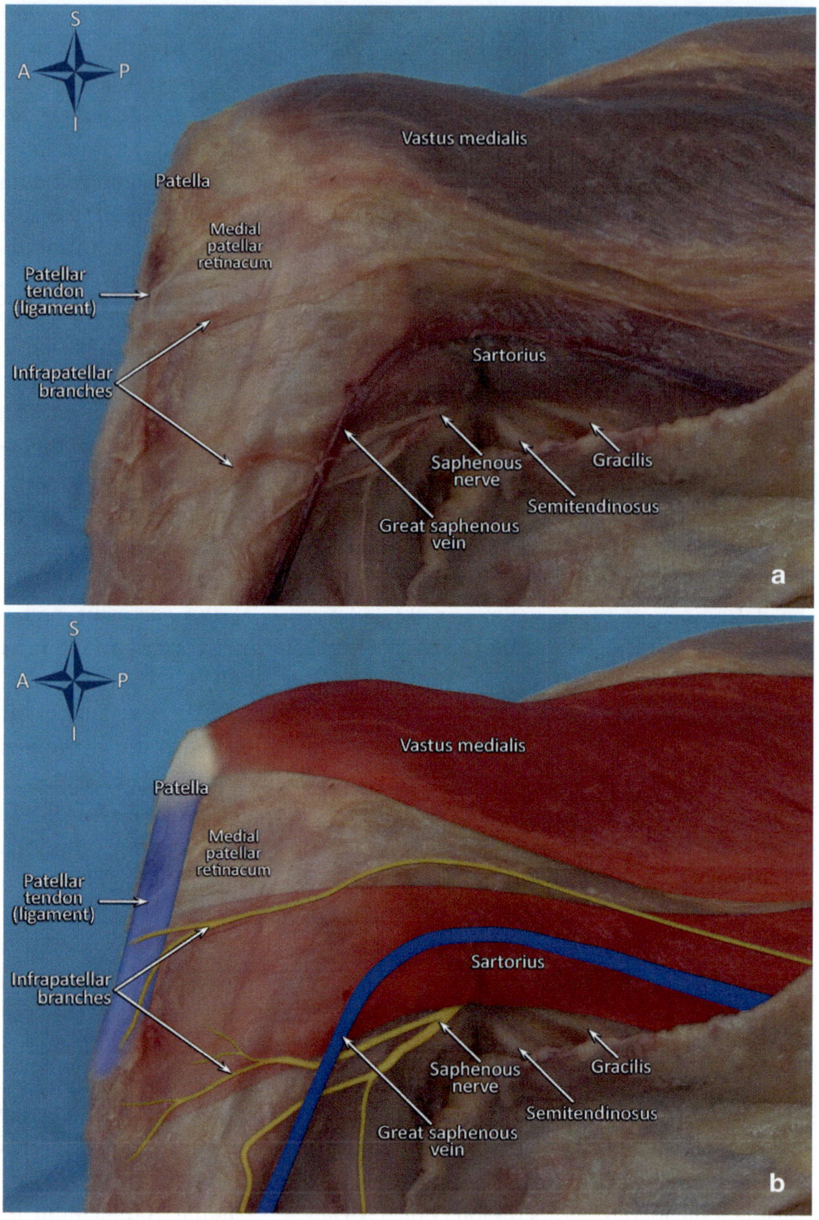

Fig. 1.14 Medial view of the right knee at 90° flexion. (**a**) Native view. (**b**) Colored view. *S* superior, *I* inferior, *A* anterior, *P* posterior, on the star showing directions

thin line, just in front of the attachment of the gracilis and semitendinosus tendons, distal to the medial condyle (Fig. 1.17). The insertion on the tibia is about the level of the tibial tuberosity or approximately 5 cm from the joint line and extends 4–5 cm distally. The more distal part of the sartorial fascia combines with the fascia in the medial of the leg. There are connections with the semitendinosus and gracilis tendons close to the attachment site to the bone [23]. The tendons of the gracilis and semitendinosus with the sartorial fascia form the "pes anserinus" [19, 23]. More posteriorly, the sartorius fascia shows continuity with the popliteal fascia covering the popliteal structures.

Medial patellar retinaculum: In front of the sartorius fascia, the aponeurotic extensions of the vastus medialis in the medial of the patella form

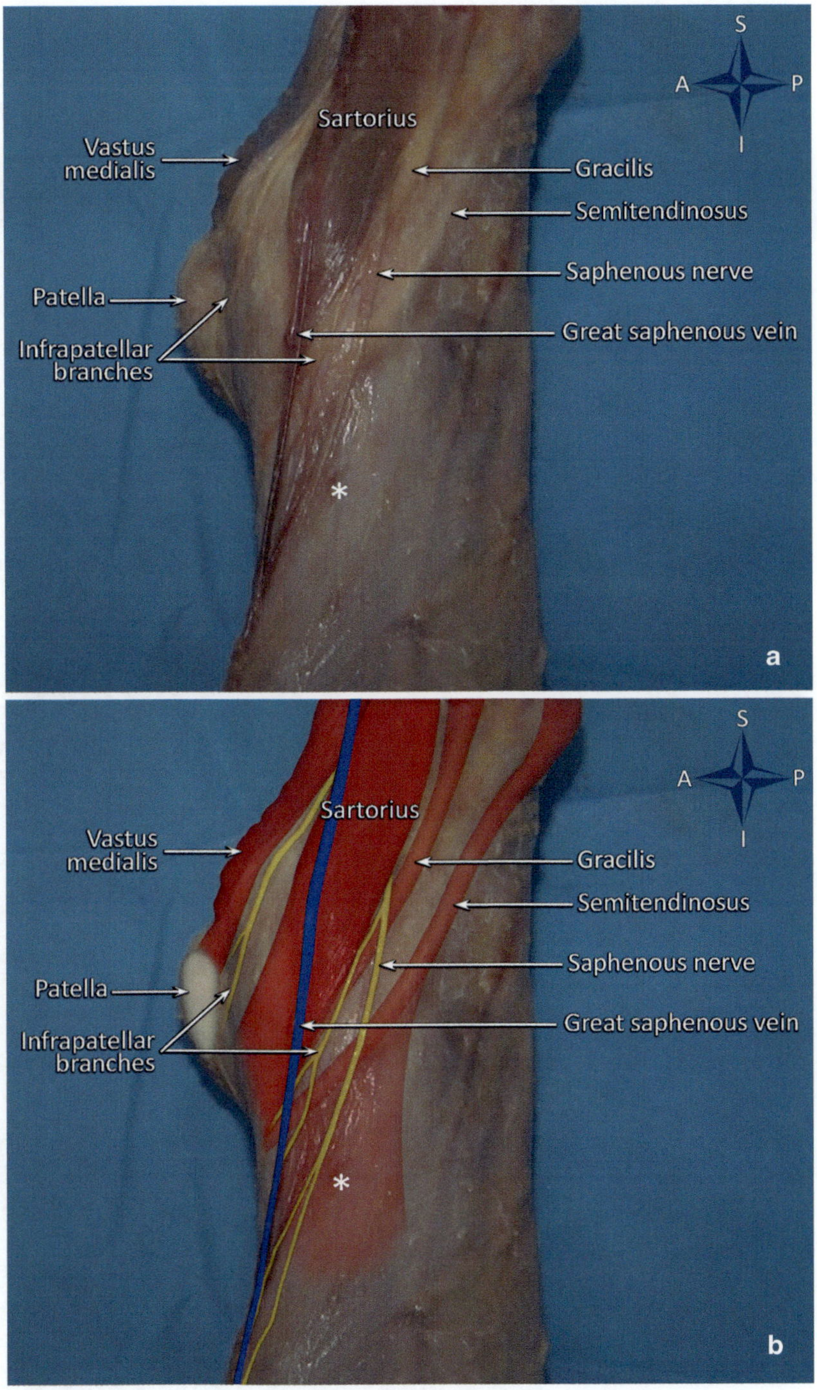

Fig. 1.15 Posteromedial view of the right knee at extension. (**a**) Native view. (**b**) Colored view. *S* superior, *I* inferior, *A* anterior, *P* posterior, on the star showing directions

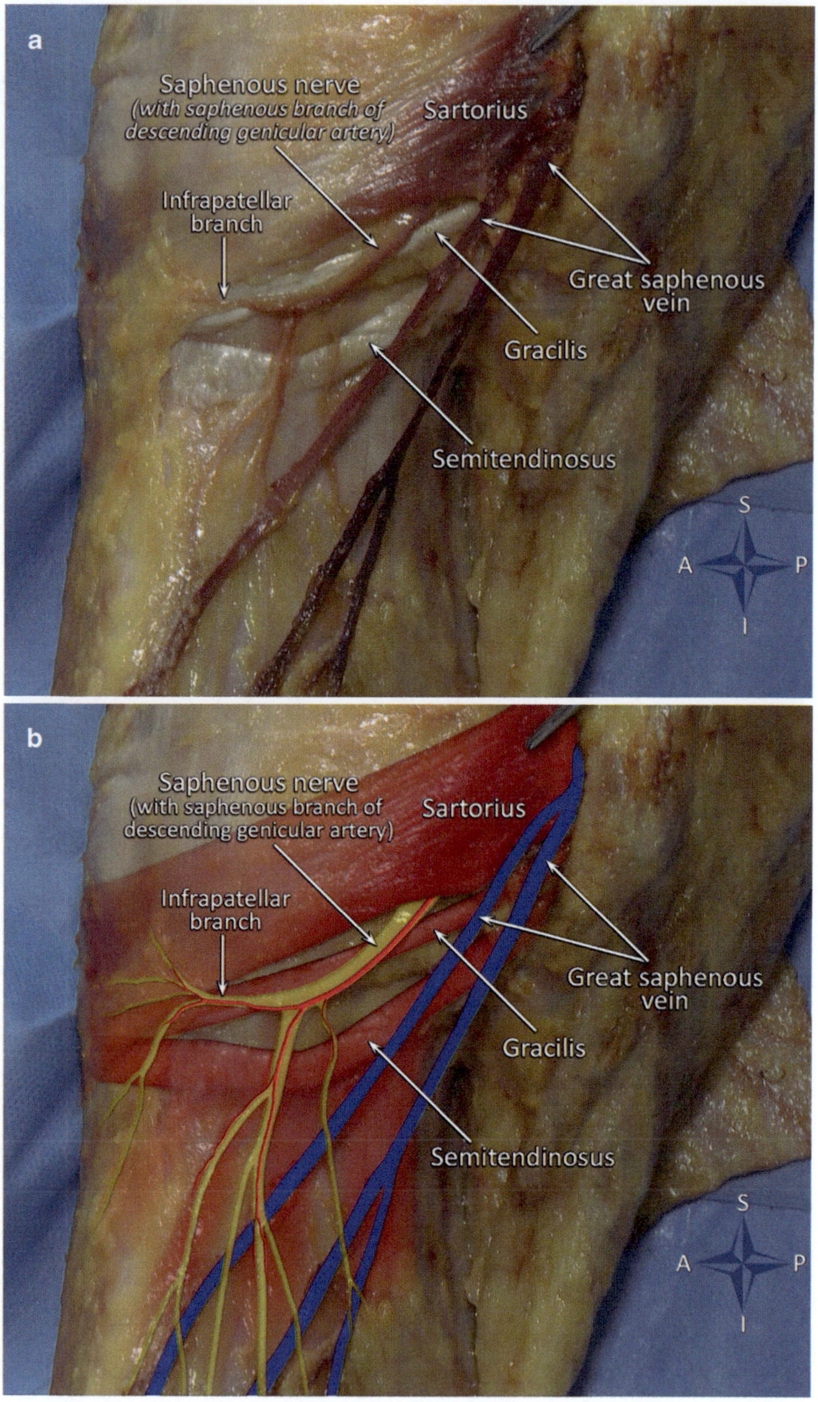

Fig. 1.16 Medial view of the right knee at 90° flexion. (**a**) Native view. (**b**) Colored view. *S* superior, *I* inferior, *A* anterior, *P* posterior, on the star showing directions

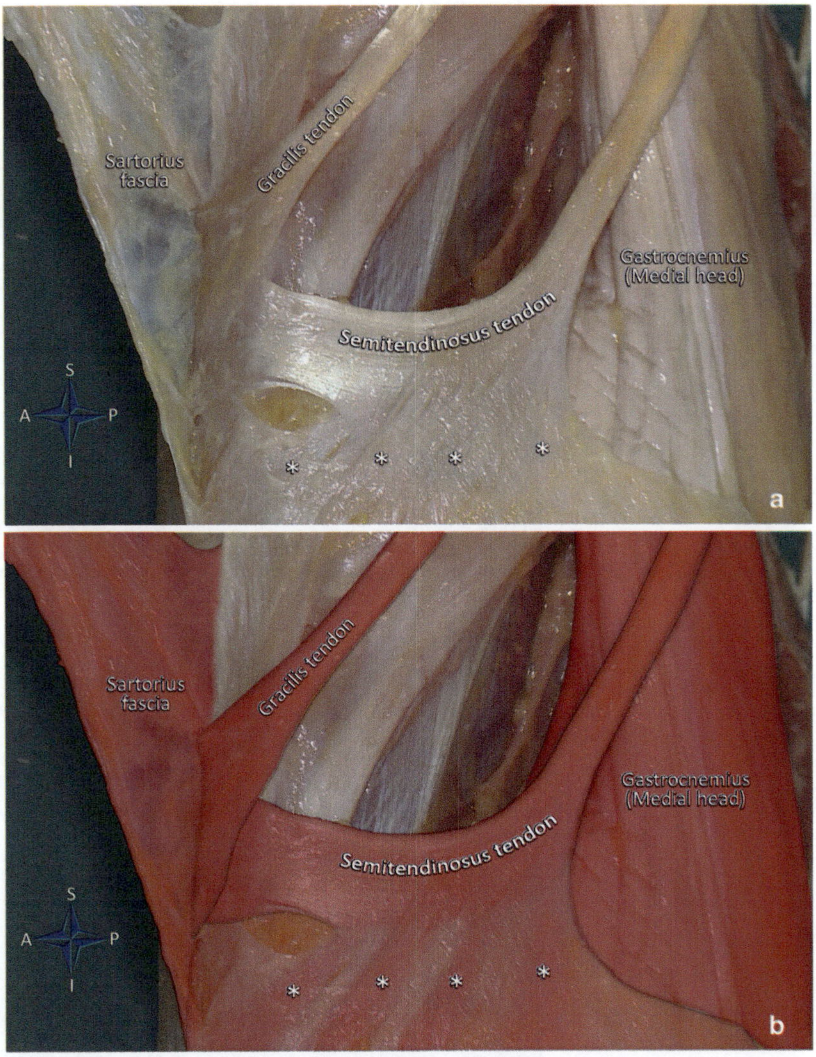

Fig. 1.17 Medial view of the right knee. (**a**) Native view. (**b**) Colored view. Pes anserinus muscles (sartorius, gracilis, and semitendinosus) are seen. Asterisks indicate accessory bands of semitendinosus blended with the fascia of medial head of gastrocnemius. *S* superior, *I* inferior, *A* anterior, *P* posterior, on the star showing directions

the medial patellar retinaculum observed in the first layer. However, most of the fibers are inserted distally underneath the sartorius fascia and attach to the anterolateral of the medial condyle in front of superficial medial collateral ligament (sMCL) (Fig. 1.18) [8, 9, 18].

Tendons of gracilis and semitendinosus: These tendons extend between the first and the second layers [18]. The tendons insert on to the anteromedial of the tibia, approximately 2 cm medial and 2 cm distal from the tibial tuberosity. The gracilis tendon is anterior to the semitendinosus, and the attachment site to the bone is more proximal. Just as there are connections with overlying the sartorius fascia; particularly, the semitendinosus has wide connections with the deep fascia of the leg covering the medial head of the gastrocnemius (Figs. 1.16 and 1.17) [8, 18, 19, 23–25]. During tendon harvesting, these connections should be cut in order to isolate the tendon. There is a high risk of early rupture of the tendon which is tried to be removed without isolating. The bursa of the pes anserinus (anserine bursa) is located between the superficial MCL and the tendons of gracilis and semitendinosus [2].

Layer 2: Most of this layer is formed by the superficial medial collateral ligament (sMCL) [18]. The medial patellofemoral ligament is another important structure in this layer (Fig. 1.18).

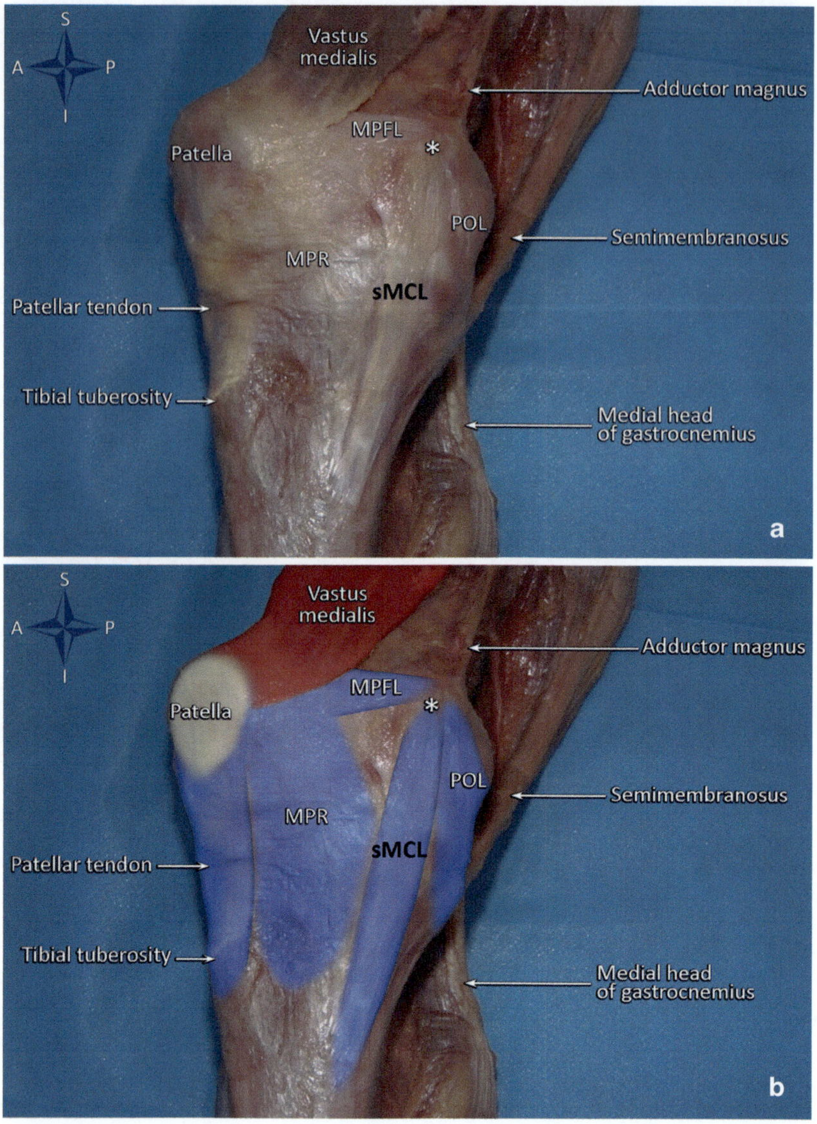

Fig. 1.18 Medial view of the right knee. (**a**) Native view. (**b**) Colored view. Pes anserinus muscles were removed. Asterisk indicates the medial epicondyle. *MPFL* medial patellofemoral ligament, *MPR* medial patellar retinaculum, *sMCL* superficial medial collateral ligament. *S* superior, *I* inferior, *A* anterior, *P* posterior, on the star showing directions

Superficial medial collateral ligament (sMCL): The medial collateral ligament (MCL) is the most frequently injured ligament in the knee [26]. It consists of two parts: the superficial and deep. While sMCL is in the second layer, the deep MCL (dMCL) is in the third layer (Figs. 1.18 and 1.19). sMCL is the primary stabilizer protecting the knee from valgus at all flexion angles starting from full extension [18]. sMCL originates from just proximal and posterior to the medial epicondyle of the femur [8, 27]. Unlike the LCL, it is a smooth and wide ligament, extending vertically under the tendons of gracilis and semitendinosus. The anserine bursa is located between these tendons and the ligament. The vertical fibers of the sMCL attach to a relatively large area extending 6–7 cm distally on the just anterior to the medial edge of the tibia [28].

Medial patellofemoral ligament (MPFL): MPFL originates from immediately posterosuperior to the medial epicondyle and anteroinferior to the adductor tubercle. The ligament courses transversely to anterolateral over the capsule and extends to the inferior edge of the vastus medialis obliquus. It enters deep into the fibers of vastus medialis obliquus (VMO) and fuses with the aponeurotic lower edge of this muscle (Fig. 1.18). The MPFL together with the distal part of VMO

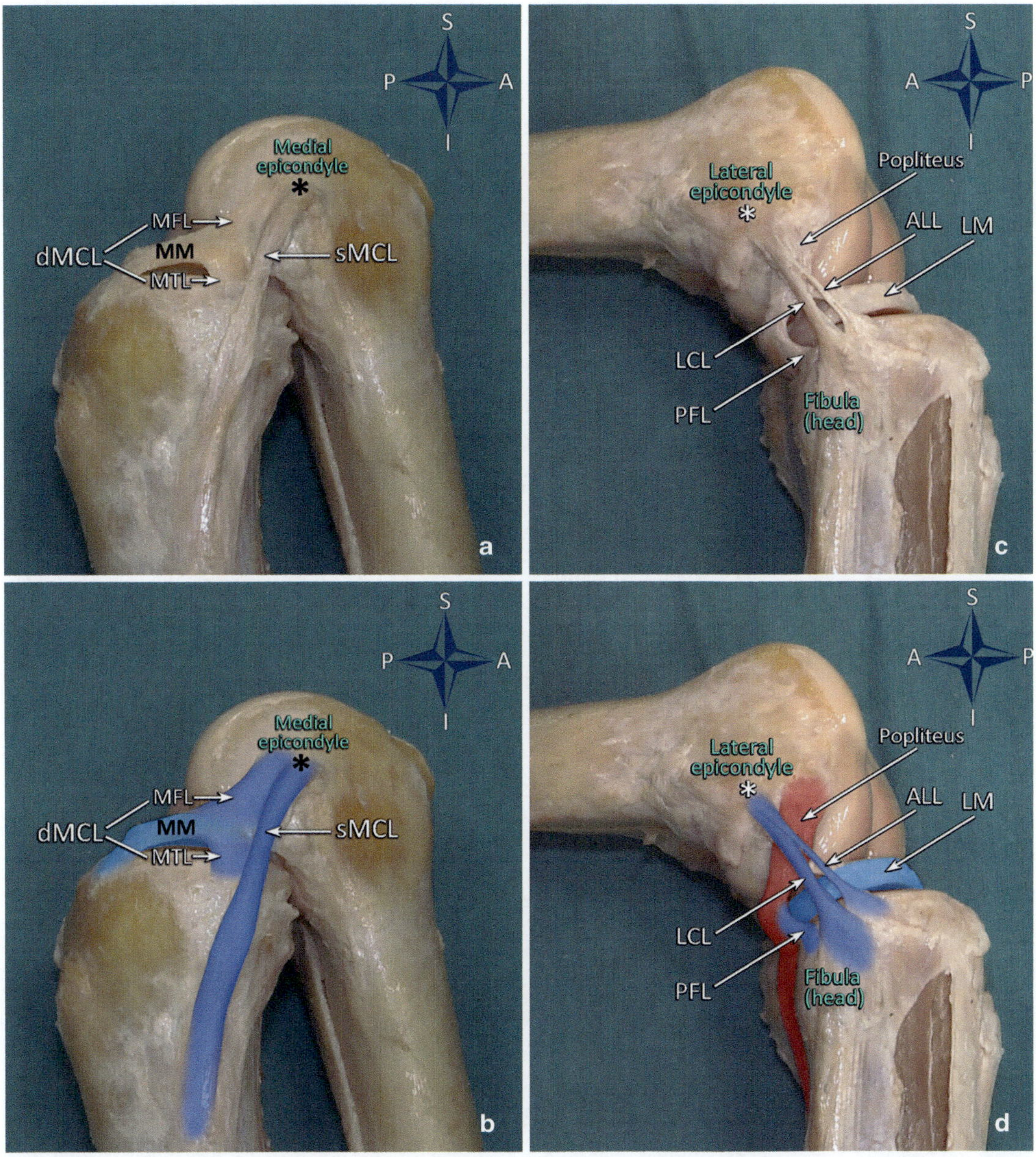

Fig. 1.19 Lateral (**a**, **b**) and medial (**c**, **d**) views of the right knee. (**a**, **c**) Native views. (**b**, **d**) Colored views. *sMCL* superficial medial collateral ligament, *dMCL* deep medial collateral ligament, *MFL* meniscofemoral ligament, *MTL* meniscotibial ligament, *MM* medial meniscus, *LCL* lateral collateral ligament, *PFL* popliteofibular ligament, *ALL* anterolateral ligament, *LM* lateral meniscus. *S* superior, *I* inferior, *A* anterior, *P* posterior, on the star showing directions

attach to the upper half of the medial edge of the patella. MPFL is one of the most important static stabilizers of the patella, especially in the knee in extension [8, 18, 27].

Layer 3: Deep part of the MCL and the posterior oblique ligament are seen in this layer. These structures are secondary stabilizers protecting the knee against valgus.

Deep medial collateral ligament (dMCL): The vertically extending fibers beneath sMCL form dMCL. The upper part of this ligament extends from the femur to the medial meniscus. This part is named as the *meniscofemoral ligament*. The inferior part of the dMCL extends from the medial meniscus to the tibia. These fibers are named the *meniscotibial or the coronary ligament* (Fig. 1.19) [8, 27, 29, 30].

Posterior oblique ligament (POL): The oblique fibers of the posteromedial joint capsule located posterior to sMCL form POL (Fig. 1.20). This ligament is a secondary stabilizer protecting the knee against valgus and, more importantly,

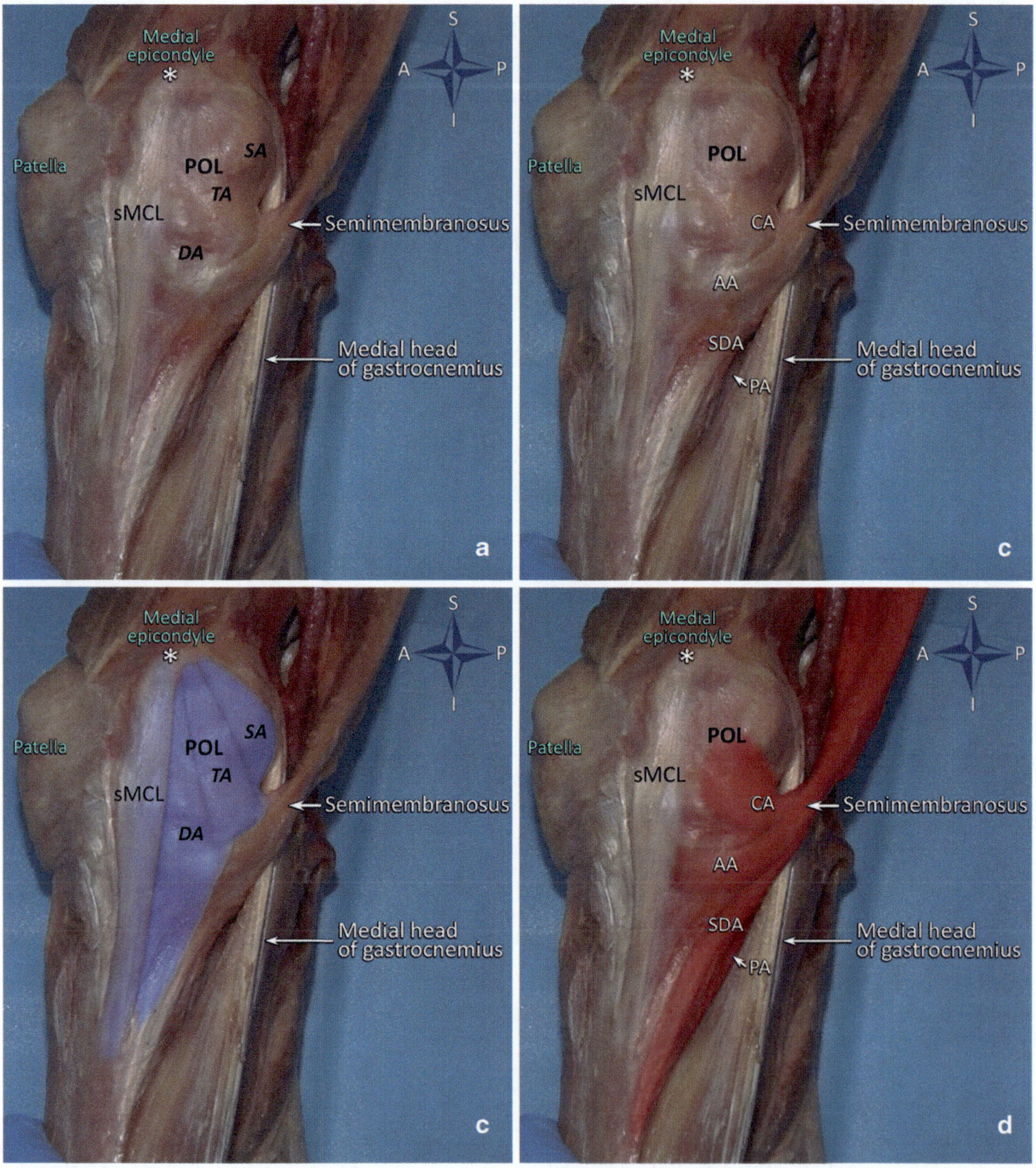

Fig. 1.20 Medial view of the right knee. (**a**, **c**) Native views. (**b**, **d**) Colored views. Pes anserinus were removed. The arms of posterior oblique ligament (POL) (**a**, **b**) and semimembranosus (**c**, **d**) in the posteromedial corner of the knee are shown. sMCL superficial medial collateral ligament. The arms of posterior oblique ligament (POL): *SA* superior (or capsular) arm, *TA* tibial (central) arm, *DA* distal (or superficial) arm. The arms of semimembranosus: *CA* capsular arm, *AA* anterior arm, *SDA* superficial direct arm, *PA* popliteal arm. *S* superior, *I* inferior, *A* anterior, *P* posterior, on the star showing directions

provides rotational stability by limiting internal rotation of the tibia in extension [8, 31, 32].

POL has three arms [8, 9, 27]. The most proximal fibers of POL extend to the posterior knee capsule. This part is called capsular or superior arm. It is weaker than other parts of the ligament. The middle and strongest part of POL is named as tibial or central arm. It inserts posterior to the medial tibial condyle, deep in the semimembranosus tendon. The most distal fibers extend immediately posterior to and parallel to the sMCL. This part, called distal or superficial arm, passes over the anterior arm of semimembranosus (Fig. 1.20).

Numerous extensions of the semimembranosus tendon that are related to POL have been described: capsular, anterior, superficial direct, popliteal, deep direct, oblique popliteal ligament arms [8, 9, 28, 33]. The most proximal capsular arm (CA) extends to the posteromedial joint capsule. Here it fuses with POL's central arm. Anterior arm (AA) enters beneath the distal arm of POL and inserts on posteromedial to the medial tibial condyle. Superficial direct arm (SDA) extends parallel to the distal arm of POL, on the medial edge of the tibia (Fig. 1.20). Popliteal arm (PA) blends with the fascia of popliteus. Deep direct arm (DDA) inserts on posterior to the medial tibial condyle, directly (Fig. 1.21).

Deep structures between sMCL and PCL is located in the posteromedial corner of the knee (Fig. 1.21) [27]. The "posteromedial corner" (PMC) structures of the knee include the posteromedial joint capsule which contains POL and the tendon of semimembranosus. In addition to limiting valgus, these structures have important functions in providing rotational stability, particularly by limiting internal rotation.

1.3.2 The Lateral Side of the Knee

Recent studies have emphasized the importance of the structures providing stability from the lateral aspect of the knee, especially "posterolateral corner (PLC)." Injuries to several structures disregarded are associated with knee instability and unsuccessful results following cruciate ligament repair [34–37].

The lateral structures of the knee limit the varus of the knee. In addition to this, posterior translation and external rotation are limited by the PLC structures. Isolated PLC damage is rarely seen. Injuries of these structures are more often (43%–80%) associated with damage to other ligamentous structures of the knee, including damage to the PCL and/or ACL [38–42].

The structures in the lateral of the knee can be identified in three layers from superficial to deep [35, 36, 38, 43, 44]. The structure described in one layer can sometimes be incorporated into the more superficial or deeper layer in different references.

Layer 1: The iliotibial tract, lateral patellar retinaculum, and biceps femoris tendon are in this layer (Fig. 1.22). The common peroneal nerve has a course between the first and the second layers.

Iliotibial tract: The lateral part of the fascia lata thickens and extends to the leg in the form of a firm band. This band is named the iliotibial tract. The tensor fasciae latae and the gluteus maximus attach to the proximal part of it anteriorly and posteriorly, respectively. A large part of the iliotibial tract extending to the leg terminates on Gerdy's tubercle on the anterolateral surface of the lateral tibial condyle. The anterior part of the iliotibial tract terminates on the lateral edge of the patella. These fibers are named the "iliopatellar band" (Figs. 1.22 and 1.23) [42, 45].

The iliotibial tract moves forward during knee extension and backward during knee flexion. It contributes to the maintenance of extension during knee extension, and also after about 30° of flexion, it passes behind the transverse axis of the knee and contributes to flexion. By preventing varus in the knee, especially in extension, it helps with stability of the knee together with the lateral ligaments and capsular structures.

Lateral patellar retinaculum: Superficially, distal aponeurotic fibers of the vastus lateralis extend at the lateral of the patella. It ends at the lateral edge of the patella together with the oblique fibers of the iliotibial tract (iliopatellar band) (Figs. 1.22 and 1.23) [45]. In the deep layer of the lateral patellar retinaculum, there are fibers

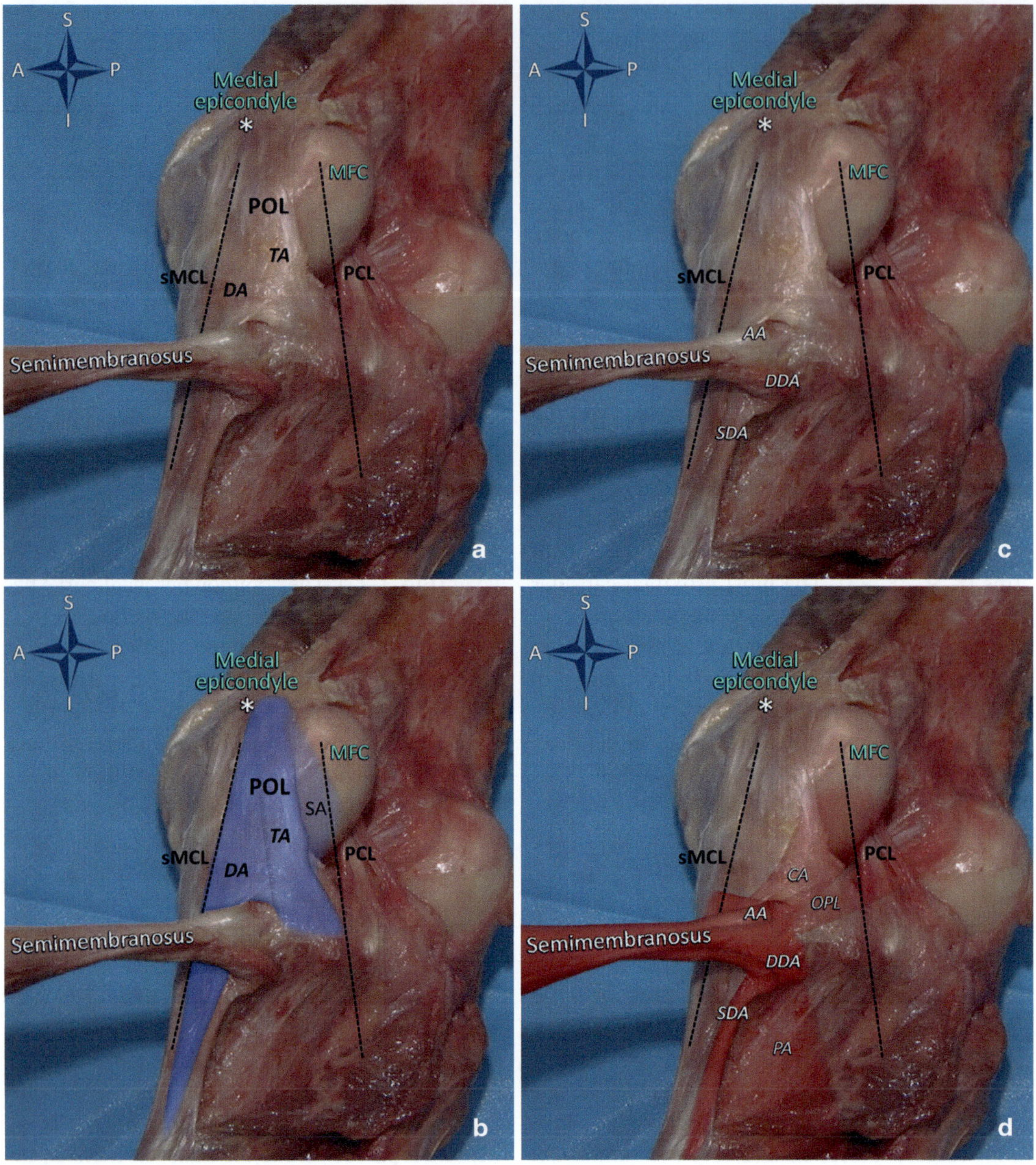

Fig. 1.21 Posteromedial corner (PMC) of the right knee. (**a**, **c**) Native views. (**b**, **d**) Colored views. The PMC of the knee is between the posterior cruciate ligament (PCL) and the superficial medial collateral ligament (sMCL). The dashed lines indicate the boundaries of this region. The arms of posterior oblique ligament (POL) (**a**, **b**) and semimembranosus (**c**, **d**) are demonstrated. The superior arm of POL, capsular arm of semimembranosus, and oblique popliteal ligament removed from the previous stages of dissection are shown as translucent in **b** and **d**. sMCL superficial medial collateral ligament. The arms of POL: *SA* superior (or capsular) arm, *TA* tibial (or central) arm, *DA* distal (or superficial) arm. The arms of semimembranosus: *CA* capsular arm, *AA* anterior arm, *SDA* superficial direct arm, *DDA* deep direct arm, *PA* popliteal arm. *S* superior, *I* inferior, *A* anterior, *P* posterior, on the star showing directions

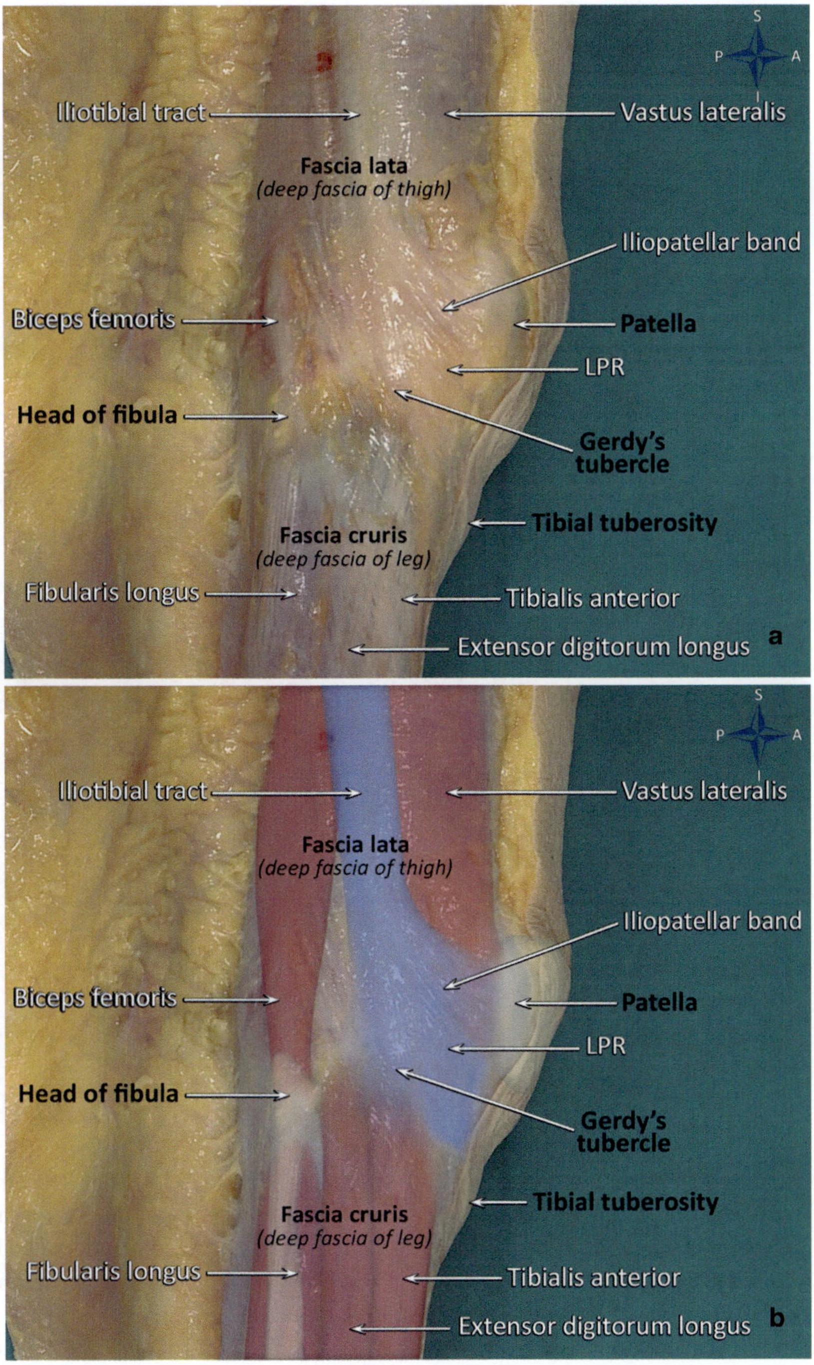

Fig. 1.22 Lateral view of the right knee at extension. (**a**) Native view. (**b**) Colored view. LPR lateral patellar retinaculum. *S* superior, *I* inferior, *A* anterior, *P* posterior, on the star showing directions

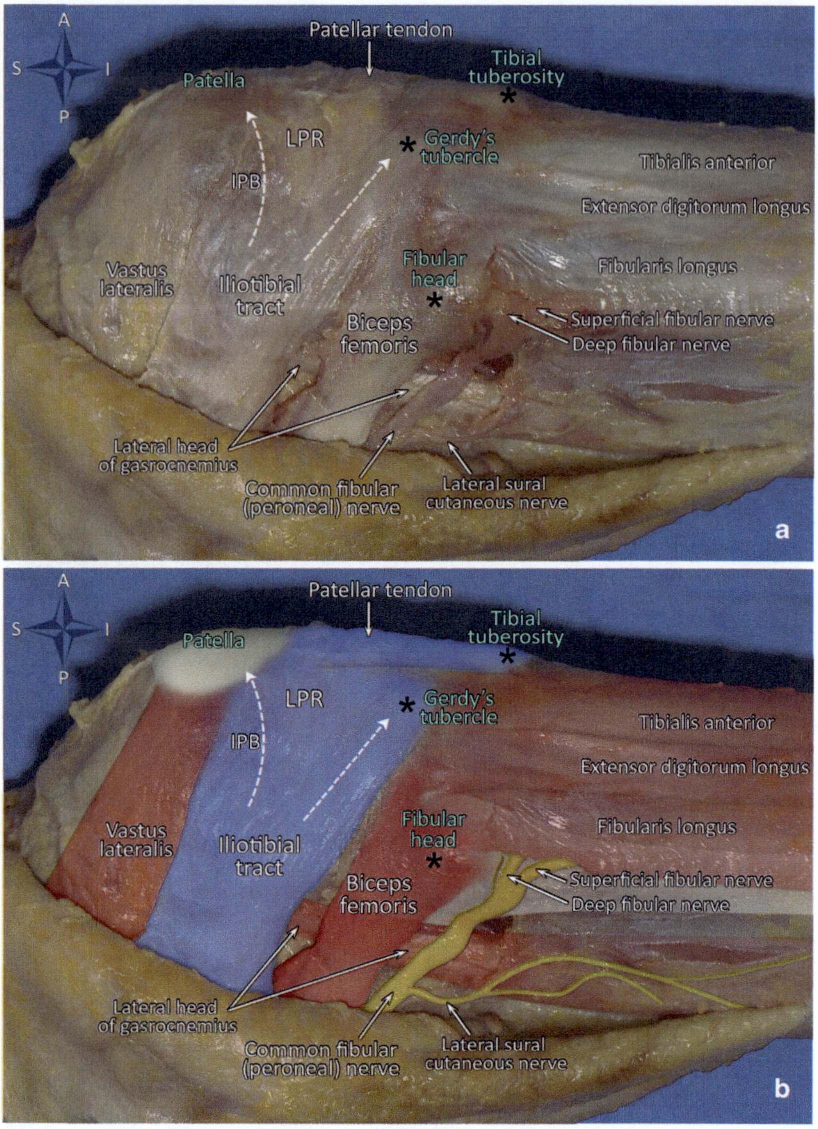

Fig. 1.23 Lateral view of the right knee at 90° flexion. (**a**) Native view. (**b**) Colored view. *LPR* lateral patellar retinaculum, *IPB* iliopatellar band. *S* superior, *I* inferior, *A* anterior, *P* posterior, on the star showing directions

extending from the lateral epicondyle to the lateral of the patella. However, these fibers are much weaker than medial. These fibers are named as the *lateral patellofemoral ligament*. The distal part of deep fibers lies between the patella and the lateral condyle of the tibia. These fibers are named as the *lateral patellotibial ligament*.

Biceps femoris: The biceps femoris is the muscle in the lateral of the posterior compartment of the thigh (lateral hamstring). It has two heads, long and short. The long head starts from the ischial tuberosity via the common hamstring tendon with other ischiocrural muscles. The short head originates from the lateral lip of the linea aspera and the lateral intermuscular septum. The long head is innervated by the tibial nerve and the short head by the common peroneal nerve [1, 2, 42].

The biceps femoris tendon separates into two, over the last part of the lateral collateral ligament (Fig. 1.24). These two parts of the tendon insert on the fibular head, anterior and posterior to the attachment site of LCL. There is a small bursa (biceps femoris bursa) between the LCL and the anterior part of the tendon (Fig. 1.25) [46]. A small part of the fibers can be attached to several

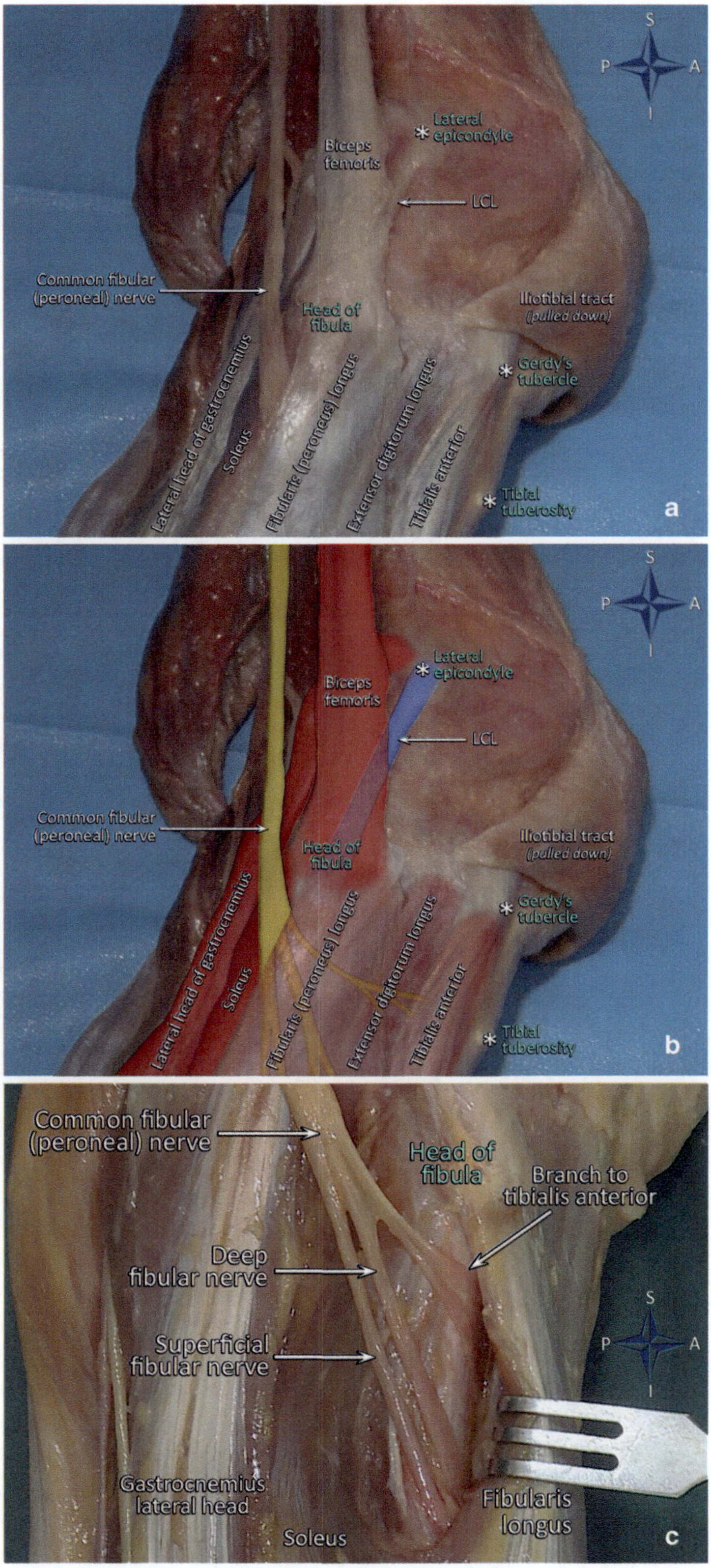

Fig. 1.24 Lateral view of the right knee at extension. (**a**) Native view. (**b**) Colored view. (**c**) Branches of common peroneal nerve and fibularis longus. The iliotibial tract was separated and pulled forward. *LCL* lateral collateral ligament. *S* superior, *I* inferior, *A* anterior, *P* posterior, on the star showing directions

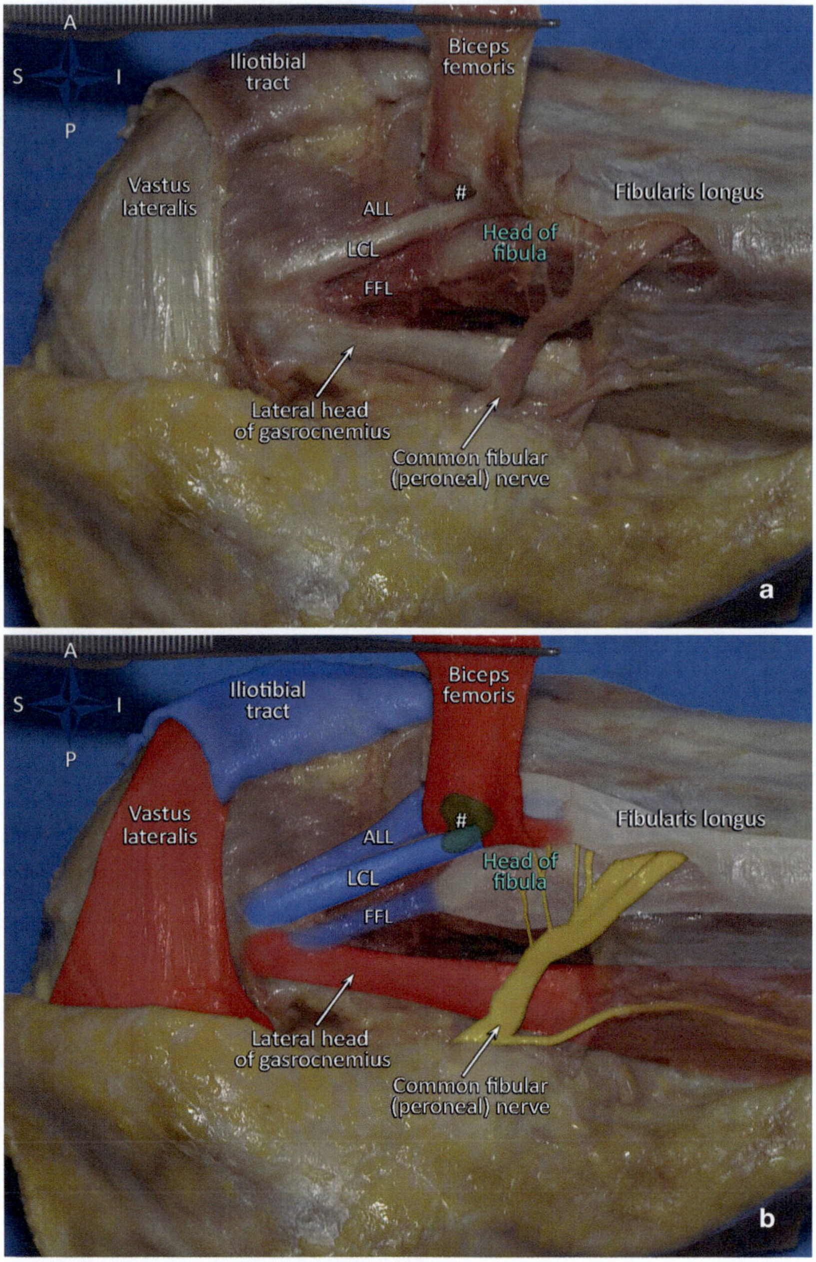

Fig. 1.25 Lateral view of the right knee at 90° flexion. (**a**) Native view. (**b**) Colored view. The iliotibial tract was pulled forward. The biceps femoris tendon was elevated from the lateral collateral ligament (LCL). # biceps femoris bursa, *ALL* anterolateral ligament, *FFL* fabellofibular ligament. *S* superior, *I* inferior, *A* anterior, *P* posterior, on the star showing directions

anatomic structures adjacent to this region, including the LCL, the lateral tibial condyle, the joint capsule, and the meniscotibial ligament, which is a capsular ligament [46–48].

The *common fibular (or peroneal) nerve* is the first structure to be detected during surgeries regarding the lateral side of knee. This nerve separated from the sciatic nerve in the proximal part of the popliteal fossa follows the posterior edge of the biceps tendon. It courses superficially just beneath the popliteal fascia. Common peroneal nerve is located between the superficial and middle layers of the lateral structures of the knee [6, 36, 49]. The nerve gives the lateral sural cutaneous branch that extends to the lateral of the leg before separated from the terminal branches (Fig. 1.23). Where it curves around the head of the fibula, it separates into the terminal branches

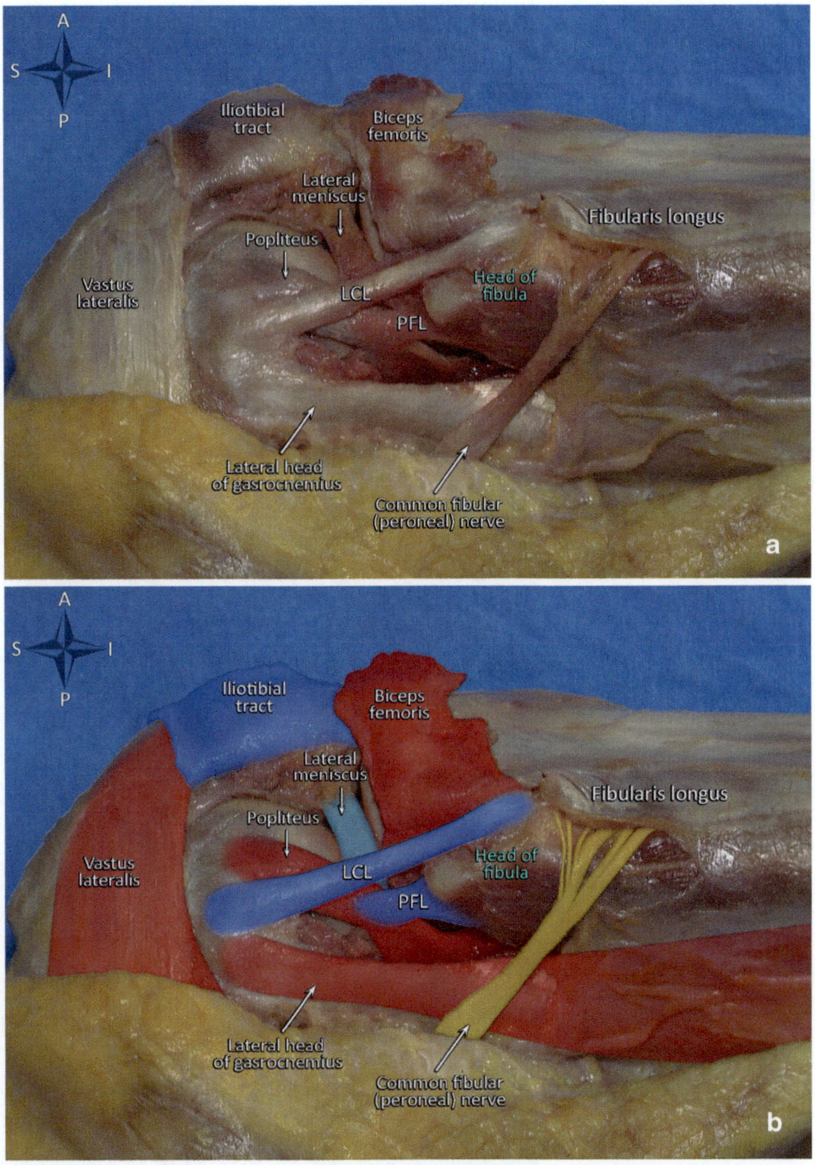

Fig. 1.26 Lateral view of the right knee at 90° flexion. (**a**) Native view. (**b**) Colored view. The iliotibial tract and biceps were pulled forward. Posterolateral capsule was removed. *LCL* lateral collateral ligament, *PFL* popliteofibular ligament. *S* superior, *I* inferior, *A* anterior, *P* posterior, on the star showing directions

(Fig. 1.26) [1, 2]. Here, it is located approximately 2 cm distal to the fibular styloid [50].

The terminal branches of the common peroneal nerve enter the tendinous arch formed by the fibularis longus, as they pass over lateral to the fibular head (Figs. 1.23, 1.24, 1.25 and 1.26) [51]. Deep fibular nerve is located more proximally, and superficial fibular nerve is located more distally (Fig. 1.24). Deep peroneal nerve innervates the anterior compartment muscles of the leg, and the superficial peroneal nerve innervates the lateral compartment muscles of the leg [1, 2]. A few sensory branches that participate to innervation of the tibiofemoral and proximal tibiofibular joints are separated from common peroneal nerve just before giving terminal branches or the first part of the deep peroneal nerve (Figs. 1.25 and 1.26). The first muscular branch separated distally to the sensory branches innervates the tibialis anterior (Fig. 1.24). Therefore, the tibialis anterior is the first muscle to be affected during a surgical procedure around the head of the fibula.

Layer 2: The lateral collateral ligament is the most significant structure in this layer. This layer

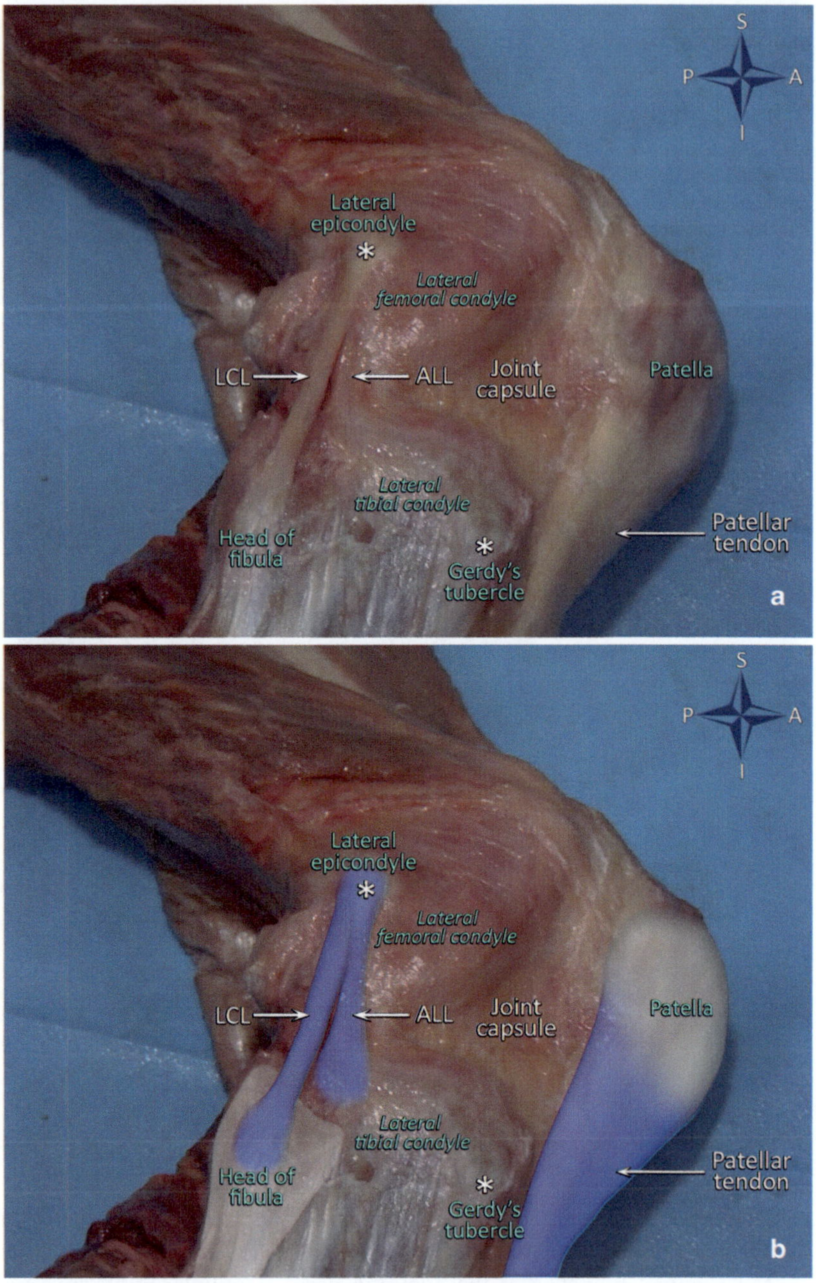

Fig. 1.27 Lateral view of the right knee at 45° flexion. (**a**) Native view. (**b**) Colored view. Iliotibial tract and biceps femoris tendon were removed. *LCL* lateral collateral ligament, *ALL* anterolateral ligament, *S* superior, *I* inferior, *A* anterior, *P* posterior, on the star showing directions

also includes the gastrocnemius lateral head and lateral patellofemoral ligament.

Lateral (fibular) collateral ligament (LCL): LCL originates from the just proximal and posterior to the lateral epicondyle of the femur (Figs. 1.27 and 1.28) [39, 44, 46, 52].

LCL is approximately 66 mm in length. Mean thickness of the LCL is 3.4 mm [5, 52]. Looking at its cross-section, LCL has a round structure, similar to the tendon. It is an extracapsular ligament. Inferior lateral genicular vessels pass through between LCL and the joint capsule. Namely, these vessels are located between the second and third layers.

Distally, LCL terminates on the lateral surface of the fibular head (Figs. 1.27 and 1.28). The liga-

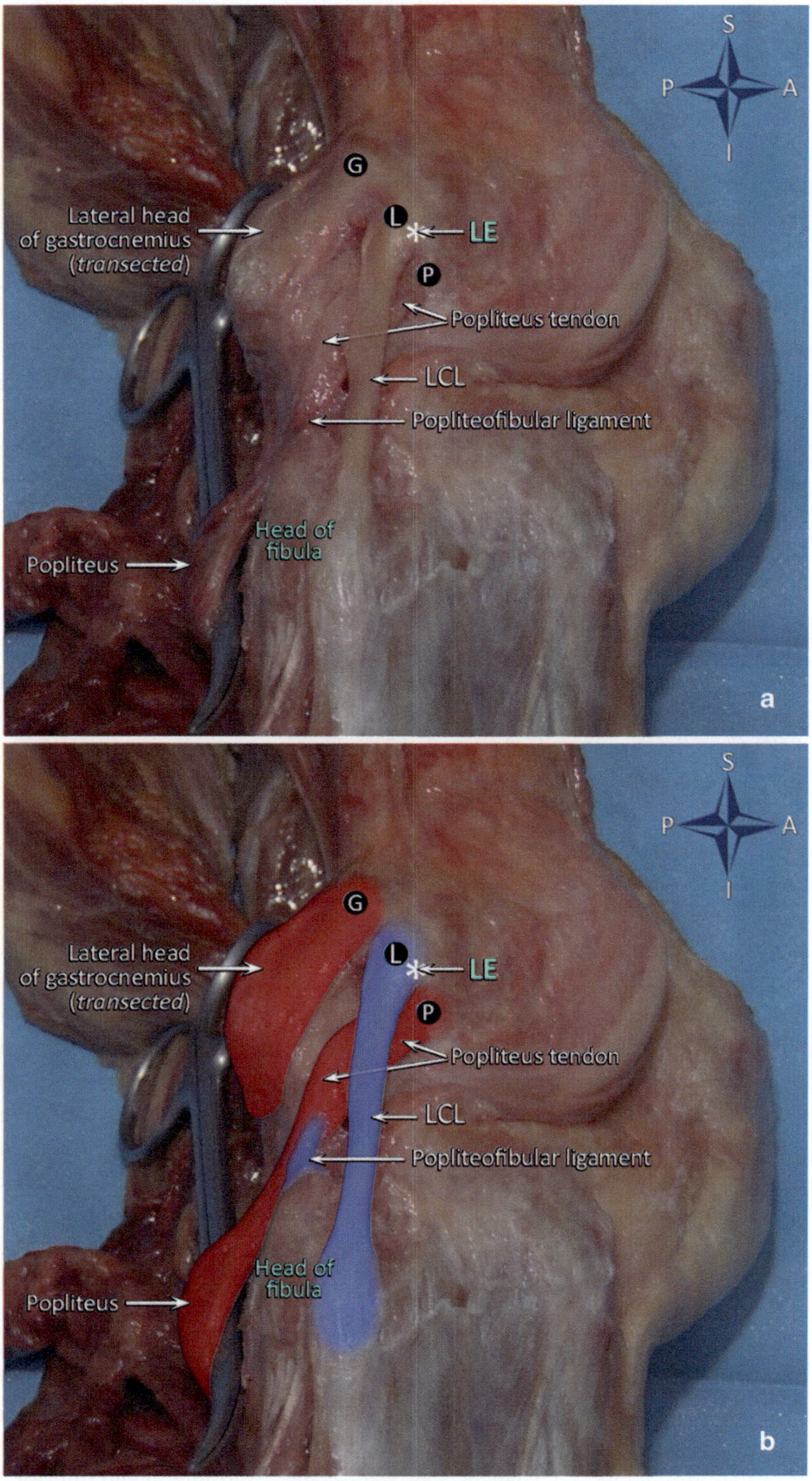

Fig. 1.28 Lateral view of the right knee at extension. (**a**) Native view. (**b**) Colored view. Iliotibial tract, biceps femoris tendon and lateral capsule were removed. Structures that attach to the lateral side of the lateral femoral condyle are seen. Attachment areas: *P* popliteus, *L* lateral collateral ligament (LCL), *G* lateral head of the gastrocnemius. *S* superior, *I* inferior, *A* anterior, *P* posterior, on the star showing directions

ment inserts 28.4 mm distal to the apex of fibular head and 8.2 mm posterior to the anterior edge of the fibula [5, 52]. It can be easily revealed with a 3 cm incision anterior and parallel to the biceps femoris tendon during surgery. The incision is made over the proximal part of the fibular head, and the anterior band of the long head of the biceps femoris should be cut.

LCL is the most important stabilizer of the knee against varus force in extension and in the first 30° of flexion. It also limits external rotation of the tibia at angles close to extension. At flexion angles above 30°, the ligament loosens slightly.

Anterolateral ligament (ALL): ALL originates from just posterior and proximal to the lateral epicondyle, beneath the iliotibial tract. It passes over the initial part of LCL and extends in an anteroinferior direction, anterior to LCL. ALL crosses the lateral meniscus and inserts on the anterolateral of the lateral tibial condyle (Figs. 1.19 and 1.27). The insertion of ALL is 21.6 mm posterior to Gerdy's tubercle and 4–10 mm distal to the tibial plateau [53]. There are connections between the ALL and the lateral meniscus.

ALL is taut in extension and in internal rotation. Injury of ALL often occurs together with ACL damage and is related with a Segond fracture [53]. Segond fracture is a small avulsion fracture immediately distal to the joint surface on the lateral tibial condyle which is caused by forcing knee into varus with internal rotation of the tibia. Generally, there is a combination of ACL (75%–100%), medial meniscus (66%–75%), and ALL damage.

Layer 3: The third layer contains the structures forming the posterolateral corner which are the arcuate ligament, popliteus tendon, popliteofibular ligament, and the joint capsule.

1.3.2.1 Popliteus and Popliteus Complex

The popliteus originates from the posterior surface of the proximal part of the tibia (Fig. 1.29). It has an oblique course laterally and superiorly, and the muscle becomes a tendon in the lateral third of the popliteal fossa. Then, the popliteus tendon becomes intra-articular by passing through the opening found in the posterolateral capsule. The arcuate ligament forms over the opening (Fig. 1.30). A recess of the joint cavity occurs between the popliteus tendon and the lateral meniscus. The opening of this recess between the lateral meniscus and the popliteus tendon is called the popliteal hiatus (Fig. 1.31) [5, 54]. Synovial membrane covers the meniscus in the popliteal hiatus. As this part of the meniscus has no lateral connection, it appears bare (*bare area of the lateral meniscus*). The popliteus tendon passes under LCL during its course within the joint. The tendon is located in a shallow groove in flexion, which can be seen on the outer surface of the lateral femoral condyle (Figs. 1.3 and 1.19). However, full seating in this groove requires approximately 110° flexion [5]. The popliteus tendon inserts immediately anteroinferior to the femoral attachment site of LCL (Fig. 1.31).

The popliteus tendon, which is approximately 55 mm long, has significant functions among PLC structures of the knee. The popliteus allows tibial internal rotation or femoral external rotation. At the same time, it is responsible for dynamic stabilization of the lateral meniscus. To be able to achieve these functions, the popliteus is connected to the several structures in the posterolateral aspect of the knee. The muscle–tendon unit of the popliteus and the connections of this unit with the fibula, tibia, and lateral meniscus form the "popliteus complex." *The* popliteofibular ligament provides its connection with the fibula. By passing over the tendon, the arcuate ligament contributes to its stability. Popliteus tendon is attached via the popliteomeniscal fascicles to the lateral meniscus (Fig. 1.31).

Popliteofibular ligament (PFL): It is the second most important structure in the posterolateral corner, after the popliteus tendon. PFL together with the popliteus tendon is the most significant stabilizer in the PLC. The ligament is separated near the musculotendinous junction of the popliteus tendon [5, 55, 56]. Coursing toward the posterolateral and inferior, it is attached to almost the top of the fibular styloid. The fibers of the ligament extend a few millimeters posterior to the fibular styloid (Figs. 1.29 and 1.30).

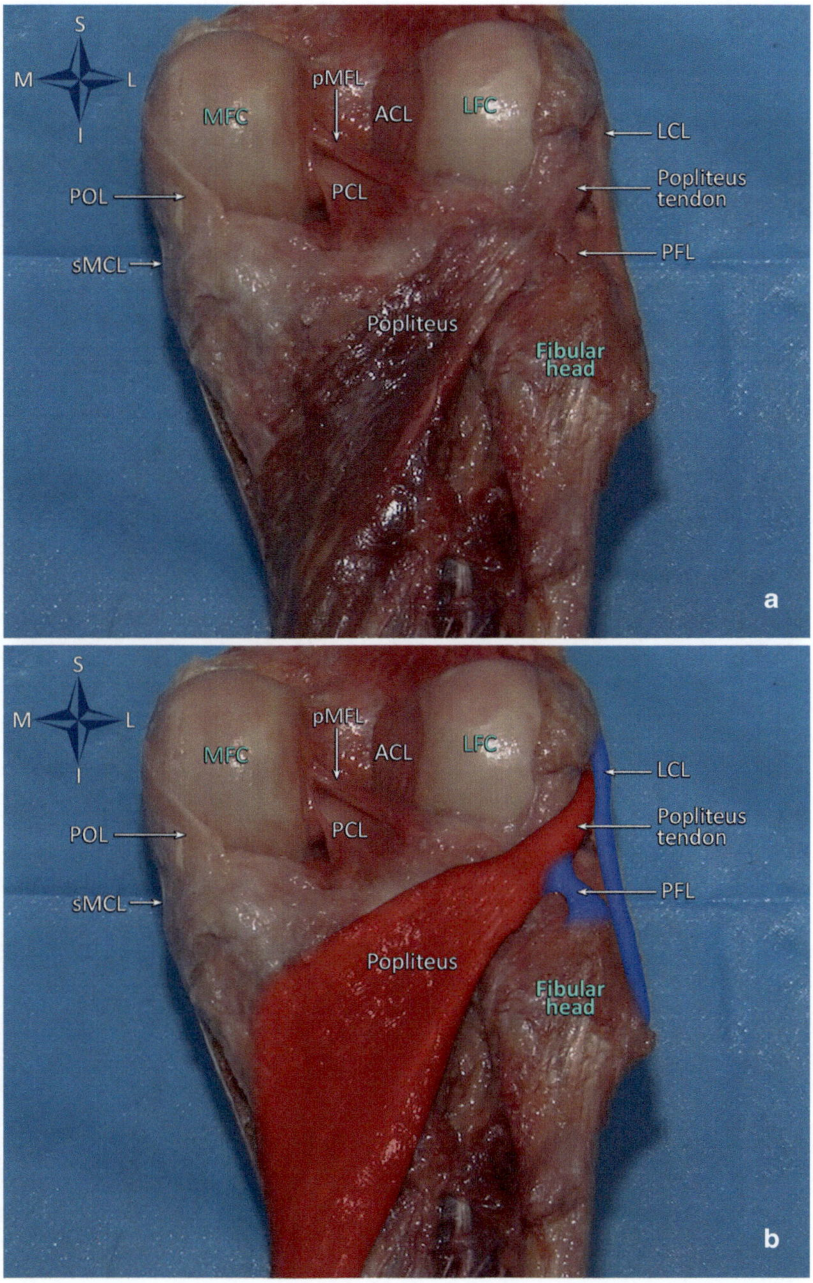

Fig. 1.29 Posterior view of the right knee. (**a**) Native view. (**b**) Colored view. All tissues behind the knee joint, including the capsule, were removed. Popliteus and its connections are demonstrated. *LCL* lateral collateral ligament, *PFL* popliteofibular ligament, *ACL* anterior cruciate ligament, *PCL* posterior cruciate ligament, *pMFL* posterior meniscofemoral ligament, *sMCL* superficial medial collateral ligament, *POL* posterior oblique ligament. *S* superior, *I* inferior, *M* medial, *L* lateral, on the star showing directions

The popliteofibular ligament limits external rotation, posterior tibial translation, and varus. It is a short, strong ligament requiring approximately 300 N for rupture and, when forced, causes avulsion in the styloid process rather than rupture (arcuate fracture). This usually occurs with cruciate ligament damage.

1.4 The Anterior Side of the Knee

The anterior structures of the knee are especially responsible for extension and patellar stabilization.

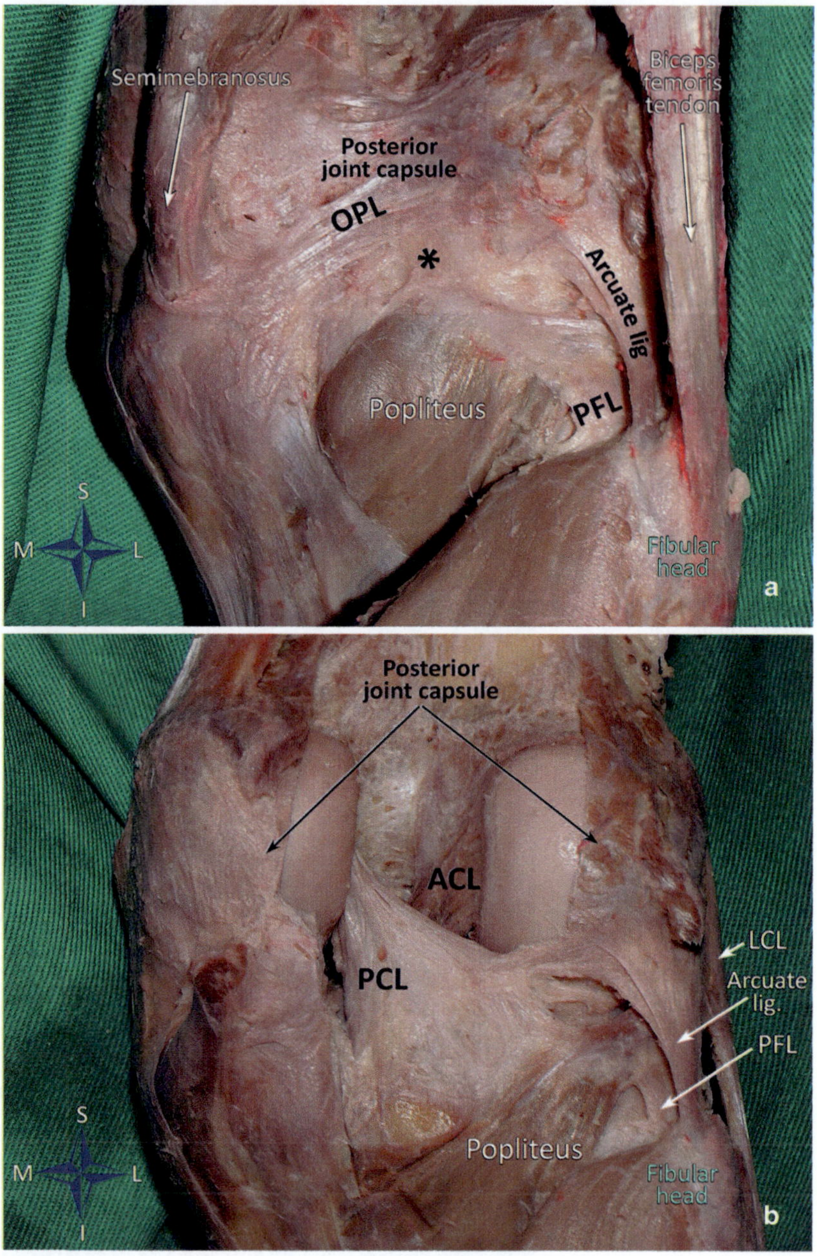

Fig. 1.30 Posterior view of the right knee. All tissues behind the knee joint were removed. Popliteus and its connections are demonstrated. The posterior capsule was preserved in **a**, while the middle part of it was removed in **b**. Oblique popliteal ligament (OPL) and arcuate ligament, which are capsular ligaments, are demonstrated in **a**. Asterisks marks the capsular connection of the popliteus. Important structures in the posterolateral corner of the knee are seen in **b**. *ACL* anterior cruciate ligament, *PCL* posterior cruciate ligament, *LCL* lateral collateral ligament, *PFL* popliteofibular ligament. *S* superior, *I* inferior, *M* medial, *L* lateral, on the star showing directions

The *patellar tendon* (or patellar ligament in anatomical terminology) is a strong tendon extending from the inferior aspect of the patella to tibial tuberosity. This ligament transmits the power of quadriceps femoris to tibia. The approximate length of the patellar tendon is about 6–8 cm [2]. The infrapatellar fat pad (Hoffa's fat pad) is located deep in these structures (Fig. 1.32) [57].

The lateral patellar retinaculum is located lateral to the patella. Superficial fibers of lateral patellar retinaculum extend from patella to anterolateral aspect of the tibia. This part is called the *lateral patellotibial ligament*. This ligament attaches just proximal to Gerdy's tubercle at the anterolateral aspect of the tibia (Fig. 1.32). Similar to the lateral side, the part of medial

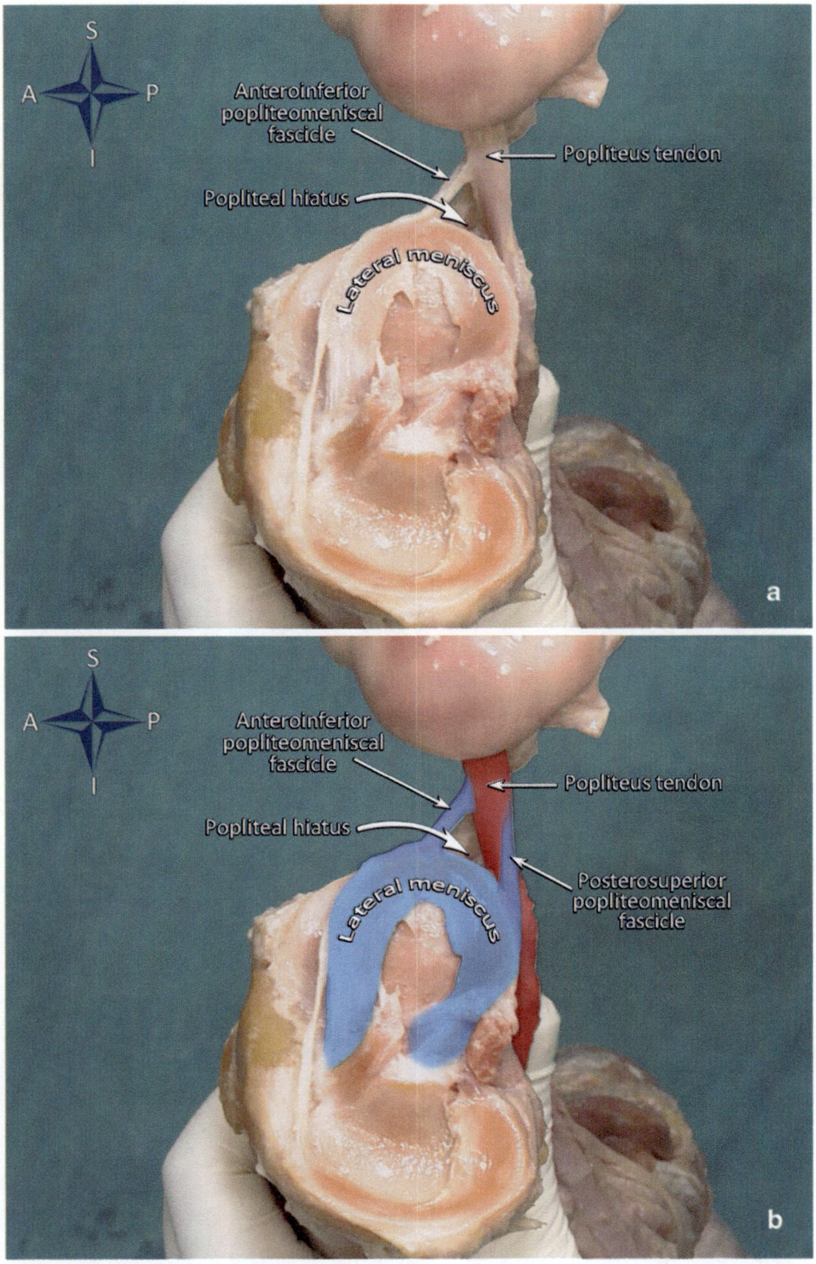

Fig. 1.31 Medial view of the popliteus tendon and its connections. (**a**) Native view. (**b**) Colored view. *S* superior, *I* inferior, *A* anterior, *P* posterior, on the star showing directions

patellar retinaculum which extends from the medial edge of patella to tibia is called *medial patellotibial ligament* (Fig. 1.32).

Patella tends to move laterally because of the existence of a 5–7 degree tibiofemoral angle. Eventually, structures that support the medial side of the patella are more prominent. Vastus medialis obliquus and medial patellofemoral ligament are the most important among these structures [58, 59]. Medial patellotibial ligament contributes to supporting the medial side of the patella (Fig. 1.33).

As it is mentioned before, superficial fibers of lateral and medial patellar retinaculum make the lateral and medial patellotibial ligaments that extend from patella to tibia. Deep fibers that extend from both the sides of the patella through distally are called patellomeniscal ligaments. As its name implies, *lateral and medial patellomeniscal ligaments* connect the patella to the

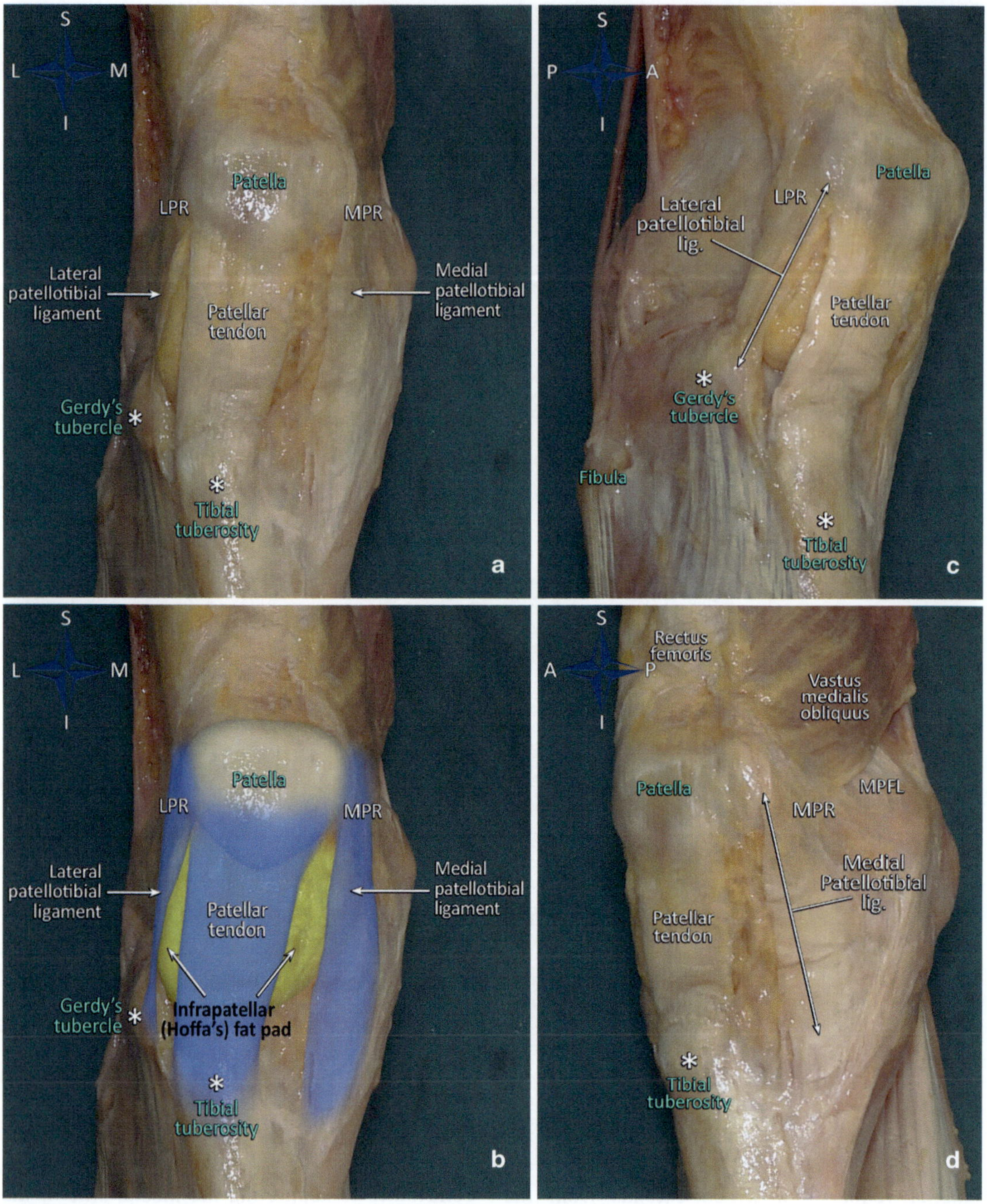

Fig. 1.32 Anterior (**a**, **b**), anterolateral (**c**), and anteromedial (**d**) views of the right knee. (**a**, **c** and **d**) Native view. (**b**) Colored view. *LPR* lateral patellar retinaculum, *MPR* medial patellar retinaculum, *MPFL* medial patellofemoral ligament. *S* superior, *I* inferior, *L* lateral, *M* medial, on the star showing directions

anterior horns of menisci both laterally and medially (Figs. 1.34 and 1.35).

The quadriceps femoris has four sections that attach to the patella proximally (Figs. 1.36 and 1.37). The middle and most superficial section is the rectus femoris (Fig. 1.36). Rectus femoris becomes tendon 3–5 cm proximal to patella. While some of the fibers are attached to the

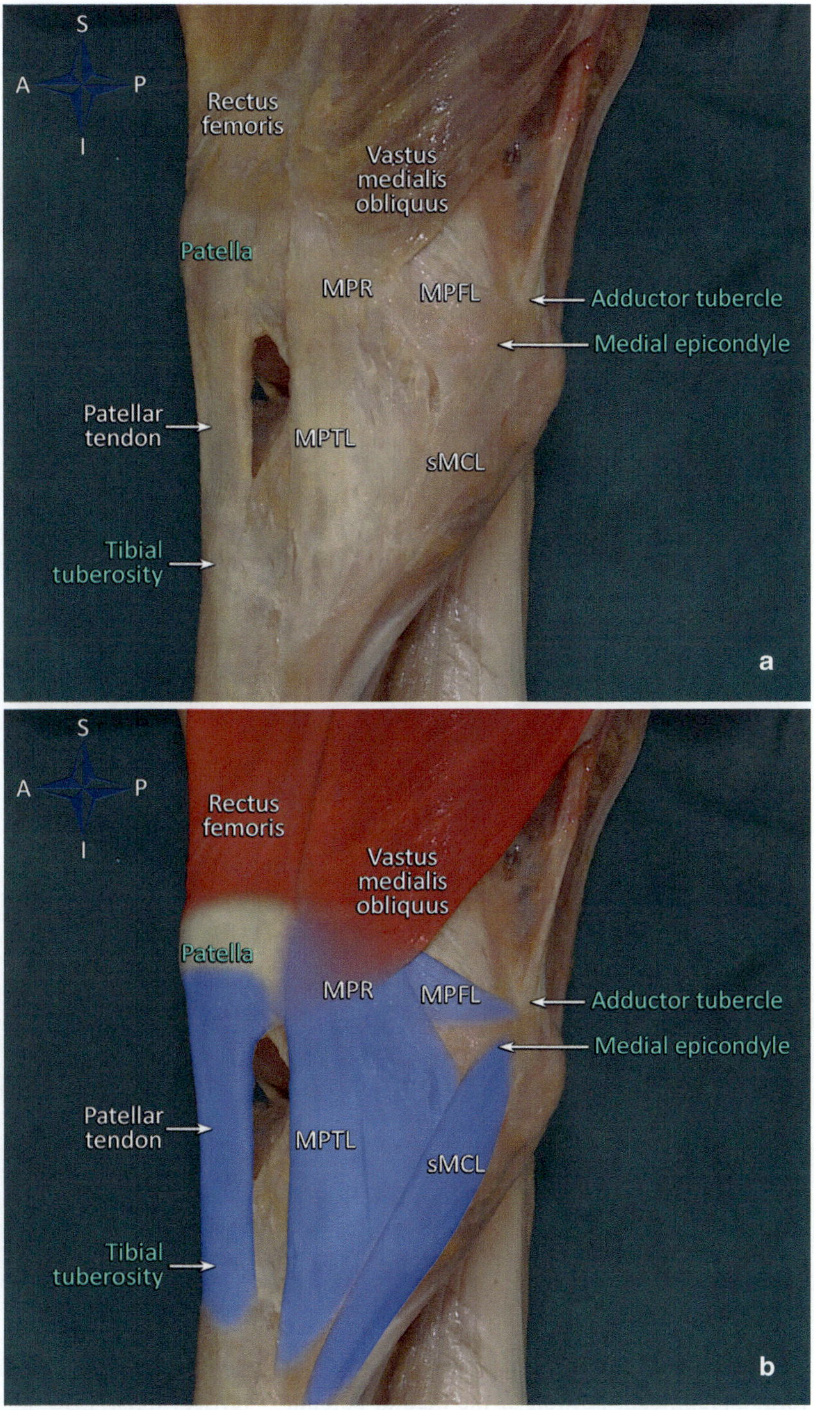

Fig. 1.33 Anteromedial view of the right knee. (**a**) Native view. (**b**) Colored view. *LPR* lateral patellar retinaculum, *MPR* medial patellar retinaculum, *MPFL* medial patellofemoral ligament, *MPTL* medial patellotibial ligament, *sMCL* superficial medial collateral ligament. *S* superior, *I* inferior, *A* anterior, *P* posterior, on the star showing directions

upper edge of patella, most of fibers pass over the patella and join the structure of the patellar tendon. Vastus medialis is the medial section of quadriceps femoris. Vastus medialis has two parts. Lateral part extending more vertical is the vastus lateralis longus. Medial part extending more oblique is the vastus medialis obliquus. It is attached to superomedial edge of patella.

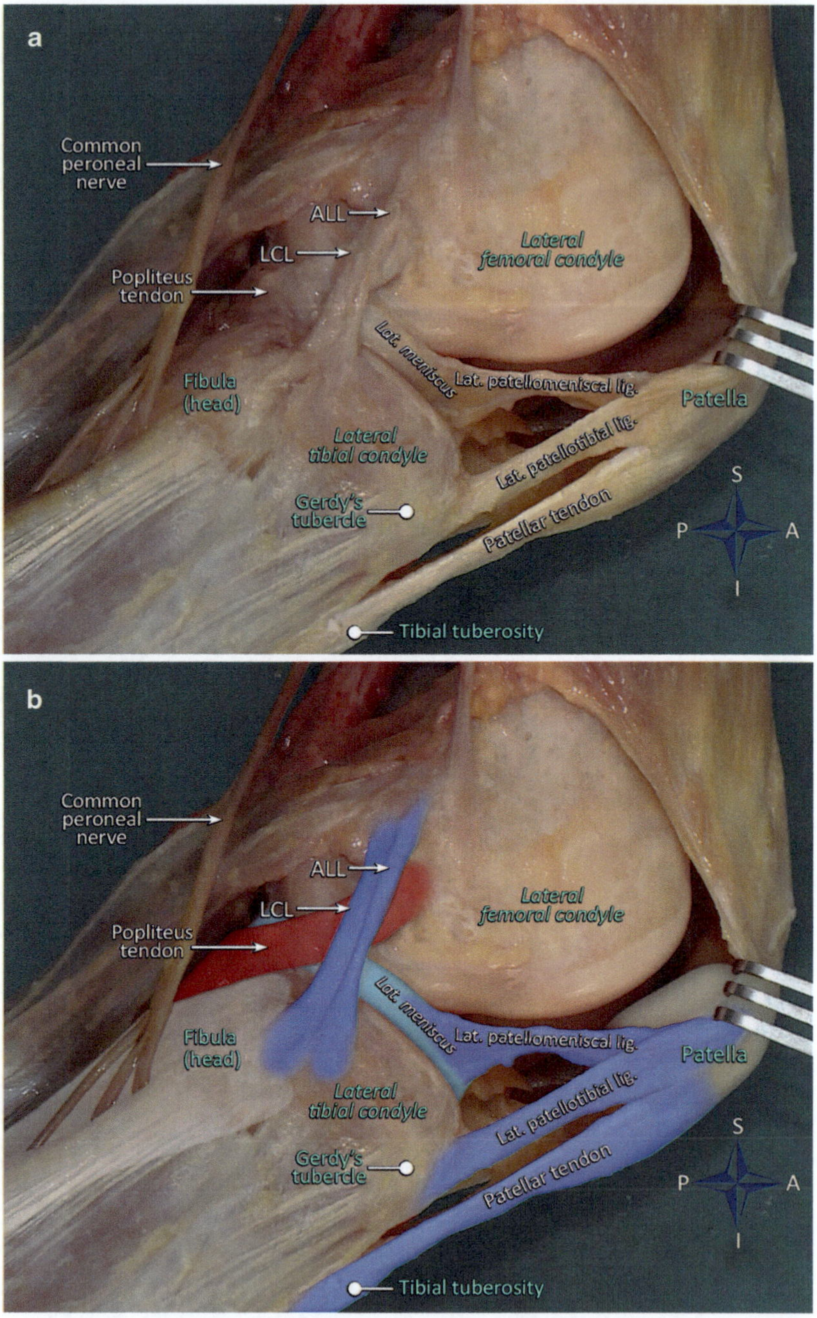

Fig. 1.34 Anterolateral view of the right knee. (**a**) Native view. (**b**) Colored view. Lateral joint capsule was removed. Lateral connections of the patella are demonstrated. *LCL* lateral collateral ligament, *ALL* anterolateral ligament. *S* superior, *I* inferior, *A* anterior, *P* posterior, on the star showing directions

Aponeurotic fibers of distal portion of VMO form a significant portion of the medial patellar retinaculum (Fig. 1.36). A small part joins patellar tendon. Vastus lateralis also includes longitudinal and oblique parts like vastus medialis. It becomes tendinous before VMO, and almost all of it is attached to the superolateral edge of patella. VMO fibers angle about 60° with the vertical axis. VLO fibers angle about 40° with the vertical axis. The vastus intermedius is the middle and deep section of quadriceps femoris. This muscle is attached to most superior and posterior parts of anterior patellar surface (Fig. 1.37) [58, 59].

Fig. 1.35 Anteromedial view of the right knee. (**a**) Native view. (**b**) Colored view. Anteromedial joint capsule was removed. Medial connections of the patella are demonstrated. *S* superior, *I* inferior, *L* lateral, *M* medial, on the star showing directions

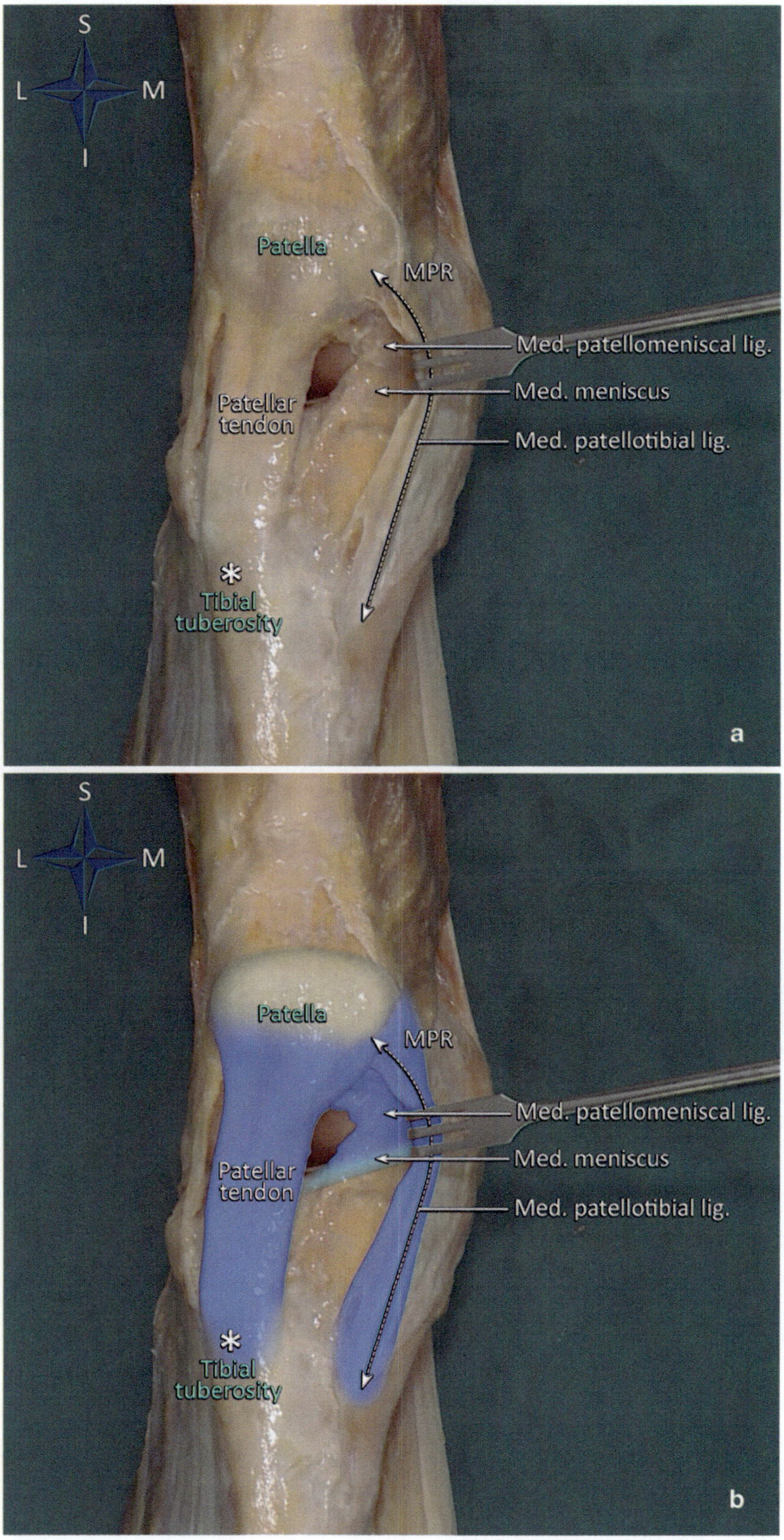

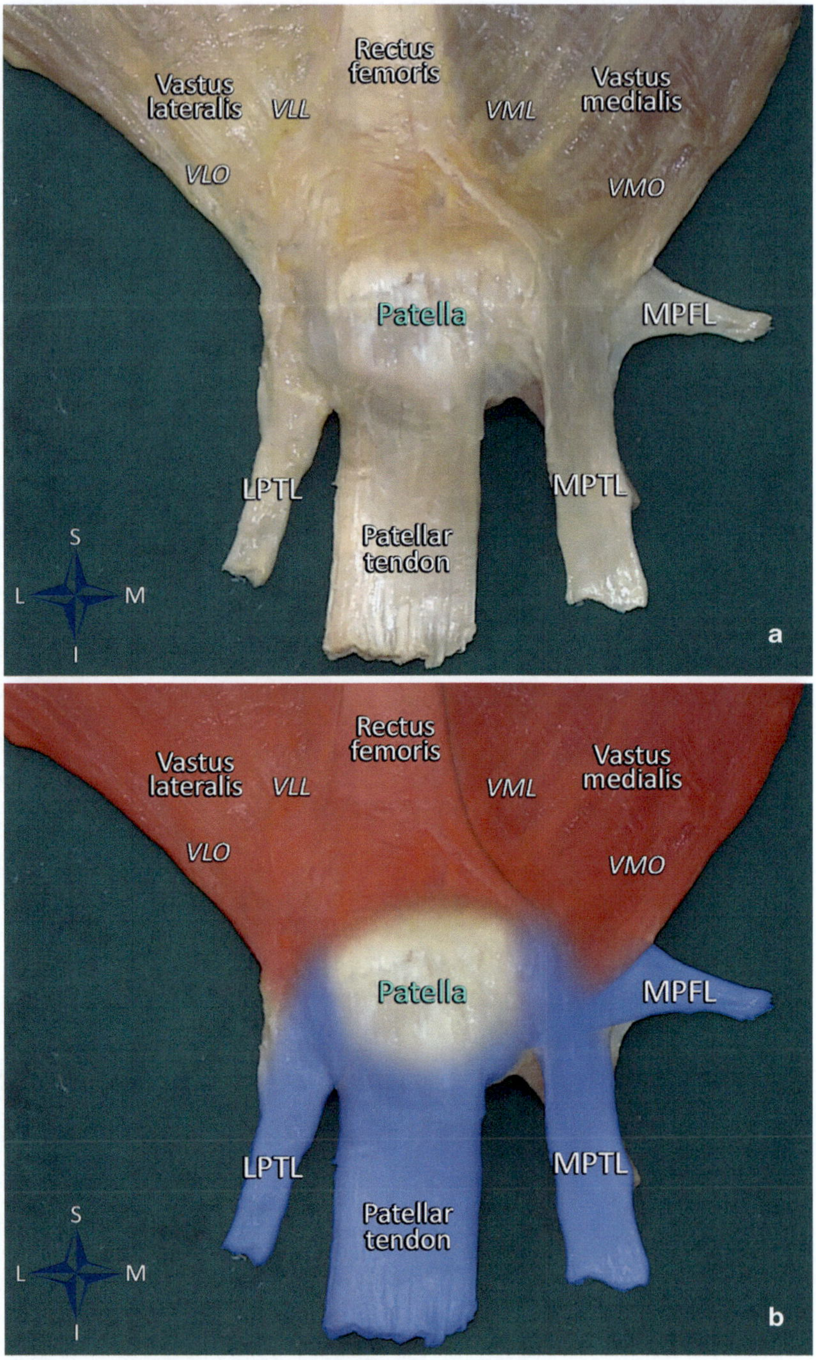

Fig. 1.36 Anterior view of patella with all structures attached to it. (**a**) Native view. (**b**) Colored view. *VLO* vastus lateralis obliquus, *VLL* vastus lateralis longus, *VMO* vastus medialis obliquus, *VML* vastus medialis longus, *MPFL* medial patellofemoral ligament, *MPTL* medial patellotibial ligament, *LPTL* lateral patellotibial ligament. *S* superior, *I* inferior, *L* lateral, *M* medial, on the star showing directions

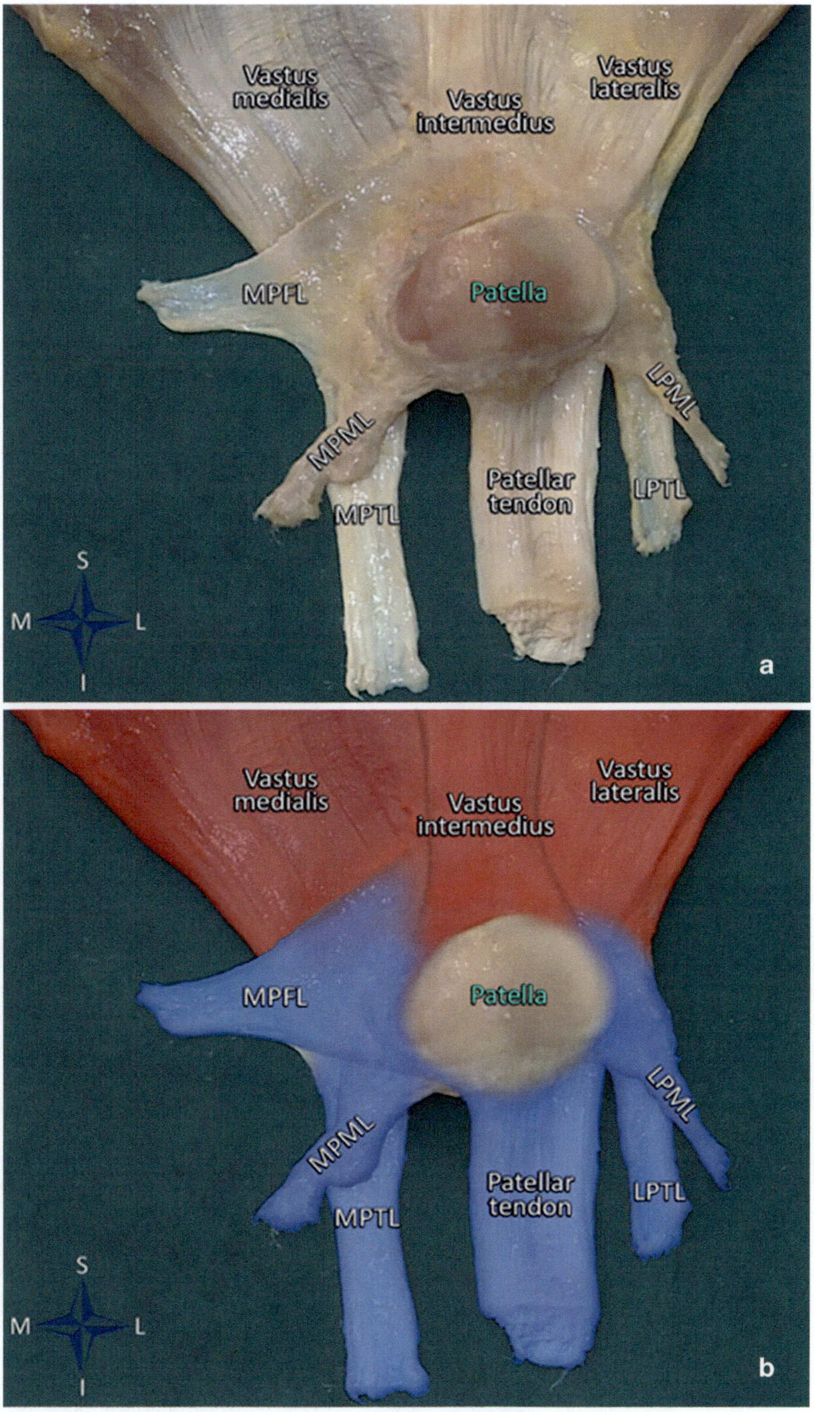

Fig. 1.37 Posterior view of patella with all structures attached to it. (**a**) Native view. (**b**) Colored view. *MPFL* medial patellofemoral ligament, *MPML* medial patellomeniscal ligament, *MPTL* medial patellotibial ligament, *LPML* lateral patellomeniscal ligament, *LPTL* lateral patellotibial ligament. *S* superior, *I* inferior, *M* medial, *L* lateral, on the star showing directions

The synovial space of the knee joint extends upward under the quadriceps muscle. This impasse of the joint is called suprapatellar recess or suprapatellar pouch. This recess extends about 5 cm above the superior pole of the patella (Fig. 1.38).

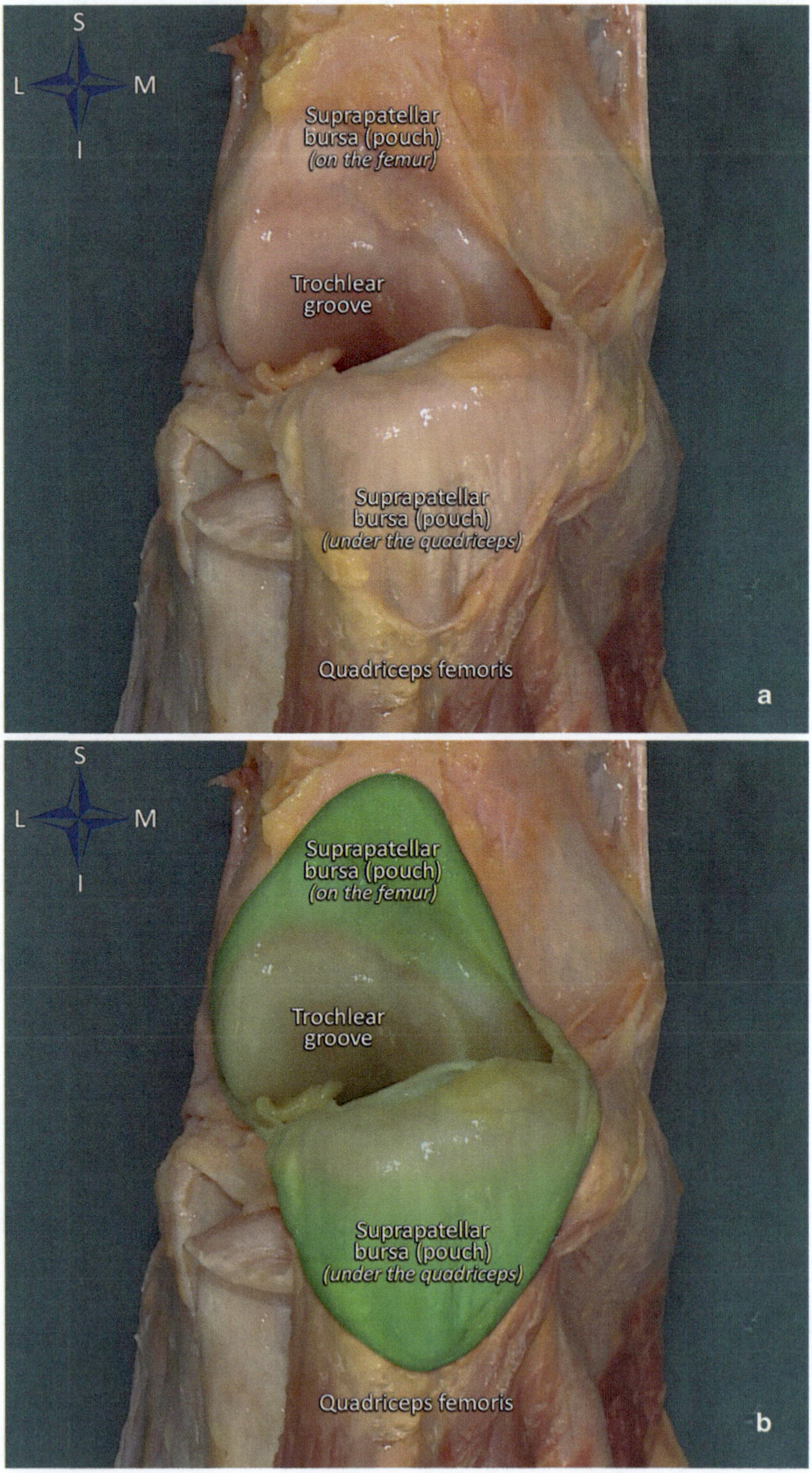

Fig. 1.38 Anterior view of the right knee. (**a**) Native view. (**b**) Colored view. The quadriceps femoris was pulled down with the patella. The boundaries of the suprapatellar pouch are seen. *S* superior, *I* inferior, *L* lateral, *M* medial, on the star showing directions

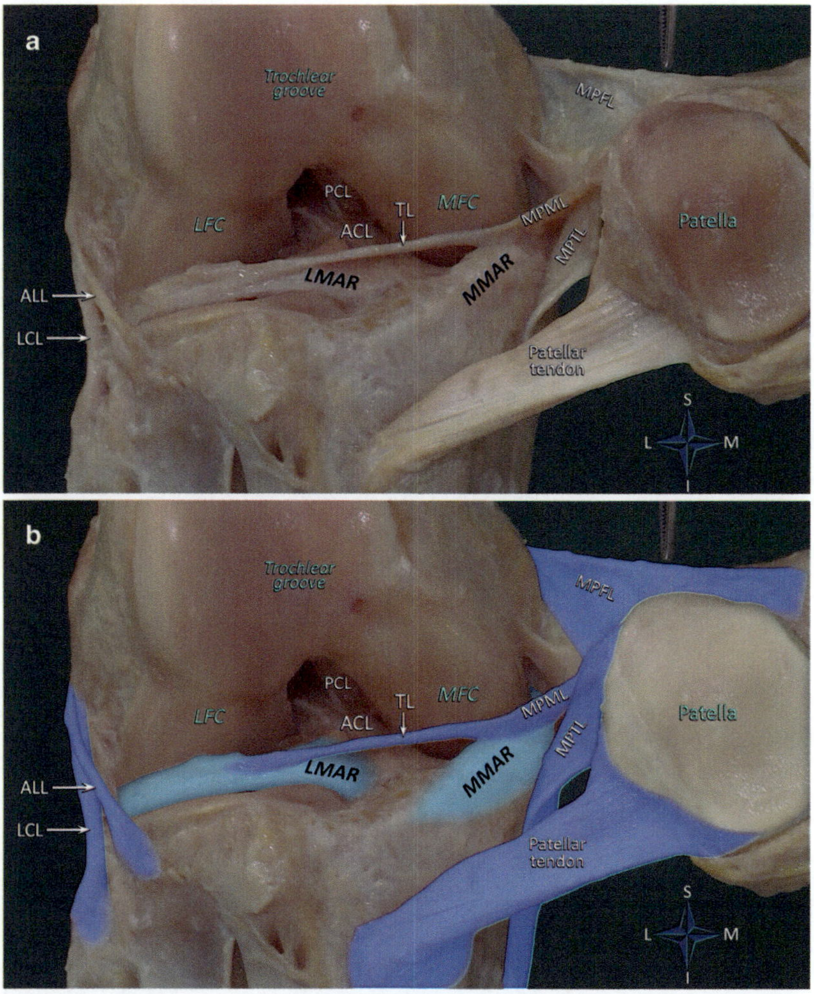

Fig. 1.39 Anterior view of the right knee. (**a**) Native view. (**b**) Colored view. The quadriceps femoris was retracted with the patella inferomedially. The joint capsule was opened. The anterior horns of menisci and anterior cruciate ligament (ACL), which attach to the anterior intercondylar area, are seen. *LCL* lateral collateral ligament, *ALL* anterolateral ligament, *PCL* posterior cruciate ligament, *TL* transverse ligament, *LMAR* lateral meniscus anterior root, *MMAR* medial meniscus anterior root, *MPFL* medial patellofemoral ligament, *MPML* medial patellomeniscal ligament, *MPTL* medial patellotibial ligament, *LFC* lateral femoral condyle, *MFC* medial femoral condyle. *S* superior, *I* inferior, *L* lateral, *M* medial, on the star showing directions

Transverse ligament is a weak ligament that connects the anterior horns of both menisci (Fig. 1.39).

1.5 The Popliteal Fossa and Posterior Side of the Knee

Popliteal fossa is a diamond-shaped area posterior to the knee joint. It is limited by the superolaterally biceps femoris, the superomedially semimembranosus and semitendinosus, the inferomedially medial head of gastrocnemius, and the inferolaterally lateral head of gastrocnemius and plantaris (Fig. 1.40). As we mentioned before in the medial side of the knee, the gracilis tendon and sartorius are located anterior to semitendinosus. Saphenous nerve emerges between these muscles. Although this nerve, which is the branch of the femoral nerve, is not located in the fossa poplitea, it can be observed medially from the posterior view [1, 2].

When the popliteal fascia is removed, the popliteal fat pad appears on the popliteal fossa (Fig. 1.41). Posterior femoral cutaneous nerve, a branch of sacral plexus, extends to the popliteal fossa with a long course under the deep fascia on posterior thigh and carries the cutaneous sense of this area. The lateral sural cutaneous nerve, which is the common fibular nerve branch, can be observed laterally to the popliteal fossa. The small saphenous vein extends from the posterior leg advances to deep and opens to the popliteal vein. This vein is accompanied by the medial

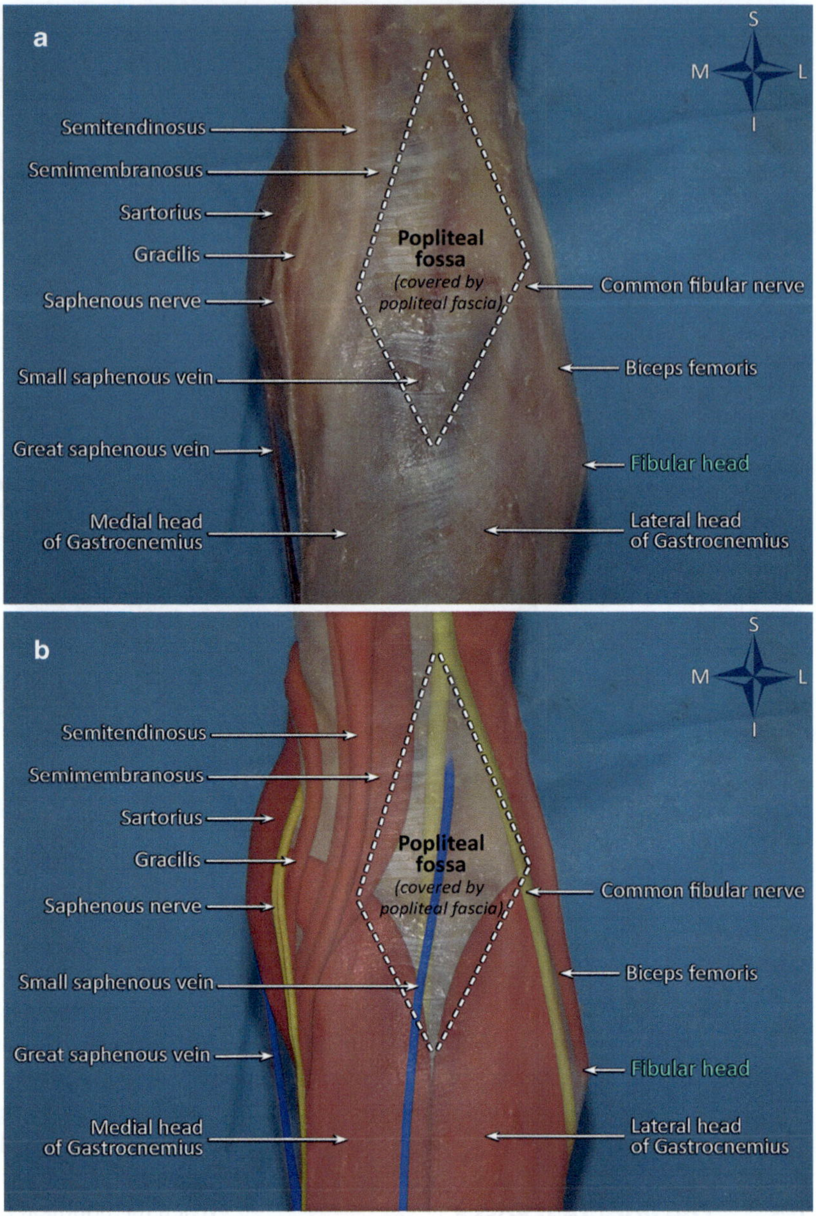

Fig. 1.40 Posterior view of the right knee. (**a**) Native view. (**b**) Colored view. Skin and subcutaneous tissue were removed. The boundaries of the popliteal fossa are shown. *S* superior, *I* inferior, *M* medial, *L* lateral, on the star showing directions

sural cutaneous nerve, which is the branch of the tibial nerve (Fig. 1.41).

Sciatic nerve and its terminal branches (tibial and common peroneal nerves) are located more superficially compared to the popliteal vessels in the popliteal fossa, just beneath popliteal fascia [1, 2]. Superolaterally, the biceps femoris tendon attaches to the fibular head. Common fibular (or peroneal) nerve courses parallel to the medial edge of the biceps femoris at the most lateral side of popliteal fossa (Fig. 1.41).

When the popliteal fat pad is removed, the muscles around the popliteal fossa and important neurovascular structures inside the fossa can be clearly visible. Popliteal vessels are located deeper in the popliteal fossa. The popliteal artery is the deepest structure, that is, closest to the posterior joint capsule. Popliteal vein is located just above it. The tibial nerve is located on the superficial and slightly lateral of the popliteal vessels. The common fibular nerve extends more laterally parallel to the biceps tendon (Fig. 1.41).

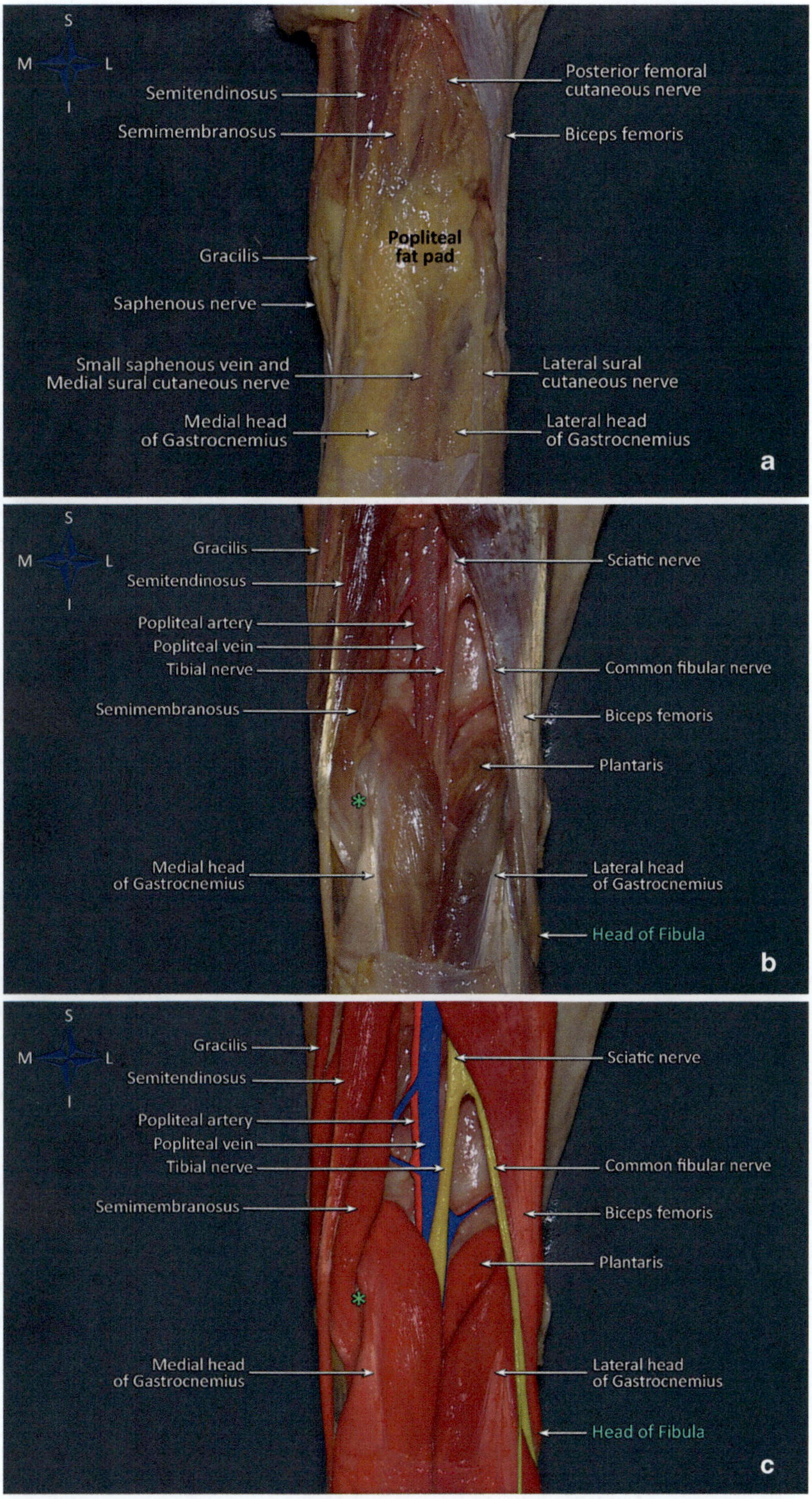

Fig. 1.41 Posterior view of the right knee. (**a**, **b**) Native view. (**c**) Colored view of **b**. Skin, subcutaneous tissue and popliteal fascia were removed in a. In addition, popliteal adipose tissue was also removed in **b** and **c**. The neurovascular structures are shown in the popliteal fossa. Green asterisk shows the location of the gastrocnemius-semimembranosus bursa. *S* superior, *I* inferior, *M* medial, *L* lateral, on the star showing directions

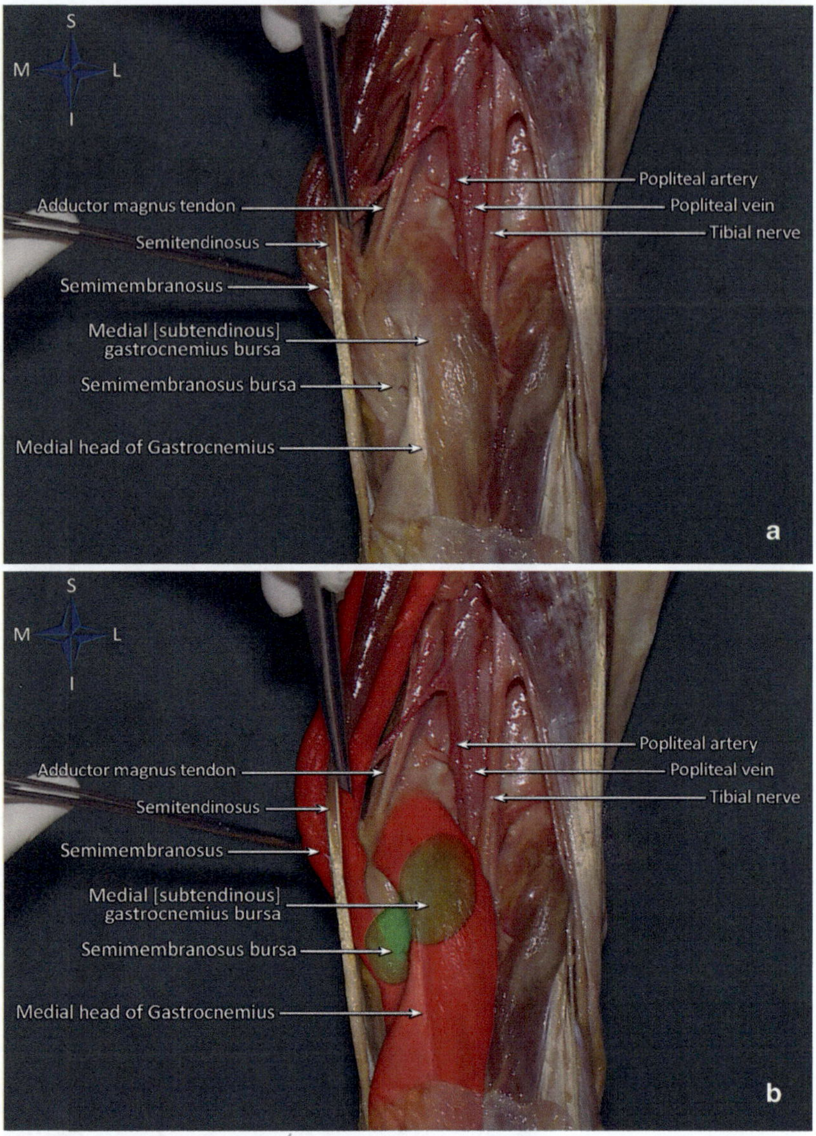

Fig. 1.42 Posterior view of the right knee. (**a**) Native view. (**b**) Colored view. The semimembranosus was pulled medially. The gastrocnemius-semimembranosus bursa is demonstrated in green. Baker's cyst originates from this bursa and extends to the popliteal fossa. *S* superior, *I* inferior, *M* medial, *L* lateral, on the star showing directions

The semimembranosus bursa is located under the semimembranosus, and the gastrocnemius bursa is under the medial head of the gastrocnemius. These bursae are generally merged and named as gastrocnemius-semimembranosus bursa. Baker's cyst originates from this bursa and extends between those two muscles to the popliteal fossa (Fig. 1.42).

When the lateral and medial heads of the gastrocnemius are pulled laterally and medially, the branches of the popliteal vessels and tibial nerve supplying to these muscles can be seen (Fig. 1.43).

The medial sural cutaneous nerve, which originates from the tibial nerve in the popliteal fossa, extends distally with the small saphenous vein. The lateral sural cutaneous nerve, which originates from the common fibular nerve in the popliteal fossa, extends distally over the lateral head of gastrocnemius (Figs. 1.41 and 1.43). This nerve gives cutaneous branches to lateral side of the leg. Terminal branch of the nerve courses toward the medial and merge with the last part of the medial sural cutaneous nerve to form the sural nerve.

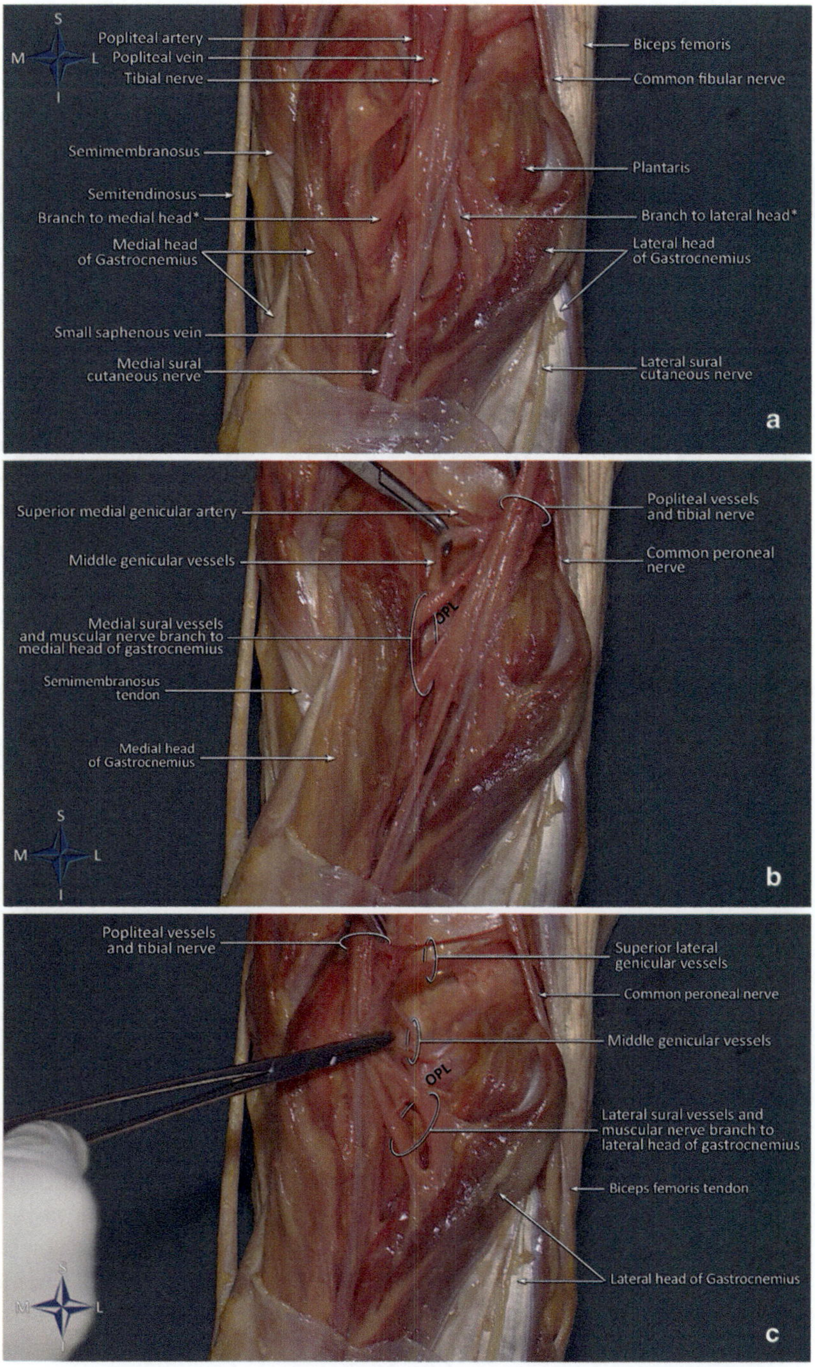

Fig. 1.43 Posterior view of the right knee. (**a**) The lateral and medial heads of the gastrocnemius were pulled laterally and medially. (**b**) The popliteal vessels and tibial nerve were pulled laterally. (**c**) The popliteal vessels and tibial nerve were pulled medially. The branches of the popliteal vessels and tibial nerve are seen in the popliteal fossa. *S* superior, *I* inferior, *M* medial, *L* lateral, on the star showing directions

When the popliteal vessels and tibial nerve are pulled laterally, superior medial genicular vessels proximally, middle genicular vessels just behind the posterior joint capsule (close to the upper edge of the oblique popliteal ligament), and the medial sural vessels extending to the medial head of gastrocnemius more distally can be seen. When the popliteal vessels are pulled medially, the superior lateral genicular vessels, middle genicular vessels, and lateral sural vessels and motor branch of tibial nerve extending to the lateral head can be seen, respectively (Fig. 1.43).

When both the heads of the gastrocnemius are separated and all structures are eliminated behind the capsule, the relationship of popliteal neurovascular structures with the posterior capsule are seen clearly (Fig. 1.44). Semimembranosus inserts at the posteromedial of the proximal tibia. Semimembranosus is one of the most important structures at the posteromedial of the knee joint. This muscle has many extensions that continue as ligaments in this region. One of these is the oblique popliteal ligament on the posterior joint capsule. The popliteus is seen that starts from the posterior surface of the tibia, proximal to soleus. Popliteus extends superolaterally. Musculotendineous junction of the muscle is connected by the popliteofibular ligament to head of fibular. Popliteus muscle also has connections with the posterior joint capsule. At more proximal, the popliteus tendon passes beneath the lateral collateral ligament and inserts lateral to the lateral femoral condyle. Popliteofibular ligament and popliteus tendon are very important structures in the posterolateral corner of the knee (Fig. 1.44).

1.6 Intra-Articular Structures of Knee

1.6.1 Menisci

The menisci are half-moon-shaped, intra-articular, and fibrocartilaginous structures. The peripheral parts of the menisci, which are largely attached to the capsule, are thick. The thickness decreases toward the central part. The free inner edges are concave. The upper surface is concave to be compatible with the femur, and the lower surface is flat in order to fit with the tibia. Thus, a depression is formed by the flat joint surface of the tibia for the placement of the femoral condyles. Blood is provided to the lateral parts by vessels coming from the capsule. Therefore, tears of the lateral parts have a chance of healing. However, in the inner part, blood vessels are not present, and there is no chance of healing when torn. The anterior and posterior ends of the meniscus are called anterior and posterior horns, respectively. The attachments of the horns to the bone occurs by anterior and posterior roots. The anterior horns of the menisci are connected with the transverse (intermeniscal) ligament (Fig. 1.45) [1, 2].

Medial meniscus: It is approximately a semicircular structure (Fig. 1.45). The posterior part is wider than the anterior. It attaches to the anterior part of the anterior intercondylar area in front of the attachment site of ACL via the anterior root (Figs. 1.45 and 1.46). It is attached to the posterior intercondylar area by the posterior root. The footprint of the posterior root is located anteromedial to the PCL attachment site. The anterior and posterior roots are penetrated into the bone in order to provide the meniscal strength. The peripheral parts is attached to the capsule with the meniscotibial and the meniscofemoral ligaments (Fig. 1.19). Due to these connections, the medial meniscus is more fixed structure than the lateral meniscus [12, 13].

Lateral meniscus: The lateral meniscus is different from the medial meniscus regarding mobility, shape, and footprint. It is about 4/5 of a circle in shape, and occupies more space over the lateral condyle. The widths of the anterior and posterior parts are approximately equal. The anterior horn of the lateral meniscus attaches to the anterior intercondylar area via the anterior root immediately lateral and posterior to ACL (Fig. 1.45). Some fibers of the anterior root show continuity with the ACL (Figs. 1.45 and 1.46). The posterior horn is attached to the posterior intercondylar area on the anterolateral side of the posterior root of the medial meniscus. The posterolateral of the lateral meniscus has a groove formed by the intra-articular extension of the popliteus tendon. The connections with popliteus and the absence of meniscofemoral ligament make the lateral meniscus more mobile. Discoid meniscus occurs approximately 5% and is often bilateral [13].

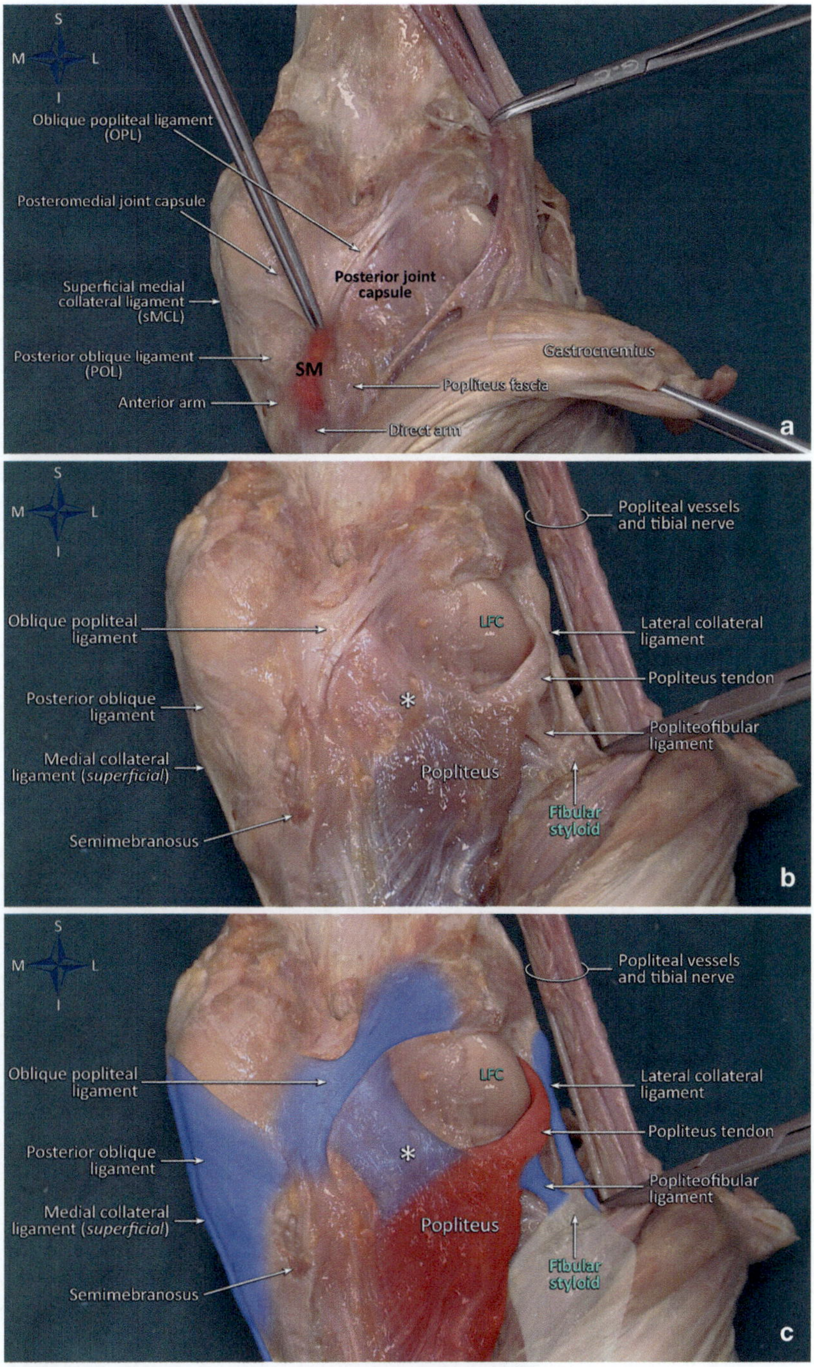

Fig. 1.44 Posterior view of the right knee. (**a**) Native view. (**b**) Colored view. All tissues behind the posterior joint capsule were removed. Insertion of semimembranosus (SM) is shown at the posteromedial of the knee in **a**. Connections of popliteus are demonstrated in **b** and **c**. Asterisk marks the capsular connection of the popliteus. *S* superior, *I* inferior, *M* medial, *L* lateral, on the star showing directions

1.6.2 Cruciate Ligaments

Cruciate ligaments are the strong intracapsular but extrasynovial ligaments of the knee. They are referred by this name as they cross each other in their courses. The anterior and posterior cruciate ligaments are named according to their attachment sites on the tibia (Fig. 1.7). They are enclosed in synovial membrane. The synovial membrane extends to the posterior by

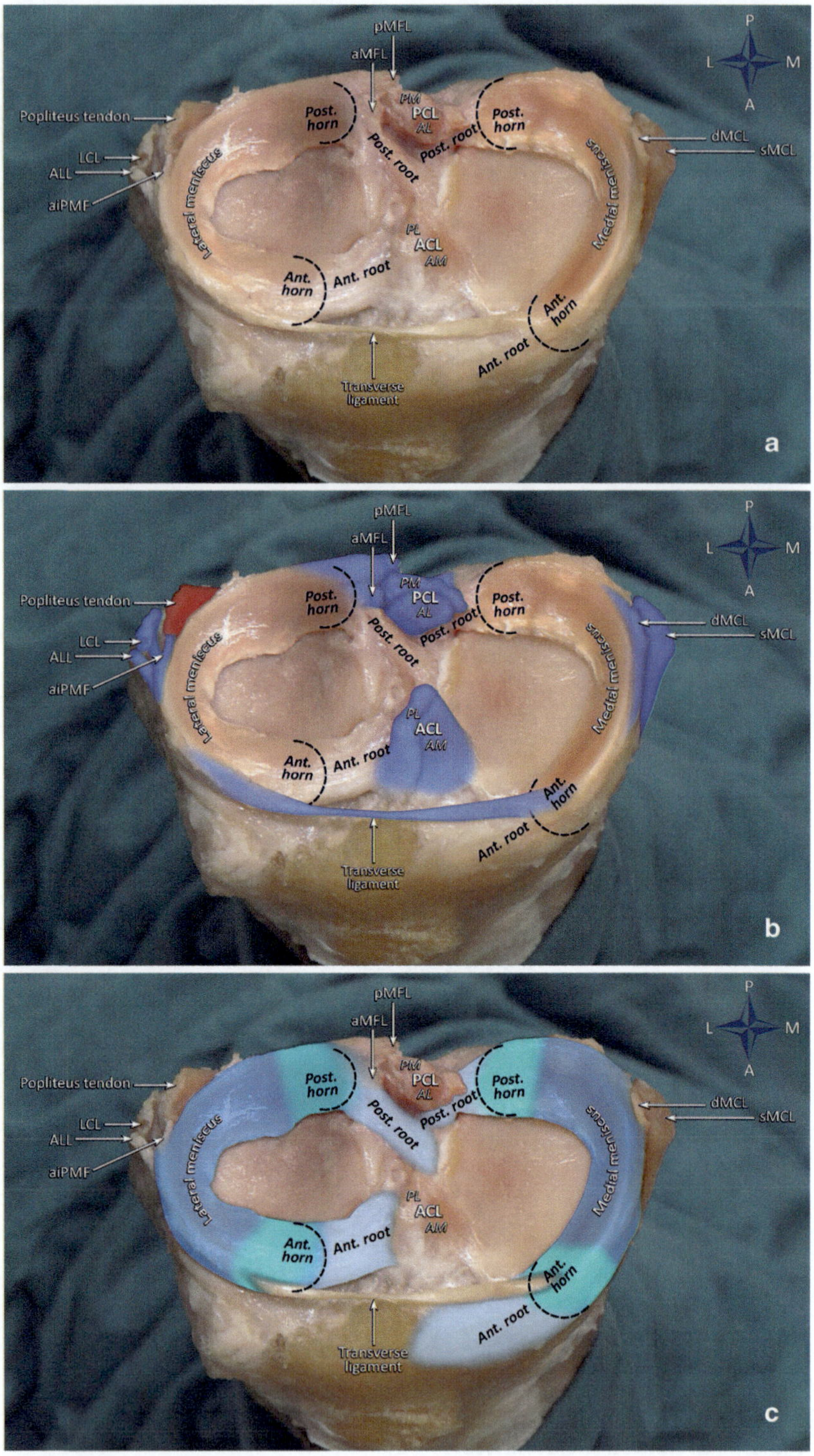

Fig. 1.45 Superior view of the proximal tibia. The structures on the tibia are shown. (**a**) Native view. (**b**) Structures other than the menisci are colored. (**c**) Parts of the menisci are colored. *ACL* anterior cruciate ligament, *AM* anteromedial bundle of ACL, *PL* posterolateral bundle of ACL, *PCL* posterior cruciate ligament, *AL* anterolateral bundle of PCL, *PM* posteromedial bundle of PCL, *LCL* lateral collateral ligament, *ALL* anterolateral ligament, *aiPMF* anterior inferior popliteal meniscal fascicle, *aMFL* anterior meniscofemoral ligament, *pMFL* posterior meniscofemoral ligament, *sMCL* superficial medial collateral ligament, *dMCL* deep medial collateral ligament. *P* posterior, *A* anterior, *L* lateral, *M* medial, on the star showing directions

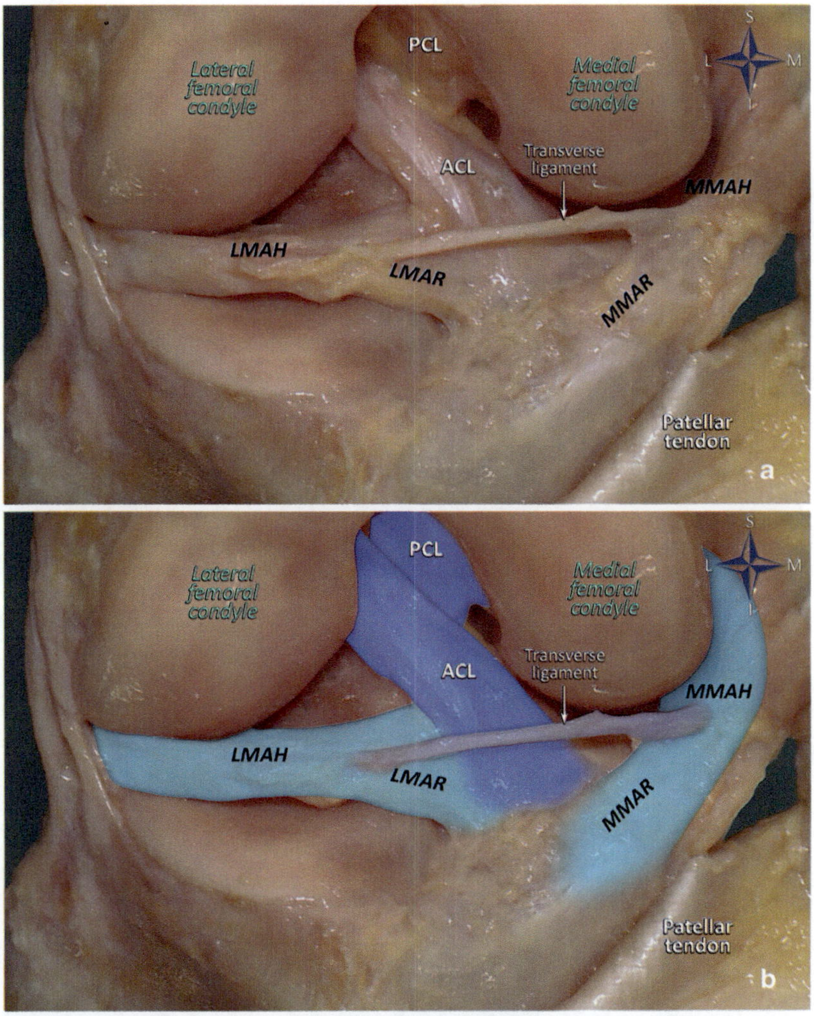

Fig. 1.46 Anterior view of the right knee. (**a**) Native view. (**b**) Colored view. The anterior joint capsule was removed. The anterior horns of menisci and anterior cruciate ligament (ACL), which attach to the anterior intercondylar area, are seen closely. *PCL* posterior cruciate ligament, *LMAH* lateral meniscus anterior horn, *LMAR* lateral meniscus anterior root, *MMAH* medial meniscus anterior horn, *MMAR* medial meniscus anterior root. *S* superior, *I* inferior, *L* lateral, *M* medial, on the star showing directions

covering the cruciate ligaments. It continues on the posterior joint capsule without fully covering the posterior surface of the posterior cruciate ligament. Thus, the cruciate ligaments are intracapsular, but they remain extrasynovial [1, 2].

1.6.2.1 Anterior Cruciate Ligament (ACL)

ACL originates from the anterior intercondylar area (Figs. 1.45 and 1.46). The attachment site is immediately anterior and slightly lateral to the medial intercondylar tubercle (Figs. 1.45 and 1.46). Some fibers of ACL blend with the anterior root of the lateral meniscus at the origin (Fig. 1.46). It extends obliquely toward the postero-supero-lateral aspect and attaches to the superomedial surface of the lateral femoral condyle (Figs. 1.47 and 1.48). It is a strong ligament with a tensile strength of 2200 N, approximately 38 mm long, 11 mm width. ACL limits anterior translation and internal rotation of the tibia. There are two bundles that are named according to the attachment site on the tibia: the anteromedial (AM) and the posterolateral (PL) (Figs. 1.48 and 1.49). The fibers of these bundles are paral-

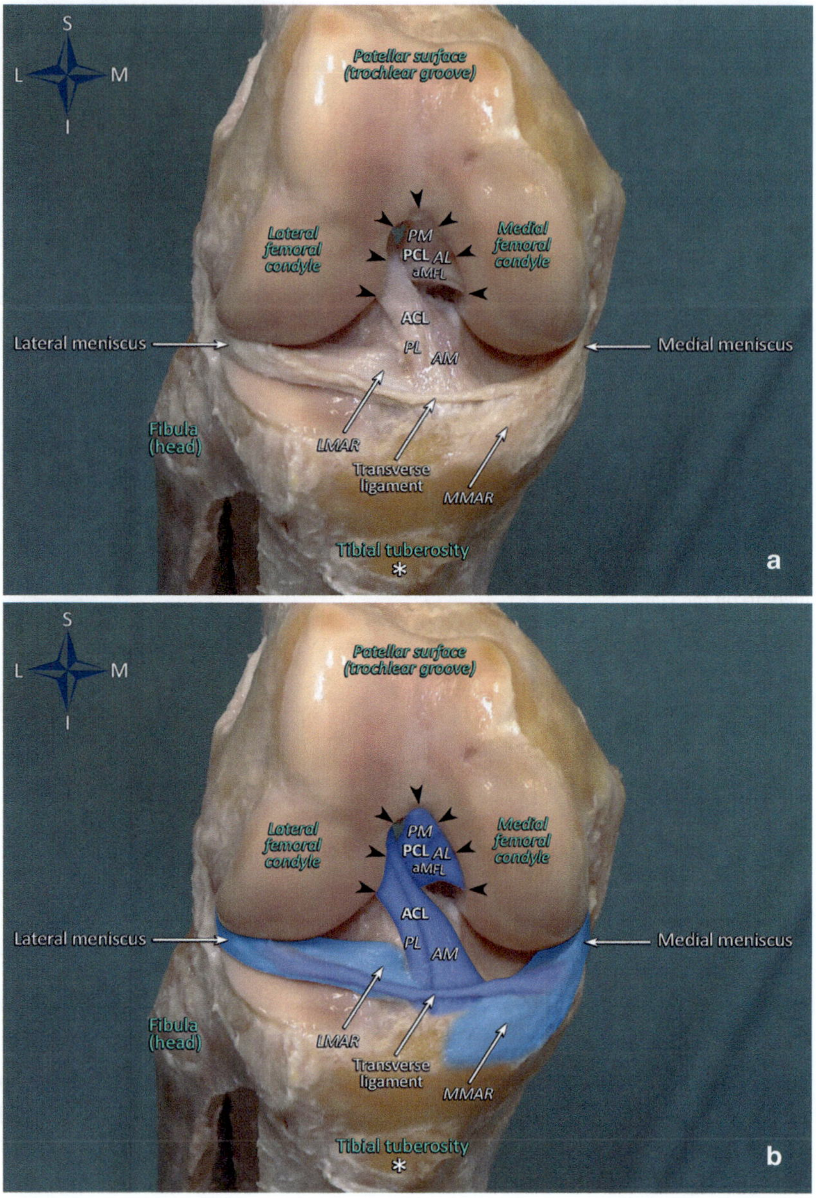

Fig. 1.47 Anterior view of the right knee at 90° flexion. (**a**) Native view. (**b**) Colored view. The anterior and posterior cruciate ligaments and their bundles are shown. Black arrowheads indicate anterior border of the intercondylar notch. *ACL* anterior cruciate ligament, *AM* anteromedial bundle of ACL, *PL* posterolateral bundle of ACL, *PCL* posterior cruciate ligament, *AL* anterolateral bundle of PCL, *PM* posteromedial bundle of PCL, *aMFL* anterior meniscofemoral ligament, *LMAR* lateral meniscus anterior root, *MMAR* medial meniscus anterior root. *S* superior, *I* inferior, *L* lateral, *M* medial, on the star showing directions

lel to each other in extension (Fig. 1.48). In flexion, they cross and twist each other (Fig. 1.49) [10, 11].

AM bundle is thicker. Its attachment site is closer to the roof of the intercondylar notch (Fig. 1.48). Although tense in both flexion and extension, it is extension, much tauter in flexion. AM bundle is the main part of ACL which prevents anterior translation of the tibia. PL bundle is taut in extension and slightly loose in flexion. In particular, PL bundle limits the internal rotation of the tibia [10, 11].

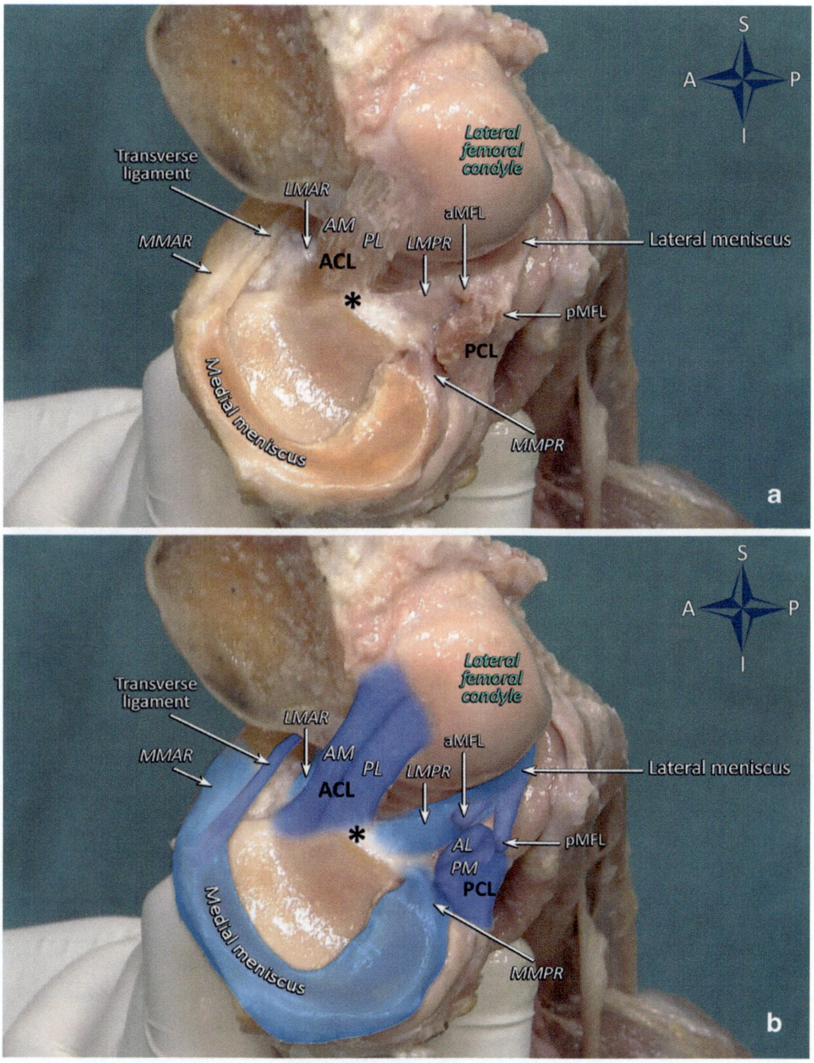

Fig. 1.48 Posteromedial view of the anterior cruciate ligament (ACL). (**a**) Native view. (**b**) Colored view. The femur was cut median and the medial half was removed. Asterisks marks the medial intercondylar tubercle. *AM* anteromedial bundle of ACL, *PL* posterolateral bundle of ACL, *PCL* posterior cruciate ligament, *AL* anterolateral bundle of PCL, *PM* posteromedial bundle of PCL, *aMFL* anterior meniscofemoral ligament, *pMFL* posterior meniscofemoral ligament, *LMAR* lateral meniscus anterior root, *MMAR* medial meniscus anterior root, *LMPR* lateral meniscus posterior root, *MMPR* medial meniscus posterior root. *S* superior, *I* inferior, *A* anterior, *P* posterior, on the star showing directions

1.6.2.2 Posterior Cruciate Ligament (PCL)

The distal end of PCL is attached to the most posterior part of the posterior intercondylar area (Figs. 1.45 and 1.50). This part of the posterior intercondylar area, which extends distal to the tibial plateau level, is in the form of a groove. PCL originates from this groove and courses slightly obliquely, antero-supero-medially (Fig. 1.50). This ligament attaches to the superolateral surface of the medial femoral condyle and the anterior aspect of the intercondylar notch (Figs. 1.47 and 1.49). It is thicker and stronger than the ACL; approximately 38 mm long, 13 mm width, and tensile strength of 2500 N. It limits external rotation of tibia with posterior translation of the tibia or anterior sliding of the femur over the tibia. There are two bundles that are named according to the attachment site on the femur: the anterolateral (AL) and the posteromedial (PM) (Figs. 1.49 and 1.50) [14–16].

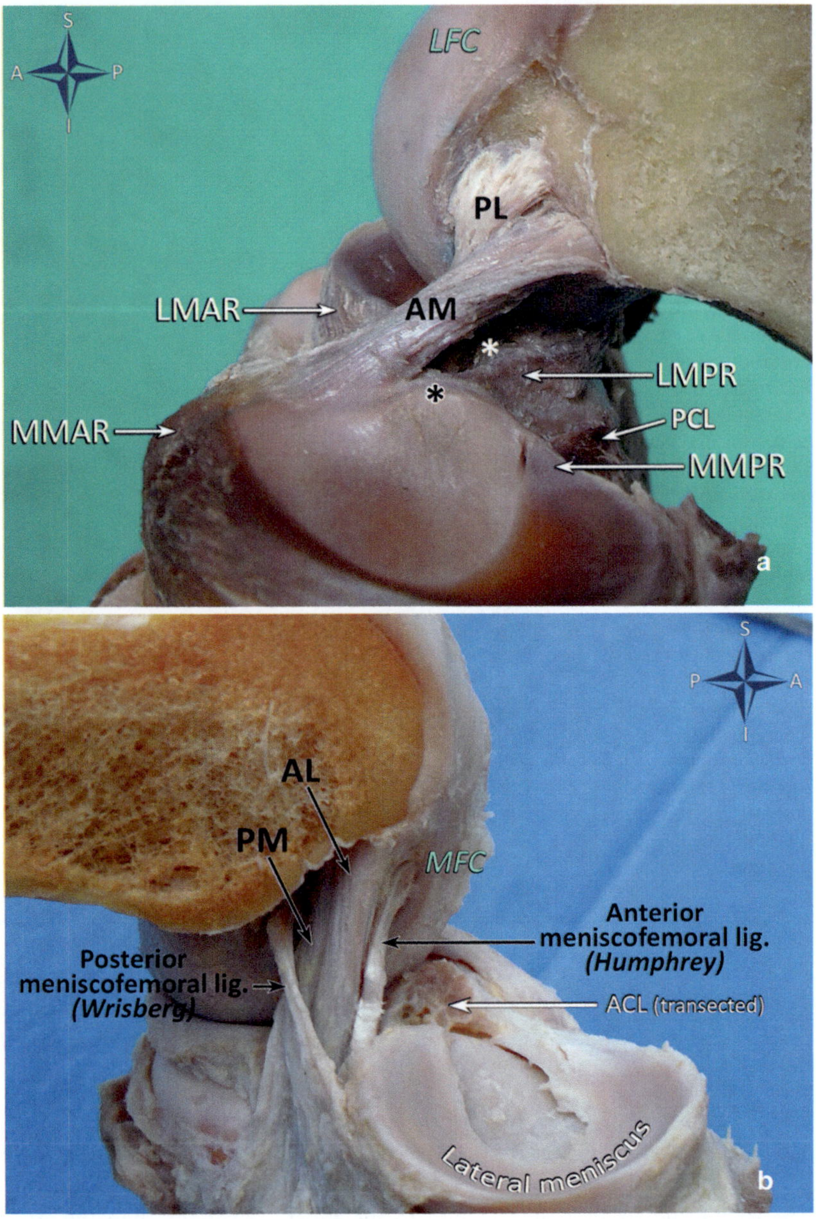

Fig. 1.49 The right knee. The distal femur was cut median. The medial half of femur was removed in **a**. The lateral half of femur was removed in **b**. Both knees are flexed at 90°. The anterior cruciate ligament (ACL) is seen from medial (**a**) and posterior cruciate ligament (PCL) is seen from lateral (**b**). White asterisk marks the medial intercondylar tubercle and black asterisk marks the lateral intercondylar tubercle in **a**. *AM* anteromedial bundle of ACL, *PL* posterolateral bundle of ACL, *AL* anterolateral bundle of PCL, *PM* posteromedial bundle of PCL, *LMAR* lateral meniscus anterior root, *MMAR* medial meniscus anterior root, *LMPR* lateral meniscus posterior root, *MMPR* medial meniscus posterior root. *S* superior, *I* inferior, *A* anterior, *P* posterior, on the star showing directions

AL bundle is shorter, thicker, and stronger. It is especially tense in half-flexion. However, PM bundle is longer, thinner, and weaker. It becomes stretched during extension and prevents hyperextension. At every angle from extension to flexion, some part of the fibers of the ligament is taut [14–16].

Anterior and posterior meniscofemoral ligaments (Humphrey's and Wrisberg's ligaments, respectively) originate from the posterior horn of the lateral meniscus (Fig. 1.48). These ligaments course anterior and posterior to PCL (Fig. 1.49). They attach to the anteroinferior and posterosuperior of PCL footprint on the medial femoral condyle. Posterior meniscofemoral ligament (pMFL) is thicker than anterior meniscofemoral ligament (aMFL). Meniscofemoral ligaments protect the posterior horn of the lateral meniscus. aMFL is stretched during flexion, whereas pMFL is stretched during extension [2, 60].

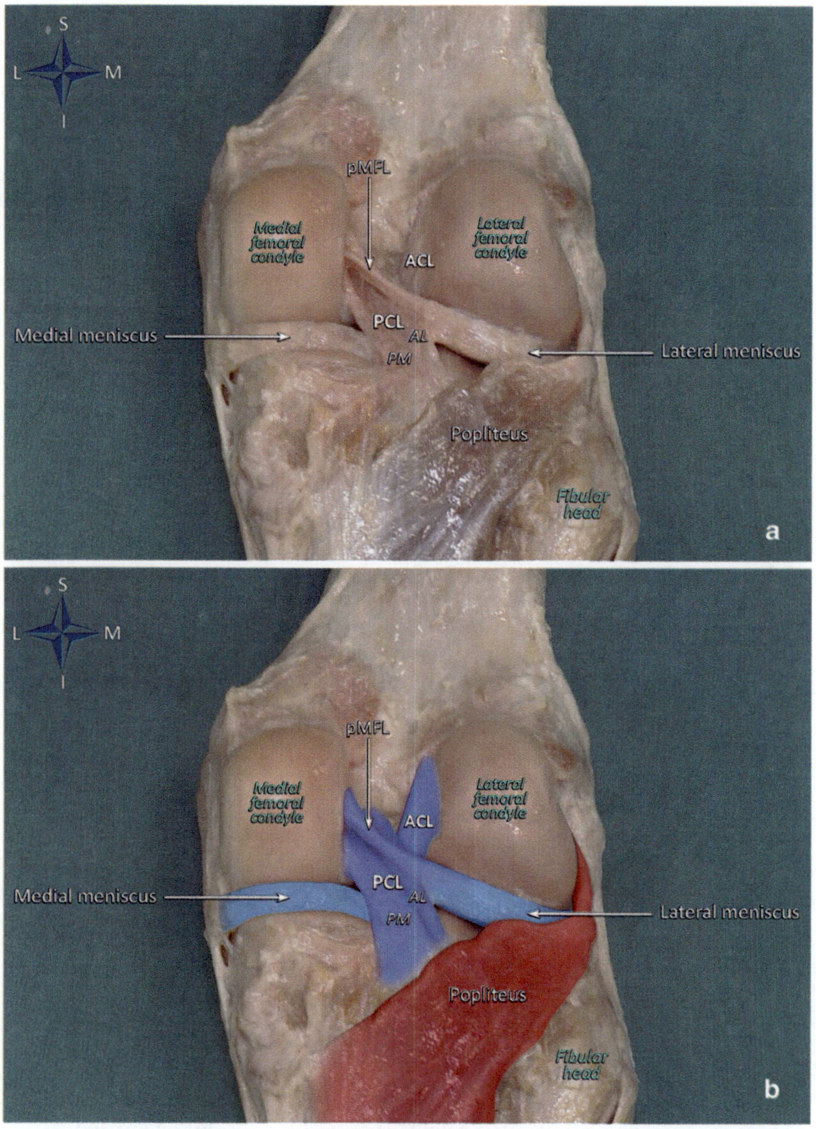

Fig. 1.50 Posterior view of the right knee. (**a**) Native view. (**b**) Colored view. All tissues behind the knee joint, including the capsule, were removed. (**b**). The menisci and the posterior cruciate ligament (PCL) are seen from posterior. *ACL* anterior cruciate ligament, *AL* anterolateral bundle of PCL, *PM* posteromedial bundle of PCL, *pMFL* posterior meniscofemoral ligament. *S* superior, *I* inferior, *L* lateral, *M* medial, on the star showing directions

References

1. Keith L, Moore AFD, Agur AMR. Lower limb. In: Moore KL, editor. Clinically oriented anatomy. Philadelphia, PA: Lippincott Williams & Wilkins; 2014. p. 508–669.
2. Standring S. Gray's anatomy: the anatomical basis of clinical practice. Amsterdam: Elsevier; 2016.
3. Cherian JJ, Kapadia BH, Banerjee S, Jauregui JJ, Issa K, Mont MA. Mechanical, anatomical, and kinematic Axis in TKA: concepts and practical applications. Curr Rev Musculoskelet Med. 2014;7(2):89–95.
4. Tang WM, Zhu YH, Chiu KY. Axial alignment of the lower extremity in Chinese adults. J Bone Joint Surg Am. 2000;82(11):1603–8.
5. LaPrade RF, Ly TV, Wentorf FA, Engebretsen L. The posterolateral attachments of the knee: a qualitative and quantitative morphologic analysis of the fibular collateral ligament, popliteus tendon, popliteofibular ligament, and lateral gastrocnemius tendon. Am J Sports Med. 2003;31(6):854–60.
6. Justin P, Strickland EWF, Noyes FR. Lateral and posterior knee anatomy. In: Noyes FR, editor. Noyes' knee disorders: surgery, rehabilitation, clinical outcomes. Amsterdam: Elsevier; 2017. p. 23–35.
7. Jadhav SP, More SR, Riascos RF, Lemos DF, Swischuk LE. Comprehensive review of the anatomy, function, and imaging of the popliteus and associated pathologic conditions. Radiographics. 2014;34(2):496–513.

8. LaPrade RF, Engebretsen AH, Ly TV, Johansen S, Wentorf FA, Engebretsen L. The anatomy of the medial part of the knee. J Bone Joint Surg Am. 2007;89(9):2000–10.
9. Alvin Detterline JB, Noyes FR. Medial and anterior knee anatomy. In: Noyes FR, editor. Noyes' knee disorders surgery, rehabilitation, clinical outcomes. Amsterdam: Elsevier; 2017. p. 2–22.
10. Dargel J, Gotter M, Mader K, Pennig D, Koebke J, Schmidt-Wiethoff R. Biomechanics of the anterior cruciate ligament and implications for surgical reconstruction. Strategies Trauma Limb Reconstr. 2007;2(1):1–12.
11. Kraeutler MJ, Wolsky RM, Vidal AF, Bravman JT. Anatomy and biomechanics of the native and reconstructed anterior cruciate ligament: surgical implications. J Bone Joint Surg Am. 2017;99(5):438–45.
12. Smigielski R, Becker R, Zdanowicz U, Ciszek B. Medial meniscus anatomy-from basic science to treatment. Knee Surg Sports Traumatol Arthrosc. 2015;23(1):8–14.
13. Bryceland JK, Powell AJ, Nunn T. Knee Menisci. Cartilage. 2017;8(2):99–104.
14. Logterman SL, Wydra FB, Frank RM. Posterior cruciate ligament: anatomy and biomechanics. Curr Rev Musculoskelet Med. 2018;11(3):510–4.
15. Arthur JR, Haglin JM, Makovicka JL, Chhabra A. Anatomy and biomechanics of the posterior cruciate ligament and their surgical implications. Sports Med Arthrosc Rev. 2020;28(1):e1–e10.
16. Voos JE, Mauro CS, Wente T, Warren RF, Wickiewicz TL. Posterior cruciate ligament: anatomy, biomechanics, and outcomes. Am J Sports Med. 2012;40(1):222–31.
17. Fulkerson JP. Disorders of the patellofemoral joint. 4th ed. Philadelphia: Lippincott Williams & Wilkins; 2004.
18. Warren LF, Marshall JL. The supporting structures and layers on the medial side of the knee: an anatomical analysis. J Bone Joint Surg Am. 1979;61(1):56–62.
19. Charalambous CP, Kwaees TA. Anatomical considerations in hamstring tendon harvesting for anterior cruciate ligament reconstruction. Muscles Ligaments Tendons J. 2012;2(4):253–7.
20. Patterson DC, Cirino CM, Gladstone JN. No safe zone: the anatomy of the saphenous nerve and its posteromedial branches. Knee. 2019;26(3):660–5.
21. Henry BM, Tomaszewski KA, Pekala PA, Ramakrishnan PK, Taterra D, Saganiak K, Mizia E, Walocha JA. The variable emergence of the infrapatellar branch of the saphenous nerve. J Knee Surg. 2017;30(6):585–93.
22. James NF, Kumar AR, Wilke BK, Shi GG. Incidence of encountering the infrapatellar nerve branch of the saphenous nerve during a midline approach for total knee arthroplasty. J Am Acad Orthop Surg Glob Res Rev. 2019;3(12):e19.
23. Olewnik L, Gonera B, Podgorski M, Polguj M, Jezierski H, Topol M. A proposal for a new classification of pes anserinus morphology. Knee Surg Sports Traumatol Arthrosc. 2019;27(9):2984–93.
24. Reina N, Abbo O, Gomez-Brouchet A, Chiron P, Moscovici J, Laffosse JM. Anatomy of the bands of the hamstring tendon: how can we improve harvest quality? Knee. 2013;20(2):90–5.
25. Yasin MN, Charalambous CP, Mills SP, Phaltankar PM. Accessory bands of the hamstring tendons: a clinical anatomical study. Clin Anat. 2010;23(7):862–5.
26. Cohen M, Astur DC, Branco RC, de Souza Campos Fernandes R, Kaleka CC, Arliani GG, Jalikjian W, Golano P. An anatomical three-dimensional study of the posteromedial corner of the knee. Knee Surg Sports Traumatol Arthrosc. 2011;19(10):1614–9.
27. Lundquist RB, Matcuk GR Jr, Schein AJ, Skalski MR, White EA, Forrester DM, Gottsegen CJ, Patel DB. Posteromedial corner of the knee: the neglected corner. Radiographics. 2015;35(4):1123–37.
28. Loredo R, Hodler J, Pedowitz R, Yeh LR, Trudell D, Resnick D. Posteromedial corner of the knee: MR imaging with gross anatomic correlation. Skeletal Radiol. 1999;28(6):305–11.
29. Cavaignac E, Carpentier K, Pailhe R, Luyckx T, Bellemans J. The role of the deep medial collateral ligament in controlling rotational stability of the knee. Knee Surg Sports Traumatol Arthrosc. 2015;23(10):3101–7.
30. Kim MS, Koh IJ, In Y. Superficial and deep medial collateral ligament reconstruction for chronic medial instability of the knee. Arthrosc Tech. 2019;8(6):e549–54.
31. Kuroda R, Muratsu H, Harada T, Hino T, Takayama H, Miwa M, Sakai H, Yoshiya S, Kurosaka M. Avulsion fracture of the posterior oblique ligament associated with acute tear of the medial collateral ligament. Arthroscopy. 2003;19(3):E18.
32. Saigo T, Tajima G, Kikuchi S, Yan J, Maruyama M, Sugawara A, Doita M. Morphology of the insertions of the superficial medial collateral ligament and posterior oblique ligament using 3-dimensional computed tomography: a cadaveric study. Arthroscopy. 2017;33(2):400–7.
33. Kim YC, Yoo WK, Chung IH, Seo JS, Tanaka S. Tendinous insertion of semimembranosus muscle into the lateral meniscus. Surg Radiol Anat. 1997;19(6):365–9.
34. Cooper JM, McAndrews PT, LaPrade RF. Posterolateral corner injuries of the knee: anatomy, diagnosis, and treatment. Sports Med Arthrosc Rev. 2006;14(4):213–20.
35. Covey DC. Injuries of the posterolateral corner of the knee. J Bone Joint Surg Am. 2001;83(1):106–18.
36. Davies H, Unwin A, Aichroth P. The posterolateral corner of the knee. Anatomy, biomechanics and management of injuries. Injury. 2004;35(1):68–75.
37. Hughston JC, Jacobson KE. Chronic posterolateral rotatory instability of the knee. J Bone Joint Surg Am. 1985;67(3):351–9.

38. Alpert JM, McCarty LP, Bach BR Jr. The posterolateral corner of the knee: anatomic dissection and surgical approach. J Knee Surg. 2008;21(1):50–4.
39. Bolog N, Hodler J. MR imaging of the posterolateral corner of the knee. Skeletal Radiol. 2007;36(8):715–28.
40. Lasmar RC, Marques de Almeida A, Serbino JW Jr, Mota Albuquerque RF, Hernandez AJ. Importance of the different posterolateral knee static stabilizers: biomechanical study. Clinics (Sao Paulo). 2010;65(4):433–40.
41. Raheem O, Philpott J, Ryan W, O'Brien M. Anatomical variations in the anatomy of the posterolateral corner of the knee. Knee Surg Sports Traumatol Arthrosc. 2007;15(7):895–900.
42. Sanchez AR II, Sugalski MT, LaPrade RF. Anatomy and biomechanics of the lateral side of the knee. Sports Med Arthrosc Rev. 2006;14(1):2–11.
43. Malone AA, Dowd GS, Saifuddin A. Injuries of the posterior cruciate ligament and posterolateral corner of the knee. Injury. 2006;37(6):485–501.
44. Seebacher JR, Inglis AE, Marshall JL, Warren RF. The structure of the posterolateral aspect of the knee. J Bone Joint Surg Am. 1982;64(4):536–41.
45. Terry GC, Hughston JC, Norwood LA. The anatomy of the iliopatellar band and iliotibial tract. Am J Sports Med. 1986;14(1):39–45.
46. LaPrade RF, Hamilton CD. The fibular collateral ligament-biceps femoris bursa. An anatomic study. Am J Sports Med. 1997;25(4):439–43.
47. Terry GC, LaPrade RF. The biceps femoris muscle complex at the knee. Its anatomy and injury patterns associated with acute anterolateral-anteromedial rotatory instability. Am J Sports Med. 1996;24(1):2–8.
48. Terry GC, LaPrade RF. The posterolateral aspect of the knee. Anatomy and surgical approach. Am J Sports Med. 1996;24(6):732–9.
49. Jia Y, Gou W, Geng L, Wang Y, Chen J. Anatomic proximity of the peroneal nerve to the posterolateral corner of the knee determined by MR imaging. Knee. 2012;19(6):766–8.
50. Rausch V, Hackl M, Oppermann J, Leschinger T, Scaal M, Muller LP, Wegmann K. Peroneal nerve location at the fibular head: an anatomic study using 3D imaging. Arch Orthop Trauma Surg. 2019;139(7):921–6.
51. Anderson JC. Common fibular nerve compression: anatomy, symptoms, clinical evaluation, and surgical decompression. Clin Podiatr Med Surg. 2016;33(2):283–91.
52. James EW, LaPrade CM, LaPrade RF. Anatomy and biomechanics of the lateral side of the knee and surgical implications. Sports Med Arthrosc Rev. 2015;23(1):2–9.
53. Sonnery-Cottet B, Daggett M, Fayard JM, Ferretti A, Helito CP, Lind M, Monaco E, de Padua VBC, Thaunat M, Wilson A, et al. Anterolateral ligament expert group consensus paper on the management of internal rotation and instability of the anterior cruciate ligament - deficient knee. J Orthop Traumatol. 2017;18(2):91–106.
54. LaPrade RF, Morgan PM, Wentorf FA, Johansen S, Engebretsen L. The anatomy of the posterior aspect of the knee. An anatomic study. J Bone Joint Surg Am. 2007;89(4):758–64.
55. LaPrade RF, Tso A, Wentorf FA. Force measurements on the fibular collateral ligament, popliteofibular ligament, and popliteus tendon to applied loads. Am J Sports Med. 2004;32(7):1695–701.
56. Wadia FD, Pimple M, Gajjar SM, Narvekar AD. An anatomic study of the popliteofibular ligament. Int Orthop. 2003;27(3):172–4.
57. Leese J, Davies DC. An investigation of the anatomy of the infrapatellar fat pad and its possible involvement in anterior pain syndrome: a cadaveric study. J Anat. 2020;237:20.
58. Andrikoula S, Tokis A, Vasiliadis HS, Georgoulis A. The extensor mechanism of the knee joint: an anatomical study. Knee Surg Sports Traumatol Arthrosc. 2006;14(3):214–20.
59. Reider B, Marshall JL, Koslin B, Ring B, Girgis FG. The anterior aspect of the knee joint. J Bone Joint Surg Am. 1981;63(3):351–6.
60. Knapik DM, Salata MJ, Voos JE, Greis PE, Karns MR. Role of the Meniscofemoral ligaments in the stability of the posterior lateral meniscus root after injury in the ACL-deficient knee. JBJS Rev. 2020;8(1):e0071.

2 Arthroscopic Anatomy of the Knee

Murat Bozkurt, Mustafa Akkaya, Mesut Tahta, Özgür Kaya, and Halil İbrahim Açar

2.1 Introduction

Normal arthroscopic anatomy should be well known for an adequate and effective surgical intervention for successful results in surgery. Successful differentiation of normal tissue and pathological tissue is the first step of arthroscopic surgery. In this context, the arthroscopic anatomy of the knee joint is essential for knee arthroscopy which is the most commonly performed orthopaedic procedure [1, 2].

Knee joint is the biggest joint in the body. It is primarily a large synovial hinge joint. Its bony structure is formed by femoral condyles, tibial plateau and patella [3]. Knee joint has flexion and extension movements and limited internal and external rotations with certain knee flexion.

In this chapter, meniscal, synovial, chondral, ligamentous and bony structures in the knee joint and normal anatomy of joint architecture have been studied regionally and systematically.

2.2 Suprapatellar Pouch

Suprapatellar pouch is located in the proximal part of femoral trochlea and contains synovial membrane (Fig. 2.1). This area contains richly vascularised synovial tissue and fat. A layer of fat separates the pouch from the distal anterior femoral shaft. The proximal border is approximately 4 cm proximal to the proximal edge of the patella [4]. While the quadriceps tendon can be seen on the top, the walls are covered with smooth and consistent synovium (Fig. 2.2). Normal synovium is generally pink and mildly villous. A suprapa-

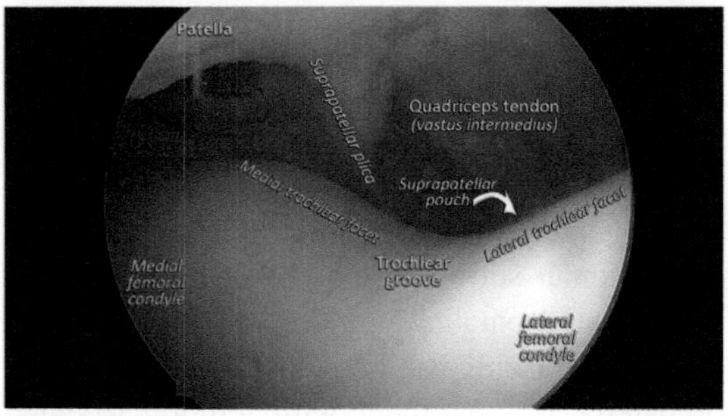

Fig. 2.1 Arthroscopic anatomy of the suprapatellar pouch

M. Bozkurt (✉) · M. Akkaya
Department of Orthopaedics and Traumatology, Faculty of Medicine, Ankara Yildirim Beyazit University, Ankara, Turkey

M. Tahta
Department of Orthopaedics and Traumatology, Ataturk Training and Research Hospital, Izmir Katip Celebi University, Izmir, Turkey

Ö. Kaya
Department of Orthopaedics and Traumatology, Etlik Lokman Hekim Hospital, Ankara, Turkey

H. İ. Açar
Department of Anatomy, Faculty of Medicine, Ankara University, Ankara, Turkey

M. Bozkurt, H. İ. Açar (eds.), *Clinical Anatomy of the Knee*, https://doi.org/10.1007/978-3-030-57578-6_2

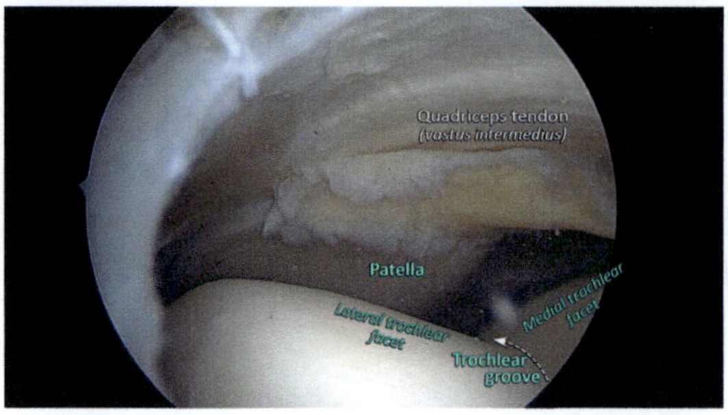

Fig. 2.2 Arthroscopic anatomy of the quadriceps tendon

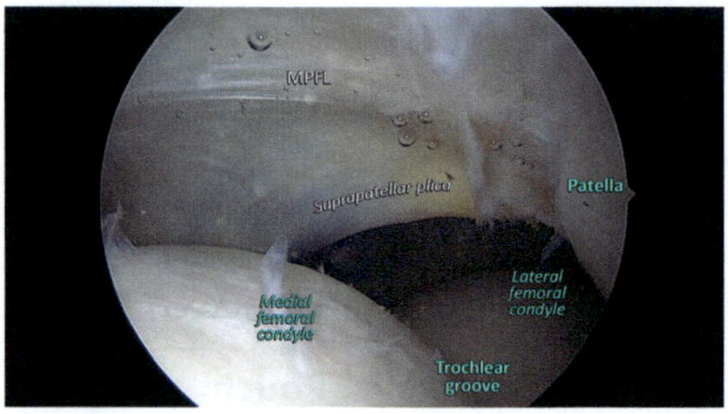

Fig. 2.3 Arthroscopic anatomy of the suprapatellar plica

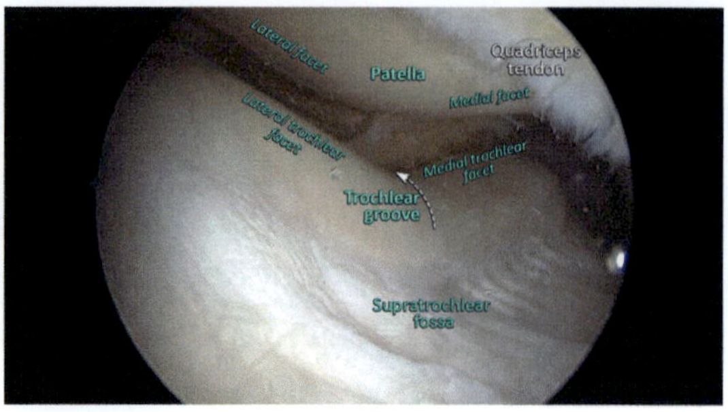

Fig. 2.4 Arthroscopic anatomy of the patellofemoral joint

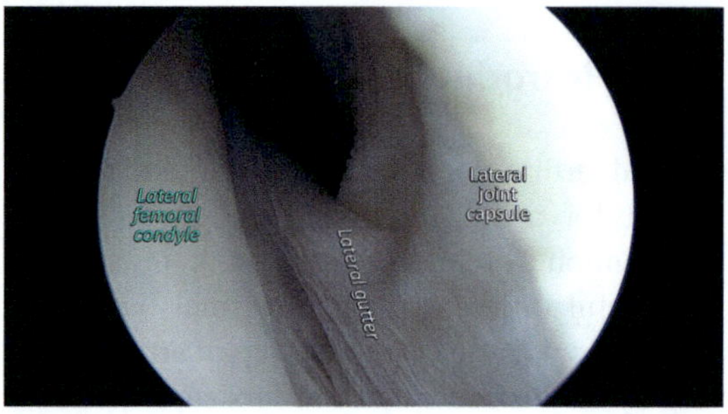

Fig. 2.5 Arthroscopic anatomy of the lateral gutter

tellar plica which can range from complete separation of the pouch from the joint to a band-shaped remnant can be observed in majority of the knees (Fig. 2.3) [5–7].

Pathological conditions in this region: plica, synovitis/inflammation, adhesions, loose bodies, crystalline deposits, traumatic rupture and neoplastic masses.

2.3 Patellofemoral Joint

It is the area between the patella (which is the largest sesamoid bone in the body) and the femoral trochlea [8]. In this region, smooth, femoral and patellar cartilage structures are observed. Normal cartilage is white, smooth and glistening. The articular surface of patella contains vertical ridge that separates lateral facet from medial facet. Central ridge and medial and lateral facets of patella are congruent with femoral trochlea (Fig. 2.4). In this context, lateral femoral condyle is higher than medial condyle; however, the medial femoral condyle is larger than the lateral from proximal to distal and anterior to posterior [9, 10]. The patella should be in the natural groove of the femoral trochlea. From full-knee extension to approximately 20° of flexion, the patella stands superior to the trochlear groove. With 25° of flexion, the patella becomes engaged in the trochlear groove, and the patella should be fully engaged in the trochlea at about 40° flexion [11–13]. The inferior pole of the patella is generally non-articulating.

Pathological conditions in this region: trochlear dysplasia, patellar maltracking, unstable bipartite patella, pathologic plica, patellar or trochlear chodromalacia.

2.4 Lateral Gutter

It is the area between lateral ridge of lateral femoral trochlea and joint capsule (Fig. 2.5). Lateral synovial folds, periphery of lateral meniscus, popliteus tendon, popliteal hiatus and margin of lateral femoral condyle can be observed.

Pathological conditions in this region: loose bodies, perimeniscal cysts, femoral osteophytes.

2.5 Medial Gutter

It is the area between medial ridge of medial femoral condyle and the joint capsule (Fig. 2.6). Synovial folds can be observed as seen in lateral gutter. A mediopatellar plica which is a remnant of embryonic development can be observed in about 40% of knees [14, 15].

Pathological conditions in this region: loose bodies, perimeniscal cysts, pathologic mediopatellar plica, femoral osteophytes.

2.6 Intercondylar Notch

It is the area between medial and femoral condyles. It contains anterior and posterior cruciate ligaments, medial and lateral tibial spines, intermeniscal ligament (Fig. 2.7). Patellar fat pad and ligamentum mucosum can be observed as normal anatomical structures. The mean intercondylar notch expands from the distal to the proximal and the average width is 1.8–2.3 cm in adult knee [16, 17]. Fat pad is primarily an adipose tissue and provides vascular supply to the anterior cruciate ligament. Ligamentum mucosum is a synovial reflection which generally covers intercondylar notch. The average length of anterior cruciate ligament is 33 mm, and the average thickness is 11 mm [18, 19]. The anterolateral and posteromedial bundles of the anterior cruciate ligament that progress towards the medial of the lateral femoral condyle can be observed (Fig. 2.8). The tibial attachment of the anterior cruciate ligament is located on the same line as the anterior horn of the lateral meniscus on the intercondylar eminence, and it is approximately 30 mm long and 10 mm wide. The femoral attachment is on the posterior aspect of the lateral femoral condyle, and it is approximately 20 mm long and 10 mm wide (Fig. 2.9).

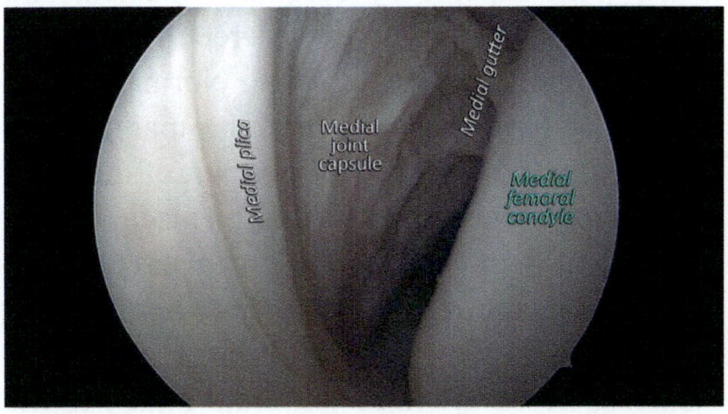

Fig. 2.6 Arthroscopic anatomy of the medial gutter

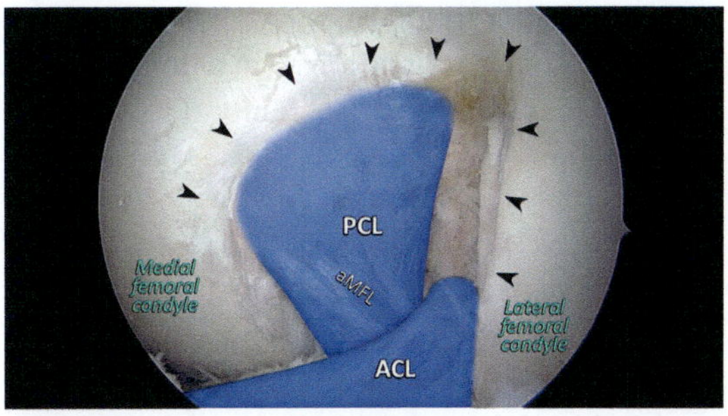

Fig. 2.7 Arthroscopic anatomy of the intercondylar notch

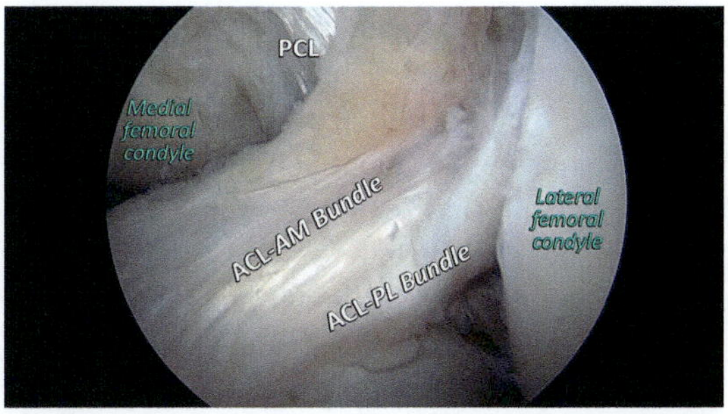

Fig. 2.8 Arthroscopic anatomy of the anterior cruciate ligament

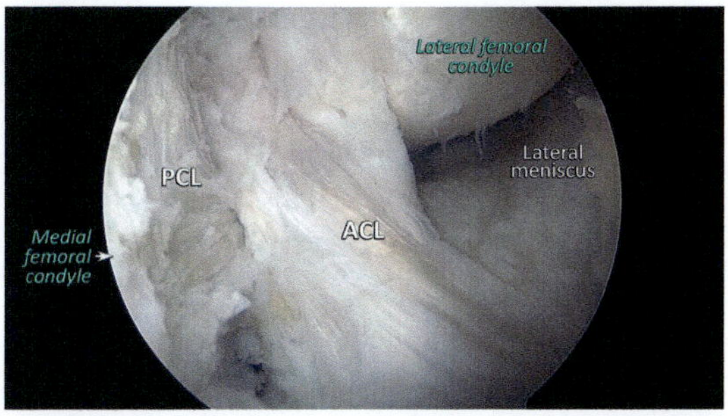

Fig. 2.9 Arthroscopic view of femoral and tibial attachments of anterior cruciate ligament

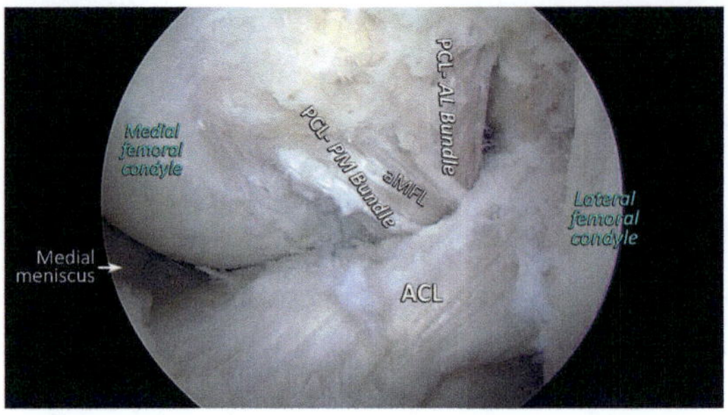

Fig. 2.10 Arthroscopic anatomy of the posterior cruciate ligament

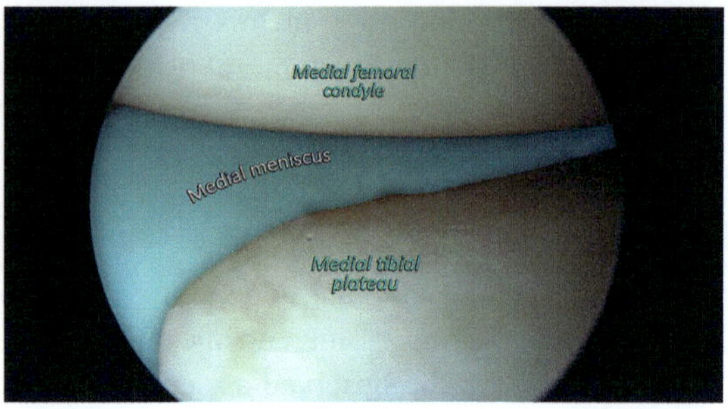

Fig. 2.11 Arthroscopic view of the medial compartment

The femoral attachment of the posterior cruciate ligament starts immediately from the posterior of the medial femoral condyle cartilage, and it is approximately 30 mm long and 5 mm wide. Posterior cruciate ligament extends distally to insert on posterior aspect of proximal tibia. The average length of posterior cruciate ligament is 38 mm, and the average thickness is 13 mm [20, 21]. Usually only the femoral attachment of the posterior cruciate ligament can be observed (Fig. 2.10).

Meniscofemoral ligaments can be observed in the posterior region: Wrisberg or Humphrey ligaments can be seen in 70% cases [22]. The Wrisberg ligament is about half the posterior cruciate ligament and extends from the lateral meniscus to the medial femoral condyle. The Humphrey ligament is thinner and extends from the lateral meniscus to the medial femoral condyle [23, 24]. The Wrisberg ligament is posterior to the posterior cruciate ligament, whereas the Humphrey ligament is anterior to the posterior cruciate ligament.

Pathological conditions in this region: loose bodies, anterior cruciate ligament tears, posterior cruciate ligament tears, trochlear chondromalacia.

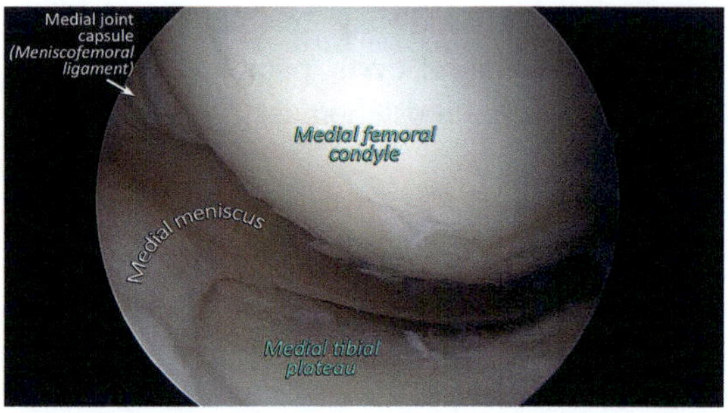

Fig. 2.12 Arthroscopic anatomy of the medial meniscus

2.7 Medial Compartment

It is the area between medial femoral condyle and medial tibia plateau and contains cartilages of both bone structures and medial meniscus (Fig. 2.11). Medial meniscus is wedge-shaped in cross-section. Medial meniscus is firmly attached to the joint capsule by coronary ligament and excursion below 5 mm is considered normal in an intact medial meniscus [25, 26]. Normal meniscus should have a smooth articular margin. It is anchored in the central posterior part of the tibial intercondylar spine. The medial meniscus has an average width of 8–10 mm and an average thickness of 4–6 mm (Fig. 2.12) [26]. Some fibres originating from the anterior horn of medial meniscus cross the knee joint and attach to the lateral meniscus. These fibres form the transverse meniscal ligament. The medial tibial plateau is larger than the lateral plateau. It is also concave in frontal and sagittal planes.

Pathological conditions in this region: medial meniscal tears, femoral or tibial chondromalacia.

2.8 Lateral Compartment

It is the area between lateral femoral condyle and lateral tibia plateau. It includes cartilages of both bone structures, lateral meniscus and popliteus tendon (Fig. 2.13). The lateral meniscus, which is

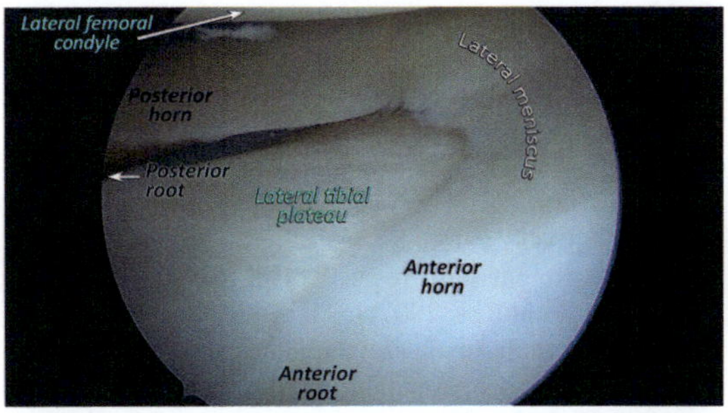

Fig. 2.13 Arthroscopic view of the lateral compartment

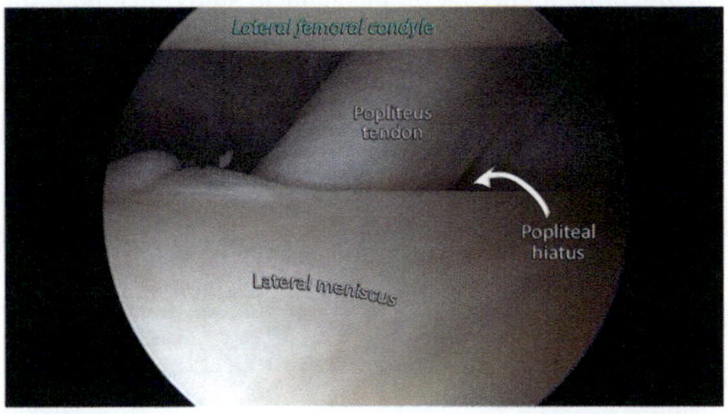

Fig. 2.14 Arthroscopic view of the poptliteus tendon and popliteal hiatus

closer to the O-shape than the C-shape form, is connected to the joint capsule with the coronary ligament, and it has a triangular-shaped cross-section. Anterior attachment of lateral meniscus combines with fibres of anterior cruciate ligament [27, 28]. There is a normal hiatus, created by the traversing popliteus tendon, and lateral meniscus is not connected with capsule in this area (Fig. 2.14). This area along the popliteal hiatus is described as the avascular region. The lateral meniscus has more excursion than the medial meniscus: Anterior horn has 9.5 mm of excursion, and posterior horn has 5.6 mm of excursion [25]. The normal depression of sulcus terminalis can be assessed in lower degrees of flexion. Lateral tibial plateau is smaller and more circular than medial tibial plateau. It is also convex in sagittal plane.

Pathological conditions in this region: lateral meniscal tears, femoral and tibial chondromalacia.

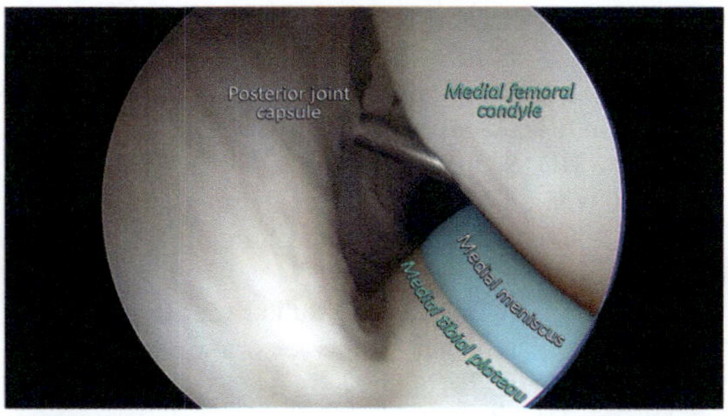

Fig. 2.15 Arthroscopic view of the posterior medial compartment

2.9 Posterior Medial Compartment

It is the area between posterior aspect of medial femoral condyle and posterior joint capsule. Posterior cruciate ligament can be observed (Fig. 2.15).

Pathological conditions in this region: posterior cruciate ligament injuries, loose bodies, medial meniscus root injuries, meniscocapsular injuries, synovitis, medial femoral condyle, posterior chondral lesions.

2.10 Posterior Lateral Compartment

It is the area between posterior aspect of lateral femoral condyle and posterior joint capsule.

Pathological conditions in this region: loose bodies, lateral meniscus root injuries, meniscocapsular injuries, synovitis, popliteomeniscal fascicle tears, lateral femoral condyle, posterior chondral lesions.

References

1. McGinty JB, Johnson LL, Jackson RW, McBryde AM, Goodfellow JW. Uses and abuses of arthroscopy: a symposium. J Bone Joint Surg Am. 1992;74:1563–77.
2. Small NC. Complications in arthroscopic surgery of the knee and shoulder. Orthopedics. 1993;16:985–8.

3. Flandry F, Hommel G. Normal anatomy and biomechanics of the knee. Sports Med Arthrosc Rev. 2011;19(2):82–92. https://doi.org/10.1097/JSA.0b013e318210c0aa.
4. Clarke HD, Scott WN, Insall JN, et al. Anatomy. In: Scott WN, editor. Insall & Scott surgery of the knee, vol. 1. 4th ed. Philadelphia: Churchill Livingstone; 2006. p. 3–66.
5. Deliwala UH, Jadeja HR, Rathod CL, Loya N. The suprapattellar pouch of the knee and its disorders. Gujarat Med J. 2010;65:47–54.
6. Dandy DJ. Anatomy of the medial suprapatellar plica and medial synovial shelf. Arthroscopy. 1990;6(2):161–76.
7. Schindler O. Synovial plicae of the knee. Curr Orthop. 2004;18(3):210–9.
8. Sherman SL, Plackis AC, Nuelle CW. Patellofemoral anatomy and biomechanics. Clin Sports Med. 2014;33(3):389–401.
9. Walsh W. Recurrent dislocation of the knee in the adult. In: Delee J, Drez D, Miller M, editors. Delee and Drez's orthopaedic sports medicine. Philadelphia: Saunders; 2003. p. 1710–49.
10. Ahmed AM, Burke DL, Hyder A. Force analysis of the patellar mechanism. J Orthop Res. 1987;5:6–85.
11. Grelsamer RP, Proctor CS, Bazos AN. Evaluation of patellar shape in the sagittal plane. A clinical analysis. Am J Sports Med. 1994;22:61.
12. White BJ, Sherman OH. Patellofemoral instability. Bull NYU Hosp Jt Dis. 2009;67:22–9.
13. Dejour D, Saggin P. Disorders of the patellofemoral joint. In: Scott N, editor. Insall & Scott surgery of the knee. Philadelphia: Elsevier; 2012. Chapter 61.
14. Ewing JW. Plica: pathologic or not? J Am Acad Orthop Surg. 1993;1:117–21.
15. Al-Hadithy N, Gikas P, Mahapatra AM, Dowd G. Review article: plica syndrome of the knee. J Orthop Surg (Hong Kong). 2011;19(03):354–35.
16. Koukoubis TD, Glisson RR, Bolognesi M, Vail TP. Dimensions of the intercondylar notch of the knee. Am J Knee Surg. 1997;10(2):83–7; discussion 87-8.
17. Farrow LD, Chen MR, Cooperman DR, Victoroff BN, Goodfellow DB. Morphology of the femoral intercondylar notch. J Bone Joint Surg Am. 2007;89(10):2150–5.
18. Girgis FG, Marshall JL, Monajem A. The cruciate ligaments of the knee joint. Anatomical, functional and experimental analysis. Clin Orthop Relat Res. 1975;106:216–31.
19. Petersen W, Zantop T. Anatomy of the anterior cruciate ligament with regard to its two bundles. Clin Orthop Relat Res. 2007;454:35–47.
20. Van Dommelen BA, Fowler PJ. Anatomy of the posterior cruciate ligament. A review. Am J Sports Med. 1989;17(1):24–9.
21. Gollehon DL, Torzilli PA, Warren RF. The role of the posterolateral and cruciate ligaments in the stability of the human knee. A biomechanical study. J Bone Joint Surg Am. 1987;69(2):233–42.
22. Cupte CM, Bull AM, Thomas RD, Amis AA. A review of the function and biomechanics of the meniscofemoral ligaments. Arthroscopy. 2003;19:161–71.
23. Kusayama T, Harner CD, Carlin GJ, Xerogeanes JW, Smith BA. Anatomical and biomechanical characteristics of human meniscofemoral ligaments. Knee Surg Sports Traumatol Arthrosc. 1994;2:234–7.
24. Wan AC, Felle P. The meniscofemoral ligaments. Clin Anat. 1995;8:323–6.
25. Gries P, Bandana D, Holstrom M, Burks RT. Meniscal injury: I. Basic science and evaluation. J Am Acad Orthop Surg. 2002;10:168–76.
26. Vedi V, Spouse E, Williams A, Tennant JJ, Hunt D, Gedroyc W. Meniscal movement: an in vivo study using dynamic MRI. J Bone Joint Surg Br. 1999;81:37–41.
27. Makris EA, Hadidi P, Athanasiou KA. The knee meniscus: structure-function, pathophysiology, current repair techniques, and prospects for regeneration. Biomaterials. 2011;32(30):7411–31. https://doi.org/10.1016/j.biomaterials.2011.06.037.
28. Fox AJ, Wanivenhaus F, Burge AJ, Warren RF, Rodeo SA. The human meniscus: a review of anatomy, function, injury, and advances in treatment. Clin Anat. 2015;28(2):269–87. https://doi.org/10.1002/ca.22456.

3 Knee Radiology

Nurdan Çay

Knee joint is the largest synovial joint of the human body, and most damage occurs in the knee joint during lower extremity injuries. It is the most frequently affected joint in motor vehicle accidents and sports-related injuries, especially in pediatric and adolescent populations. Bone fractures, meniscus, ligament, and intraarticular and extraarticular soft tissue injuries are common. Although history and physical examination are essential for clinical diagnosis, the accurate diagnosis is usually made with the help of imaging modalities. Plain film radiography is the first imaging method to be used in traumatic knee pain. Plain radiography allows the evaluation of the medial femorotibial, lateral femorotibial, and patellofemoral compartments forming the knee joint, and femur, tibia, fibula, and patella fractures can also be evaluated. However, it is recommended that radiography should be obtained in accordance with the Ottawa Knee Rule and the Pittsburgh Decision Rule to avoid unnecessary exams that would not be useful [1, 2]. Computed tomography (CT) with three-dimensional reconstructions is useful for evaluating tibial plateau fractures, loose bodies in the joint, and other complex knee injuries especially for preoperative planning [3]. CT arthrogram may also be a good alternative to assess soft tissues of the joint where magnetic resonance imaging (MRI) cannot be used. Lipohemarthrosis, acute anterior cruciate ligament injuries, and acute meniscal injuries can be evaluated with ultrasonography (US) and MRI [1, 2]. MRI is a useful tool in the evaluation of occult fractures and intra- and extra-articular soft tissues. Also, it has an important place in the planning of treatment with the help of early and accurate diagnosis.

Plain radiography should be the first imaging tool in the evaluation of nontraumatic knee pain. Joint space narrowing associated with osteoarthritis, osteophytes, subchondral cysts, and sclerosis can be easily visualized in elderly patients. Plain radiography alone may not be sufficient in the diagnosis of osteoarthritis, and MRI can be used in patients with unexplained symptoms [4]. Patients with a normal plain radiograph but suspicion of internal derangement or persistent knee pain should also have a subsequent MRI [5]. Joint effusion, synovial membrane, articular cartilage, bone marrow, meniscal/ligamentous pathologies, and friction syndromes can be evaluated with MRI. US can be used for evaluating the popliteal cyst.

Anatomical knowledge forms the basis for the interpretation of radiological images. The aim in this chapter is to review the radiological anatomy of the knee joint with different imaging modalities. Radiographic anatomy with knee radiographs and cross-sectional anatomy with CT and MRI are demonstrated.

N. Çay (✉)
Department of Radiology, Ankara Yıldırım Beyazıt University, School of Medicine, Ankara, Turkey

M. Bozkurt, H. İ. Açar (eds.), *Clinical Anatomy of the Knee*,
https://doi.org/10.1007/978-3-030-57578-6_3

3.1 Plain Film Radiography of the Knee

Plain radiography is the preferred imaging modality in abnormalities of the knee joint. Routine radiographic projections are anteroposterior (Fig. 3.1), lateral (Fig. 3.2), and axial (skyline/tangential) views (Fig. 3.3). Also, tunnel (or notch) projection can sometimes be added to routine imaging (Fig. 3.4). In evaluating arthritis, weight-bearing radiographs are preferred to assess the joint space and bone wear. Varus-valgus stress radiographs can be used to evaluate the collateral ligaments and physeal fractures (in children). Posterior stress radiographs can be used to evaluate the posterior cruciate ligament and posterolateral corner. However, both the stress radiographs are rarely preferred nowadays.

3.2 Computed Tomography of the Knee

Computed tomography with 3D reconstruction is more sensitive than radiography for evaluating fractures. Obtained 3D rendered images provide fast and accurate results especially in tibial plateau fractures, avulsion fractures, and complex injuries of the knee [2]. Although computed tomography seems even better than MRI for demonstrating the cortex and trabecular structure of the bone, MRI has better soft tissue spatial resolution than computed tomography. However, when the MRI evaluation is contraindicated (e.g., claustrophobia, MRI incompatible implanted devices), computed tomography can also be used for soft tissue evaluation.

After the patient is placed in the supine position, scout images in the two orthogonal planes

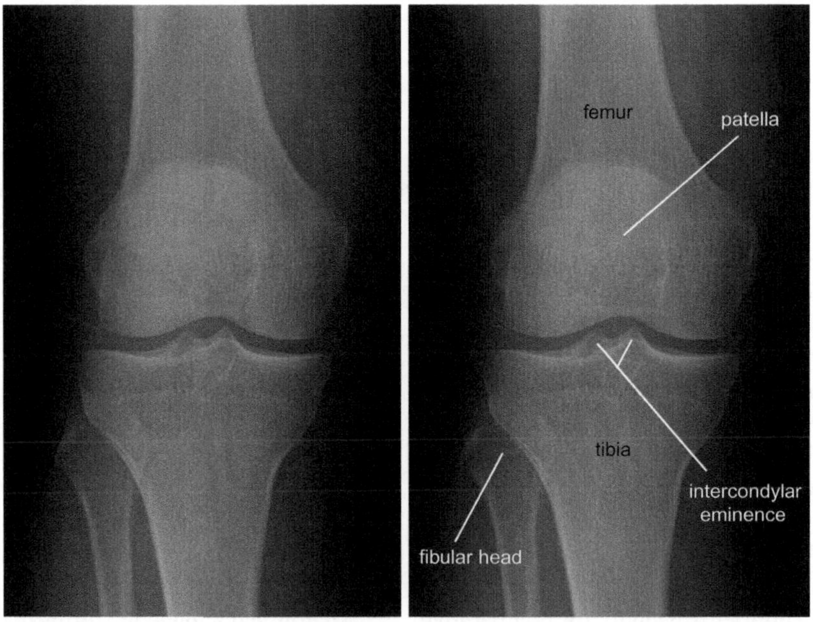

Fig. 3.1 Anteroposterior knee radiograph. While achieving anteroposterior knee X-ray, the patient is in the supine position on the table, leg is in the neutral position, and the knee is fully extended. Central X-ray should be directed vertically to the center of the knee (1.5 cm distal to the apex of the patella) with 5–7° cephalad angulation. Anteroposterior knee view demonstrates the distal femoral condyles, the proximal tibia/fibula, the patella, and the medial/lateral femorotibial joint compartments

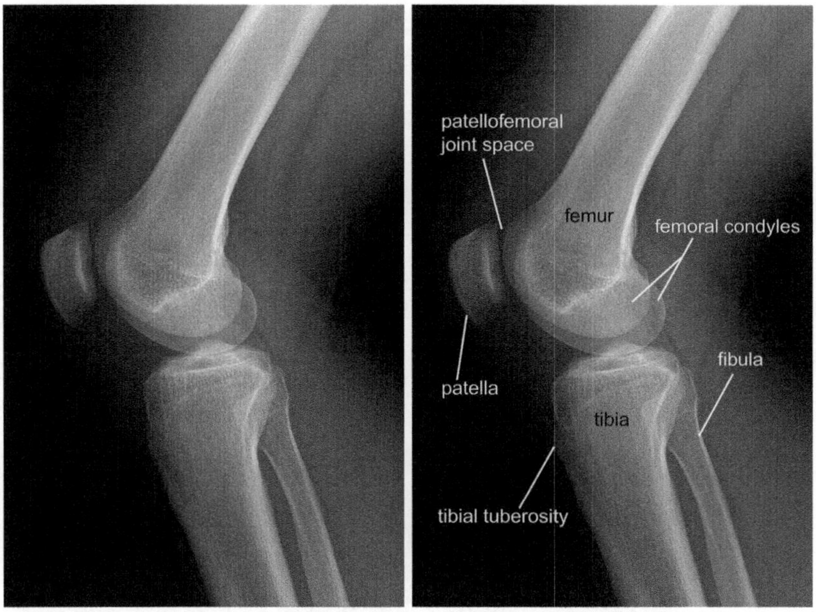

Fig. 3.2 Lateral knee radiograph. While achieving lateral knee X-ray, the patient is lying flat on the affected knee side on the table with 25–30° of flexion of the knee. Central X-ray should be directed vertically from medial to lateral with 5–7° cephalad angulation. Lateral view demonstrates the patellofemoral joint and patella in profile. In this view, femoral condyles project over each other. In trauma patients, horizontal beam lateral view may be preferred to demonstrate lipohemarthrosis

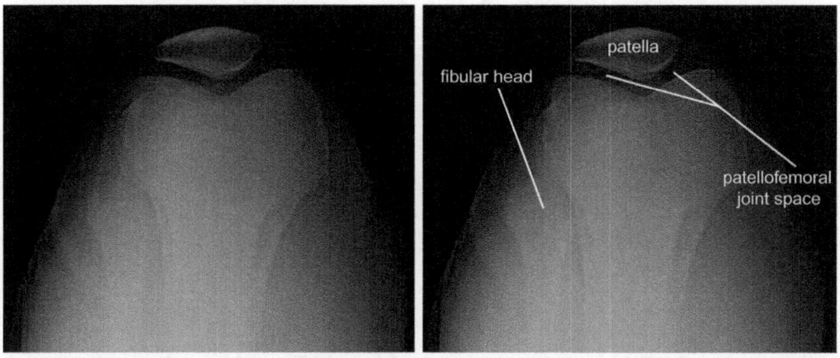

Fig. 3.3 Axial (sunrise/tangential) knee radiograph. While achieving axial knee X-ray, the patient is in the prone position with 115° of flexion of the knee toward the patella with approximately 15° cephalad angulation. Central X-ray should be directed vertically. This view demonstrates axial view of the patella and patellofemoral joint compartment

are obtained. Only the side of interest is focused using the smallest possible field of view (FOV) to reduce radiation risk. Axial images must be sufficient to cover the whole knee joint from the distal femoral metadiaphysis to the proximal tibial metadiaphysis. The coronal reformatted images are formed in the plane parallel to the line tangential to the posterior of the condyles at the level of the femoral condyles in the axial plane. The sagittal reformatted images are created in the plane vertical to the coronal reformatted images in the same axial slice.

The following CT images demonstrate the important anatomical structures of the knee joint in the axial, coronal, and sagittal planes (Figs. 3.5a–p, 3.6a–l, and 3.7a–n).

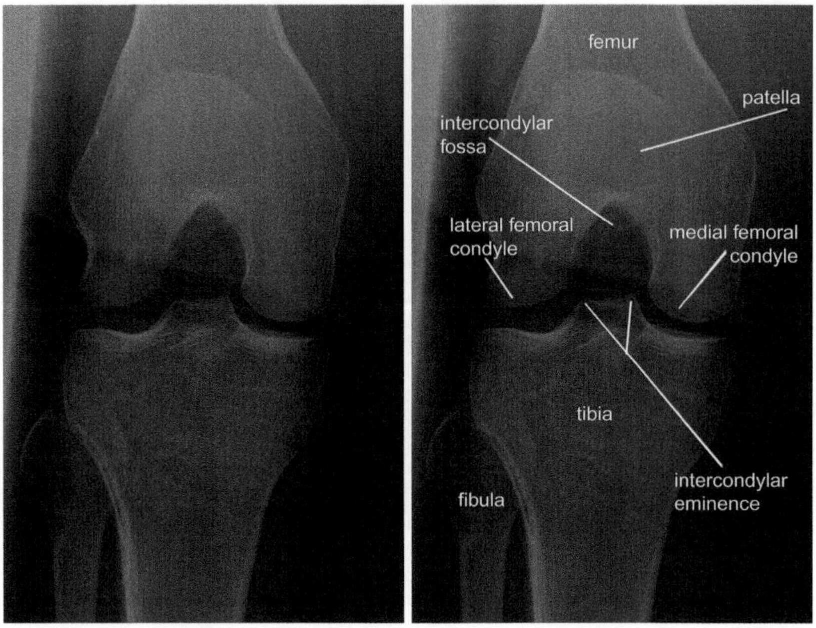

Fig. 3.4 Tunnel (notch) radiograph. For this view, the patient is in the prone position with approximately 40–45° of flexion of the knee. Central X-ray should be directed vertically toward the knee joint with 40° caudally. Tunnel view demonstrates the intercondylar notch and intercondylar eminence of the tibia

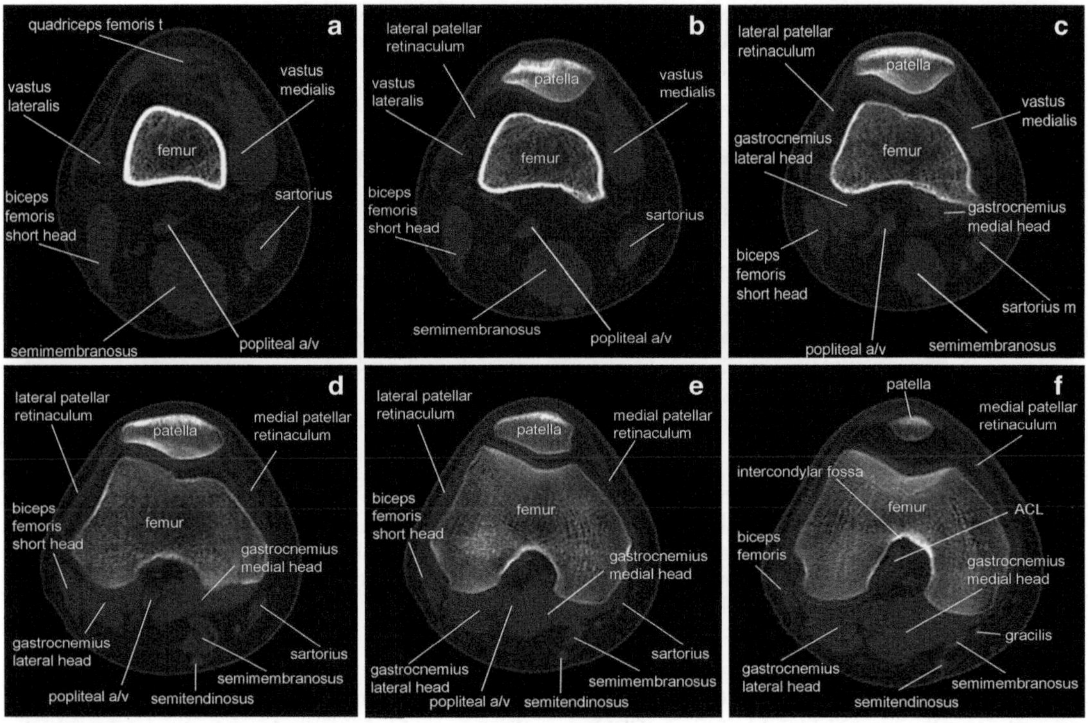

Fig. 3.5 (**a–p**) Axial CT images; bone window (*t* tendon, *a* artery, *v* vein, *ACL* anterior cruciate ligament, *PCL* posterior cruciate ligament)

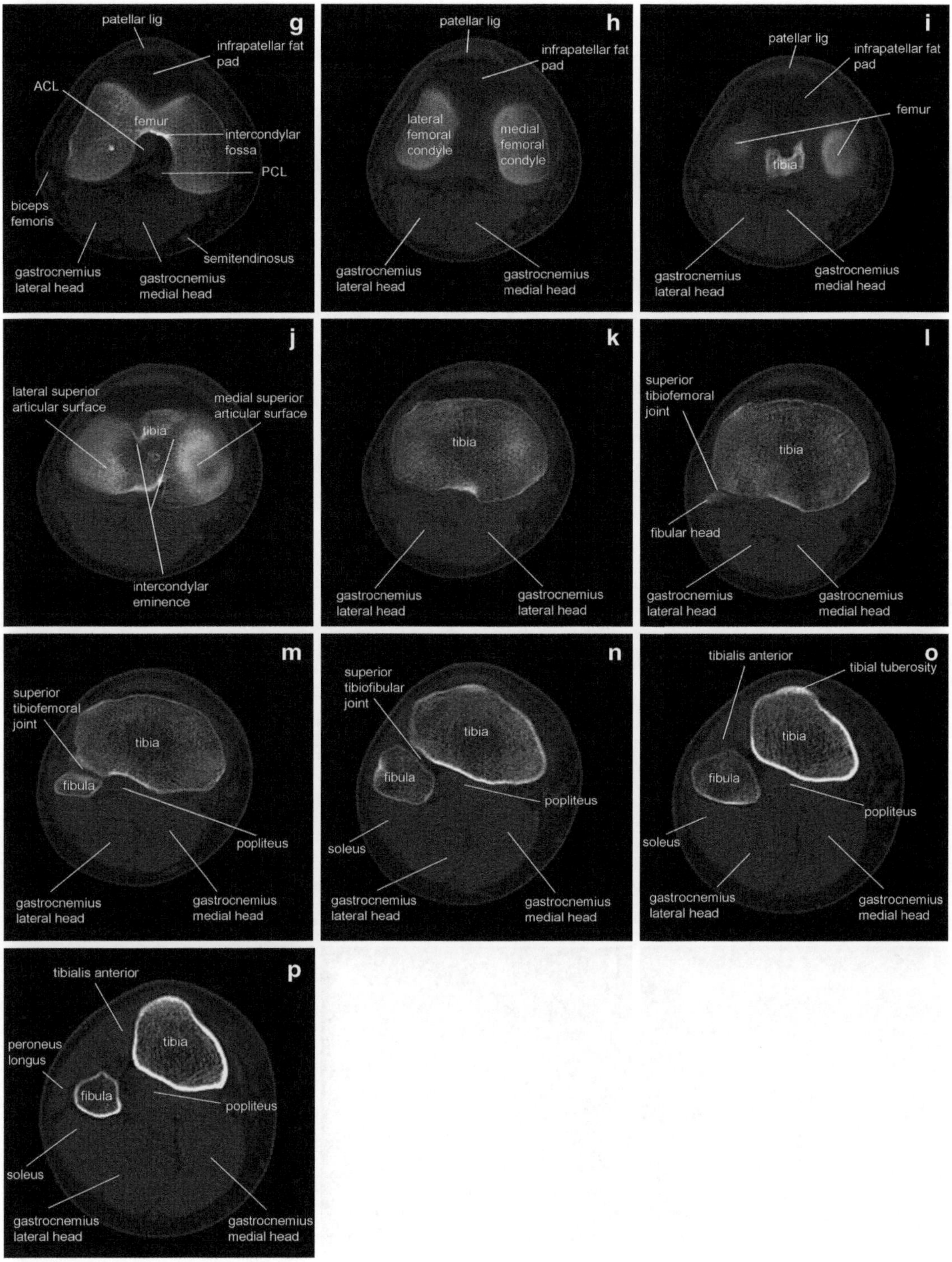

Fig. 3.5 (continued)

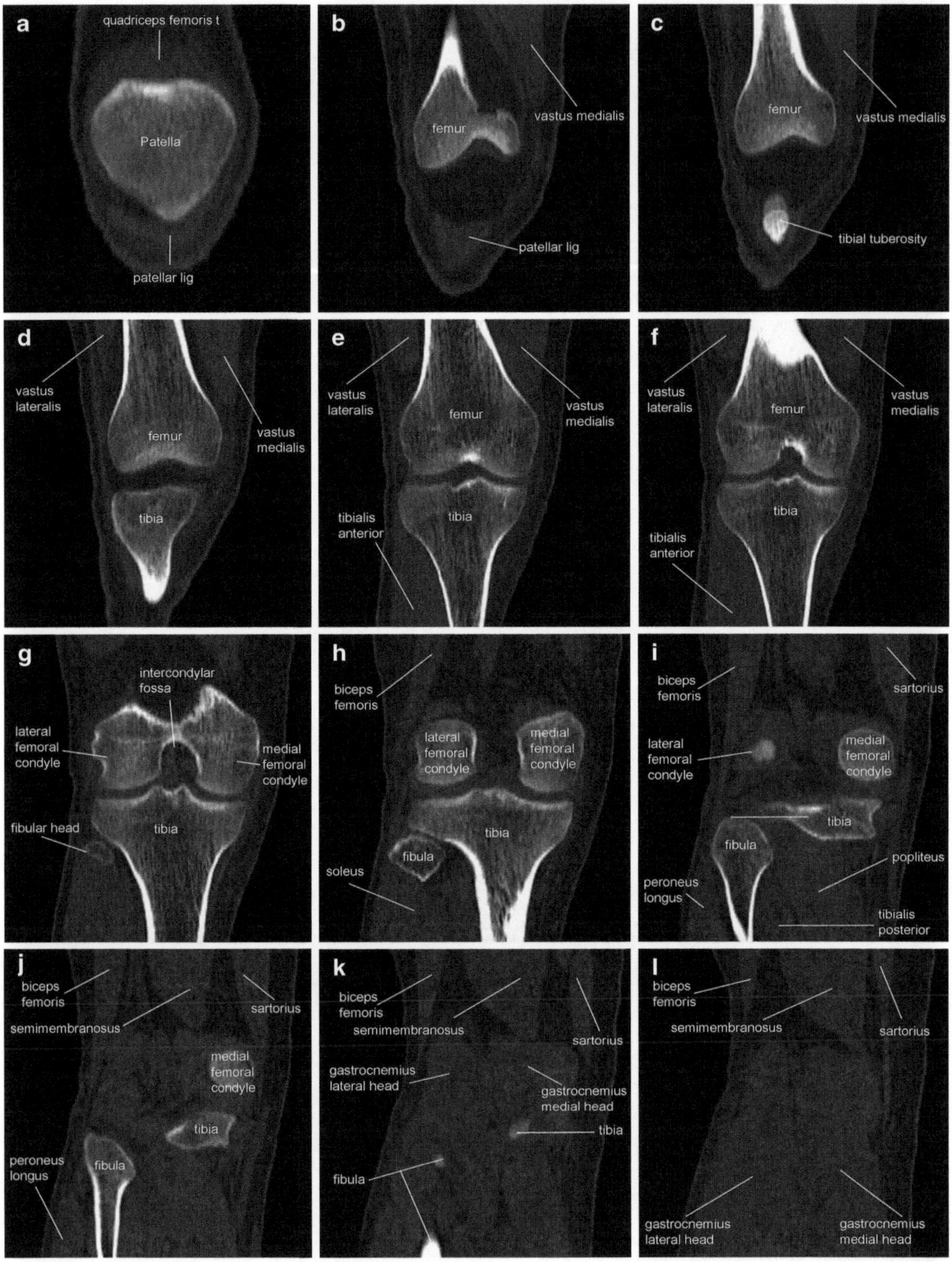

Fig. 3.6 (**a**–**l**) Coronal CT images; bone window (*t* tendon)

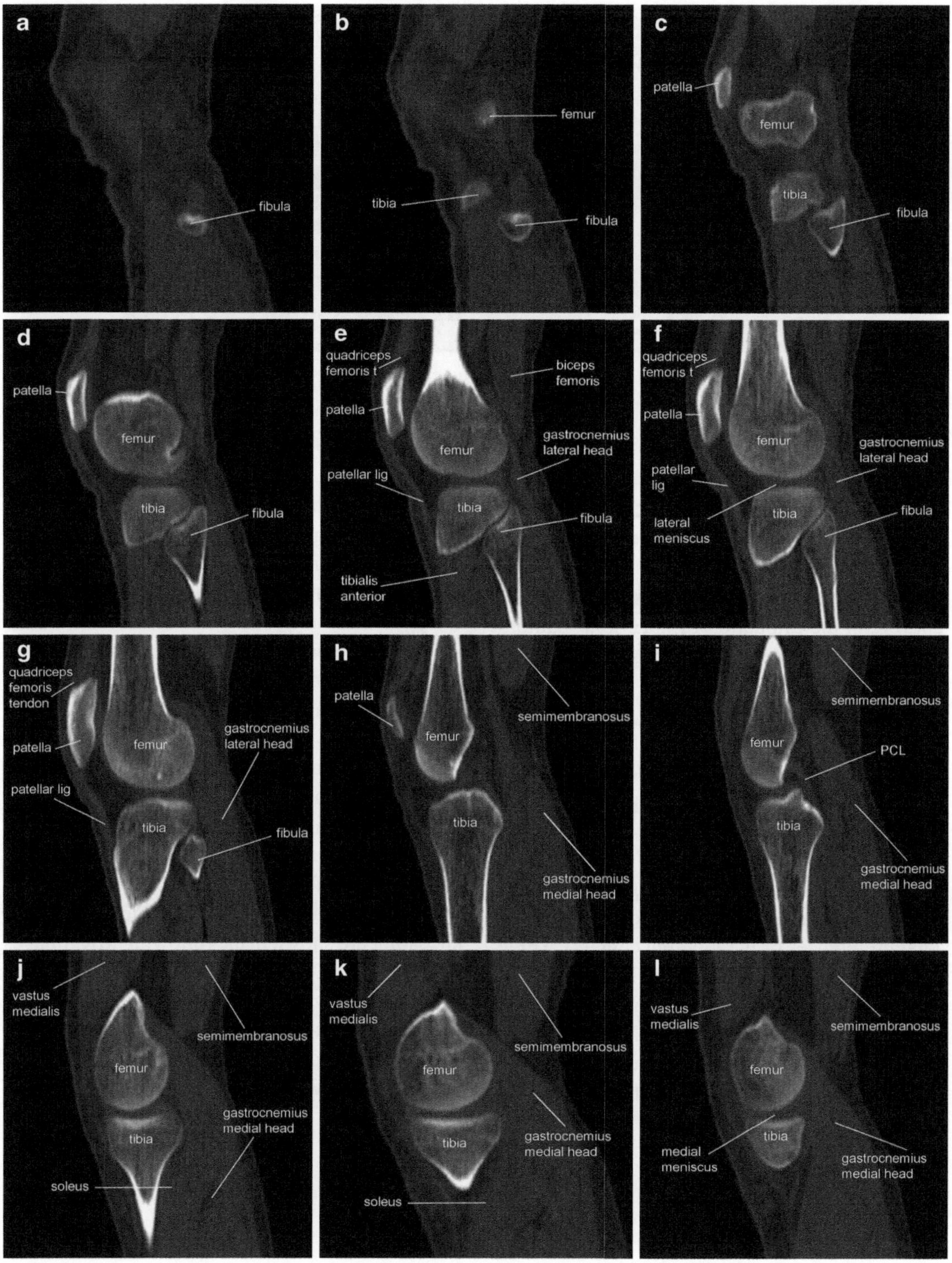

Fig. 3.7 (**a**–**n**) Sagittal CT images; bone window (*t* tendon, *PCL* posterior cruciate ligament)

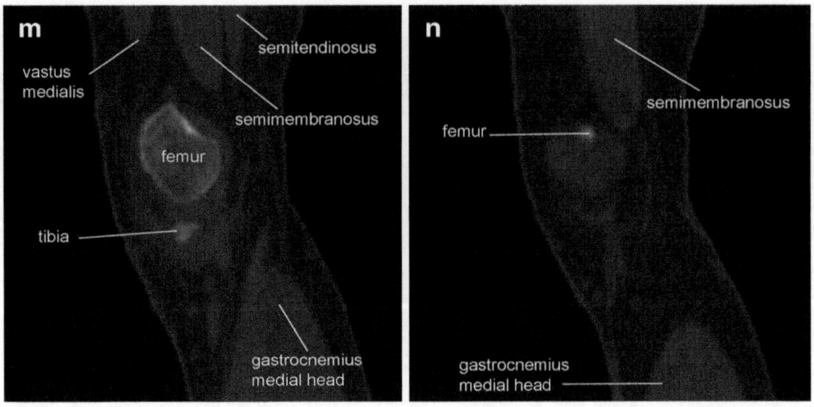

Fig. 3.7 (continued)

3.3 Magnetic Resonance Imaging of the Knee

Magnetic resonance imaging has a very important role in the evaluation of knee joint pathologies using dedicated extremity coils and high field systems [6]. Multi-planar high-resolution imaging capability of the cortex, bone marrow, cartilage, menisci, ligaments, tendons, synovium, and surrounding soft tissues without joint movement is the superiority of MRI compared to other imaging modalities [7, 8].

Routine MRI examination of the knee joint consists of axial, coronal, and sagittal images in different sequences (changes according to personal preferences of the imaging centers). Three-plane scout images must be obtained to localize and plan the sequences. The axial slices are planned on the coronal plane localizer. They must be sufficient to cover the all -knee joint from the tibial tuberosity up to the line of the superior border of the patella. The axial slices are particularly useful for imaging of the retropatellar cartilage and in evaluating fluid collections. The coronal slices are planned on the axial plane localizer. They must be sufficient to cover the whole-knee joint from the patella down to the line of the popliteal artery. The coronal slices are useful in evaluating collateral ligaments and meniscocapsular separation. The sagittal slices are planned on the axial plane localizer. They must be sufficient to cover the all-knee joint from the medial condyle up to the lateral condyle. The sagittal slices are useful in evaluating menisci, cruciate ligaments, and especially femoral cartilage.

The important anatomical structures of the knee joint are shown in the following MRI images (Figs. 3.8a–o, 3.9a–h and 3.10a–i). On the right side of the view, T1-weighted images and, on the left side, fat-saturated proton density images are found. The important anatomical structures are marked on T1-weighted images.

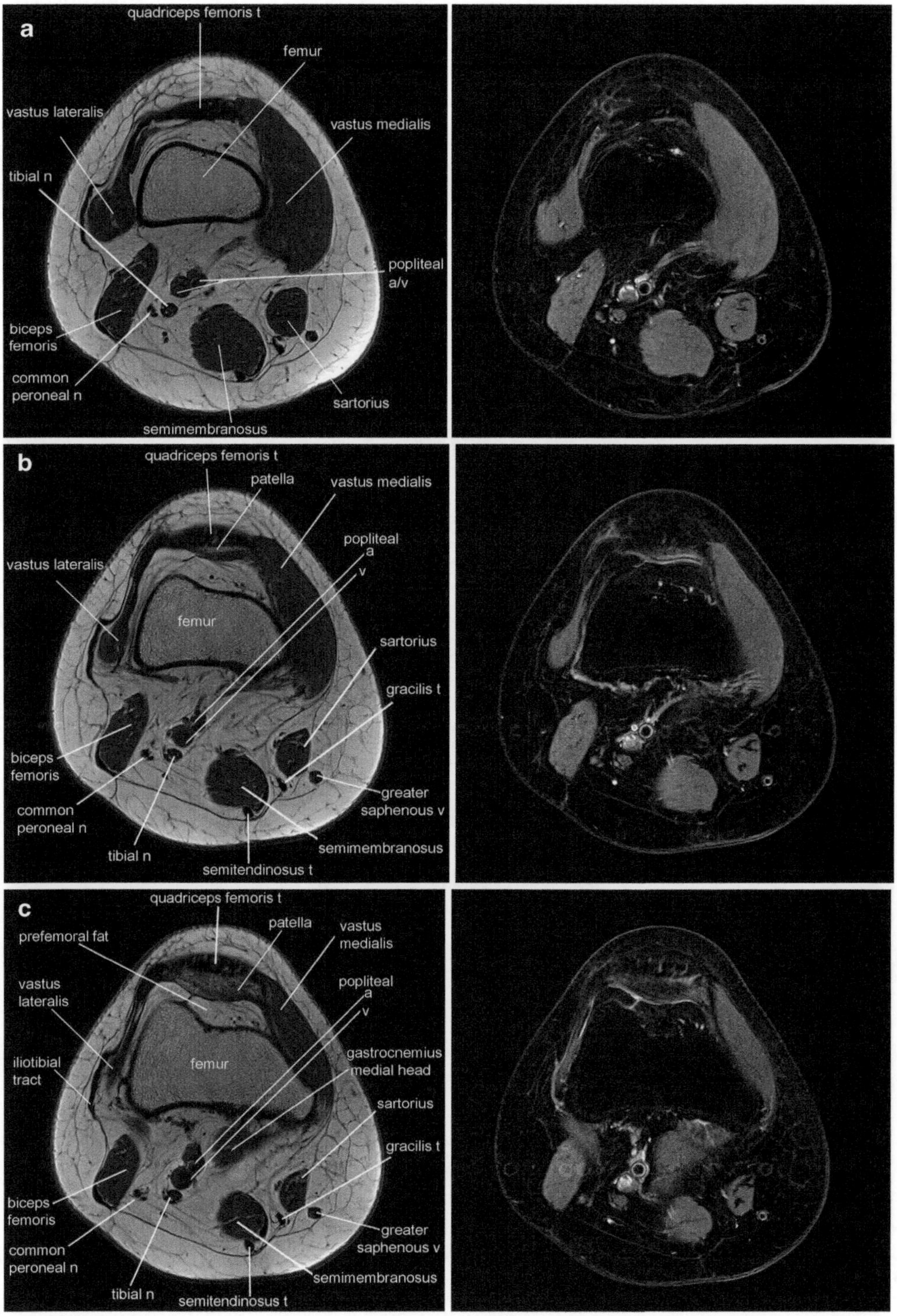

Fig. 3.8 (**a**–**o**) Axial MRI images (*n* nerve, *t* tendon, *a* artery, *v* vein, *lig* ligament, *ACL* anterior cruciate ligament, *PCL* posterior cruciate ligament, *MCL* tibial collateral ligament, *LCL* fibular collateral ligament)

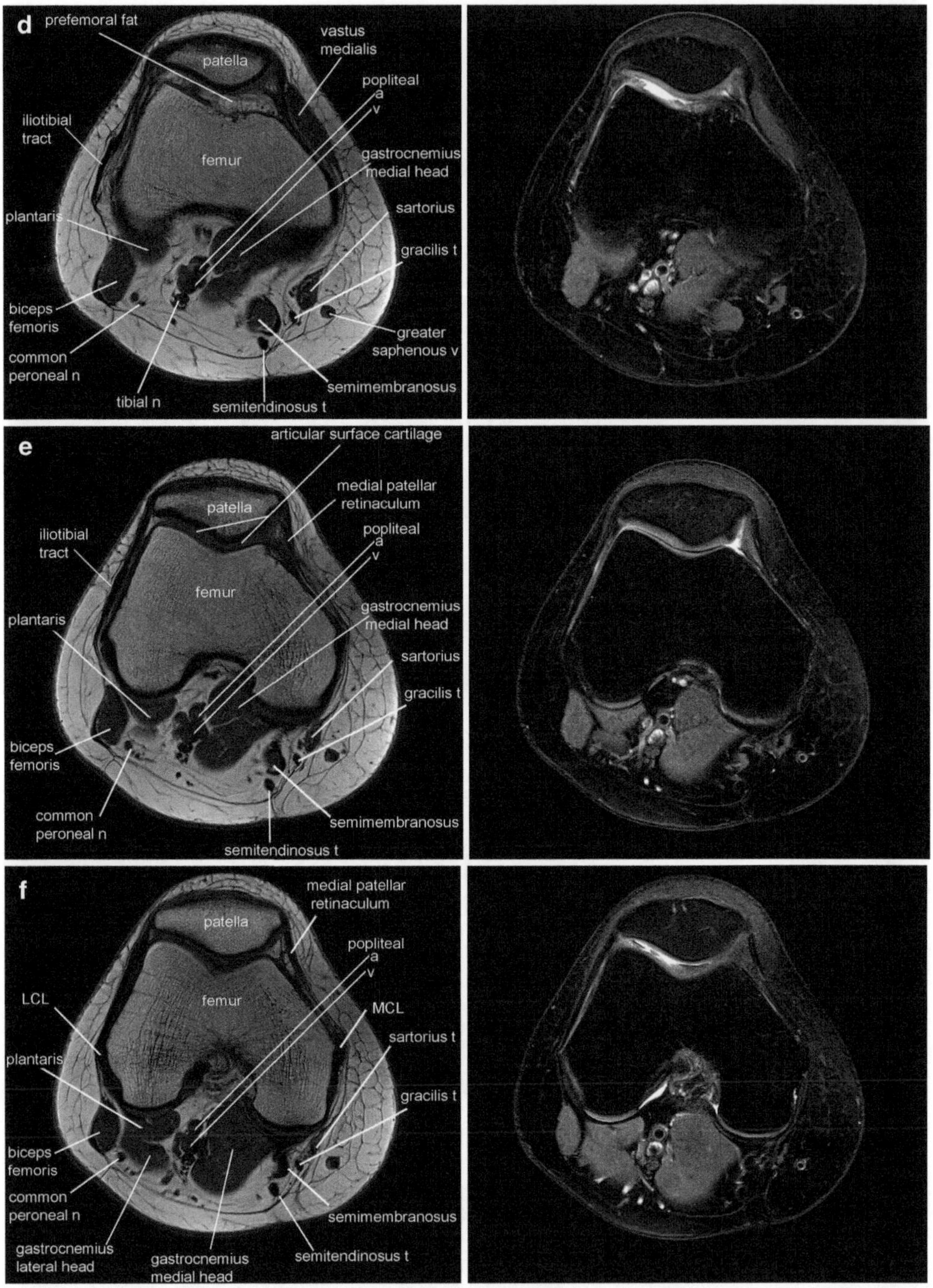

Fig. 3.8 (continued)

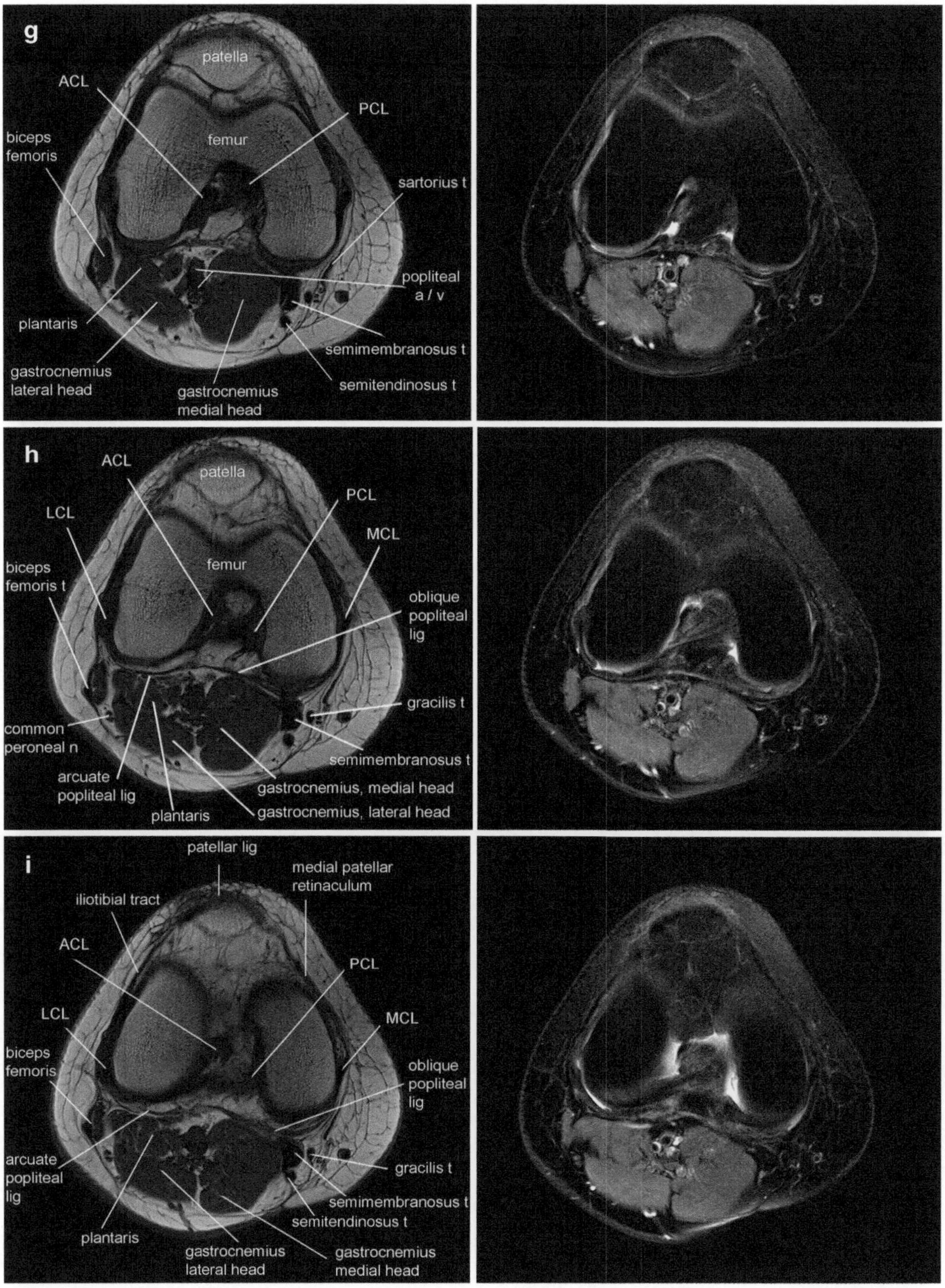

Fig. 3.8 (continued)

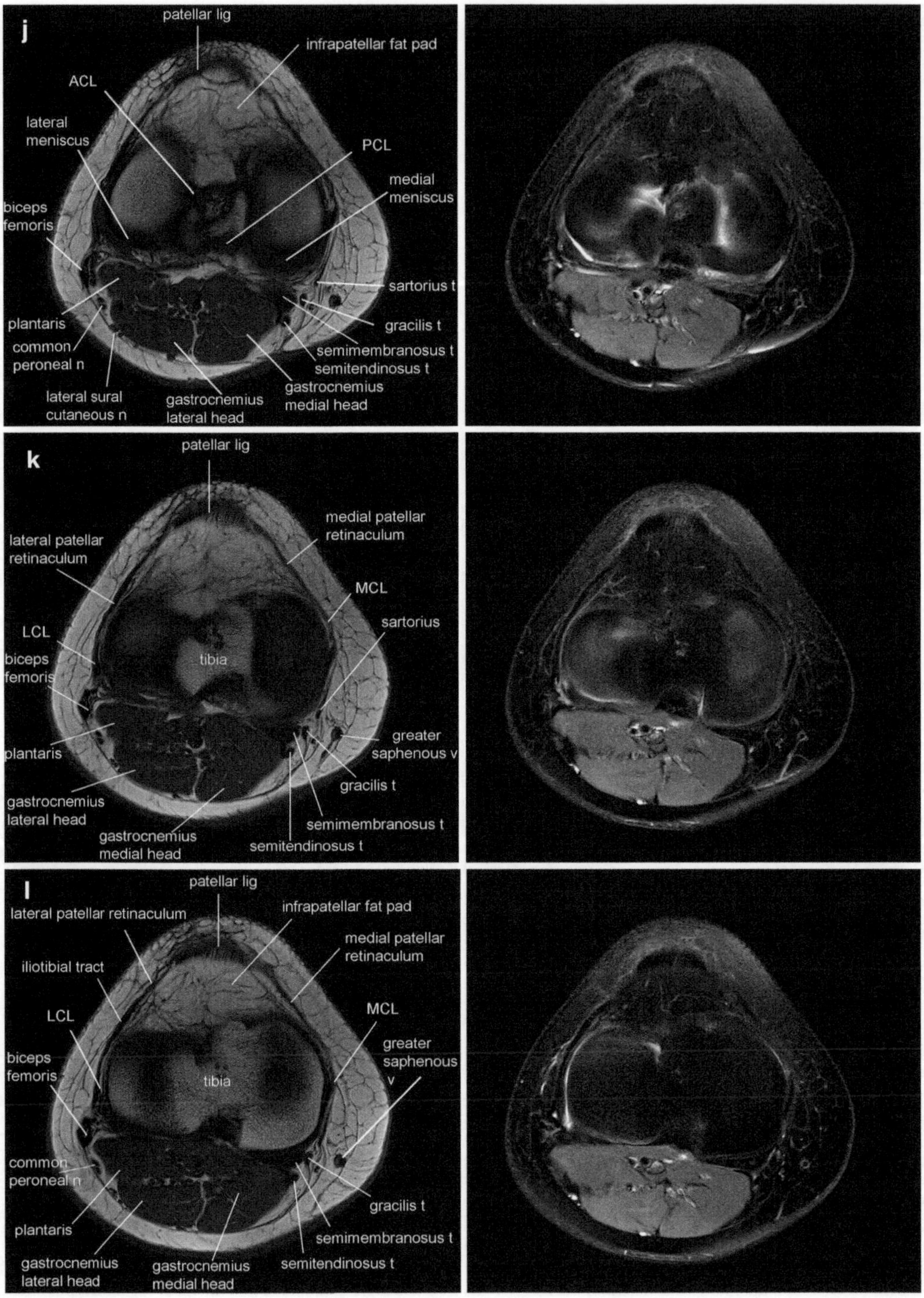

Fig. 3.8 (continued)

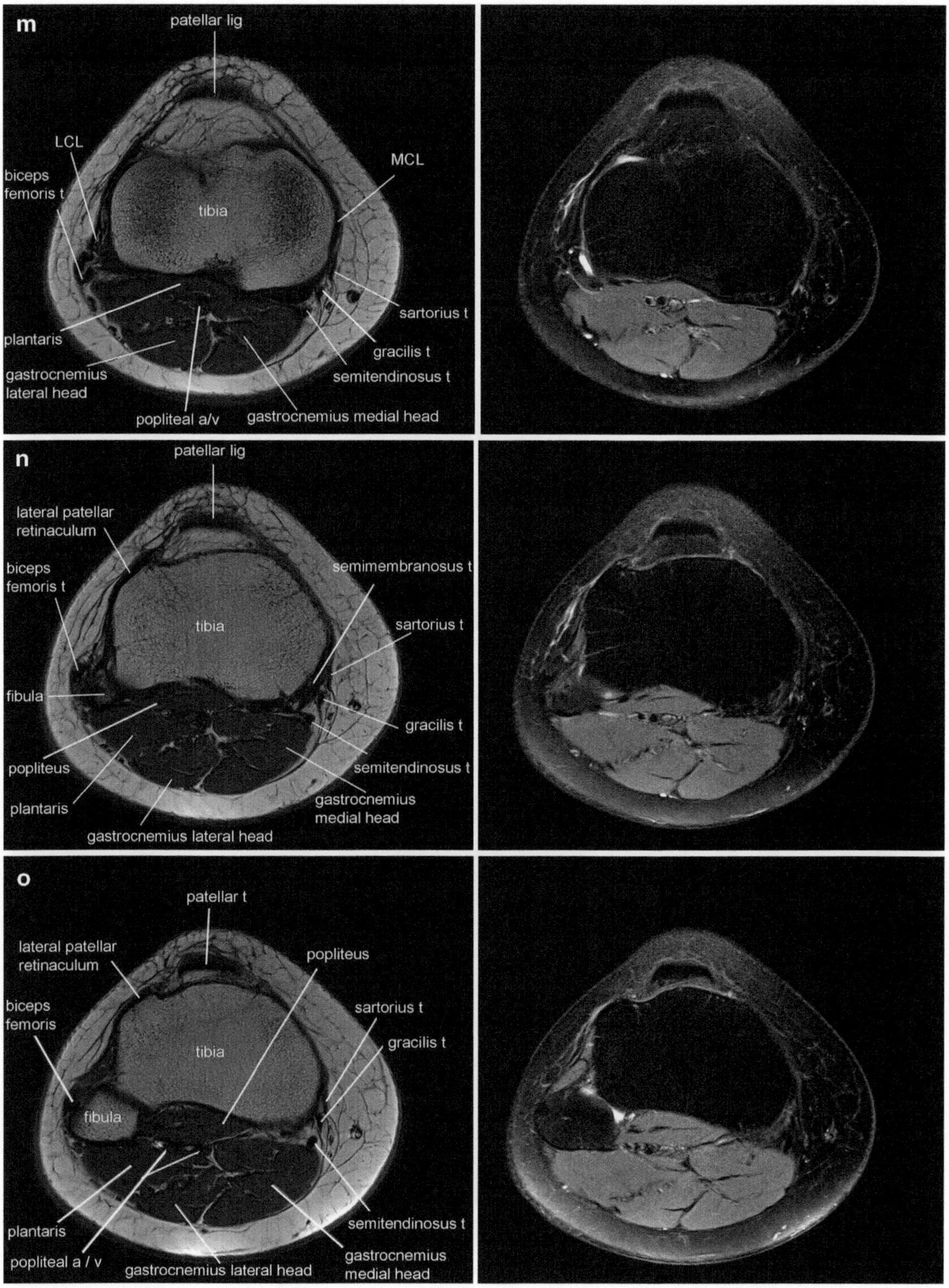

Fig. 3.8 (continued)

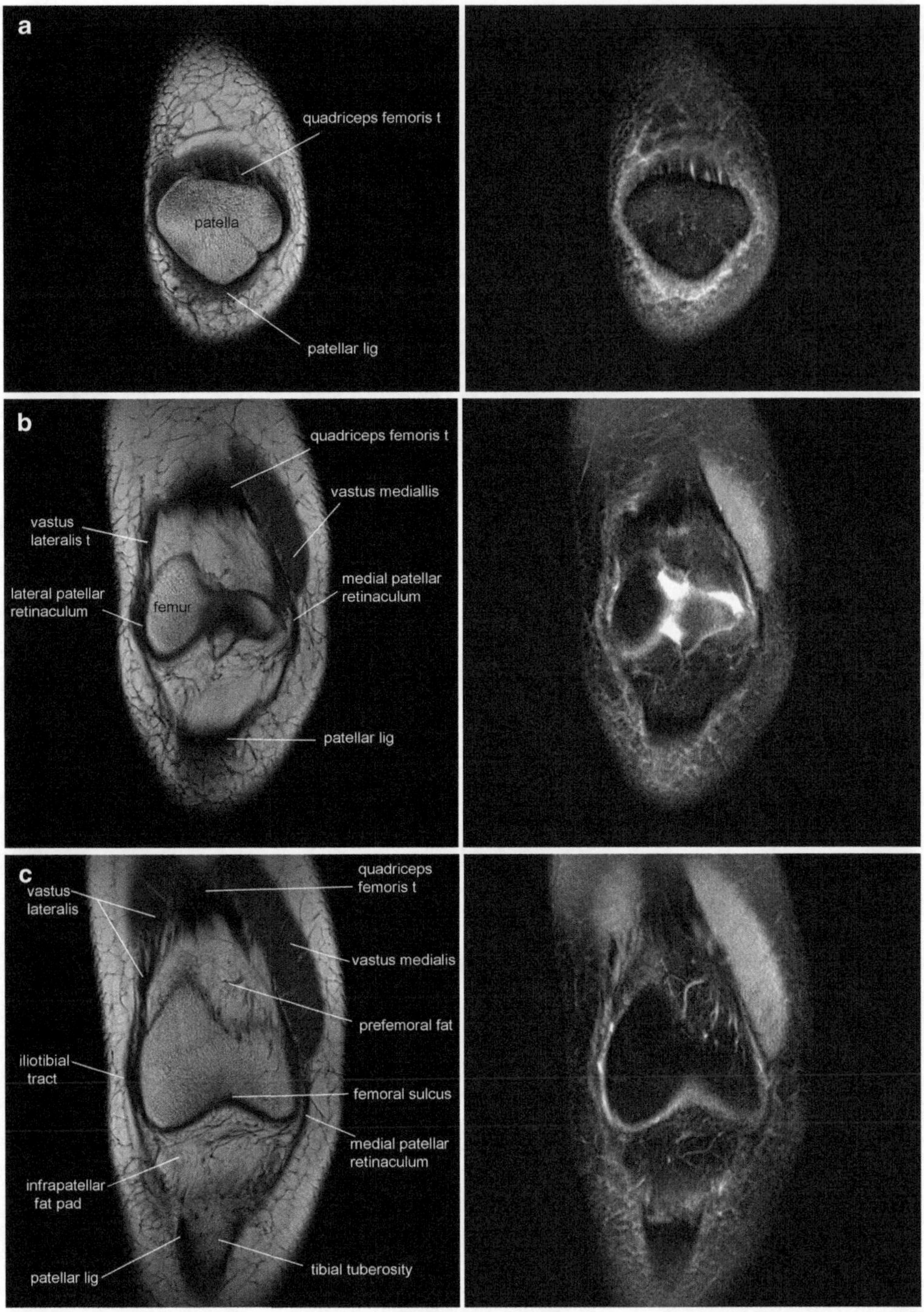

Fig. 3.9 (**a**–**h**) Coronal MRI images (*n* nerve, *t* tendon, *a* artery, *v* vein, *lig* ligament, *ACL* anterior cruciate ligament, *PCL* posterior cruciate ligament)

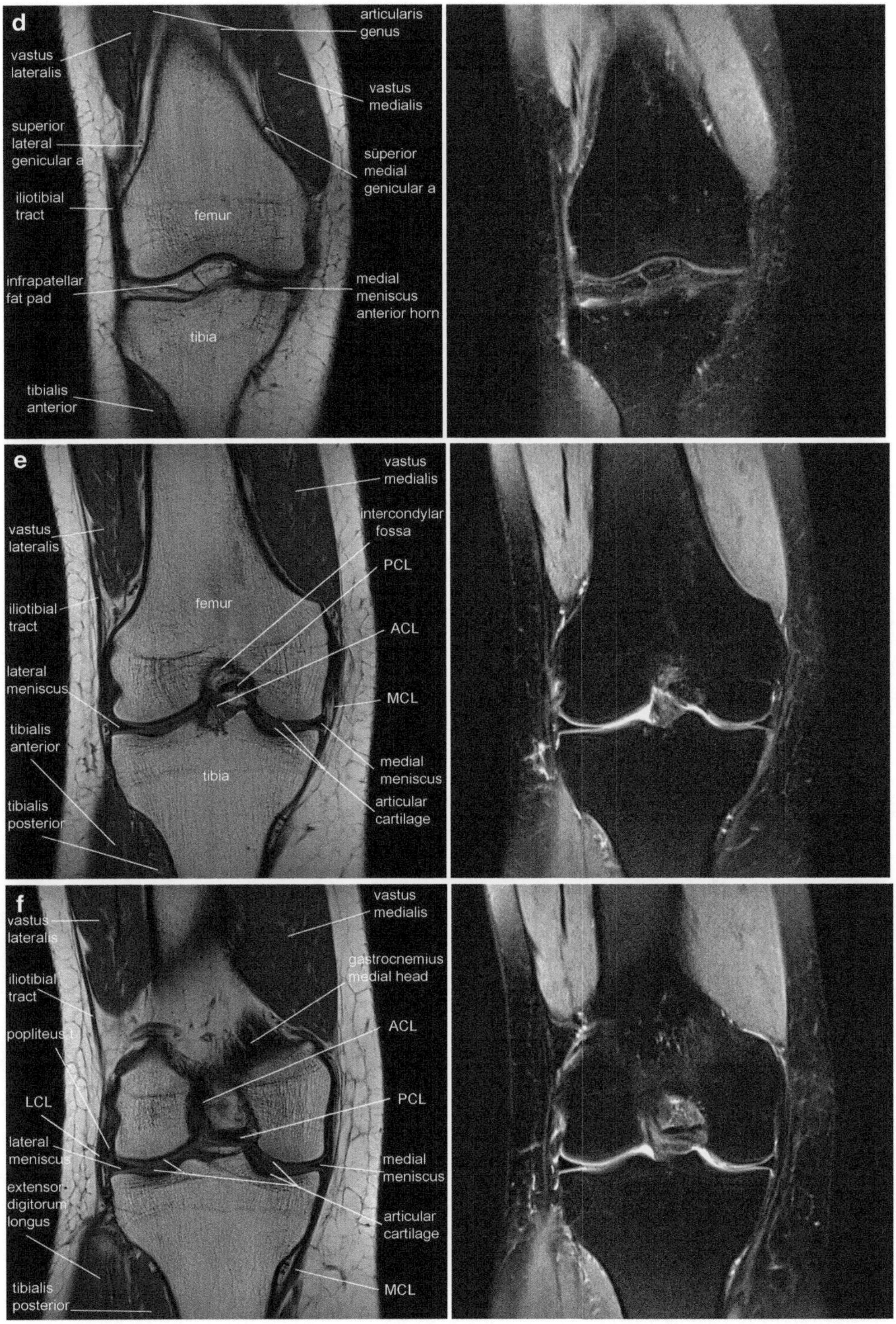

Fig. 3.9 (continued)

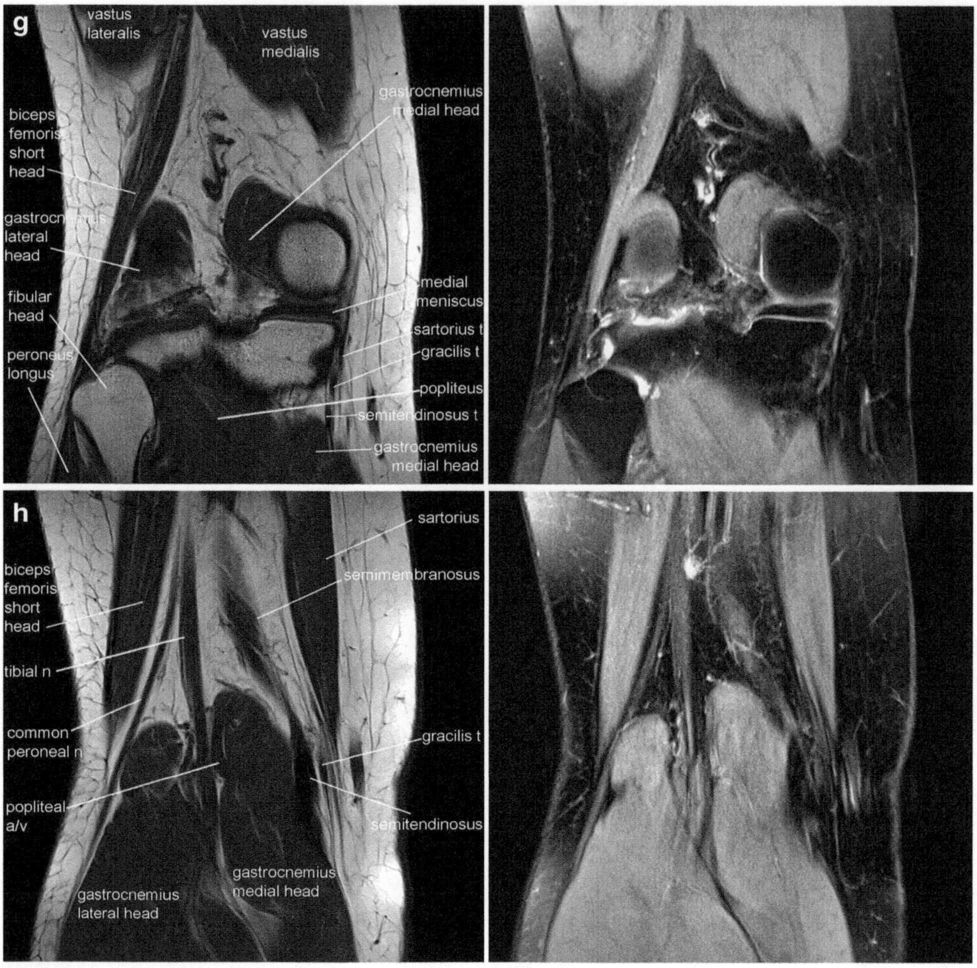

Fig. 3.9 (continued)

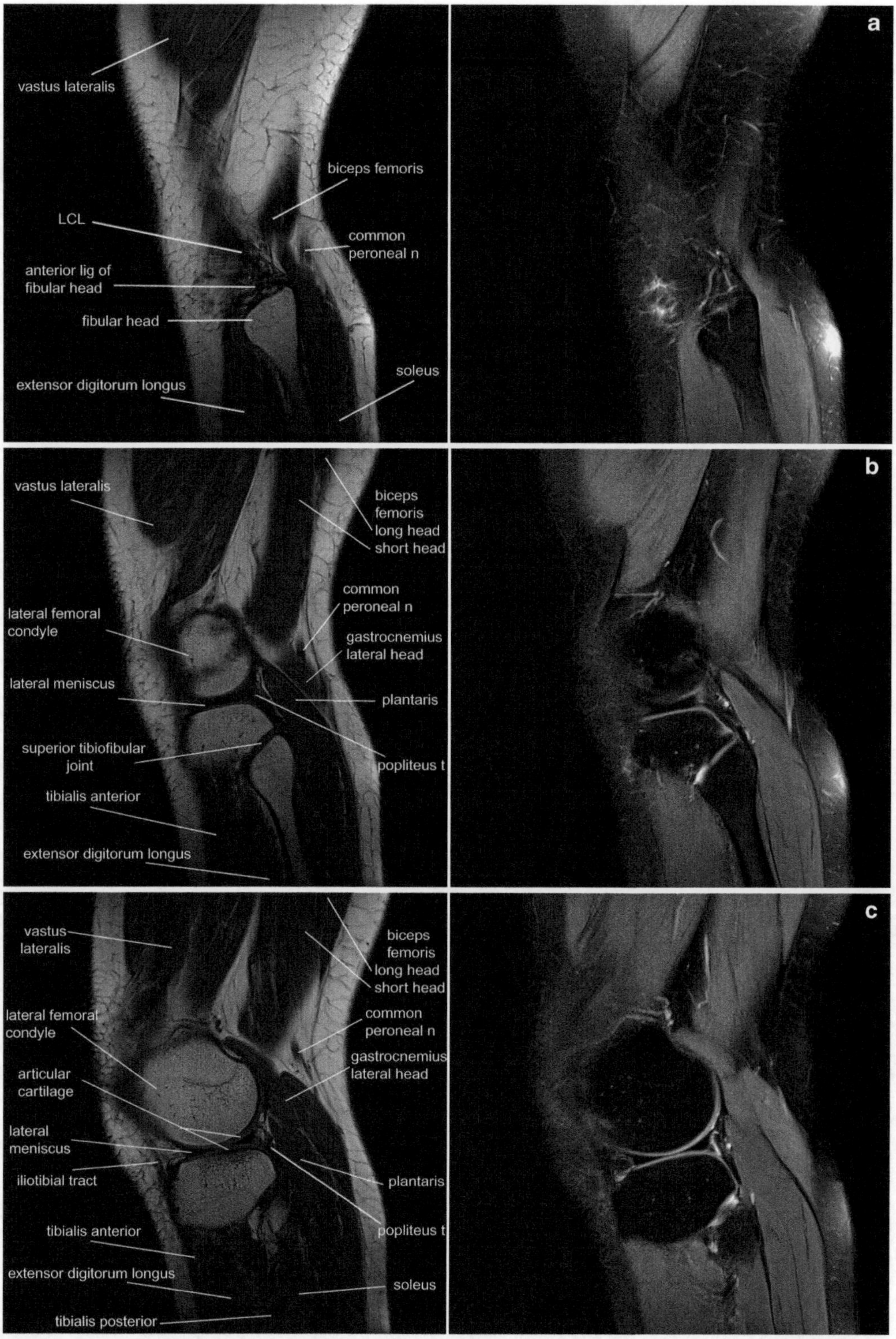

Fig. 3.10 (**a–i**) Sagittal MRI images (*n* nerve, *t* tendon, *a* artery, *v* vein, *lig* ligament, *ACL* anterior cruciate ligament, *PCL* posterior cruciate ligament)

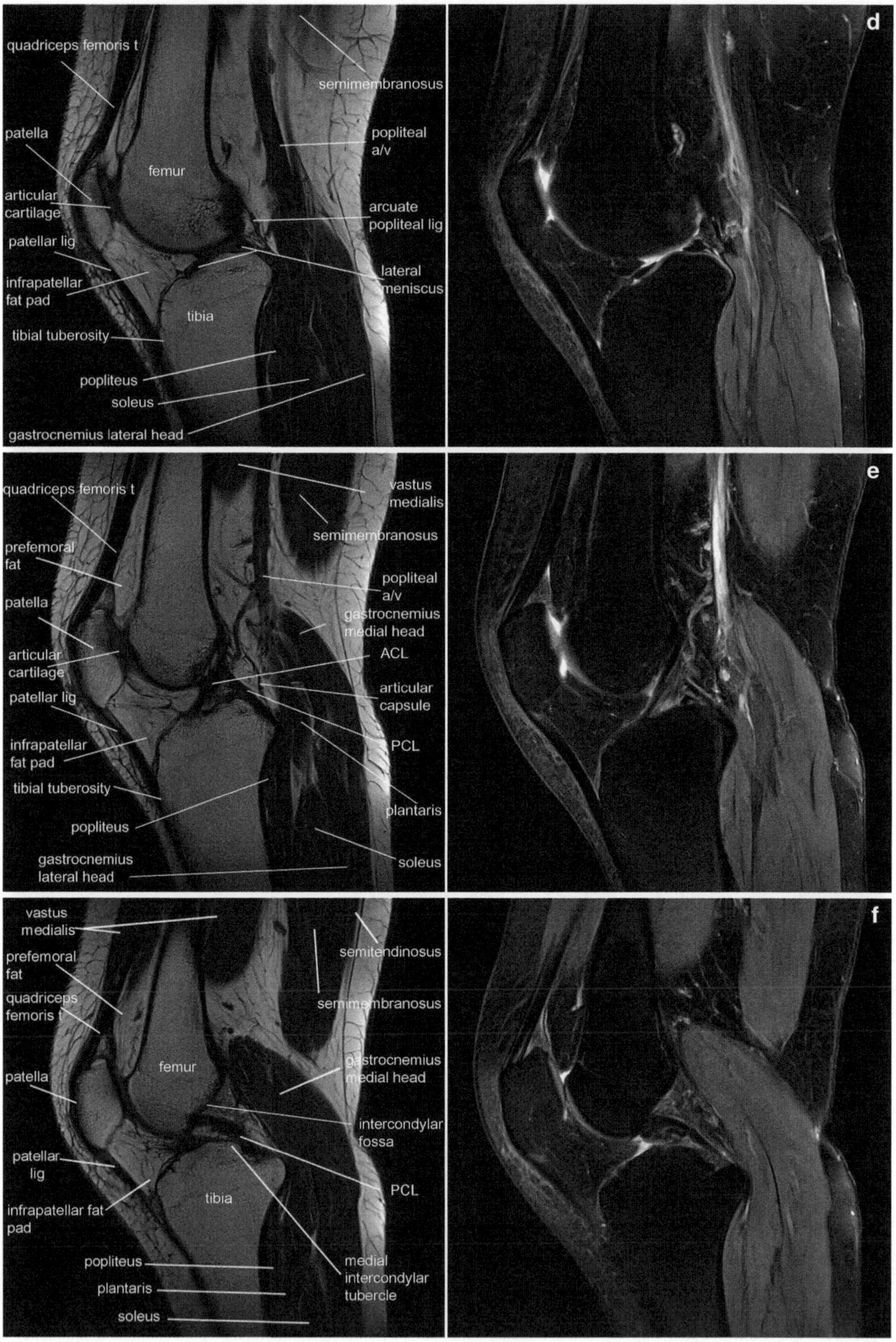

Fig. 3.10 (continued)

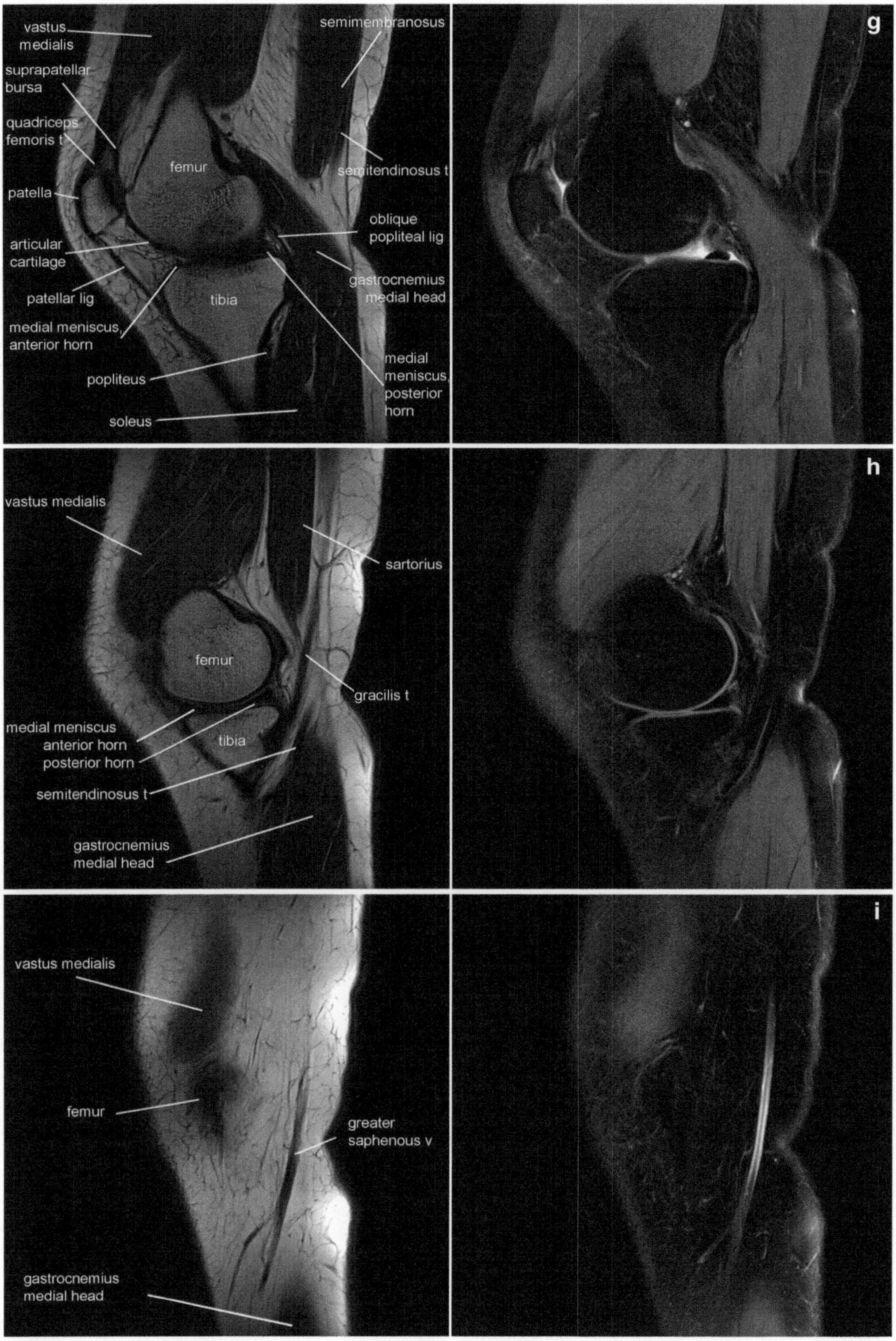

Fig. 3.10 (continued)

References

1. Cheung TC, Tank Y, Breederveld RS, Tuinebreijer WE, de Lange-de Klerk ES, Derksen RJ. Diagnostic accuracy and reproducibility of the Ottawa knee rule vs the Pittsburgh decision rule. Am J Emerg Med. 2013;31(4):641–5.
2. Tuite MJ, Kransdorf MJ, Beaman FD, Adler RS, Amini B, Appel M, Bernard SA, Dempsey ME, Fries IB, Greenspan BS, Khurana B, Mosher TJ, Walker EA, Ward RJ, Wessell DE, Weissman BN. ACR appropriateness criteria acute trauma to the knee. J Am Coll Radiol. 2015;12(11):1164–72.
3. Mustonen AO, Koskinen SK, Kiuru MJ. Acute knee trauma: analysis of multidetector computed tomography findings and comparison with conventional radiography. Acta Radiol. 2005;46(8):866–74.
4. Lo GH, McAlindon TE, Niu J, et al. Bone marrow lesions and joint effusion are strongly and independently associated with weight-bearing pain in knee osteoarthritis: data from the osteoarthritis initiative. Osteoarthr Cartil. 2009;17(12):1562–9.
5. Vincken PW, ter Braak AP, van Erkel AR, et al. MR imaging: effectiveness and costs at triage of patients with nonacute knee symptoms. Radiology. 2007;242(1):85–93.
6. Miller TT. MR imaging of the knee. Sports Med Arthrosc. 2009;17(1):56–67.
7. Prickett WD, Ward SI, Matava MJ. Magnetic resonance imaging of the knee. Sports Med. 2001;31(14):997–1019. Review.
8. Bennett DL, Nelson JW, Weissman BN, et al. ACR Appropriateness Criteria®; nontraumatic knee pain 2012. http://www.acr.org/~/media/ACR/Documents/AppCriteria/Diagnostic/NontraumaticKneePain.pdf. Accessed 31 Mar 2014.

Physical Examination of the Knee

4

Safa Gursoy

4.1 Introduction

Knee joint pathologies are the most frequently encountered orthopedic problems. The appropriate treatment for knee joint pathologies is determined not only by the correct diagnosis but also by the history and the application of a complete physical examination.

Several factors such as patient age, activity level, and type of trauma sustained will provide extremely important information in the differential diagnosis. In young patients or sports-related injuries, meniscus or ligament injuries related to the knee are often seen, whereas in elderly patients, degenerative joint diseases are the most frequent. In adolescents presenting with anterior knee pain, the diagnosis may often be osteochondritis; in young or middle-aged patients who are sports active, it may be tendinitis; and in middle-aged or elderly women, it is often patellofemoral arthritis.

Even if the basic structure of the knee is known and the functional anatomy of the ligamentous structures that provide stability, predicting the structures that could potentially be damaged by the trauma mechanism of a known injury will correctly direct the physical examination.

A complete anamnesis and a correct physical examination are key in reaching the correct diagnosis. Several authors have described how a correct knee joint physical examination should be applied. Just as there is no single correct way, it is undisputed that physical examination of the knee joint must be made systematically [1]. Following a careful inspection of the knee joint, palpation is made and then the joint range of movement is evaluated. Then tests are applied to the patellofemoral joint, meniscus, and knee stability.

4.2 Examination

4.2.1 Inspection

Gait is an important component of knee joint inspection. The physician must always evaluate the gait and weight-bearing capability of the patient, because these findings can be helpful in differentiating knee pathology from pain in the hip, lower back, or foot. When walking, knee joint-related pathologies can be observed in the nature of typical walking forms such as varus or valgus tilt in the coronal plane, antalgic gait in the sagittal plane, stiff knee gait, and gait with the knee in flexion.

Then, with the patient standing, the varus-valgus position of the lower extremity and the patella alignment can be observed from opposite. When observed from the side, conditions such as

S. Gursoy (✉)
Department of Orthopaedics and Traumatology, Ankara Yildirim Beyazit University, Ankara, Turkey

M. Bozkurt, H. İ. Açar (eds.), *Clinical Anatomy of the Knee*, https://doi.org/10.1007/978-3-030-57578-6_4

recurvatum or extension loss can be determined. Swelling of the knee can be best evaluated with the patient lying down. When there is a large amount of swelling or conditions accompanying advanced arthritis, the knee is in partial flexion. Redness concomitant to swelling may be seen in pathologies such as sepsis or gout. Localized swelling is seen more in bone dislocation, cyst, or bursitis.

4.2.2 Range of Joint Movement

Knee joint range of movement is limited to flexion and extension measurements. In the presence of laxity, rotational movements can be evaluated as explained in the next section on knee stability.

Extension is generally examined with the patient in a supine position (Fig. 4.1). The ankle of the patient is held and raised in the air, and the knee extension status is observed. With this movement, it is expected that the femur and tibia will reach a neutral position in the same linear direction. In some cases, hyperextension of the knee (genu recurvatum) can be seen up to 10°. When the knee cannot reach full extension (excessive effusion, advanced arthrosis, displaced meniscus tear, etc.), flexion contracture can be seen.

Flexion is generally examined with the patient in a supine position (Fig. 4.2). When a normal knee is in flexion, the heel is expected to touch the hip, and this generally indicates 140°–150° flexion. Knee joint pain during flexion can be a sign of many conditions, which can be determined with additional information such as pain localization. Knee flexion may be restricted earlier because of extra-articular thigh thickness and the calf muscle structure rather than by the bone edge.

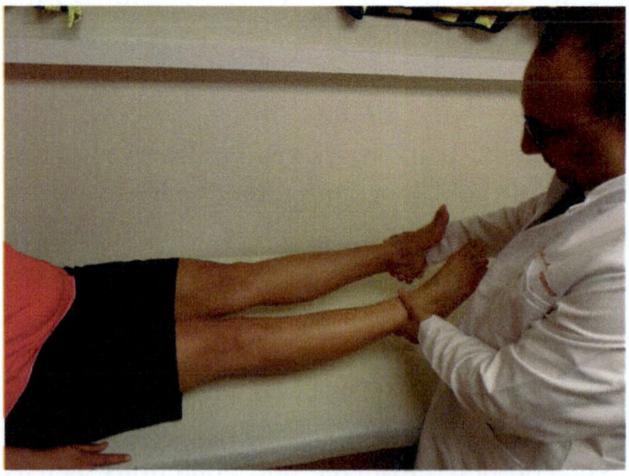

Fig. 4.1 Range of movement examination: extension of the knee

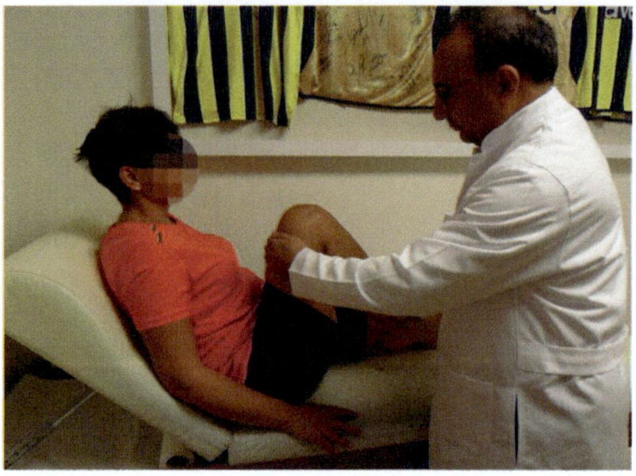

Fig. 4.2 Range of movement examination: flexion of the knee

4.2.3 Palpation

Palpation of the knee joint should be started on the unaffected side. This method allows the patient to feel safe, and the healthy side can be a reference for the affected knee. Palpation of the knee allows the doctor making the examination to become familiarized with the joint orientation and provides advantages in cases such as obesity or edema when pathologies may not be able to be determined with inspection. Palpation is extremely valuable in inflammatory or septic conditions when there is an increase in temperature.

In palpation of the joint, the anterior structures are evaluated first (Fig. 4.3). Moving upwards from the tibia tuberositas, the patellar tendon is identified and continues to the whole quadriceps tendon and the sensitive points of this region. The presence of anteromedial and medial patellar plica can create sensitivity in this region. When there has been trauma, a fracture that could have occurred in the patella can be easily determined with palpation.

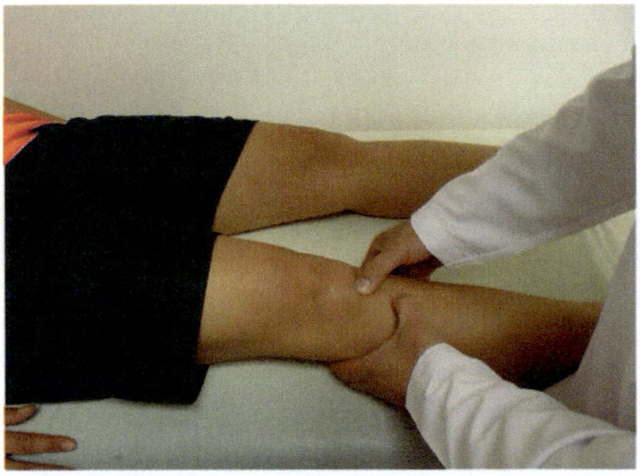

Fig. 4.3 Palpation of anterior structures of the knee

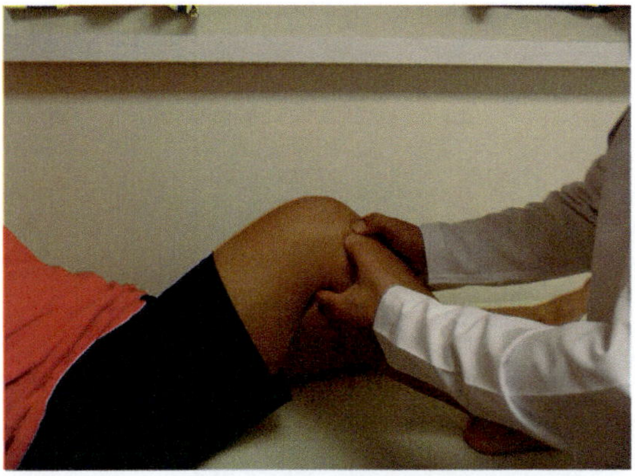

Fig. 4.4 Palpation of lateral structures of the knee

The palpation then moves sequentially to the medial section. First, defining the medial joint line in this region is important in respect of orientation. In patients with sensitivity in the medial joint line, examination is made with the knees at 90° and is usually positive in pathologies such as meniscus tear and osteoarthritis. Other important structures in this region that require palpation are the medial collateral ligament, semi-membranous, and other pes anserinus tendons and sensitivity in the bursae of these.

Palpation on the lateral side again starts with the identification of the joint line (Fig. 4.4). Sensitivities are observed in the meniscus and the lateral side arthrosis in this region just as on the medial side. Then the lateral ligament can be easily palpated in the position in Fig. 4.4 (Fig. 4.5). Of the other anatomic structures, knee problems

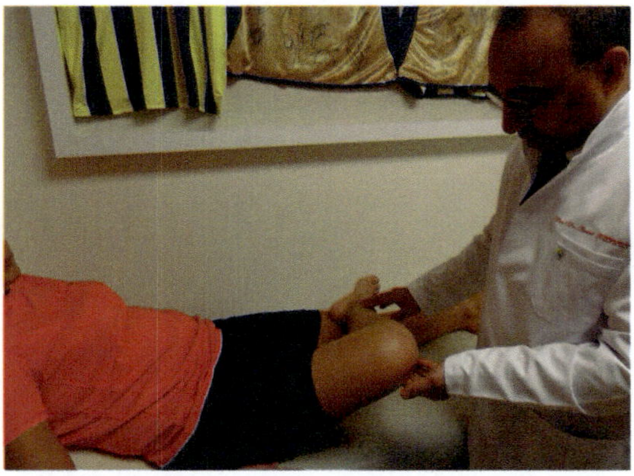

Fig. 4.5 Palpation of lateral structures of the knee in figure of four position

that are often seen in runners originate from the iliotibial band. The proximal tibiofibular joint is one of the less important structures in knee problems. Congenital or traumatic dislocations of this joint in the posterolateral region can be determined by examining the knee at 90°.

Posterior popliteal cysts or acute ruptures of these can be evaluated with palpation of this section. Examination of neurovascular structures in the posterior can be made with the patient in the prone or supine position.

4.2.4 Specific Pathologies and Examination Tests

4.2.4.1 Patellofemoral Joint and Extensor Mechanism

Q Angle

Patellar alignment is an important form of measurement. Q angle is the angle between the line drawn from the spina iliaca anterior superior to the center of the patella and the line drawn from the patella center to the tibia tuberositas (Fig. 4.6). It is recommended that it is measured with the knee in full extension or in 30° flexion. The normal value of the Q angle is 15° (men 14°, women 17°). An increased Q angle is related to patella inversion or outward movement of the tibial tubercle. Although a high Q angle is a risk factor for lateral tilt or patellar instability [2],

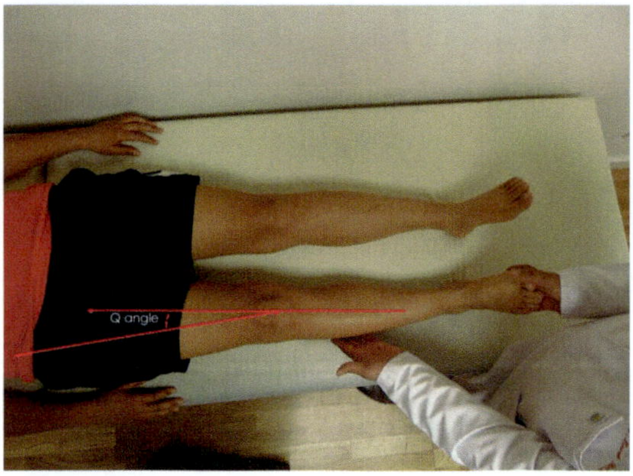

Fig. 4.6 Measurement of the "Q" angle

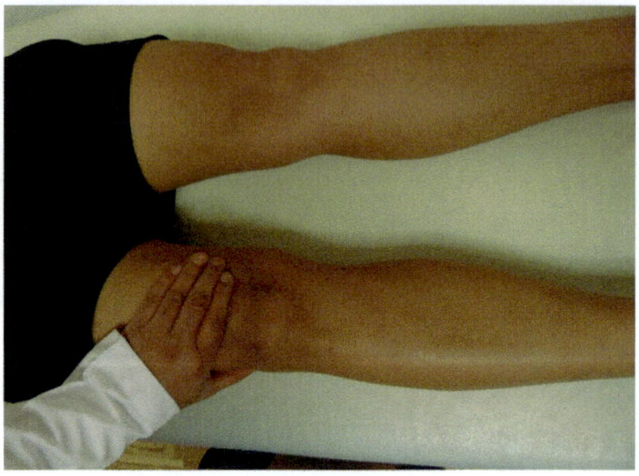

Fig. 4.7 Patellofemoral grinding test

the clinical benefit of the Q angle is debatable. It has been reported that there is no relationship between clinical symptoms and the Q angle and measurements of patellofemoral pain [3].

Patellofemoral Grinding Test

This test was first described by Owre in 1936 [4]. The currently used form was described by Soloman et al. [5]. With different modifications, this test is used for the conditions of patellofemoral pain. In the test, the patient is positioned supine, and the knee is brought into full extension. The thumb is placed above the edge of the inner section of the patella, the patient is instructed to contract the quadriceps muscle, and pressure is applied slowly and downwards with the thumb (Fig. 4.7). Pain with movement of the patella or inability to complete the test is a sign of patellofemoral dysfunction.

Patellar Glide Test (Sage Sign)

The glide test is applied with the knee in 30° flexion. By moving the patella being examined medially and laterally in sequence, the distance from the normal position is evaluated (Fig. 4.8). A movement of >1 cm in any direction, although not definitive for patellar instability, is accepted as a symptom. The rate of movement can be used in the patellar quadrants. In contrast to excessive movement, <25% movement of the patella laterally suggest lateral pressure syndrome.

Fig. 4.8 Patellar Glide Test (Sage Sign)

Patellar (Fairbanks) Apprehension Test

This test was first described by Fairbanks in 1936 [6]. With the patient in a supine position on the examination table, the patient's ankle is held and abduction is applied until the knee to be evaluated comes into flexion away from the table. Then with the other hand, pressure is applied from the medial side of the patella to move it to the lateral side. At the same time, the knee is slowly brought into flexion (Fig. 4.9). Positivity in the test is shown by the development of anxiety in patients with a tendency or history of instability during flexion. In a study by Sallay et al., only 39% of patients with a history of dislocation showed a positive apprehension sign [7].

4.2.4.2 Meniscal Tests

McMurray Test

McMurray test is the primary of the clinical tests in the evaluation of meniscus tears and was first described by McMurray in 1940 [8]. While the knee is in flexion, the leg is turned in external rotation, and the joint line is palpated with the other hand. Then the knee is brought slowly into extension (Fig. 4.10a). The lateral meniscus test is applied by turning the leg in internal rotation (Fig. 4.10b). A meniscus lesion within the condyle is felt with pain or sound.

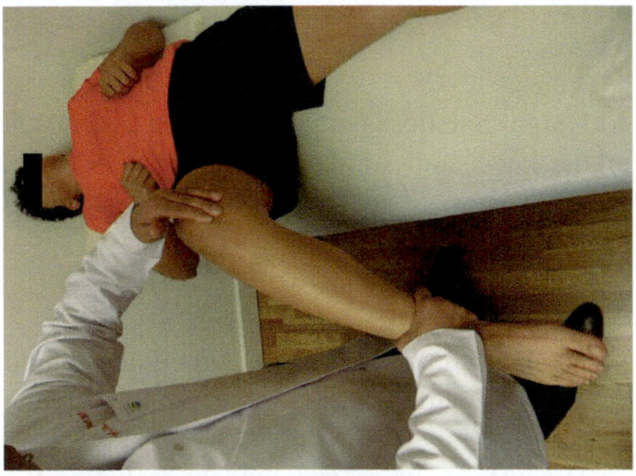

Fig. 4.9 Patellar (Fairbanks) apprehension test

Apley Test

This test was first described in 1947 by Apley [9]. The test has two phases: distraction and compression. With the patient in the prone position, first the knee is brought into 90° flexion for the distraction phase. Pressure is placed on the back of the patient's thigh, and distraction is applied to the knee by pulling the patient's foot upwards with the hands and the knee is moved to internal and external rotation. The distraction phase is expected to be painless even if there is a meniscus lesion (Fig. 4.11a). If the patient experiences pain at this stage, the test is evaluated as not safe and is terminated. If there is no pain, with the patient in the same position, compression is applied to the knee by pressing on the sole of the foot, and then in sequence moved in external and

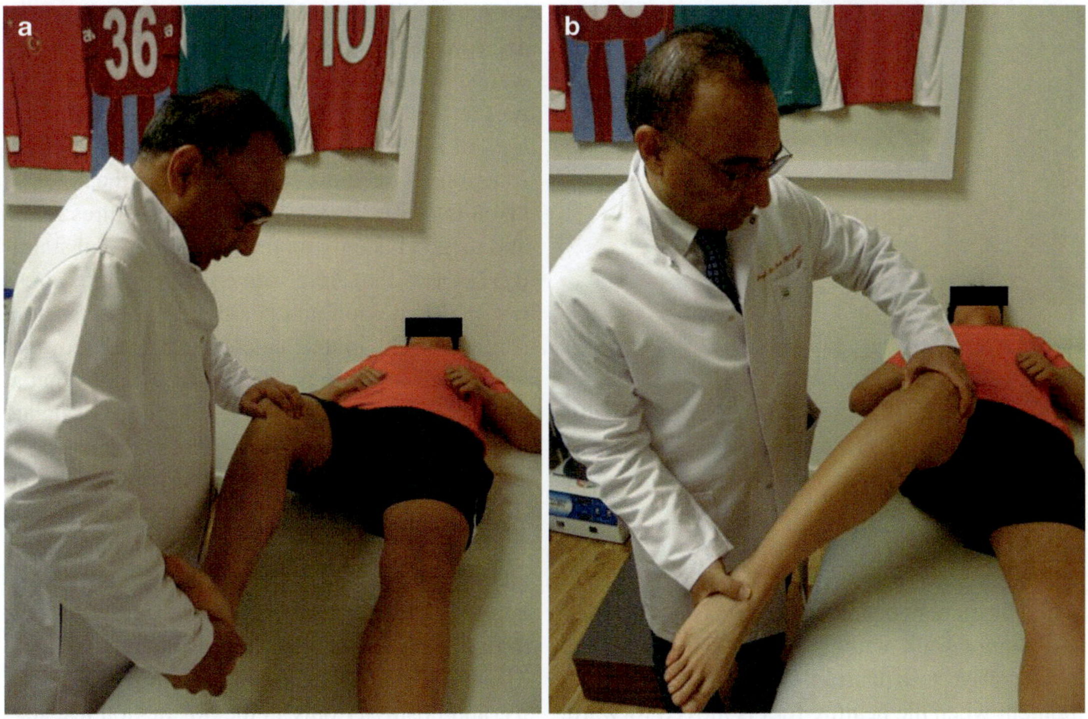

Fig. 4.10 (**a**) Medial meniscus McMurray test in external rotation. (**b**) Lateral meniscus McMurray test in internal rotation

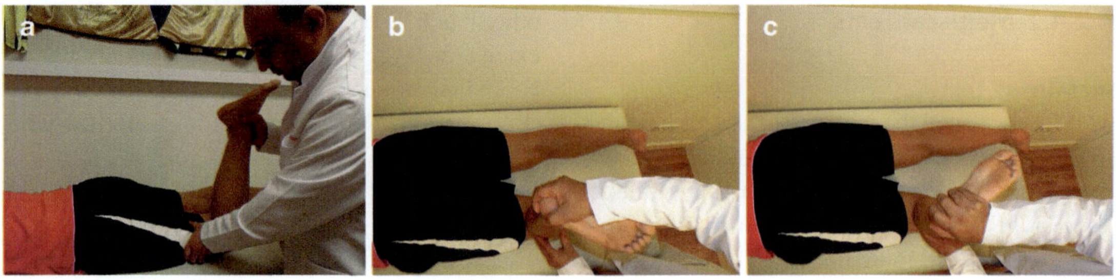

Fig. 4.11 (**a**) Apley test distraction phase. (**b**) Medial meniscus Apley test in external rotation. (**c**) Lateral meniscus Apley test in internal rotation

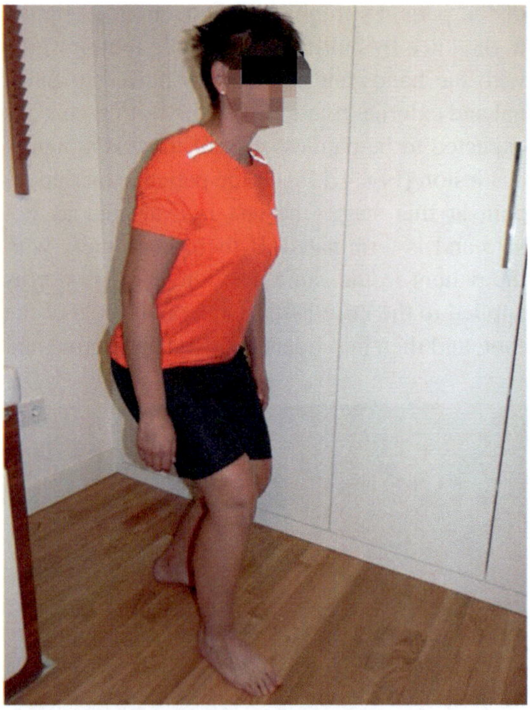

Fig. 4.12 Childress' sign (squat test)

internal rotation. In patients with a medial meniscus lesion, pain generally emerges in external rotation (Fig. 4.11b) and in those with a lateral meniscus lesion, in internal rotation (Fig. 4.11c).

Childress' Sign (Squat Test)

When the patient bends over while trying to walk and cannot walk without pain, this is accepted as a positive test. During this movement, at the moment when the healthy foot is moved forward, all the load is transferred to the affected side, and this causes compression on the knee (Fig. 4.12). Although patellofemoral pathologies, effusion in the knee, and similar findings can be seen, pain localization in the joint line is helpful in diagnosis.

4.2.4.3 Varus and Valgus Stability Tests

Varus and valgus stress tests of the knee should be repeated with the knee in full extension and in 30° flexion (Fig. 4.13a–d).

The valgus stress test, which evaluates the medial structures, is applied with the patient in a supine position. While one hand pushes the knee inwards, the ankle is held in the other hand and force is applied in the opposite direction outwards. When the knee is in full extension, the posteromedial capsule and cruciate ligaments, as well as the medial collateral ligament (MCL), function in medial stability. When the knee is taken into 30° flexion, the other structures relax, and the load is only on the MCL in medial stability. In cases that are positive for the valgus stress test in 30° flexion and healthy in full extension, isolated MCL injury should be considered, and when both tests are positive, this indicates that at least one of the posteromedial capsule or cruciate ligaments is injured.

The varus stress test is applied with the patient in the same position, but in this test, the force is applied to the ankle internally and the knee externally. The status of the valgus test is valid for the lateral side. Test positivity in full extension shows injury in structures other than the lateral collateral ligament (LCL), which functions in lateral stability, and positivity only in the 30° flexion test suggests an isolated LCL injury. Unlike the valgus test, laxity is more widespread in healthy indi-

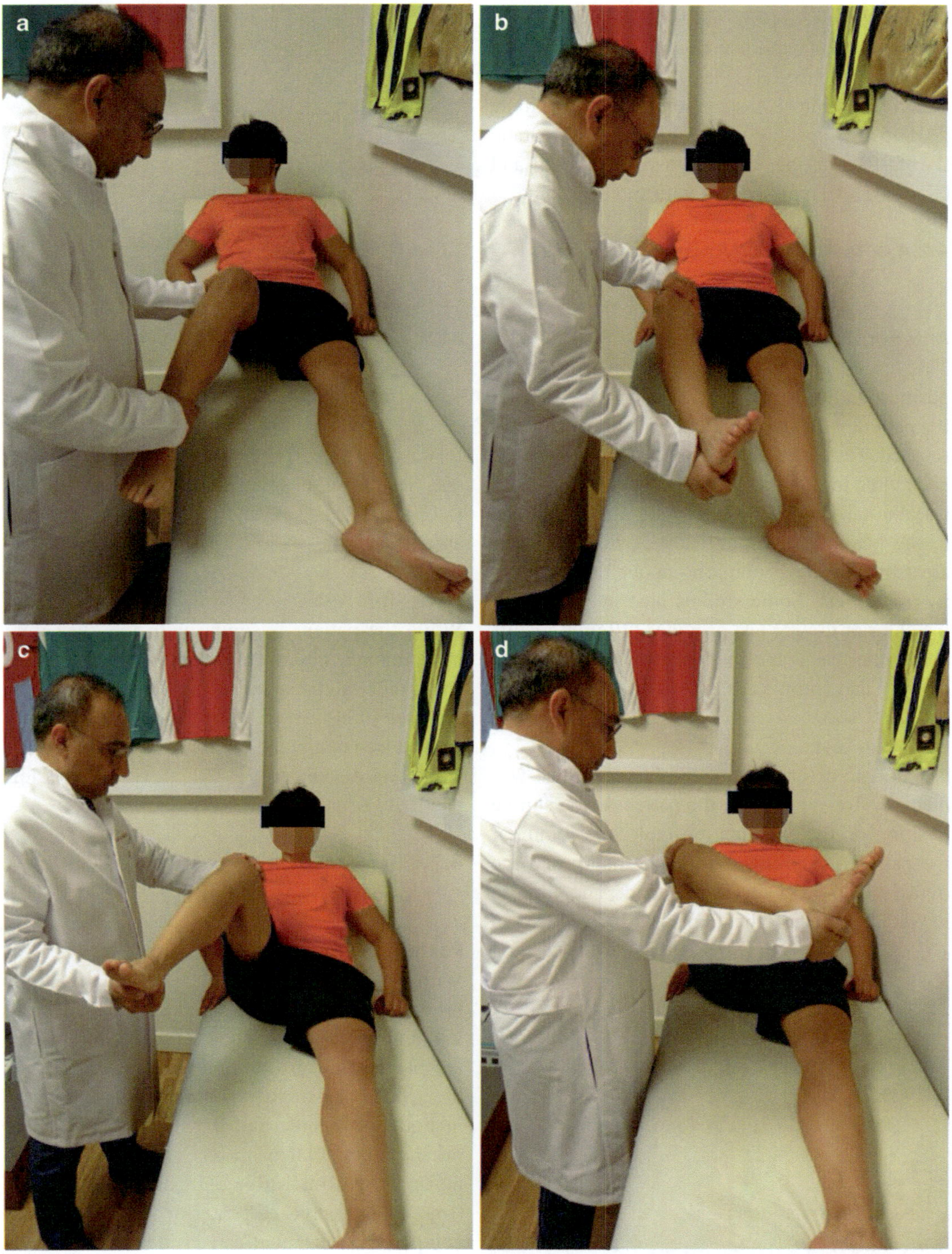

Fig. 4.13 (**a**) Valgus stability test in knee extension. (**b**) Varus stability test in knee extension. (**c**) Valgus stability test in knee flexion. (**d**) Varus stability test in knee flexion

viduals in the varus test. This should be confirmed by comparison with the contralateral side. The amount of angulation is classified according to the American Medical Association (AMA) as Grade I = 0–5 mm opening with hard end point, Grade II = 5–10 mm opening with hard end point, and Grade III ≥10 mm opening with soft end point.

4.2.4.4 Tests for the Anterior Cruciate Ligament

Anterior Drawer Test

This test is applied with the patient in a supine position and the knee in 90° flexion. By placing the patient's feet flat on the table, this position is secured. The calf is then grasped in both the hands and pulled toward the front of the tibia (Fig. 4.14). Greater anterior movement of the tibia compared to the healthy side shows test positivity. There are some negative opinions about the sensitivity of the test. Some studies have reported that the anterior drawer test provides different results under anesthesia and in awake patients [10] and in acute and chronic cases [11].

Lachman Test

The classic description of the Lachman test was made by Torg et al. With the patient in the supine position on the examination table, the femur of the patient is stabilized with one hand, and the tibia is held with the other hand. The tibia is moved in front of the femur with the knee in 15°–20° flexion (Fig. 4.15). The tibia coming forward with a soft end point indicates ACL damage. Several studies have shown the Lachman test to be more sensitive and more specific than the anterior drawer test [12, 13].

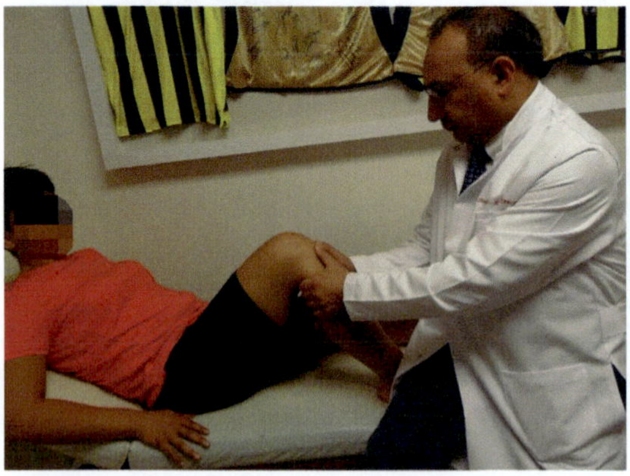

Fig. 4.14 Anterior drawer test

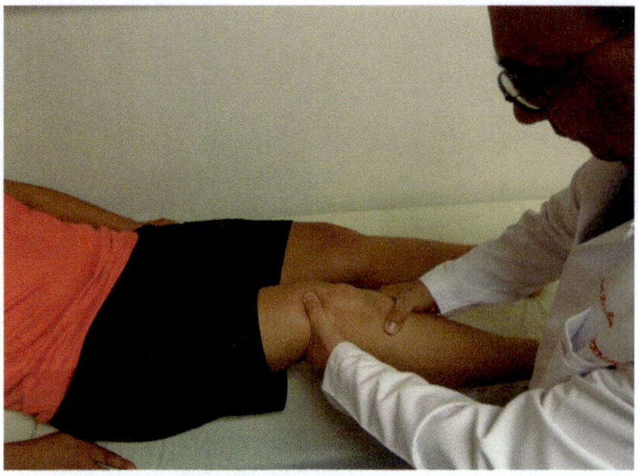

Fig. 4.15 Lachman test

Pivot Shift Test

While the patient is supine with the knee joint in extension, internal rotation is applied to the ankle and valgus stress to the tibia. During this, the knee joint is moved into flexion, and anterior subluxation below the femoral condyle of the lateral tibial plateau in 20°–30° flexion shows test positivity (Fig. 4.16).

4.2.4.5 Tests for the Posterior Cruciate Ligament

Posterior Drawer Test

With the patient in the supine position, the hip is brought into 45° flexion, the knee into 90° flexion, and the foot into a neutral position. The feet are placed flat on the table to secure the position. The back of the proximal tibia is grasped in both the hands, and the thumbs are placed on the tibial plateau. Force is applied to the posterior proximal tibia compared to the femur (Fig. 4.17). Increased posterior tibial movement compared to the unaffected side is a sign of a partial or full tear [14].

Quadriceps Active Test

The patient is positioned supine with the knee in 90° flexion and the feet flat on the examination

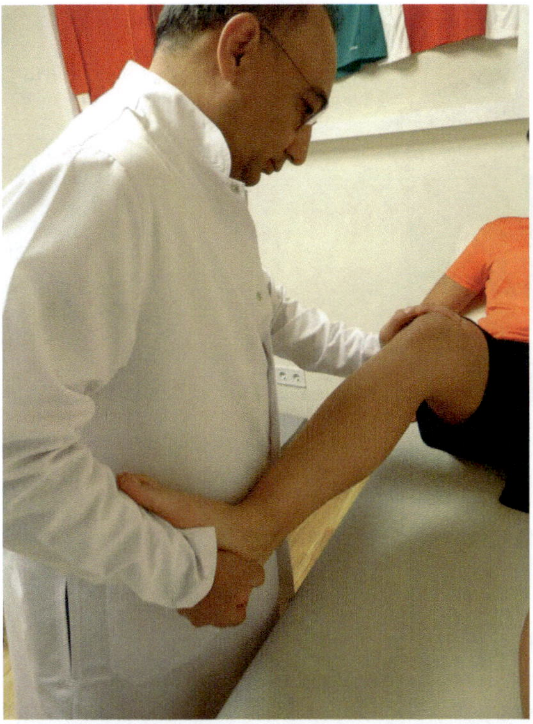

Fig. 4.16 Pivot shift test

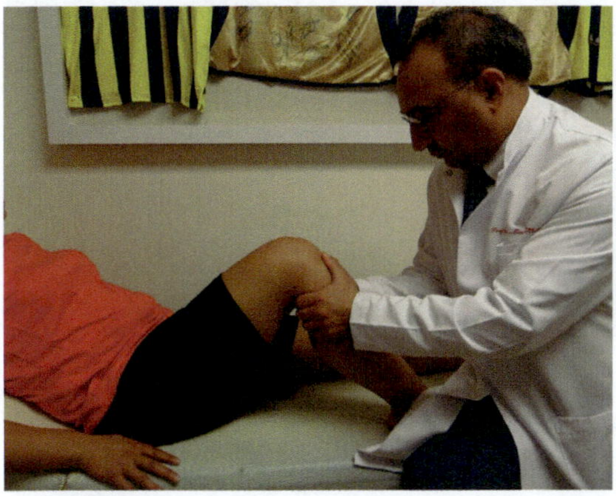

Fig. 4.17 Posterior drawer test

table. The patient is instructed to bring the knee into extension, and resistance to this is applied (Fig. 4.18). During this movement in a normal knee, the quadriceps muscle contracts, no anterior shift is seen in the tibia, and there may even be an amount of posterior shift. In a patient with

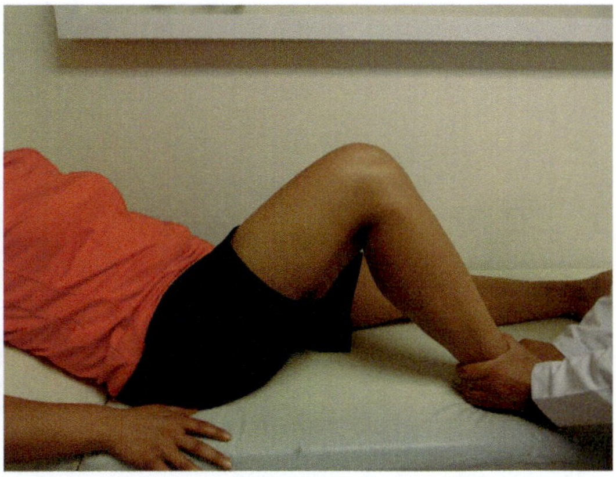

Fig. 4.18 Quadriceps active test

PCL damage, a shift of ≥2 mm of the tibia to the anterior together with quadriceps contraction shows test positivity [15].

4.2.4.6 Posterolateral Corner Tests

External Rotation Test (Dial Test)

With the patient in the supine position and the knees in flexion, the feet are held and taken into external rotation. The test result is evaluated by comparing the amounts of external rotation between the affected and the unaffected sides. A difference of >10° between the sides is evaluated as positive. The test is repeated with the knees in 30° and 90° flexion (Fig. 4.19a, b). In an isolated posterolateral corner injury, increased external rotation is seen at 30° but not at 90°. If increased external rotation is observed at both the degrees, a combined posterior and posterolateral injury should be considered.

Varus Recurvatum Test

With the patient relaxed and supine and the knee in extension, the big toes of the feet are held and raised (Fig. 4.20). When there is posterior, posterolateral, and LCL combined injury, the knee is expected to come into recurvatum and varus position.

Posterolateral Drawer Test

Increased positivity in the posterior drawer test while the foot is in external rotation is significant in respect of posterolateral injury (Fig. 4.21a, b).

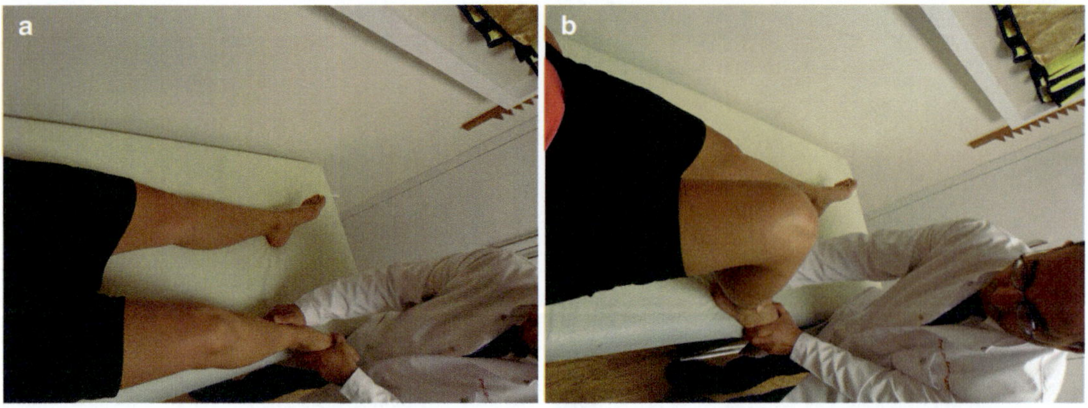

Fig. 4.19 (**a**) External rotation test (dial test) in 30° knee flexion. (**b**) External rotation test (dial test) in 90° knee flexion

Fig. 4.20 Varus recurvatum test

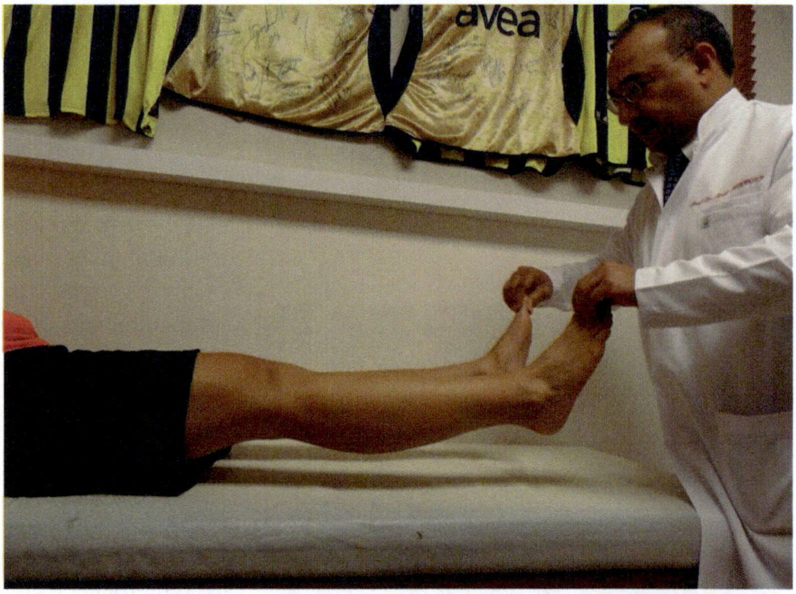

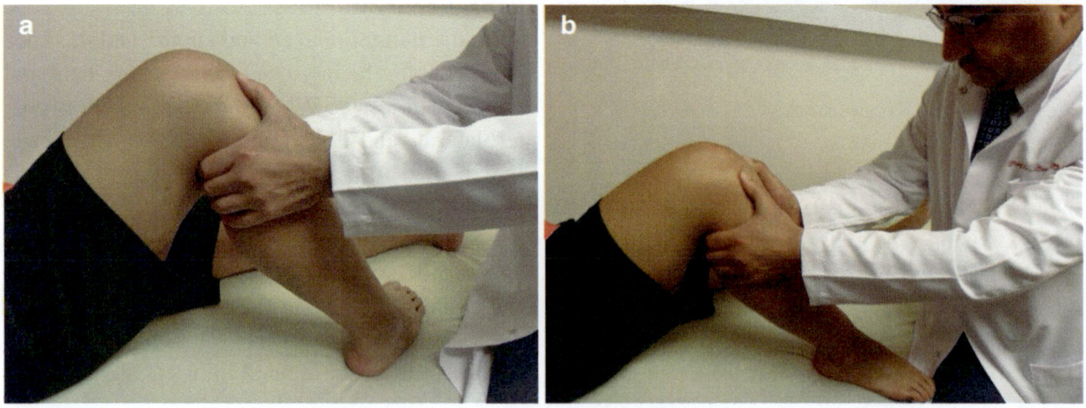

Fig. 4.21 (**a**) Posterolateral drawer test in internal rotation. (**b**) Posterolateral drawer test in external rotation

References

1. Magee DJ. Shoulder. Orthopedic physical assessment. 3rd ed. Philadelphia: W.B. Saunders; 1997. p. 5.
2. Fredericson M, Yoon K. Physical examination and patellofemoral pain syndrome. Am J Phys Med Rehabil. 2006;85:234–43.
3. Post WR, Teitge R, Amis A. Patellofemoral malalignment: looking beyond the viewbox. Clin Sports Med. 2002;21(3):521–46.
4. Owre A. Chondromalacia patellae. Acta Chir Scand. 1936;77(Suppl 41):1–159.
5. Solomon DH, Simel DL, Bates DW, Katz JN, Schaffer JL. Does this patient have a torn meniscus or ligament of the knee? Value of the physical examination. JAMA. 2001;286:1610–20.
6. Fairbank HA. Internal derangement of the knee in children and adolescents. Proc R Soc Med. 1936;30:427–32.
7. Sallay PI, Poggi J, Speer KP, Garrett WE. Acute dislocation of the patella. A correlative pathoanatomic study. Am J Sports Med. 1996;24:52–60.
8. McMurray TP. The semilunar cartilages. Br J Surg. 1942;29:407–14.
9. Gillis L. Diagnosis in orthopaedics. Toronto: Butterworth; 1969.
10. Harilainen A. Evaluation of knee instability in acute ligamentous injuries. Ann Chir Gynaecol. 1987;76:269–73.
11. Konin JG. Special tests for orthopedic examination. Thorofare, NJ: SLACK; 1997.
12. Mitsou A, Vallianatos P. Clinical diagnosis of ruptures of the anterior cruciate ligament: a comparison between the Lachman test and the anterior drawer sign. Injury. 1988;19:427–8.
13. Torg JS, Conrad W, Kalen V. Clinical diagnosis of anterior cruciate ligament instability in the athlete. Am J Sports Med. 1976;4:84–93.
14. Hughston JC. The absent posterior drawer test in some acute posterior cruciate ligament tears of the knee. Am J Sports Med. 1988;16:39–43.
15. Daniel DM, Stone ML, Barnett P, Sachs R. Use of the quadriceps active test to diagnose posterior cruciate-ligament disruption and measure posterior laxity of the knee. J Bone Joint Surg Am. 1988;70:386–91.

5 Patient Position and Setup

Özgür Kaya and Mehmet Emin Şimşek

5.1 Introduction

Arthroscopy or open surgery are among the commonly practiced surgical procedures in the treatment of meniscus and cartilage lesions, PCL rupture, and ACL rupture, which may cause symptomatic instability or symptoms such as knee locking and catching. Therefore, various surgical techniques have been developed for each procedure, and accordingly, several intraoperative patient positions have been used. There are two major intraoperative patient positions, i.e., supine position and with using a leg holder on the operating table [1, 2]. Although it is known that the patient position can affect the success of surgery, there are very few publications on the relationship between intraoperative patient position and surgical success in the literature [3]. Throughout the last century, technological advancements in the optical systems for imaging have facilitated the developments in arthroscopic surgery. Today, a majority of articular pathologies in the knee joint can be successfully treated with arthroscopic techniques.

Arthroscopic surgeries require small incisions, thereby enabling rapid recovery and early mobility of the patient. It prevents patients from developing muscle atrophy and reduces complications as a result of early mobility and rapid rehabilitation. Arthroscopy also facilitates reaching the posterior structures in the knee without the need to dislocate the knee. In addition, arthroscopy may help periarticular fracture fixation. The first challenge in arthroscopic surgeries, which are relatively more common than open surgery in ligament, meniscus, and cartilage injuries in the knee joint, is the learning curve. In order to carry out the surgery, it is of utmost importance to select patient position according to the surgical procedure after learning the use of portals [4].

5.2 Operating Room Setup

5.2.1 Patient and Operating Table Position

It is ensured that the room provides maximum efficiency and ease. Anesthetic is delivered to the patient at the bedside. Side control and marking are performed before the patient is anesthetized. The monitor and other devices are placed contralaterally. A Mayo stand/table can also be placed contralaterally. The patient is shaved and prepared. The surgical technician is generally posi-

Ö. Kaya
Department of Orthopedics and Traumatology, Ankara Lokman Hekim University, Etlik Hospital, Ankara, Turkey

M. E. Şimşek (✉)
Department of Orthopedics and Traumatology, Ankara Lokman Hekim University, Sincan Hospital, Ankara, Turkey

M. Bozkurt, H. İ. Açar (eds.), *Clinical Anatomy of the Knee*,
https://doi.org/10.1007/978-3-030-57578-6_5

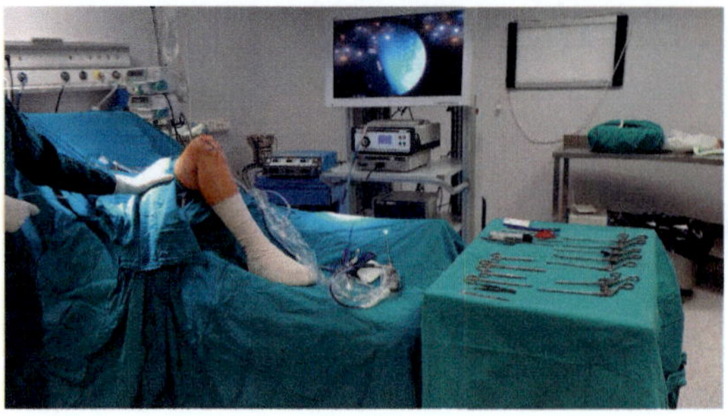

Fig. 5.1 Patient and operating table position

tioned contralaterally at the foot side of the bed. Suction canisters and the fluids that will be used in surgery may be placed on the side of the bed. The monitor should be on, fluid line cleaned and suction switched on before starting the surgery. The camera should be set up for an appropriate number of photos, and a recording device such as a CD-ROM device should be in place. In order to prevent lens fogging, a suitable amount of time should pass between equipment sterilization by the surgical personnel and the start of surgery [5] (Fig. 5.1).

5.2.2 Anesthesia

Spinal or epidural anesthesia can be used in arthroscopic surgery. Spinal anesthesia has a more predictable onset of action, despite its adverse effects on circulation and its potential to cause postoperative spinal headache and urinary bladder dysfunction. Although an anesthesiologist should always be present, local anesthetics can be administered by the surgeon. After establishing anesthesia, 1% lidocaine with epinephrine is administered to portal areas. The patient is placed in supine position on the operating table. For routine arthroscopy, a side support should be placed on the lateral side of the femur as well as a sandbag or a foot support that can be attached under the foot. This side support should facilitate two things, i.e., allowing the placement of the knee with a 90° angle and applying a valgus force on the knee in order to open the medial joint when necessary and to move perpendicular to the knee joint [6, 7].

5.2.3 Tourniquet

Tourniquet control and limb exsanguination are controversial issues in arthroscopic surgery. An exsanguinated area provides the highest intraarticular imaging quality. All surgical manipulations in the knee joint can be performed without significantly blurred view, and inconvenient bleeding can be eliminated. If the surgery is performed without using a tourniquet, the resulting bleeding can significantly disrupt the image. However, this does not mean that every arthroscopic surgery requires tourniquet inflation or long-term changes. A differential approach is necessary. A pneumatic tourniquet is placed on the proximal third of the thigh or on the junction of the proximal and middle third of the thigh. The tourniquet should be placed more proximally for a shorter and thicker femur, especially when an ACL reconstruction is planned. If the arthroscopy is going to be performed on a bent knee in a leg holder, the tourniquet should be placed in coordination with the position of the leg holder. The tourniquet should be applied right before general anesthesia induction, regardless of whether it will be inflated during surgery [8].

The tourniquet is applied on the proximal aspect of the thigh. An appropriately applied tourniquet would significantly facilitate the use of the surgical site during the procedure and particularly during femoral drilling for anterior cruciate ligament (ACL) reconstruction. First, a thick piece of cotton should be placed on the thigh, and the cotton should be wider than the width of the tourniquet. Otherwise, the tourniquet could put pressure on the skin around the thigh during inflation and cause injuries. The cotton also provides equal distribution of the pressure from the tourniquet to the thigh. Then, the tourniquet should be covered with a bandage with sides rolled under the tourniquet [8] (Figs. 5.2 and 5.3).

5.2.4 Supports

If a post is going to be used for arthroscopy, it should be placed at the correct level. A tourni-

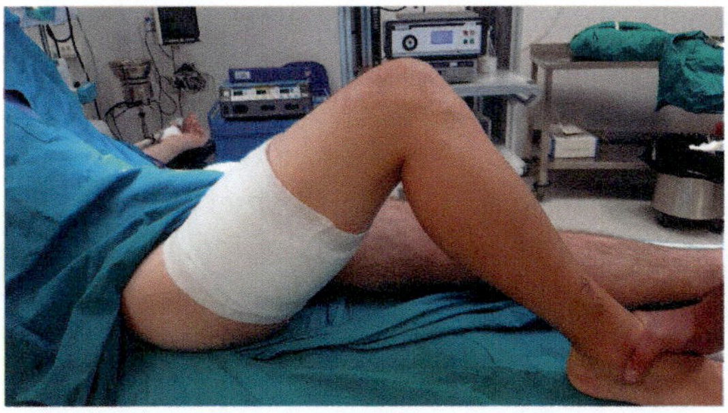

Fig. 5.2 The cotton also provides equal distribution of the pressure from the tourniquet to the thigh

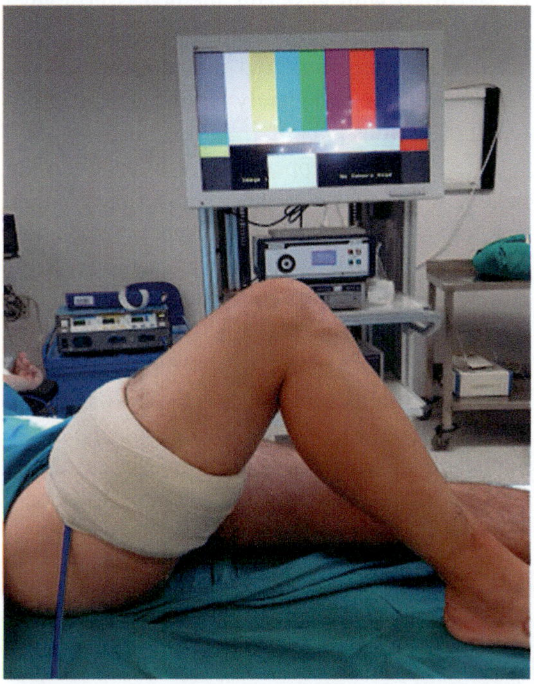

Fig. 5.3 The tourniquet should be covered with a bandage

quet should be applied before placing the leg in the leg holder, if desired. The leg holder should be as close as possible to the knee joint to provide the maximum mechanical advantage. Moreover, it should be ensured that there is enough space to move the surgical tools around the knee in all directions, if the surgery will involve procedures more complicated than a simple meniscectomy such as meniscus repair

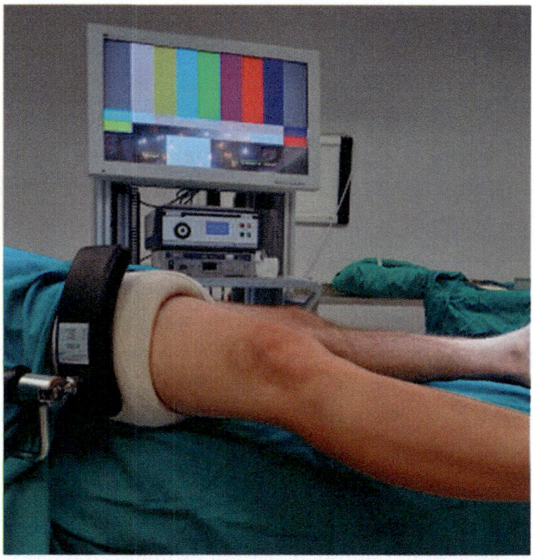

Fig. 5.4 This proximal foot support should be positioned in a manner to ensure a flexion that is slightly lower than full knee flexion

and ACL reconstruction. In such cases, there should be one hand breadth between the upper part of the patella and the leg holder [9].

The surgeon is positioned next to the knee that will be operated. An assistant surgeon, if present, is positioned at the proximal side of the surgeon, and a nurse is positioned further at the distal side of the surgeon with a surgical instrument trolley. Arthroscopic towers are placed on the side of the other knee on the operating table. All arthroscopic tools that are near the tower should be on the Mayo table. If the surgery will involve graft use or preparation, the surgical nurse may use another table that is connected to the instrument trolley. The operating table and the surgeon should be inside laminar flow. The surgeon may need to lower the patient's leg from the operating table during surgery. Therefore, the surgeon should check the side support for both the positions. An additional foot support can be used to achieve maximum knee flexion in surgeries such as ACL reconstruction. This proximal foot support should be positioned in a manner to ensure a flexion that is slightly lower than full knee flexion [10] (Fig. 5.4).

5.2.5 Equipment

The primary goal was to view the interior of joints as clearly and accurately as possible in the beginning of the arthroscopic era. However, high-performance optical systems that provide optimum image quality cannot be achieved even in the era of arthroscopic surgery, and setting up an effective arthroscopic system requires diligent work. The heart of the arthroscopic system is the arthroscope (telescope) itself. It consists of an eyepiece, a connection piece for the light cable, and a series of lenses and optics to transmit the light to the joint.

The fiberoptic and metal casing of the lens constitutes the arthroscopic barrel. Older arthroscopes were equipped with achromatic lens systems that only provided a relatively smaller visual field. Modern arthroscopes are based on the Hopkins rod lens system that combines a smaller overall diameter with a considerably larger visual field and a brighter field. Arthroscopes offer various view angles: 0°, 30° wide angle, and 70° wide angle. The standard instrument recommended for knee arthroscopy is a 30° wide-angle arthroscope. A 70° optics should be included only when sufficient view is achieved by 30° arthroscopes, since the indications for 70° optics are limited. Arthroscopes also have various trocar lengths depending on the manufacturer. A trocar length of 18 cm is recommended for knee arthroscopy [10].

Standard arthroscopes designed for various joints have diameters ranging between 1.7 and 4 mm. The 4-mm arthroscope has become the standard arthroscope which is commonly used for knee arthroscopy. Even pediatric and adolescent knees can be diagnosed and treated with a 4-mm scope. Only very small knee joints in children younger than 5 years require the use of a 2.4-mm scope. However, barrel length is a more important element than the scope diameter in smaller joints. When a long arthroscope is used, the surgeon is forced to work almost freehanded because he/she cannot stay stable by pushing a finger or hand against the patient's knee in order to enhance the control and coordination of fine arthroscope movements. Unfortunately, image quality is an important criterion that is hard to evaluate and is frequently neglected. In most of the cases, arthroscopes are selected solely based on the view angle and outer diameter. On the other hand, it is possible to observe significant differences in optical quality, comparing the scopes of different manufacturers. The image transmitted by the scope should have sharp edges (if the camera is suitably focused) and sufficient brightness. The scope should have satisfactory resolution, i.e., should be able to distinguish fine surface details. Historically, arthroscopes had to be sterilized with gas. This method is no longer used today due to environmental concerns, and arthroscopes are sterilized in steam autoclaves. They cannot be adequately sterilized with disinfectant solutions [11].

The sheath, equipped with a blunt obturator inside, is inserted into the joint during preparation for arthroscopy. We do not recommend using a sharp trocar since it may cause irreversible damage to the articular cartilage by plunging into the joint space. Once the sheath is inside the joint, the obturator is replaced with the arthroscope. The sheath consists of three parts: coupler (to secure the obturator or scope), spigot plane (to connect the inflow and outflow tubes), and sheath barrel with suction openings and an inflow channel for the distention medium [12].

To illuminate the interior part of the joint, light from a light source is transmitted via a light cable and glass fibers integrated into the arthroscope. A cold light source or xenon source can be used. The light provided to the arthroscope is transmitted to the arthroscope through a light cable (Fig. 5.5).

5.2.6 Imaging Systems

The main elements of a video system are a video camera and monitor. Digital video additionally offers an image processing option, and picture-in-a-picture is a feature that enables displaying and comparing adjacent images. The system can be expanded by adding recording devices such as a VCR, video writer, or digital image recorder. Arthroscopic surgery may include procedures

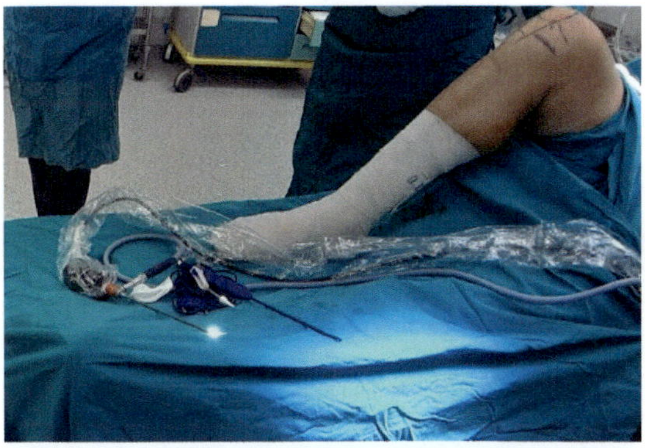

Fig. 5.5 To illuminate the interior part of the joint, light from a light source is transmitted via a light cable and glass fibers integrated into the arthroscope. A cold light source or xenon source can be used. The light provided to the arthroscope is transmitted to the arthroscope through a light cable

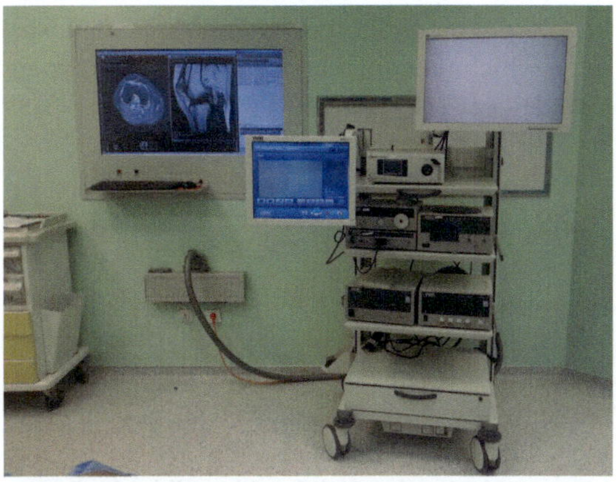

Fig. 5.6 The main elements of a video system are a video camera and monitor

concerning the meniscus, ligaments, or bones. Generally, it is required to detach tissue fragments and remove them from the joint (Fig. 5.6).

5.2.7 Punches

Mechanical instruments used in knee arthroscopy have a relatively uniform design that consists of jaws, a shaft, and handle. Manual movement of the handle transmits the cutting or grasping force to the instrument jaws. The shaft can be straight or curved. Punches may be straight or angled.

The basket forceps, also known as a punch or cutting forceps, is the most commonly used mechanical instrument. There are various types and degrees of angulation; however, the most commonly used ones are the straight or right-angled instruments. Basket forceps with an approximately 10° up-curved shaft and 10° up-angled jaw are useful for cutting the posterior horn of the medial meniscus.

It can be very hard to reach the posterior horn of the medial meniscus with a straight instrument in a very narrow medial compartment, especially due to the fact that the medial tibial plateau is convex, and the medial meniscus is positioned relatively higher on the back of the plateau. While using a punch, tissue fragments initially remain in the joint and are removed from the joint with a shaver or a large-bore irrigation cannula at the end of the resection. Special basket forceps types are as follows: a basket forceps connected to vacuum suction that almost entirely eliminates the intraarticular retention of tissue fragments when they are excised. This requires a significantly higher shaft diameter than regular basket forceps; retrograde basket forceps—tissue structures just below or adjacent to the instrument portal—are hard to reach with forward-cutting basket forceps.

The main instrument for arthroscopic surgery is narrow and angled grasping forceps, which combines the advantages of small size and tip angulation. It is inserted in the closed position, moved toward the target structure, and is slightly rotated and opened to grasp the tissue fragment that has been detached. Grasping forceps is necessary in order to remove meniscal fragments (partially detached), loose bodies, cartilage flaps, and osteophytes, to perform a synovial biopsy, to grasp the sutures for arthroscopic repair and for reconstruction (cruciate ligament reconstruction, medial retinacular repair, reconstruction, and meniscus repair). Arthroscopic scissors were used in the early days of arthroscopic surgery to detach the meniscal fragments. The problem with scissors was that they required a considerable amount of force in order to divide hard or scarred areas such as meniscal tissue. Therefore, scissors

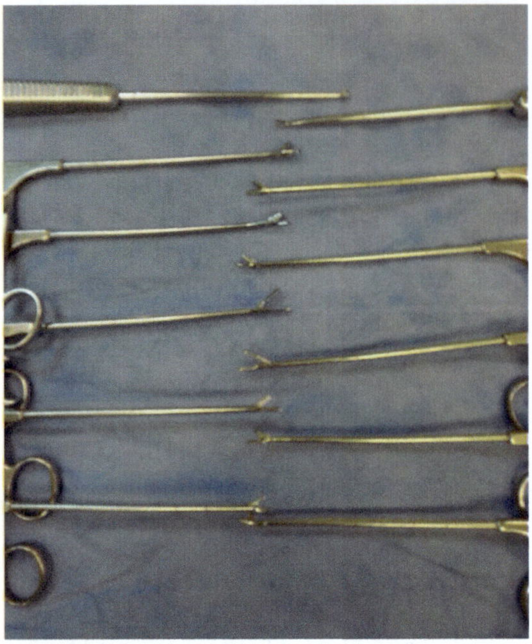

Fig. 5.7 Different punches and graspers

are rarely used today, and they were replaced by electrocautery hooks or thin basket forceps (Fig. 5.7).

5.2.8 Shavers and Electrosurgical Instruments

Motorized instruments or shavers have become a standard and established part of arthroscopic surgery routine. A motorized instrument set consists of a control unit, a connection cable between the handpiece and the control unit, a handpiece, blades, and suction.

Electrosurgical instruments have been used in urology and general surgery for decades. They are generally used to provide hemostasis in parenchymatous organs. Electrocautery devices were adapted for use in arthroscopic surgery starting from 1981. Electrosurgical techniques are entirely based on the thermal effect produced by an electric current. Electrosurgical instruments or electrocautery devices can be operated in coagulation mode or cutting mode. Some arthroscopic surgeries require special tools. A wide range of these devices are available in the market, and various surgical methods have been developed for their use. In most cases, conventional instruments can be slightly modified for use in arthroscopic surgery, which renders it unnecessary to use expensive devices [10] (Fig. 5.8).

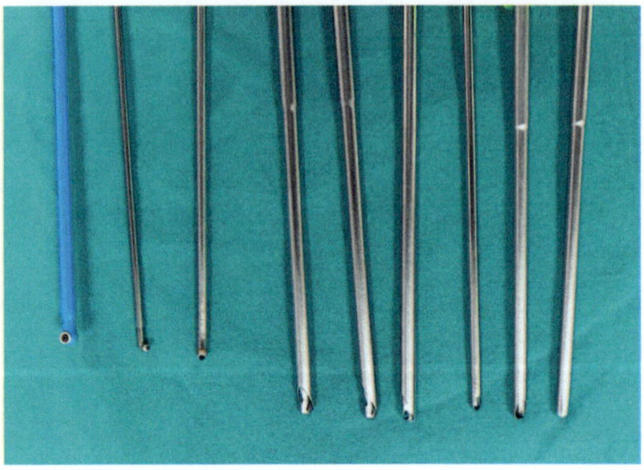

Fig. 5.8 Different types of shaver head and electrosurgical instruments for some special usage

5.2.9 Chisels

It is difficult to insert traditional chisels into the joints, and they may accidentally cause deep chondral lesions by digging into the articular cartilage. A chisel with rounded edges not only protects the cartilage but is also highly easy to pass, because sharp edges tend to “hang” in the instrument portal or during insertion. In addition to the rounded chisel, a curved chisel is a very useful tool for notchplasty. After removal of the osteophytes or other bone tissues, the resection side should be smoothed to create a homogenous surface.

5.2.10 Curettes

Sharp spoons are generally used to correct errors. However, this is a dull procedure, and it is easier to smooth and round off bone tunnels with curettes that have a special bone gradient. These instruments can also be used to debride bone surfaces covered with scars or soft tissue.

5.3 Portals

5.3.1 Anterolateral Portal

Anterolateral portal is the primary imaging portal for knee arthroscopy. When the knee is bent 90°, the inferior patellar pole, lateral patellar margins, and the lateral joint line are palpated.

The portal is created nearly 1 cm above the joint line and at the same level as the lateral line of the patella using a no. 11 blade. As long as a horizontal portal is not preferred, the incision is made perpendicular to the intercondylar notch. It both cases, the meniscus and intraarticular structures are preserved with great care. After cutting the joint capsule, a blunt trocar is advanced into the notch. Then, the knee is extended while carefully advancing into the suprapatellar bursa. The trocar is removed, and a camera is inserted into the knee at 30° angle [4].

A preliminary examination can be made before establishing the anteromedial portal. Patella and trochlea can be investigated for cartilage wear or damage. Medial and lateral grooves can be investigated to prevent loose bodies and osteophytes. Medial synovium is checked for a large plica. After completion of this initial examination, the arthroscope is advanced to the medial compartment through the medial femoral condyle.

Anteromedial portal is established when the knee is in 30° flexion. A 30° scope should be rotated to obtain an unhindered view of the anterior aspect of the medial meniscus and anterior capsule. The soft spot medial to the medial border of the patellar tendon is palpated. An 18-gauge spinal needle is used to find the most suitable spot for this portal. Pathology of the medial meniscus requires a portal that enables free access to the posterior horn. The lateral meniscal pathology requires cleaning the tibial spikes to reach the lateral compartment. An 18-gauge spinal needle is very useful in establishing this portal. The portal is established with a no. 11 blade. It should be positioned in parallel to the tibial plateau [13]. Diagnostic arthroscopy begins with establishing the anteromedial and lateral portals. The probe is brought to the medial compartment. The leg is slightly bent (10–30°), and a valgus force is exerted on the knee joint. The tibia should be in external rotation. The entire medial meniscus is probed. If it is difficult to reach the posterior horn of the meniscus, the knee might be slightly extended with great attention in order to prevent scratching the condyle with the camera. This generally opens the posterior part of the joint and enables probing the horn and posterior meniscus. If it is difficult to evaluate the posterior horn of the meniscus, a posteromedial portal can be established and a 70° camera can be used. After detailed examination of the posterior meniscus, the remaining parts of the meniscus are also examined. The medial femoral condyle and tibial plateau are examined while probing the meniscus. The condyle is visible when the knee is brought through a full range of motion. The arthroscope is advanced into the intercondylar notch once the medial compartment is examined. In the notch, ACL, PCL, meniscofemoral ligaments, and ligamentum mucosum are identified and probed. Notch morphology, depth and width are noted, particularly when there is ligament injury. It is noted whether the ligamentum mucosa flows from the upper part of the notch toward the fat pad. It can be excised if it entirely prevents evaluating the ACL, PCL, or other intraarticular structures. In general, arthroscope should be moved above the ligamentum to create an image of the ACL. A 30° intraoperative Lachman test can be performed to investigate ACL pathology [14, 15]. In patients who have a history of ACL injury or loss of extension after ACL reconstruction, ACL footprint should be investigated for debris from the ACL (cyclops lesion). After examining these structures, the probe should be moved along the lateral aspect of the ACT, and the knee should be brought to the figure four position or the varus position, if a post was used.

The lateral compartment is checked in figure four/varus position. A varus force applied right above the knee may open the lateral compartment. As in the medial side, the probe is used to examine and test the lateral meniscus. The lateral meniscus is generally easier to examine than the medial meniscus. If the anterior horn cannot be entirely examined due to the fat pad, ligamentum mucosum or portal placement, the arthroscope can be directed toward the medial portal or anterior horn [14] (Figs. 5.9 and 5.10).

5.3.2 Superomedial Portal

This portal is established on anterior knee and with the camera in anterolateral portal. The area 2–3 cm proximal to the superior pole of the patella and 1 cm medial to the midline is identified first with palpation and then with an 18-gauge spinal needle. The skin is incised with a no. 11 blade. The portal is established under direct visualization with a blunt trocar or a mosquito hemostat. Then, a switching stick is placed in a suprapatellar pouch, and a sheath is placed over it. After this point, articular surfaces and the tracking of the patella can be easily visualized. Upon completing the examination of the meniscus, articular surfaces, and ligaments, posteromedial and posterolateral compartments can be examined [16].

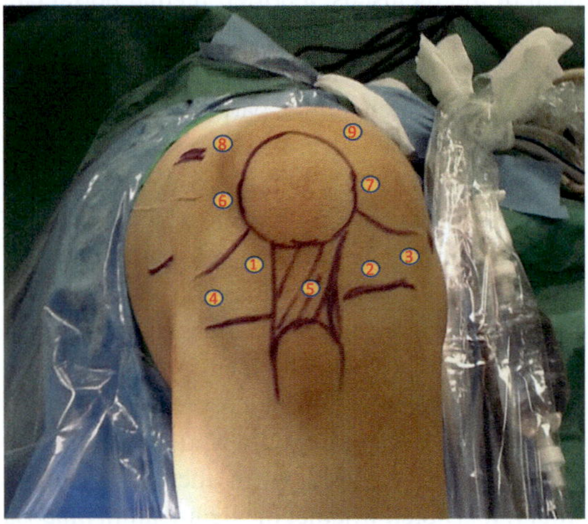

Fig. 5.9 Knee portals; 1 anterolateral portal, 2 anteromedial portal, 3 far-medial portal, 4 far-lateral portal, 5 central portal, 6 lateral midpatellar portal, 7 medial midpatellar portal, 8 superolateral portal, 9 superomedial portal, 10 posteromedial portal

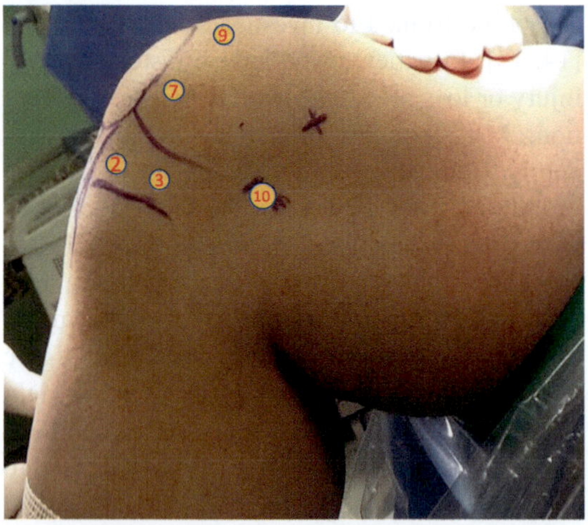

Fig. 5.10 Knee portal view medial side; 2 anteromedial portal, 3 far-medial portal, 7 medial midpatellar portal, 9 superomedial portal, 10 posteromedial portal

The ability to examine these compartments is essential for arthroscopy, especially when a posterior portal needs to be established or a loose body needs to be removed. The arthroscope has to be advanced through the ACL and PCL side in order to reach the posterior compartments in a knee with intact connective tissue. A probe can be passed through the medial portal between the medial femoral condyle and PCL, with the arthroscope in the anterolateral compartment and knee in 90° flexion (Figs. 5.9 and 5.10).

5.3.3 Posteromedial Portal

The posteromedial portal should be established nearly 1 cm posterior to the medial femoral condyle and 1 cm proximal to the joint line. The knee should be in 90° flexion, abduction, and external rotation. In general, the position can be palpated and then identified with an 18-gauge spinal needle. In PCL reconstruction, it would be useful to place a cannula in this portal. When an arthroscope is inserted through the posteromedial portal, the posterior horn of the medial meniscus, posterior medial femoral condyle, and the synovial lining of the posteromedial compartment can be examined. To assist with the inspection of this area of the knee, a probe can be provided through the anterolateral portal between the PCL and the condyle. Upon finishing the posteromedial compartment, the camera is withdrawn into the intracondylar notch. At this point, it can be possible to move the arthroscope between the ACL and lateral femoral condyle into the posterolateral compartment. The knee should be kept at 90° flexion. Generally, it is necessary to use a switching stick to enter this compartment. It may be required to turn the scope back to the anterolateral portal before entering this area and passing the switching stick to the anteromedial portal [17].

Once in the posterolateral compartment, the posterior horn of the lateral meniscus, meniscofemoral ligament, and synovial folds can be examined. With the camera facing the lateral condyle, it can be moved toward the popliteal hiatus. It may be possible to trace the popliteus up to the hiatus and view the femoral placement of the tendon with the knee in 70° flexion and under a valgus force. Most of the time, the space is too narrow to view the tendon entirely and a posterolateral portal is necessary [18] (Figs. 5.9 and 5.10).

5.3.4 Posterolateral Portal

Similar to the posteromedial side, posterolateral portal site can be palpated before establishing the portal. The knee is kept at 90° flexion. The portal site is approximately 1 cm posterior to the lateral femoral condyle and 1 cm proximal to the joint line. The surgeon must be aware of the position of the biceps femoris and the common peroneal nerve when establishing this portal. As with the medial side, upon determining the site, an 18-gauge spinal needle is used to mark the portal, and a skin incision is made. As mentioned above, the arthroscope can be passed along the posterior side of the meniscus and the condyle to the popliteal hiatus. Again, a probe can be brought from the anteromedial portal and used to help examine this compartment [19, 20] (Figs. 5.9 and 5.10).

5.3.5 Accessory Anterior Medial and Lateral Portals

According to the observed pathology, accessory anterior portals may be necessary. Accessory medial and lateral portals are established under direct visualization. Accessory medial portal is more medial and inferior to the standard portal, whereas the accessory lateral portal is more lateral and inferior to the standard portal. An 18-gauge spinal needle is used to identify the right part for the portal. It is important to visualize the needle while entering the joint in order to make sure that the portal will clear the meniscus and articular cartilage. Upon identifying the proper track, the skin is incised with a no. 11 blade and portal created with a blunt trocar. If necessary, a transpatellar portal can be established similarly. Once the case is finished, the knee is abundantly irrigated arthroscopically [21, 22]. The portal can be closed with simple nylon sutures or Steri-Strips, 3 M (St. Paul, MN). The patient is placed in a dry and sterile compression dressing, extubated by anesthesia and brought to the recovery room (Figs. 5.9 and 5.10).

References

1. Arthroscopy Association of C, Wong I, Hiemstra L, et al. Position Statement of the Arthroscopy Association of Canada (AAC) concerning arthroscopy of the knee joint-September 2017. Orthop J Sports Med. 2018;6:2325967118756597. https://doi.org/10.1177/2325967118756597.
2. Kim SJ, Kim HJ. High portal: practical philosophy for positioning portals in knee arthroscopy. Arthroscopy. 2001;17:333–7. https://doi.org/10.1053/jars.2001.21507.
3. Arthroscopy Association of C, Kopka M, Sheehan B, et al. Arthroscopy Association of Canada position statement on intra-articular injections for knee osteoarthritis. Orthop J Sports Med. 2019;7:2325967119860110. https://doi.org/10.1177/2325967119860110.
4. Hussein R, Southgate GW. Management of knee arthroscopy portals. Knee. 2001;8:329–31.
5. Stetson WB, Morgan SA, Hung NJ, et al. Knee arthroscopy: a diagnostic and therapeutic tool for management of ochronotic arthropathy. Arthrosc Tech. 2018;7:e1097–101. https://doi.org/10.1016/j.eats.2018.07.004.
6. Steiner SRH, Cancienne JM, Werner BC. Narcotics and knee arthroscopy: trends in use and factors associated with prolonged use and postoperative complications. Arthroscopy. 2018;34:1931–9. https://doi.org/10.1016/j.arthro.2018.01.052.
7. Gebhardt V, Hausen S, Weiss C, et al. Using chloroprocaine for spinal anaesthesia in outpatient knee-arthroscopy results in earlier discharge and improved

operating room efficiency compared to mepivacaine and prilocaine. Knee Surg Sports Traumatol Arthrosc. 2019;27:3032–40. https://doi.org/10.1007/s00167-018-5327-2.
8. Hoogeslag RA, Brouwer RW, van Raay JJ. The value of tourniquet use for visibility during arthroscopy of the knee: a double-blind, randomized controlled trial. Arthroscopy. 2010;26:S67–72. https://doi.org/10.1016/j.arthro.2009.12.008.
9. Howard DH. Trends in the use of knee arthroscopy in adults. JAMA Intern Med. 2018;178:1557–8. https://doi.org/10.1001/jamainternmed.2018.4175.
10. Gross RM. Arthroscopy. Basic setup and equipment. Orthop Clin North Am. 1993;24:5–18.
11. Lill H, Frosch KH, Voigt C. Recommendations of the German Working Party for Arthroscopy (Section of the German Society for Orthopedics and Trauma Surgery) on equipment of facilities, process quality and qualification of operators by arthroscopic interventions: special features from the perspective of trauma surgery. Unfallchirurg. 2010;113:964–5. https://doi.org/10.1007/s00113-010-1862-0.
12. Kowalski JM, Monica JTA. Novel method of patient positioning using shoulder arthroscopy equipment for elbow arthroscopy. Orthopedics. 2018;41:e158–60. https://doi.org/10.3928/01477447-20171102-04.
13. Jennings JK, Leas DP, Fleischli JE, et al. Transtibial versus anteromedial portal ACL reconstruction: is a hybrid approach the best? Orthop J Sports Med. 2017;5:2325967117719857. https://doi.org/10.1177/2325967117719857.
14. Ye SM, Jing JH, Lv H, et al. Accessory anteromedial portal may not provide clinically superior results compared with the anteromedial portal in anterior cruciate ligament reconstruction. J Knee Surg. 2018;31:716–22. https://doi.org/10.1055/s-0037-1607074.
15. Eysturoy NH, Nielsen TG, Lind MC. Anteromedial portal drilling yielded better survivorship of anterior cruciate ligament reconstructions when comparing recent versus early surgeries with this technique. Arthroscopy. 2019;35:182–9. https://doi.org/10.1016/j.arthro.2018.08.030.
16. Sekiya H, Takatoku K, Kimura A, et al. Arthroscopic fixation with EndoButton for tibial eminence fractures visualised through a proximal superomedial portal: a surgical technique. J Orthop Surg (Hong Kong). 2016;24:417–20. https://doi.org/10.1177/1602400329.
17. McGinnis MD, Gonzalez R, Nyland J, et al. The posteromedial knee arthroscopy portal: a cadaveric study defining a safety zone for portal placement. Arthroscopy. 2011;27:1090–5. https://doi.org/10.1016/j.arthro.2011.02.031.
18. Lanham NS, Tompkins M, Milewiski M, et al. Knee arthroscopic posteromedial portal placement using the medial epicondyle. Orthopedics. 2015;38:366–8. https://doi.org/10.3928/01477447-20150603-03.
19. Alentorn-Geli E, Stuart JJ, Choi JH, et al. Inside-out antegrade tibial tunnel drilling through the posterolateral portal using a flexible reamer in posterior cruciate ligament reconstruction. Arthrosc Tech. 2015;4:e537–44. https://doi.org/10.1016/j.eats.2015.05.016.
20. Alentorn-Geli E, Stuart JJ, James Choi JH, et al. Posterolateral portal tibial tunnel drilling for posterior cruciate ligament reconstruction: technique and evaluation of safety and tunnel position. Knee Surg Sports Traumatol Arthrosc. 2017;25:2474–80. https://doi.org/10.1007/s00167-015-3958-0.
21. Tompkins M, Milewski MD, Brockmeier SF, et al. Anatomic femoral tunnel drilling in anterior cruciate ligament reconstruction: use of an accessory medial portal versus traditional transtibial drilling. Am J Sports Med. 2012;40:1313–21. https://doi.org/10.1177/0363546512443047.
22. Tompkins M, Cosgrove CT, Milewski MD, et al. Anterior cruciate ligament reconstruction femoral tunnel characteristics using an accessory medial portal versus traditional transtibial drilling. Arthroscopy. 2013;29:550–5. https://doi.org/10.1016/j.arthro.2012.10.030.

Anatomical Meniscal Repair

6

Robbert van Dijck

6.1 Introduction

The menisci are crescent-shaped fibrocartilageous structures primarily composed of collagen type I with important biomechanical functions such as load transmission, shock absorption, stability, nutrition, joint lubrication and proprioception. The menisci have an important role in preventing osteoarthritic changes [1].

The medial meniscus is a semilunar C-shaped structure and covers 60% of the medial compartment measuring approximately 45.7 mm in length and 27.4 mm in width. The lateral meniscus is semicircular U-shaped structure, covering 80% of the lateral compartment measuring 35.7 mm in length and 29.3 mm in width and has greater variability in shape, size and mobility than the medial meniscus [2]. The menisci are stabilized by their anterior and posterior roots, the anterior intermeniscal (transverse) ligament, the medial collateral ligament, the meniscofemoral ligaments and the coronary ligaments [3]. The vascularity of the peripheral menisci is primarily derived from the superior and inferior medial and lateral geniculate arteries (Arnozcksy). Radial branches from a perimeniscal plexus enter the meniscus at intervals, with a richer supply to the anterior and posterior horns. The peripheral 20–30% of the medial meniscus and the peripheral 10–25% of the lateral meniscus are vascularized [4]. The popliteal hiatus creates a relatively hypovascular area in the posterior horn of the lateral meniscus. Arnozcksy and Warren classified the location of the tear into three zones: Zone 0 represents the peripheral meniscosynovial junction; zone 1, the red-red zone; zone 2, the red-white zone and zone 3, the white-white zone [5]. Another classification of DeHaven [6] classified tears into a peripheral 3 mm vascular zone (red-red zone), tears greater than 5 mm from the meniscocapsular junction as avascular (white-white zone) and tears in between as variable (red-white zone). The meniscus has limited healing capacity, the tears in the red-red and the red-white zones are repairable, whereas the meniscus repair for tears in the white-white zone has poor healing potential. A few studies reported good results of meniscus repair in the white-white zone [7, 8]. Several classifications of meniscal injuries have been proposed over time. Meniscal tears are often classified according to their orientation. Meniscal tear patterns can be radial, oblique, flap or parrot peak, vertical longitudinal, vertical radial, bucket handle or complex (degenerative) [9, 10]. A recent and reliable classification system for meniscal tears is the International Society of Arthroscopy, Knee Surgery and Orthopaedic Sports Medicine (ISAKOS) classification; important factors to consider are tear depth, tear

R. van Dijck (✉)
Bergman Clinics Breda, Breda, The Netherlands
e-mail: r.vandijck@bergmanclinics.nl

M. Bozkurt, H. İ. Açar (eds.), *Clinical Anatomy of the Knee*,
https://doi.org/10.1007/978-3-030-57578-6_6

pattern, tear length, tear location/rim width, radial location, location according to the popliteal hiatus and quality of the meniscal tissue. Imaging techniques like MRI can be useful to see the characteristics of a tear. However, a meniscal tear is best assessed arthroscopically using a probe to determining the type, location, stability and length of the tear.

Studies have demonstrated excellent healing of peripheral tears, because of the high vascularity of these tears. They have better healing response than other meniscal tears [11–14]. Radial, oblique and horizontal cleavage tears involve the avascular zone and result in a poorer healing rate [15]. Important prognostic factor for meniscal repair healing is the distance of the tear from the meniscocapsular junction (0–2 mm). This has been identified as the greatest predictor for healing; a greater distance from the meniscocapsular junction results in poorer healing [16]. Also tear length affects the healing rate, greater lesions (extension of the lesion from anterior to posterior, bucket handle tears), and BMI >25 kg/m^2 have a greater risk of failure of healing [17, 18]. Some studies also reported that age was an important factor for meniscal healing; however, more recent studies showed that meniscal repair failure rate was not different in patients 40 years or older in comparison with younger patients [19]. Cannon and Vittori compared the healing rate of menisci repaired in association with an anterior cruciate ligament (ACL) reconstruction with that of menisci undergoing isolated meniscal repair [17]. Patients with anterior cruciate ligament reconstruction and meniscal repair did better than those with isolated meniscal repair. Lateral meniscal repair had better results in comparison with medial meniscal repair. Also acute repairs were more successful than repairs of chronic tears. Other studies reported better results of meniscal repair in combination with an ACL reconstruction [20, 21]. Bone drilling could result in the release of growth factors and pluripotent cells which results in biologic augmentation at the repair site [22].

6.2 Meniscal Repair

An important and necessary step for all meniscal repair is tear debridement and perisynovial tissue abrasion (Fig. 6.1). This stimulates a proliferative fibroblastic healing response [23].

Trephination can also promote healing of some kind of meniscal lesions. Trephination is a technique introduced to create perforations at the peripheral aspect of the meniscus rim to stimulate bleeding through vascular channels. In small stable tears located on the outer area near the meniscus and joint capsule junction, trephination promotes bleeding and could enhance vascular ingrowth and healing process. In a study of Fox, trephination of symptomatic incomplete meniscal tears demonstrated >90% good to excellent results [24]. In a goat model, improved healing (even in the avascular zone) was demonstrated when trephination was added to meniscal repair [25]. Other healing-stimulating techniques are marrow-stimulating techniques like microfracture of the intercondylar notch which can improve meniscal healing at time of repair by release of marrow elements into the knee [26, 27]. In a study of Dean et al. [28], there was no difference in outcomes in

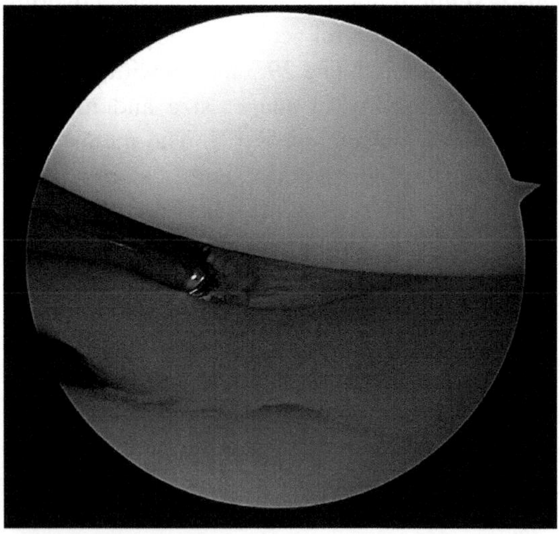

Fig. 6.1 Arthroscopic view of abrasion of a meniscal peripheral tear with a rasp, which is an important step in meniscal suturing

meniscal repair performed with biological augmentation using an marrow-stimulating technique in comparison with a meniscal repair with an ACL reconstruction. Similar outcomes may be partly attributed to biological augmentation.

Sutures of meniscal repair should be nonabsorbable or slowly absorbable [29, 30]. Vertically oriented suturing is the gold standard. The pull-out strength of vertical sutures is stronger in comparison with horizontal sutures. According to different models, the strength of vertical sutures were found to be in a range from about 60 N to more than 200 N [31, 32].

Horizontal sutures lie in between the circumferential fibre bundles and yield a lower failure load because they are pulled through those fibres as they are loaded [33]. When using horizontal sutures, it is better to place them slightly farther away from the meniscus lesion. This results in a better repair fixation in comparison with sutures placed closer to the lesion [34]. Also large diameter sutures increase fixation strength [35].

Different meniscal repair techniques have been described and are divided into inside-out, outside-in, all inside or combined.

6.3 The All-Inside Technique

There are several all-inside arthroscopic meniscus repair devices [30, 36–38]. Standard inside-out suture repairs remain the gold standard against which other techniques are compared. All-inside repairs have benefited from improvements in device and technique since their introduction in 1991, the fourth-generation devices are flexible, safe and suture based, and they allow for variable compression and retensioning cross the meniscal tear. An intact meniscal rim is needed for these devices because the meniscal rim act as an anchor for repair devices.

The all-inside technique can best be used for posterior horn meniscal lesions with an intact meniscal rim. Anterior horn tears are a relative contraindication due to difficulty in access.

6.3.1 Technique

The suture device with self-locking knot system is passed through the meniscal tear site. After fixing the first bar on the joint capsule extra-articularly behind the peripheral meniscus on the capsular surface, the suture device is passed to fix a second bar. This delivery needle is positioned at least 5 mm from the first implant in a vertical, horizontal, or oblique manner which is possible or desirable. The needle is removed from the joint, leaving the free end of the suture out of the knee. When deployed, the suture is tensioned to close the gap in the meniscus, and a pretied, sliding, self-locking knot is tightened to compress the meniscus tear. When the knot is tightened appropriately, the suture is cut.

The all-inside technique is easy to use, decreases the surgical time and avoids an accessory incision, and there is less risk to neurovascular complications. Despite the advances of these fourth-generation devices, they are not without complications. Devices can misfire, break or get tangled. Also iatrogenic chondral damage, soft tissue penetration and entrapment can occur. There is still a small risk for neurovascular damage. The risk of injury to the popliteal artery or to the peroneal nerve during all-inside repair of the posterior half of the lateral meniscus is lower at 90° of flexion and increases with knee extension to 45° and 0° [39].

Improvement in meniscal repair devices have reported equivalent biomechanical properties and success rates to those of the inside-out technique. The all-inside suture-based repairs and inside-out repairs did not differ in load-to-failure values [31, 40, 41]. A systematic review of Fillingham et al. [42] reported no difference in functional outcomes, failure rates and complications between the inside-out and all-inside meniscal repair techniques on isolated meniscal tears. However, the level of evidence of the studies of this review is low.

6.4 Inside-Out Technique

The inside-out technique is the gold standard for meniscal repair and can be applied to most meniscus tear types along the middle third and posterior horns. Because of the precise placement of the sutures, it can also be used for more complex tears. Also tears of the peripheral rim and capsular attachment can be treated with the inside-out technique.

6.4.1 Technique

Equipment needed for an inside-out suture: preloaded needles 2.0 braided, nonabsorbable, specific cannulas with various flexion angles and retractors (spoon). The standard anterolateral and anteromedial arthroscopy portals are created, and a diagnostic arthroscopy is performed. Depending on the location of the tear, the posterolateral or posteromedial capsule must be exposed: Introduction of a curved cannula with reduction of the tear; passing the first and second flexible needle through the cannula and retrieving outside the capsule; and reduction of the lesion by pulling at both ends of the suture and tying the knot outside the capsule at 90 degrees of flexion of the knee. When performing an inside-out meniscus repair, a 2.5–4 cm incision can be made on the appropriate side (posteromedial or posterolateral) of the knee with 90° of knee flexion. Another procedure is to pass the sutures through the skin first and make a skin incision in-between the sutures. Dissect onto the capsule carefully and tie down the sutures on the capsule.

With the inside-out technique, sutures can be placed with good precision and versatility in either a horizontal or vertical mattress configuration. Knots are tied on the capsule, so no chondral injury, intra-articular irritation or impingement with motion can happen.

Complications like nervus saphenous neuropathy [43], arteria poplitea lesion and nerve common peroneal injury have been reported, but their incidence is low [44].

Other complications are repair failure (retear, non-healing, persistent symptoms) and general knee arthroscopy complications (infection, deep vein thrombosis, haemarthrosis).

Inside-out repair has a success rate of 60–80% for isolated meniscal repairs and 85–90% when performed with an ACL reconstruction [45]. Because several studies reported no differences in clinical failure rate or subjective outcome between inside-out and all-inside meniscus repair techniques, the inside-out technique has become less popular, and the all-inside technique gained popularity [12, 30, 46]. Furthermore, there has been no evident difference in meniscal healing between the all-inside technique and inside-out technique found on magnetic resonance imaging in patients with a meniscal tear in combination with an ACL reconstruction. Nelson et al. [47] reported that, despite the advantages of the all-inside technique, it is still important to be familiar with the inside-out technique because it can be quite useful for a subset of meniscus tear patterns.

6.5 Outside-In Technique

The outside-in technique was first described by Warren [48] in 1985 and Morgan in 1986 [49]. The outside-in technique has lower neurovascular risks. However, the repair of posterior meniscal tears is difficult [50] because it is difficult to achieve perpendicular orientation of sutures at the posterior horn.

The anterior horn of the medial meniscus has been reported to be particularly important for stabilizing external rotation when the knee is fully extended [51] and also in preventing anterior femoral displacement [52]. An anterior horn lateral meniscus lesion was reported to significantly increase tibiofemoral contact pressures in both compartments of the knee [53]. So surgical repair is indicated when possible. The outside-in repair technique is ideal for middle and anterior horn tears.

6.5.1 Technique

The equipment that is needed for an outside-in suture: two spinal needles, suture material (2.0 ethibond, 2.0/#0/#1 PDS (absorbable), 2.0 prolene (nonabsorbable)), suture grasper.

Procedure: The standard anterolateral and anteromedial arthroscopy portals are created, and a diagnostic arthroscopy is performed. After confirmation of the anterior horn tear, the arthroscope should be placed through the contralateral portal of the compartment of the involved meniscus to visualize the extent and characteristics of the tear. A 2–3 cm vertical incision is made in line with the portal on the same side of the knee as the anterior meniscal tear. Dissection is made onto the anterior joint capsule. First needle loaded with a shuttle suture is pierced through the capsule to the desired area of the lesion. Second needle with the repair suture is passed parallel to the shuttle suture through the capsule ideally passing the loop. End of the suture is caught with a suture grasper and retrieved anteriorly. The suture is pulled back out of the knee creating a suture construct. Reduction of the lesion by pulling at both the ends of the suture outside the joint and tied over the capsule. Horizontal or vertical mattress suture configuration can be utilized. The sutures are tied in 90° of flexion of the knee.

The outside-in technique is a simple, inexpensive technique with small incisions which can provide a stable suture construct with low neurovascular risk and good clinical outcomes.

A potential disadvantage of the outside-in technique is difficulty in reducing the tear and opposing the edges while passing the sutures. It is a safe procedure; however, chondral damage and synovitis has been reported [49]. The overall results with use of the outside-in technique are comparable with those reported with use of the inside-out technique.

Morgan and Casscells [54] were the first to report excellent results with the outside-in technique. A 98% healing rate in an 18-month follow-up was described. Van Trommel [55] reported a 76% success rate in 51 patients treated with the outside-in technique. Also other studies reported high success rates with the outside-in techniques for anterior meniscal lesions. Posterior horn tears treated with the outside-in technique have a higher failure rate [56].

6.6 Peripheral Meniscal Tears

Peripheral meniscal tears are the most common meniscal tears [57–59] (Fig. 6.2) and results from disruption of the superficial radial collagen fibres in line with the circumferential collagen fibres in the red-red or red-white zone. Because of high vascularity, peripheral tears have the greatest potential of healing [11]. Larger peripheral longitudinal tears can allow the inner meniscus to flip on itself, known as a 'bucket-handle' meniscal tear [60]. Peripheral tears are believed to partially preserve the load distribution function of the meniscus, whereas other tears such as radial tears or more complex tears do not preserve the load distribution function due to the disruption of the large circumferential fibre bundles [61, 62]. Partial or subtotal meniscectomy of peripheral tears results in an increasing contact pressure, so preserve the meniscus as much as possible [63].

Peripheral meniscus tears can be repaired by an inside-out, outside-in or all-inside technique. The most used techniques for peripheral tears are the inside-out and all-inside techniques (Figs. 6.3 and 6.4). The outside-in technique has a limited

Fig. 6.2 Arthroscopic image of a medial peripheral tear

access to tears in the posterior third of both menisci and has more complications.

Peripheral longitudinal meniscus repair outcomes (with or without ACL tear) are now well-established and lead to excellent and good clinical mid-term results. For vertical peripheral longitudinal tears, the rate of failure is acceptable (6%–28%), and repair leads to a better long-term clinical outcomes [64, 65]. Biomechanical studies demonstrated similar loads to failure using the all-inside technique compared with the inside-out technique for peripheral meniscal lesions [66]. Also failure rates, functional outcomes and complications rate were comparable with these two techniques [42]. Repairs of the medial meniscus resulted in higher reoperation rates than repairs of the lateral meniscus. Meniscal repairs at the time of anterior cruciate ligament reconstruction had a lower failure rate than isolated [67]. Patient age, gender, chronicity, compartment involved (medial vs. lateral), and concurrent ACL reconstruction do not influence healing rates [68]. However, most studies reported better results of meniscal repair in association with ACL reconstruction. Peripheral meniscal lesions in the red-red zone have inherently good healing rates because of the blood supply. Lateral meniscus lesions of <10 mm in length and not extending >1 cm anterior to the popliteus can be left in situ during ACL reconstructions [20].

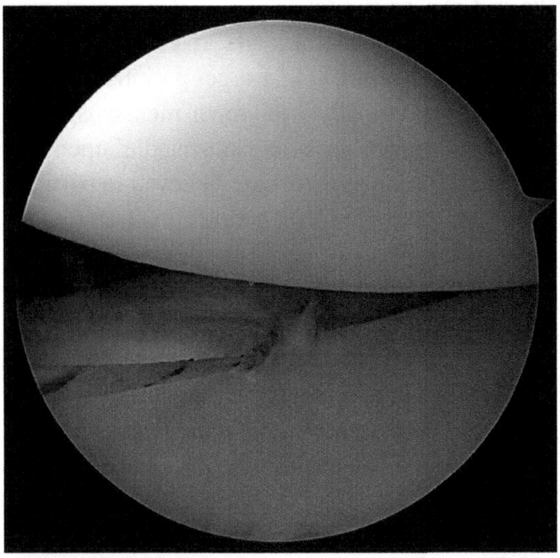

Fig. 6.3 Meniscal repair with vertical sutures by an all inside technique

6.7 Ramp Lesions

A ramp lesion was defined as a longitudinal tear of the peripheral attachment of the posterior horn of the medial meniscus at the meniscocapsular junction of less than 2.5 cm in length. However, there is still no consensus regarding the defini-

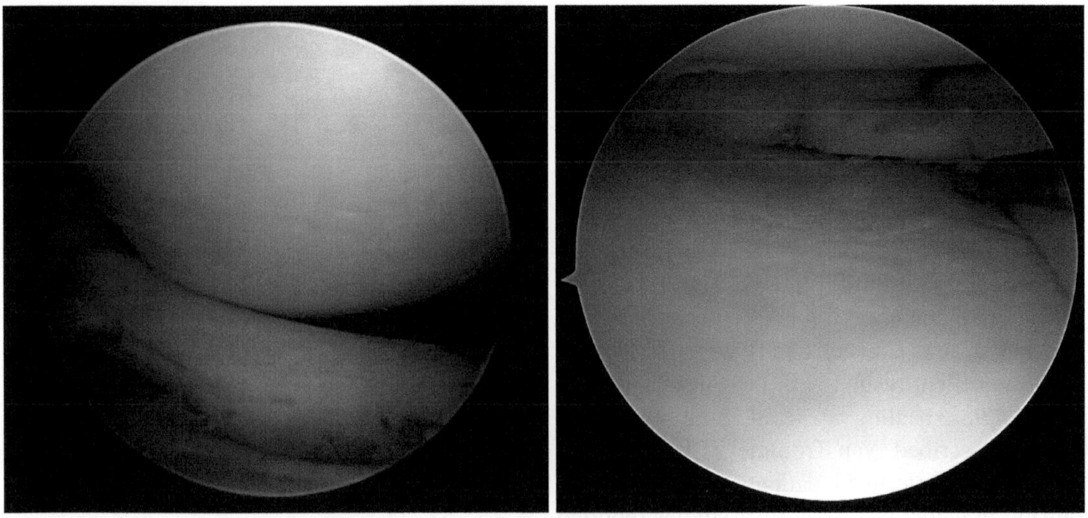

Fig. 6.4 Arthroscopic view of a bucket-handle tear with an inside-out repair

tion, and a ramp lesion is associated with an ACL lesion [69, 70]. Ramp lesions are often missed and are called 'hidden lesions' because the lesion is commonly located posteromedial and missed with standard anteromedial and anterolateral arthroscopic portals [71]. Bollen et al. [69] reported that MRI has a low sensitivity for detecting ramp lesions. A more recent study demonstrated high sensitivity and specificity in detecting ramp lesions on MRI [72]. If posteromedial tibial bone marrow oedema is present, a ramp lesion can be suspected.

The posterior horn of the medial meniscus plays a fundamental role in knee stability, particularly in limiting anterior tibial translation. An association of an ACL tear with a ramp lesion resulted in a further 30% increase in external rotation and anterior translation laxity compared to a single ACL tear [73].

No clear consensus exists on the appropriate treatment of meniscal ramp lesions. Surgical treatment of ramp lesions in the setting of an acute ACL reconstruction is controversial. In chronic ACL deficiency, ramp lesions should be treated operatively [74]. All-inside and inside-out techniques have reported good results to treat ramp lesions [75]. The all-inside technique may be insufficient to fix the gap between the posterior meniscal wall and the capsule, especially if the latter remains retracted in the extended knee. Choi et al. [43] reported that the all-inside technique cannot provide sufficient fixation strength in ramp lesions. The inside-out technique allows for a greater versatility in suture placement and increased number of sutures, thereby potentially providing a stronger construct. A posteromedial approach is performed. To visualize the meniscocapsular ramp, the knee needs to be flexed at 90°. In this position, the posteromedial capsule gets slack, in extension tight. Mostly, the inspection of the posteromedial ramp with a 30° arthroscope is sufficient to see a ramp lesion. Sometimes it is necessary to use a 70° arthroscope or posteromedial portal. Different inside-out techniques are described for ramp lesion repair. A ramp lesion repair can be difficult. Morgan et al. [76] and Ahn et al. [77] described a good technique with a posteromedial approach (Fig. 6.5). Repair occurs with curved and inclined suture passing instruments after a thorough debridement of the synovial membrane. Other techniques are using single- or double-lumen cannulas and flexible needles with preloaded nonabsorbable or absorbable sutures. The first needle is passed through the superior or inferior aspect of the posterior horn of the medial meniscus and the second nee-

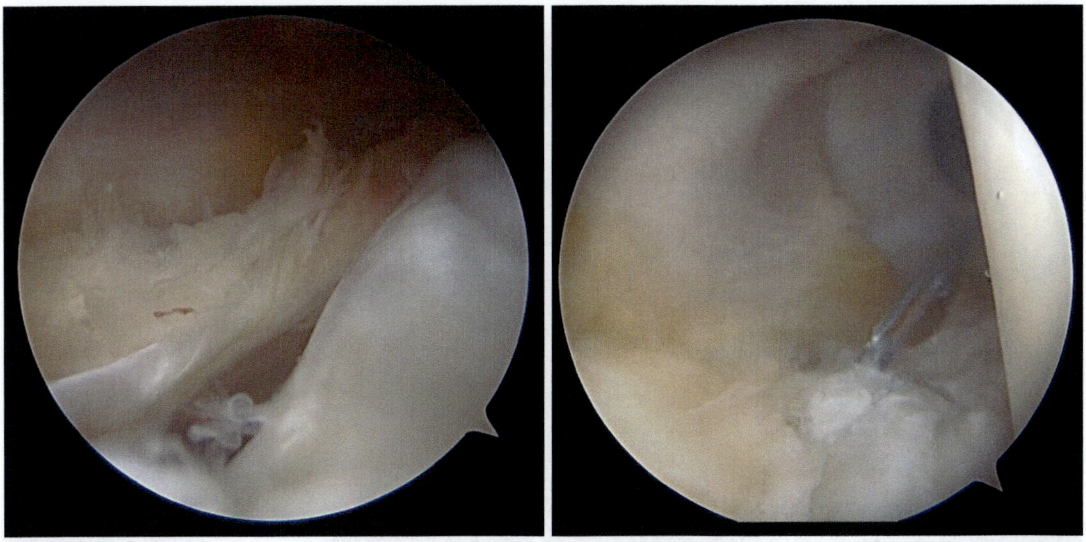

Fig. 6.5 Arthroscopic view of a ramp lesion and repair with the first posteromedial suture in the posteromedial meniscocapsular junction of the medial meniscus

dle through the meniscofemoral or meniscotibial capsule. The needles are retrieved through the posteromedial surgical interval. The needles are cut from the sutures and tied with the knee at 90° of flexion.

Inside-out repair offers a success rate of 60–80% for isolated meniscal repairs and 85–90% when performed at the time of ACL reconstruction [47]. Mentioned before the all-inside and inside-out techniques are good options to treat ramp lesions [12, 75]. If a ramp lesion is present, it is highly recommended to repair these lesions to avoid the anterior tibial translation and external rotational instability in ACL-deficient knees [78]. It is necessary to repair unstable ramp lesions. Stable ramp lesions can be treated conservatively or with abrasion/trephination.

6.8 Radial Tears

Radial meniscal tears are common in active individuals and are frequently associated with ACL and multi-ligament knee injuries. Radial tears are vertically orientated and transect the circumferential collagen fibres of the central meniscus. Normally the meniscus have the ability to transmit circumferential hoop stresses during load bearing and shock absorption. Small radial tears do not result in tibiofemoral biomechanic changes, and large radial tears (involving 90% of the meniscus) resulted in significant increase in peak compartment pressures because of impairment of transmitting circumferential hoop stresses [79, 80].

Partial meniscectomy of radial tears have negative biomechanical and worse long-term clinical outcomes. In recent years, there has been an increased interest in repairing radial meniscus tears. Repair decreases peak pressures to near-normal levels [81].

Different radial repair techniques (inside-out, all-inside and transtibial techniques) have been described in literature [82–85]. The golden standard for radial tear repair is the inside-out technique (Fig. 6.6). This technique generates tension against the periphery of the meniscus or capsule, creating single, double or crossed horizontal mattress sutures above and below the tear, approximately 5 and 10 mm from the meniscal rim. An additional incision is necessary for retrieval of the sutures. The all-inside technique has been reported to be less technically challenging; however, proper tensioning and securing sutures are still a challenge. The all-inside technique can place horizontally and vertically oriented sutures, applying direct compression at the tear site. New techniques have

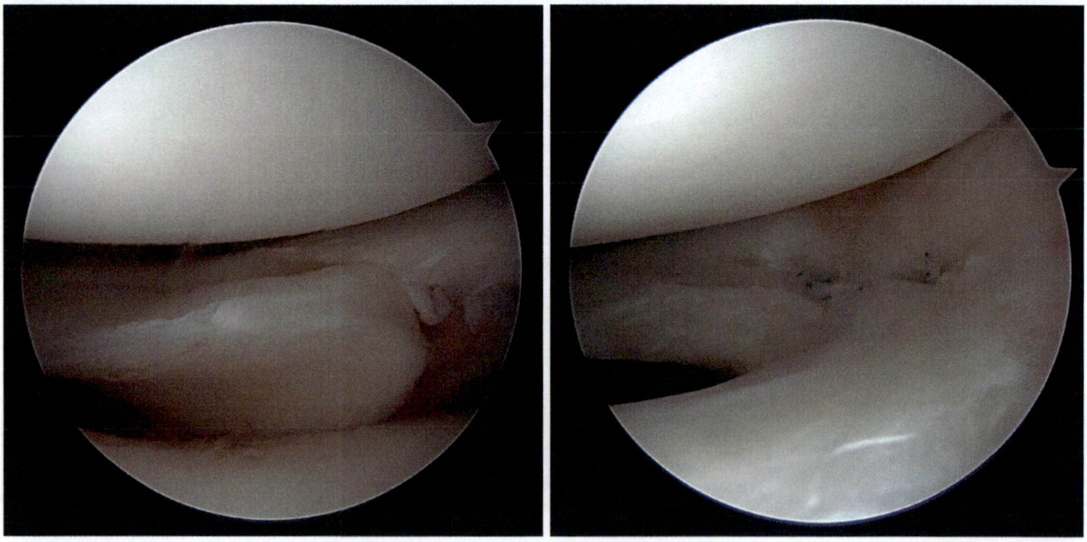

Fig. 6.6 Arthroscopic view of a radial tear with a side to side repair

been developed to augment horizontal suture repair constructs with transosseous tunnels restoring the meniscus to a more anatomically position [86]. This anatomic repair technique increases the stability of the repair construct [85]. A one or two transtibial tunnel is created at the meniscocapsular region of the tibia. Each torn edge of the meniscus is sutured superoinferiorly at the posterior corner of the tear edge. The sutures are shuttled through transtibial tunnels and tied over a button. After the transtibial repair, two inside-out horizontal mattress sutures are additionally placed on both the superior and inferior portions of the meniscus.

Meniscal preservation with repair of radial tears results in improved short-term clinical and subjective outcomes; however, long-term outcomes remain unknown [40]. No difference was seen regarding clinical outcome and clinical failure for the all-inside and inside-out techniques for radial tear repair [12]. However, the all-inside technique with a vertical suture configuration demonstrated lower displacement, higher load to failure and greater stiffness compared with the inside-out technique [87]. Transtibial repair techniques are increasing in popularity. Transtibial techniques demonstrated significantly less gapping distance and higher load to failure when compared with an inside-out technique [85]. The two-tunnel transtibial pull-out technique for the repair of radial meniscus tears reported similar clinical outcomes when compared with the repair of vertical meniscus tears [88].

6.9 Horizontal Tears

Horizontal tears are first described as 'intrasubstance tears' [89]. Horizontal cleavage in young athletes are often extensive and located in vascular and avascular zones. The aetiology of cleavage tears remains unknown but may be due to an overuse mechanism and are not traumatic. When functional treatment fails, meniscus repair can be considered. Because of the idea of minimal healing capacity of horizontal cleavage lesions, these lesions are used to be treated with partial/total menisectomy or nonoperative treatment [11]. Also partial menisectomy with single-leaflet resection has been described in literature [90]. However, studies reported minimal biomechanical benefit for single-leaflet resection [91]. By repairing the horizontal cleavage tears, the contact pressures restore to near normal levels. The study by Koh et al. [92] results in increasing interest for meniscal repair of horizontal cleavage tears. Different techniques for repair of horizontal cleavage tears have been described [93, 94].

The all-inside meniscus repair is the preferred technique for cleavage tears (Fig. 6.7). After resection of unstable meniscal fragments and/or fibrous tissue, the lesion is abraded by rasp or curette, causing capsular bleeding and providing biologic augmentation. Vertical circumferential compression stitches are placed perpendicular to the lesion, resulting in uniform compression on the superior and inferior leaflets with 5 mm intervals. Biologic augmentation could enhance the healing process [26, 95].

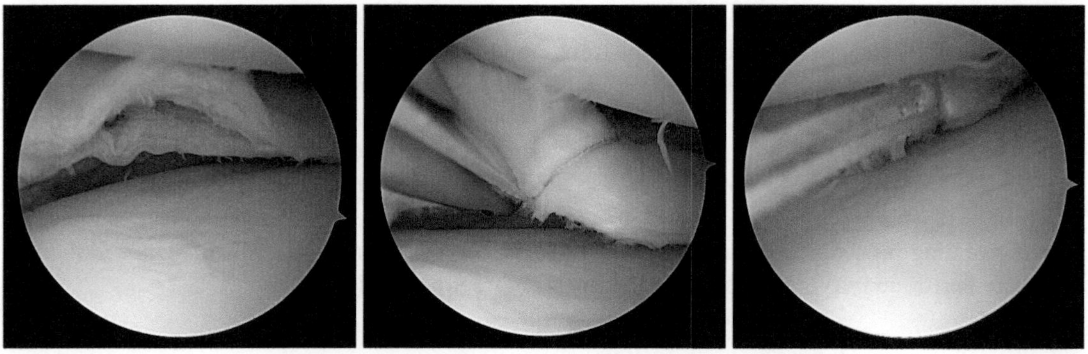

Fig. 6.7 Arthroscopic view of a horizontal cleavage tear. Repair with all-inside circumferential compression sutures

Horizontal cleavage tear repair resulted in good clinical results in literature [96].

A systematic review reported a healing rate of 78.6%. The healing rate of repair of horizontal cleavage tears is similar in comparison with other repairable tears [97]. A study of Woodmass et al. [98] showed the technique of circumferential compression suture formation with a self-retrieving suture passing device, which has been reported to have the highest load to failure of all repair patterns [99].

6.10 Biologic Augmentation

There is an increasing interest in biologic augmentation and repair enhancement to promote chemotaxis, cellular proliferation and/or matrix production at the site of meniscal repair to stimulate healing. Several adjuncts may be used to enhance meniscal healing including mechanical stimulation, marrow venting procedures, use of fibrin clots, platelet-rich plasma injections and stem cell–based therapies.

6.10.1 Mechanical Stimulation

Meniscal/synovial rasping is routinely used to stimulate bleeding and generate a healing response by promoting neovascularization. Trephination is used to improve short-term vascular access between a region of increased vascularity and an avascular region of the meniscus. The degree of mechanical stimulation achieved through trephination is balanced by the recognition that normal circumferential fibres are disrupted, which can affect the hoop-stress distribution properties of the meniscus.

6.10.2 Marrow Venting Procedures

Marrow venting procedures of the intercondylar notch are performed, trying to replicate the biologic environment when performing an ACL reconstruction.

A study of Dean et al. reported similar outcomes for meniscal repair with a marrow venting procedure and meniscal repair with ACL reconstruction. These results may be partly attributed to biological augmentation [28].

6.10.3 Use of Fibrin Clots

Exogenous fibrin clot may be useful in the setting of isolated meniscal repair [100], low level clinical studies showed improved meniscal healing using fibrin clots [101–103].

Fibrin clot enhances the local healing environment by placing peripheral blood factors, such as growth factors, fibrin and platelets, at the site of repair. This produces a healing milieu similar to the setting of concurrent ACL reconstruction. Some clinical studies have demonstrated the effectiveness of the use of a fibrin clot at the site of meniscal repair [104–106]. Comparative studies are needed to show superiority of a adding fibrin clot use to meniscal repair.

6.10.4 Stem Cell–Based Therapy

MSCs are of special interest for meniscal repair because of their multilineage potential [107], immunomodulatory and anti-inflammatory properties and extensive proliferative ability [108]. MSCs can also migrate to the site of meniscal injury [109, 110] and exert their reparative effects (132). The use of MSCs in meniscus repair is promising; however, only a few clinical studies and techniques are described in literature. Most studies are limited to preclinical animal studies. Vangsness et al. [111] reported evidence of meniscal regeneration after MSC injections following partial meniscectomy.

Further preclinical and clinical studies are needed to determine the role of stem cell therapy in treating meniscal repairs.

6.10.5 Platelet-Rich Plasma Injections

A less invasive method of treatment involves the use of PRP, an autologous, platelet-rich substance. Higher levels of platelets can release various growth factors that promote healing by enhancing meniscus cell proliferation, vascularization, and recruitment of fibroblasts and bone marrow–derived stem cells. Animal studies showed promising results of enhancing meniscal repair [112–114]. However, further studies are needed to determine if PRP and other biologics may benefit complex meniscal tear types. Only two level III studies reported the effectiveness of PRP injections for meniscal lesions [115, 116]. Both the studies found no difference in reoperation rate between meniscal repairs performed in isolation and repairs performed with platelet-rich plasma. Recently, studies showed improved results of application of PRP in meniscal repair [117, 118].

References

1. Englund M, Lohmander LS. Risk factors for symptomatic knee osteoarthritis fifteen to twenty-two years after meniscectomy. Arthritis Rhem. 2004;50(9):2811–9.
2. McDermott ID, Sharifi F, Bull AM, Gunte CM, Thomas RW, Amis AA. An anatomical study of meniscal allograft sizing. Knee Surg Sports Traumatol Arthrosc. 2004;12(2):130–5.
3. Kusayama T, Harner CD, Carlin GJ, Xerogeanes JW, Smith BA. Anatomical and biomechanical characteristics of human meniscofemoral ligaments. Knee Surg Sports Traumatol Arthrosc. 1994;2(4):234–7.
4. Richmond JC. Arthroscopy classics. Vascularity for healing of meniscus repairs. Arthroscopy. 2010;26(10):1368–9.
5. Arnoczky SP, Warrren RF. Microvasculature of the human meniscus. Am J Sports Med. 1982;10:90–5.
6. DeHaven KE. Decision-making factors in the treatment of meniscal lesions. Clin Orthop Relat Res. 1990;252:49–54.
7. Han JH, Song JG, Kwon JH, Kang KW, Sha D, Nha KW. Spontaneous healing of a displaced bucket-handle tear of the lateral meniscus in a child. Knee Surg Relat Res. 2015;27:65–7.
8. Noyes FR, Barber-Westin SD. Arthroscopic repair of meniscal tears extending into the avascular zone in patients younger than twenty years of age. Am J Sports Med. 2002;30(4):589–600.
9. Dorfmann H, Juan LH, Bonavarlet JP, Boyer T. Arthroscopy of degenerative lesions of the internal meniscus. Classification and treatment. Rev Rhum Mal Oteo Articul. 1987;54(4):303–10.
10. Klimkiewicz JJ, Shaffer B. Meniscal surgery 2002 update; indications and techniques for resection repair, regeneration. Arthroscopy. 2002;18:14–25.
11. Johnson D, Weiss B. Meniscal repair using the inside-out suture technique. Sports Med Arthrosc. 2012;20:68–76.
12. Grant JA, Wilde J, Miller BS, Bedi A. Comparison of inside-out and all-inside techniques fort the repair of isolated meniscal tears: a systematic review. Am J Sports Med. 2012;40(2):459–6.
13. Fillingham YA, Riboh JC, Erickson BJ, Bach BR Jr, Yanke AB. Inside-out versus all-inside repair of isolated meniscal tears. Am J Sports Med. 2017;45(1):234–42.
14. Krych AJ, McIntosh AL, Voll AE, Stuart MJ, Dahm DL. Arthroscopic repair of isolated meniscal tears in patients 18 years and younger. Am J Sports Med. 2008;36(7):1283–9.
15. Yim JH, Seon JK, Song EK, Choi JI, Kim MC, Lee KB, Seo HY. A comparative study of meniscectomy and nonoperative treatment for degenerative horizontal tears of the medial meniscus. Am J Sports Med. 2013;41(7):1565–70.
16. Scott GA, Jolly BL, Henning CE. Combined posterior incision and arthroscopic intra-articular repair of the meniscus. An examination of factors affecting healing. J Bone Joint Surg Am. 1986;68(6):847–61.
17. Cannon WD Jr, Vittori JM. The incidence of healing in arthroscopic meniscal repairs in anterior cruciate ligament-reconstructed knees versus stable knees. Am J Sports Med. 1992;20(2):176–81.
18. Laurendon L, Neri T, Farizon F, Philippot R. Prognostic factors for all-inside meniscal repair. A 87-case series. Orthop Traumatol Surg Res. 2017;103(7):1017–20.
19. Steadman JR, Matheny LM, Singleton SB, Johnson NS, Rodkey WG, Crespo B, Briggs KK. Meniscus suture repair: minimum 10-year outcomes in patients younger than 40 years compared with patients 40 and older. Am J Sports Med. 2015;43(9):2222–7.
20. Duchman KR, Westermann RW, Spindler KP, Reinke EK, Huston LJ, Amendola A, Wolf BR, MOON Knee Group. The fate of meniscus tears left in situ at the time of anterior cruciate ligament reconstruction: a 6-year follow-up study from the MOON cohort. Am J Sports Med. 2015;43(11):2688–95.
21. Noyes FR, Barber-Westin SD. Arthroscopic repair of meniscus tears extending into the avascular zone with or without anterior cruciate ligament reconstruction in patients 40 years of age and older. Arthroscopy. 2000;16(8):822–9.
22. Hutchinson ID, Moran CJ, Potter HG, Warren RF, Rodeo SA. Restoration of the meniscus: form and function. Am J Sports Med. 2014;42(4):987–98.

23. Canale ST, Beaty JH. Campbell's operative orthopaedics. 12th ed. St Louis, MO: Mosby; 2012. p. 2075–8.
24. Fox JM, Rintz KG, Ferkel RD. Trephination of incomplete meniscal tears. Arthroscopy. 1993;9(4):451–5.
25. Zhang Z, Arnold JA, Williams T, McCann B. Repairs by trephination and suturing of longitudinal injuries in the avascular area of the meniscus in goats. Am J Sports Med. 1995;23(1):35–41.
26. Ahn JH, Kwon OJ, Nam TS. Arthroscopic repair of horizontal meniscal cleavage tears with marrow-stimulating technique. Arthroscopy. 2015;31:92–8.
27. Howarth WR, Brochard K, Campbell SE, Grogan BF. Effect of microfracture on meniscal tear healing in a goat (Capra hircus) model. Orthopedics. 2016;39(2):105–10.
28. Dean CS, Chahla J, Matheny LM, Mitchell JJ, LaPrade RF. Outcomes after biologically augmented isolated meniscal repair with marrow venting are comparable with those after meniscal repair with concomitant anterior cruciate ligament reconstruction. Am J Sports Med. 2017;45(6):1341–8.
29. Barber FA, Herbert MA, Bava ED, Drew OR. Biomechanical testing of suture-based meniscal repair devices containing ultrahigh-molecular-weight polyethylene suture: update 2011. Arthroscopy. 2012;28:827–34.
30. Buckland M, Sadoghi P, Wimmer MD, Vavken P, Pagenstert GI, Valderrabano V, Rosso C. Meta analysis on biomechanical properties of meniscus repairs: are devices better than sutures? Knee Surg Sports Traumatol Arthrosc. 2015;23:83–9.
31. Dervin GF, Downing KJ, Keene GC, McBride DG. Failure strengths of suture versus biodegradable arrow for meniscal repair: an in vitro study. Arthroscopy. 1997;13(3):296–300.
32. Rankin CC, Lintner DM, Noble PC, Paravic V, Greer E. A biomechanical analysis of meniscal repair techniques. Am J Sports Med. 2002;30(4):492–7.
33. Rimmer MG, Nawana NS, Keene GC, Pearcy MJ. Failure strengths of different meniscal suturing techniques. Arthroscopy. 1995;11(2):146–50.
34. Kocabey Y, Taser O, Nyland J, Ince H, Sahin F, Sunbuloglu E, Baysal G. Horizontal suture placement influences meniscal repair fixation strength. Knee Surg Sports Traumatol Arthrosc. 2013;21:615–9.
35. Kocabey Y, Taşer O, Hapa O, Güçlü A, Bozdag E, Sünbüloglu E, Doral M. Meniscal repair using large diameter horizontal sutures increases fixation strength: an in vitro study. Knee Surg Sports Traumatol Arthrosc. 2011;19(2):202–6.
36. Goradia VK. All-inside arthroscopic meniscal repair with meniscal cinch. Arthrosc Tech. 2013;2:171–4.
37. Pujol N, Tardy N, Boisrenoult P, Beaufi ls P. Long-term outcomes of all-inside meniscal repair. Knee Surg Sports Traumatol Arthrosc. 2015;23:219–24.
38. Rosso C, Muller S, Buckland DM, Schwenk T, Zimmermann S, de Wild M, Valderrabano V. All-inside meniscal repair devices compared with their matched inside-out vertical mattress suture repair: introducing 10,000 and 100,000 loading cycles. Am J Sports Med. 2014;42:2226–33.
39. Cuéllar A, Cuéllar R, Cuéllar A, Garcia-Alonso I, Ruiz-Ibán MA. The effect of knee flexion angle on the neurovascular safety of all-inside lateral meniscus repair: a Cadaveric Study. Arthroscopy. 2015;31(11):2138–44.
40. Moulton SG, Bhatia S, Civitarese DM, Frank RM, Dean CS, RF LP. Surgical techniques and outcomes of repairing radial meniscal tears: a systematic review. Arthroscopy. 2016;32(9):1919–25.
41. Masoudi A, Beamer BS, Harlow ER, Manoukian OS, Walley KC, Hertz B, Haeussler C, Olson JJ, Zurakowski D, Nazarian A, Ramappa AJ, DeAngelis JP. Biomechanical evaluation of an all-inside suture-based device for repairing longitudinal meniscal tears. Arthroscopy. 2015;31(3):428–34.
42. Fillingham YA, Riboh JC, Erickson BJ, Bach BR Jr, Yanke AB. Inside-out versus all-inside repair of isolated meniscal tears: an updated systematic review. Am J Sports Med. 2017;45(1):234–42.
43. Choi NH, Kim TH, Victoroff BN. Comparison of arthroscopic medial meniscal suture repair techniques: inside-out versus all-inside repair. Am J Sports Med. 2009;37:2144–50.
44. Erduran M, Hapa O, Sen B, Kocabey Y, Erdemli D, Aksel M, Havitcioglu H. The effect of inclination angle on the strength of vertical mattress configuration for meniscus repair. Knee Surg Sports Traumatol Arthrosc. 2015;23:41–4.
45. Turman KA, Diduch DR, Miller MD. All-inside meniscal repair. Sports Health. 2009;1:438–44.
46. Espejo-Baena A, Golano P, Meschian S, Garcia-Herrera JM, Serrano Jimenez JM. Complications in medial meniscus suture: a cadaveric study. Knee Surg Sports Traumatol Arthrosc. 2007;15:811–6.
47. Nelson CG, Bonner KF. Inside-out meniscus repair. Arthrosc Tech. 2013;2(4):453–60.
48. Warren RF. Arthroscopic meniscal repair. Arthroscopy. 1985;1:170–2.
49. Morgan CD, Casscells SW. Arthroscopic meniscus repair: a safe approach to the posterior horns. Arthroscopy. 1986;2(1):3–12.
50. Roos EM, Ostenberg A, Roos H, Ekdahl C, Lohmander LS. Long-term outcome of meniscectomy: symptoms, function, and performance tests in patients with or without radiographic osteoarthritis compared to matched controls. Osteoarthr Cartil. 2001;9:316–24.
51. Chen LX, Linde-Rosen M, Hwang SC, Zhou JB, Xie Q, Smolinski P, Fu FH. The effect of medial meniscal horn injury on knee stability. Knee Surg Sports Traumatol Arthrosc. 2015;23(1):126–31.
52. Walker PS, Arno S, Bell C, Salvadore G, Borukhov I, Oh C. Function of the medial meniscus in force transmission and stability. J Biomech. 2015;48(8):1383–8.

53. Prince MR, Esquivel AO, Andre AM, Goitz HT. Anterior horn lateral meniscus tear, repair, and meniscectomy. J Knee Surg. 2014;27:229–34.
54. Kelly JD, Ebrahimpour P. Chondral injury and synovitis after arthroscopic meniscal repair using an outside-in mulberry knot suture technique. Arthroscopy. 2004;20:49–52.
55. Van Trommel MF, Simonian PT, Potter HG, Wickiewicz TL. Different regional healing rates with the outside-in technique for meniscal repair. Am J Sports Med. 1998;26(3):446–52.
56. Morgan CD, Wojtys EM, Casscells CD, Casscells SW. Arthroscopic meniscal repair evaluated by second-look arthroscopy. Am J Sports Med. 1991;19(6):632–7.
57. Ahn JH, Lee YS, Yoo JC, Chang MJ, Koh KH, Kim MH. Clinical and second-look arthroscopic evaluation of repaired medial meniscus in anterior cruciate ligament reconstructed knees. Am J Sports Med. 2010;38:472–7.
58. Metcalf MH, Barrett GR. Prospective evaluation of 1485 meniscal tear patterns in patients with stable knees. Am J Sports Med. 2004;32:675–80.
59. Weiss CB, Lundberg M, Hamberg P, DeHaven KE, Gillquist J. Non-operative treatment of meniscal tears. J Bone Joint Surg Am. 1989;71:811–22.
60. Dandy DJ. The bucket handle meniscal tear: a technique detaching the posterior segment first. Orthop Clin North Am. 1982;13(2):369–85.
61. Fithian DC, Kelly MA, Mow VC. Material properties and structure-function relationships in the menisci. Clin Orthop Relat Res. 1990;(252):19–31.
62. Messner K, Gao J. The menisci of the knee joint. Anatomical and functional characteristics, and a rationale for clinical treatment. J Anat. 1998;193(Pt2):161–78.
63. Baratz ME, Fu FH, Mengato R. Meniscal tears: the effect of meniscectomy and of repair on intraarticular contact areas and stress in the human knee. A preliminary report. Am J Sports Med. 1986;14(4):270–5.
64. Pujol H, Lorbach O. Meniscal repair: results. In: Hulet C, Pereira H, Peretti G, Dent M, editors. Surgery of the meniscus. Berlin, Heidelberg: Springer Verlag; 2016. p. 343–55.
65. Paxton ES, Stock MV, Brophy RH. Meniscal repair versus partial meniscectomy: a systematic review comparing reoperation rates and clinical outcomes. Arthroscopy. 2011;27(9):1275–88.
66. Barber FA, Herbert MA, Richards DP. Load to failure testing of new meniscal repair devices. Arthroscopy. 2004;20(1):45–50.
67. Espejo-Reina A, Serrano-Fernández JM, Martín-Castilla B, Estades-Rubio FJ, Briggs KK, Espejo-Baena A. Outcomes after repair of chronic bucket-handle tears of medial meniscus. Arthroscopy. 2014;30(4):492–6.
68. Barber-Westin SD, Noyes FR. Clinical healing rates of meniscus repairs of tears in the central-third (red-white) zone. Arthroscopy. 2014;30:134–46.
69. Bollen SR. Posteromedial meniscocapsular injury associated with rupture of the anterior cruciate ligament: a previously unrecognised association. J Bone Joint Surg Br. 2010;92:222–3.
70. Liu X, Feng H, Zhang H, Hong L, Wang XS, Zhang J. Arthroscopic prevalence of ramp lesion in 868 patients with anterior cruciate ligament injury. Am J Sports Med. 2011;39:832–7.
71. Strobel MJ. Manual of arthroscopic surgery. Berlin: Springer; 2013.
72. Arner JW, Herbst E, Burnham JM, Soni A, Naendrup JH, Popchak A, Fu FH, Musahl V. MRI can accurately detect meniscal ramp lesions of the knee. Knee Surg Sports Traumatol Arthrosc. 2017;25(12):3955–60. https://doi.org/10.1007/s00167-017-4523-9.
73. Stephen JM, Halewood C, Kittl C, Bollen SR, Williams A, Amis AA. Posteromedial meniscocapsular lesions increase tibiofemoral joint laxity with anterior cruciate ligament deficiency, and their repair reduces laxity. Am J Sports Med. 2016;44:400–8.
74. Sonnery-Cottet B, Conteduca J, Thaunat M, Gunepin FX, Seil R. Hidden lesions of the posterior horn of the medial meniscus: a systematic arthroscopic exploration of the concealed portion of the knee. Am J Sports Med. 2014;42:921–6.
75. Li WP, Chen Z, Song B, Yang R, Tan W. The FasT-fix repair technique for ramp lesion of the medial meniscus. Knee Surg Relat Res. 2015;27:56–60.
76. Morgan CD. The "all-inside" meniscus repair. Arthroscopy. 1991;7(1):120–5.
77. Ahn JH, Kim SH, Yoo JC, Wang JH. All-inside suture technique using two posteromedial portals in a medial meniscus posterior horn tear. Arthroscopy. 2004;20:101–8.
78. Chahla J, Dean CS, Moatshe G, Mitchell JJ, Cram TR, Yacuzzi C, et al. Meniscal ramp lesions: anatomy, incidence, diagnosis, and treatment. Orthop J Sports Med. 2016;4(7):459.
79. Bedi A, Kelly NH, Baad M, et al. Dynamic contact mechanics of the medial meniscus as a function of radial tear, repair, and partial meniscectomy. J Bone Joint Surg Am. 2010;92(6):1398–408.
80. Bedi A, Kelly N, Baad M, Fox AJ, Ma Y, Warren RF, Maher SA. Dynamic contact. mechanics of radial tears of the lateral meniscus: implications for treatment. Arthroscopy. 2012;28(3):372–81.
81. Zhang AL, Miller SL, Coughlin DG, Lotz JC, Feeley BT. Tibiofemoral contact pressures in radial tears of the meniscus treated with all-inside repair, inside-out repair and partial meniscectomy. Knee. 2015;22(5):400–4.
82. Matsubara H, Okazaki K, Izawa T. New suture method for radial tears of the meniscus: biomechanical analysis of cross-suture and double horizontal suture techniques using cyclic load testing. Am J Sports Med. 2012;40:414–8.
83. Choi NH, Kim TH, Son KM, Victoroff BN. Meniscal repair for radial tears of the midbody of the lateral meniscus. Am J Sports Med. 2010;38:2472–6.

84. James EW, LaPrade CM, Feagin JA, LaPrade RF. Repair of a complete radial tear in the midbody of the medial meniscus using a novel crisscross suture transtibial tunnel surgical technique: a case report. Knee Surg Sports Traumatol Arthrosc. 2015;23:2750–5.
85. Bhatia S, Civitarese DM, Turnbull TL, LaPrade CM, Nitri M, Wijdicks CA, LaPrade RF. A novel repair method for radial tears of the medial meniscus: biomechanical comparison of transtibial 2-tunnel and double horizontal mattress suture techniques under cyclic loading. Am J Sports Med. 2016;44(3):639–45.
86. Nitri M, Chahla J, Civitarese D, Bhatia S, Moulton SG, La Prade CM, La Prade RF. Medial meniscus radial tear: a transtibial 2-tunnel technique. Arthrosc Tech. 2016;5(4):889–95.
87. Beamer BS, Masoudi A, Walley KC, Harlow ER, Manoukian OS, Hertz B, Haeussler C, Olson JJ, Deangelis JP, Nazarian A, Ramappa AJ. Analysis of a new all inside versus inside-out technique for repairing radial meniscal tears. Arthroscopy. 2015;31(2):293–8.
88. Cinque ME, Geeslin AG, Chahla J, Dornan GJ, LaPrade RF. Two-tunnel transtibial repair of radial meniscus tears produces comparable results to inside-out repair of vertical meniscus tears. Am J Sports Med. 2017;45(10):2253–9.
89. Biedert RM. Intrasubstance meniscal tears. Clinical aspects and the role of MRI. Arch Orthop Trauma Surg. 1993;112:142–7.
90. Kim JG, Lee SY, Chay S, Lim HC, Bae JH. Arthroscopic meniscectomy for medial meniscus horizontal cleavage tears in patients under age 45. Knee Surg Relat Res. 2016;28:225–32.
91. Haemer JM, Wang MJ, Carter DR, Giori NJ. Benefit of single-leaf resection for horizontal meniscus tear. Clin Orthop Relat Res. 2007;457:194–202.
92. Koh JL, Yi SJ, Ren Y, Zimmerman TA, Zhang LQ. Tibiofemoral contact mechanics with horizontal cleavage tear and resection of the medial meniscus in the human knee. J Bone Joint Surg Am. 2016;98:1829–36.
93. Pujol N, Bohu Y, Boisrenoult P, Macdes A, Beaufils P. Clinical outcomes of open meniscal repair of horizontal meniscal tears in young patients. Knee Surg Sports Traumatol Arthrosc. 2013;21:1530–3.
94. Saliman JD. The circumferential compression stitch for meniscus repair. Arthrosc Tech. 2013;2:257–64.
95. Kamimura T, Kimura M. Repair of horizontal meniscal cleavage tears with exogenous fibrin clots. Knee Surg Sports Traumatol Arthrosc. 2011;19(7):1154–7.
96. Rubman MH, Noyes FR, Barber-Westin SD. Arthroscopic repair of meniscal tears that extend into the avascular zone. A review of 198 single and complex tears. Am J Sports Med. 1998;26:87–95.
97. Kurzweil PR, Lynch NM, Coleman S, Kearney B. Repair of horizontal meniscus tears: a systematic review. Arthroscopy. 2014;30:1513–9.
98. Woodmass JM, Joshua DJ, Wu IT, Saris DBF, Stuart MJ, Krych AJ. Horizontal cleavage meniscus tear treated with All-inside circumferential compression stitches. Arthrosc Tech. 2017;6(4):1329–33.
99. Asík M, Sener N. Failure strength of repair devices versus meniscus suturing techniques. Knee Surg Sports Traumatol Arthrosc. 2002;10:25–9.
100. Schmitz MA, Rouse LM Jr, DeHaven KE. The management of meniscal tears in the ACL-deficient knee. Clin Sports Med. 1996;15:573–93.
101. Henning CE, Lynch MA, Yearout KM, et al. Arthroscopic meniscal repair using an exogenous fibrin clot. Clin Orthop Relat Res. 1990;(252):64–72.
102. Van Trommel MF, Simonian PT, Potter HG, et al. Arthroscopic meniscal repair with fibrin clot of complete radial tears of the lateral meniscus in the avascular zone. Arthroscopy. 1998;14:360–5.
103. Kamimura T, Kimura M. Meniscal repair of degenerative horizontal cleavage tears using fibrin clots: clinical and arthroscopic outcomes in 10 cases. Orthop J Sports Med. 2014;2(11):2325967114555678.
104. Ra HJ, Ha JK, Jang SH, Lee DW, Kim JG. Arthroscopic inside-out repair of complete radial tears of the meniscus with a fibrin clot. Knee Surg Sports Traumatol Arthrosc. 2013;21(9):2126–30. Epub 2012 Sep 22.
105. Jang SH, Ha JK, Lee DW, Kim JG. Fibrin clot delivery system for meniscal repair. Knee Surg Relat Res. 2011;23(3):180–3.
106. Henning CE, Lynch MA, Yearout KM, Vequist SW, Stallbaumer RJ, Decker KA. Arthroscopic meniscal repair using an exogenous fibrin clot. Clin Orthop Relat Res. 1990;252:64–72.
107. Pittenger MF, Mackay AM, Beck SC, Jaiswal RK, Douglas R, Mosca JD, et al. Multilineage potential of adult human mesenchymal stem cells. Science. 1999;284:143–7.
108. Da Silva ML, Caplan AI, Nardi NB. In search of the in vivo identity of mesenchymal stem cells. Stem Cells. 2008;26:2287–99.
109. Chavakis E, Urbich C, Dimmeler S. Homing and engraftment of progenitor cells: a prerequisite for cell therapy. J Mol Cell Cardiol. 2008;45:514–22.
110. Fong EL, Chan CK, Goodman SB. Stem cell homing in musculoskeletal injury. Biomaterials. 2011;32:395–409.
111. Vangsness CT Jr, Farr J 2nd, Boyd J, Dellaero DT, Mills CR, Le Roux-Williams M. Adult human mesenchymal stem cells delivered via intra-articular injection to the knee following partial medial meniscectomy: a randomized, double-blind, controlled study. J Bone Joint Surg Am. 2014;96(2):90–8.
112. Forriol F, Longo UG, Duart J, Ripalda P, Vaquero J, Loppini M, Romeo G, Campi S, Khan WS, Muda AO, Denaro V. VEGF, BMP-7, Matrigel(TM), hyaluronic acid, in vitro cultured chondrocytes and trephination for healing of the avascular portion of the meniscus. An experimental study in sheep. Curr Stem Cell Res Ther. 2015;10(1):69–76.

113. Parrish WR, Byers BA, Su D, Geesin J, Herzberg U, Wadsworth S, Bendele A, Story B. Intra-articular therapy with recombinant human GDF5 arrests disease progression and stimulates cartilage repair in the rat medial meniscus transection (MMT) model of osteoarthritis. Osteoarthr Cartil. 2017;25(4):554–60.
114. Cucchiarini M, McNulty AL, Mauck RL, Setton LA, Guilak F, Madry H. Advances in combining gene therapy with cell and tissue engineering-based approaches to enhance healing of the meniscus. Osteoarthr Cartil. 2016;24(8):1330–9.
115. Griffin JW, Hadeed MM, Werner BC, Diduch DR, Carson EW, Miller MD. Platelet rich plasma in meniscal repair: does augmentation improve surgical outcomes? Clin Orthop Relat Res. 2015;473(5):1665–72.
116. Pujol N, Salle De Chou E, Boisrenoult P, Beaufils P. Platelet-rich plasma for open meniscal repair in young patients: any benefit? Knee Surg Sports Traumatol Arthrosc. 2015;23(1):51–8.
117. Kemmochi M, Sasaki S, Takahashi M, Nishimura T, Aizawa C, Kikuchi J. The use of platelet-rich fibrin with platelet-rich plasma support meniscal repair surgery. J Orthop. 2018;15:711–20.
118. Kaminiski R, Kuliniski K, Kozar-Kaminska K, Wielgus M, Langner M, Wasko MK, Kowalczewski J, Pomianowski S. A prospective, randomized, double-blind, parallel-group, placebo-controlled study evaluating meniscal healing, clinical outcomes and safety in patients undergoing meniscal repair of unstable, complete vertical meniscal tears (bucket handle) augmented with platelet-rich plasma. Biomed Res Int. 2018;11:1–9.

Arthroscopic Anterior Cruciate Ligament Reconstruction: Six Bundle Hamstring Tendon Autograft for Anterior Cruciate Ligament Reconstruction

7

Nader Darwich and Ashraf Abdelkafy

7.1 Introduction

Over the past eight decades, sports knee surgeons kept on developing new techniques and studying and enhancing older techniques in order to improve the results and boost the performance of anterior cruciate ligament (ACL) reconstructed patients [1–10]. The ultimate goal has been always to allow the patients to return to their pre-injury level of sports activities and the activities of daily living [11–13].

Several factors affect the final outcome of the ACL reconstruction surgery [14]. Of these factors, the most noteworthy are the tensile properties of the graft tissue, initial fixation strength of the graft, graft-tunnel healing, biologic remodeling of the graft, and the type of postoperative rehabilitation program [14, 15].

The gold standard graft is the patellar tendon bone-tendon-bone (BTB) autograft. It has been used for many years and still being used till now [16].

Because of the better donor site morbidity, improvements of soft tissue graft fixation techniques and excellent clinical outcomes, we use hamstring tendons autograft for ACL reconstruction [17]. A six-stranded hamstring tendon autograft technique for ACL reconstruction is our preferred technique.

In this chapter, we describe our surgical technique for ACL reconstruction using triple gracilis and triple semitendinosus (TGST) autograft.

7.2 Diagnosis

ACL injuries are fairly common, and the incidence increases in contact sports [18]. History taking should include a detailed mechanism of injury as well as a detailed analysis of symptoms such as pain, swelling, catching, locking, and instability. Lachman, anterior drawer, and pivot shift tests are mandatory. KT-2000 measurements comparing both knees are very helpful.

It is important to detect concomitant injuries such as meniscal, osteochondral, medial collateral ligament, and posterolateral corner injuries. Failure to identify the concomitant injuries and treat them properly might lead to failure of the ACL reconstruction and poor functional outcome results.

N. Darwich (✉)
Burjeel Orthopaedics and Sports Medicine Center, Abu Dhabi, United Arab Emirates

A. Abdelkafy
Orthopaedic Surgery Department, Faculty of Medicine, Suez Canal University, Ismailia, Egypt

Burjeel Orthopaedics and Sports Medicine Center, Burjeel Royal Hospital, Al-Ain, United Arab Emirates

M. Bozkurt, H. İ. Açar (eds.), *Clinical Anatomy of the Knee*,
https://doi.org/10.1007/978-3-030-57578-6_7

7.3 Imaging

A preoperative antero-posterior, lateral, Merchant, and full-length standing radiographs of both the lower extremities are of paramount importance for the assessment of the patellar alignment, patellar tilt, trochlear dysplasia, and varus-valgus alignments. MRI is the gold standard diagnosis method for visualizing and providing detailed images of the structures within the knee joint, including menisci, cartilage, tendons, ligaments, as well as muscles, blood vessels, and bones.

7.4 Graft Choice

The ACL graft used for ACL reconstruction can be harvested from the patient's hamstring tendons (HT), patellar tendon (PT), or quadriceps tendon. Allografts are used for special conditions. Hamstring tendon autografts are indicated for any acute or chronic ACL reconstruction. ACL reconstruction using hamstring tendon grafts has shown to result in faster recovery of quadriceps muscle strength, lower incidence of donor site pain, and less interference with kneeling compared to patellar tendon autografts [17]. Hamstring tendon graft is our graft of choice for patients whose occupation, lifestyle, or religion requires knee walking, crawling, or kneeling. Hamstring tendon grafts are also our preferred graft for patients with a history of a patellofemoral pain or patellar tendinopathy. Finally, hamstring tendon graft is the graft of choice when ACL reconstruction is indicated in patients with open growth plates [19]. The only absolute contraindication for the use of homolateral hamstring graft is previous knee surgery performed using the hamstring tendons.

7.5 Surgical Technique

7.5.1 Anesthesia and Positioning

Patients are admitted to the hospital and undergo evaluation by the anesthesiology team for determining the proper anesthesia procedure to be performed. Most patients receive regional anesthesia. We use femoral nerve blocks in order to achieve postoperative pain control. A thigh-length anti-embolism stocking and a foam rubber heel pad are applied to the non-operated leg. A padded pneumatic tourniquet is applied high on the thigh of the operative leg but is rarely inflated during the operation. The patient is positioned supine on the operating room table and given 1 g of the first-generation cephalosporin intravenous. We position the lower extremity, so that a full, free range of motion can be performed during the procedure (Figs. 7.1, 7.2, 7.3, and 7.4). We need full flexion of the knee because for drilling the femoral tunnel we use the anteromedial portal, and we need this position to avoid the risk of damage of the cartilage of the medial femoral condyle. We continuously change the position of the knee during the procedure in order to work comfortably on the entire knee. The padded hip positioner sta-

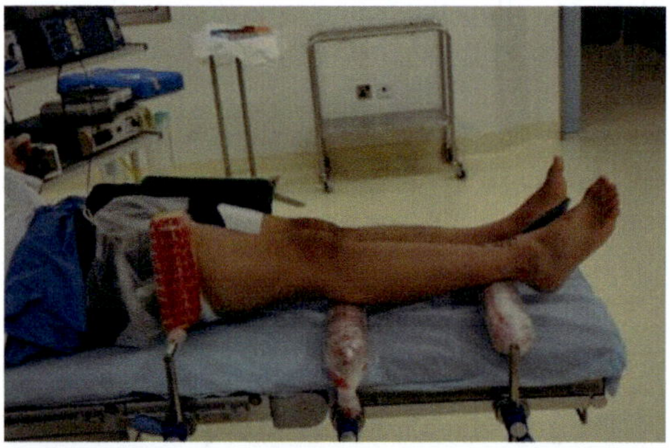

Fig. 7.1 Extension position

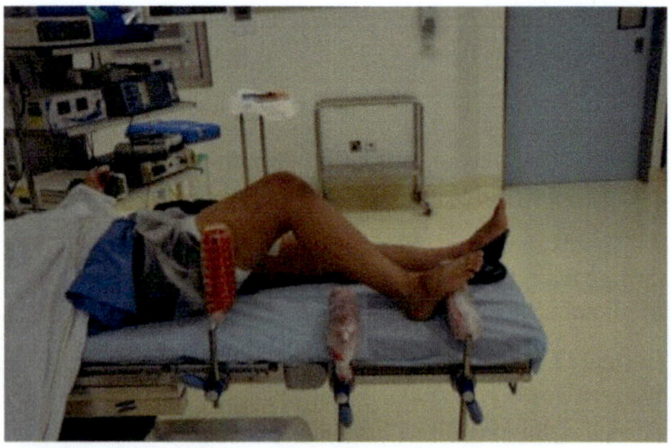

Fig. 7.2 90°position

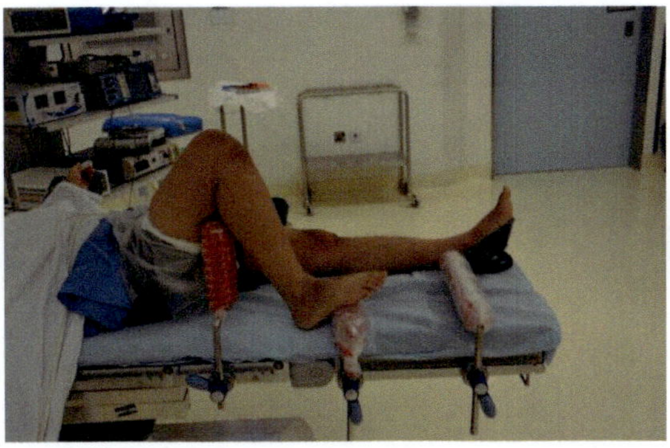

Fig. 7.3 120° position

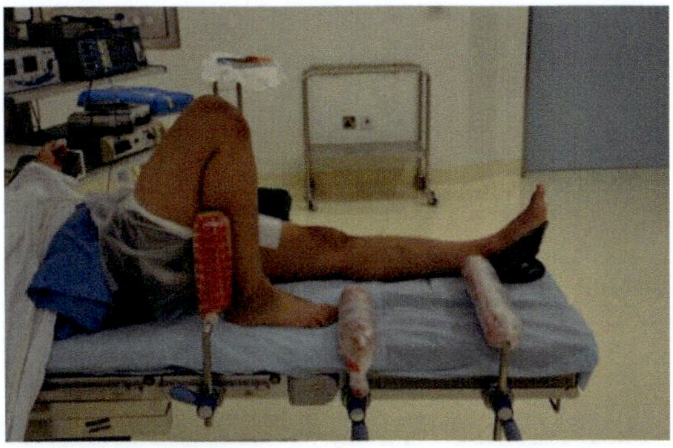

Fig. 7.4 Full flexion position

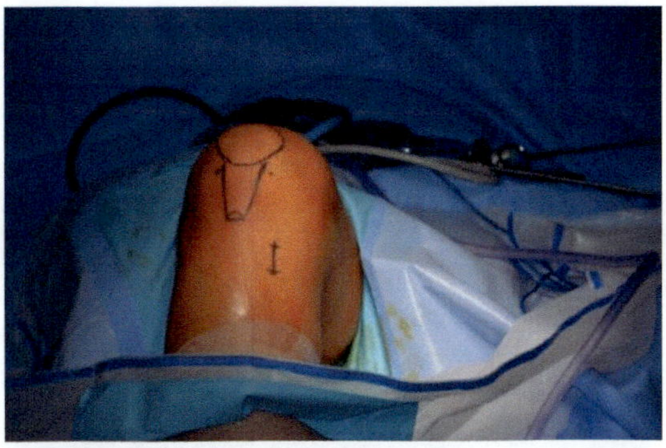

Fig. 7.5 Preoperative skin markings

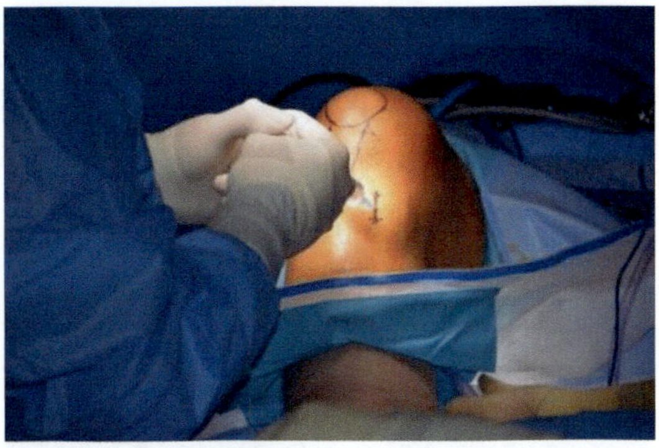

Fig. 7.6 Local anesthesia

bilizes the patient's pelvis and the padded thigh post acts as a fulcrum to allow the application of valgus force to the knee, allowing the medial compartment to be opened for the performance of any concomitant meniscus surgery.

Preoperative skin markings are crucial. We carefully mark externally the boundaries of patella, patellar tendon, anterolateral portal, anteromedial portal, and Hamstring harvesting incision site (Fig. 7.5).

Our team uses iodine skin preparation as a routine. Sterile draping is applied. A solution of 5 mg morphine sulfate, 20 mL 0.25% bupivacaine, and 1:100,000 epinephrine is injected into the supra-patellar pouch for pre-emptive analgesia. We use a pump infusion for joint distension which improves visualization and allows the procedure to be performed without tourniquet. The skin incision and subcutaneous tissues are infiltrated with a solution of 0.25% bupivacaine and 1:100,000 epinephrine for hemostasis and pre-emptive analgesia [20] (Fig. 7.6).

7.5.2 Hamstring Tendon Graft Harvest

Harvest of the hamstring tendons is performed using a vertical skin incision centered over the tibial insertion of the hamstrings tendons (Fig. 7.7).

The vertical skin incision is positioned closer to the anterior crest of the tibia, in this way this incision can be easily extended to harvest a patellar tendon graft in the case of premature amputation of the semitendinosus tendon graft. The main complication of hamstring tendon harvesting is the damage to the infrapatellar branches of the saphenous nerve. This risk can be minimized by avoiding sharp dissection of the soft tissue under the skin using the scalp. We use a scissor divulging the tissue until the fascia. Another complication is the premature amputation of the hamstring tendons where harvesting of the patellar tendon is required. The superior border of the sartorius ten-

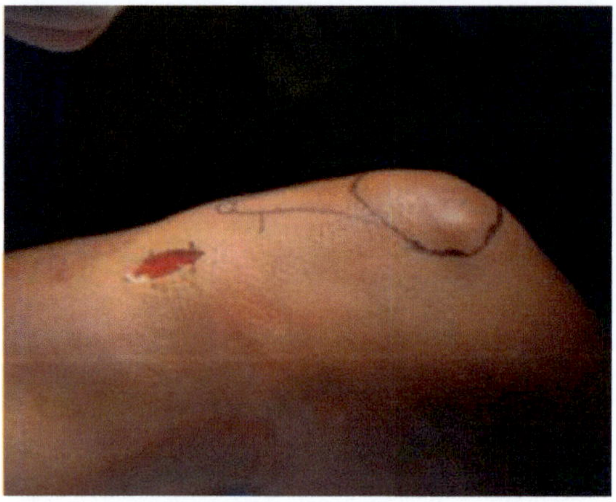

Fig. 7.7 Vertical incision

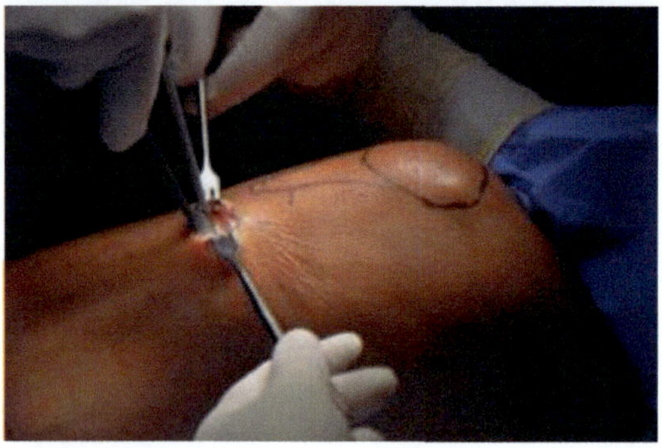

Fig. 7.8 Blunt dissection with the scissors

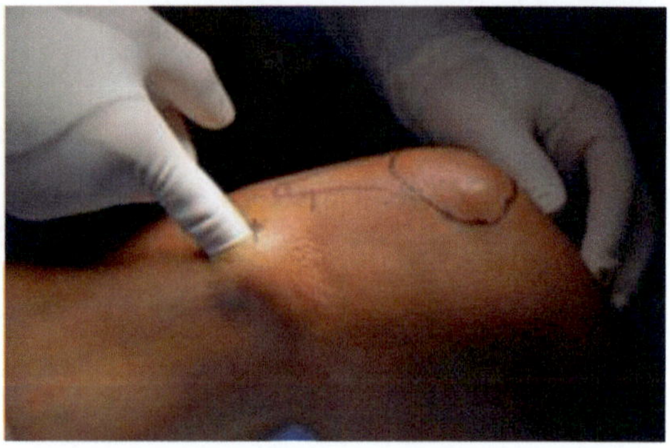

Fig. 7.9 Blunt dissection with the finger

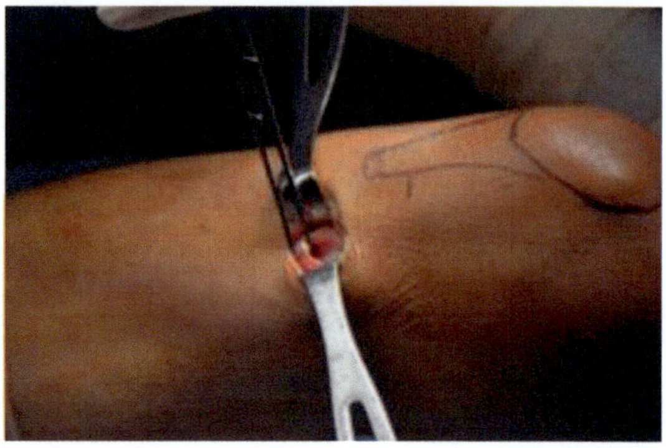

Fig. 7.10 Press distal tendons with Allis-Adair tissue forceps

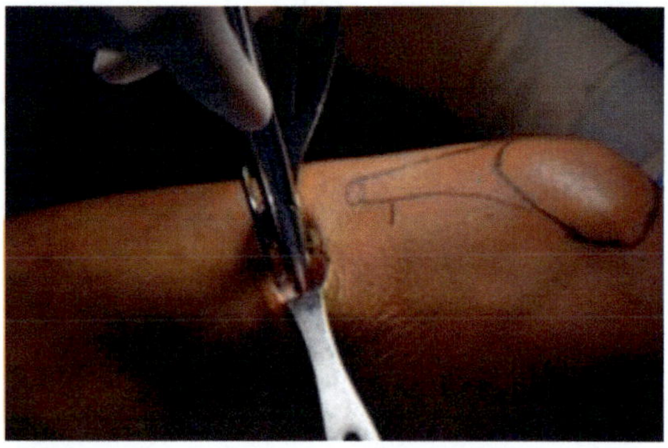

Fig. 7.11 Extension the incision longitudinally with a scissor

don is approximately one-finger width below the tibial tubercle or three-finger widths below the medial joint line. In revision cases of BTB grafts, we extend the previous patellar tendon incision distally 2–3 cm below the tibial tubercle, and we harvest the hamstring tendons. Then removal of the hardware is performed. We do not inflate the tourniquet routinely during the harvesting the grafts. The sartorius fascia is exposed by sharp and blunt dissection (Figs. 7.8 and 7.9).

We grasp the distal tendons with an Allis-Adair tissue forceps (Fig. 7.10). A 1-cm incision is made over the fascia above the superior border of the sartorius tendon. After that we extend the incision longitudinally with a scissor (Fig. 7.11). This technique gives an excellent view of the internal aspect of the pes anserine and allows the surgeon to better visualize and to identify any of the associated anatomic variations or variable tendon attachments to the tibia.

The conjoined tibial insertion of the two tendons is detached from the tibia by making an inverted L-shaped incision through the sartorius

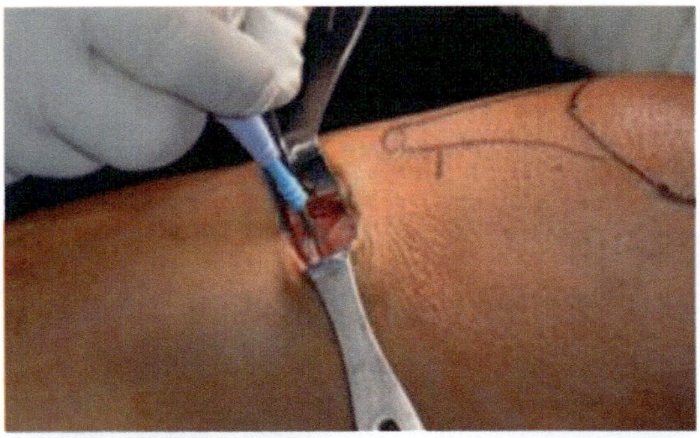

Fig. 7.12 Inverted L-shaped incision through the sartorius using the electrocautery

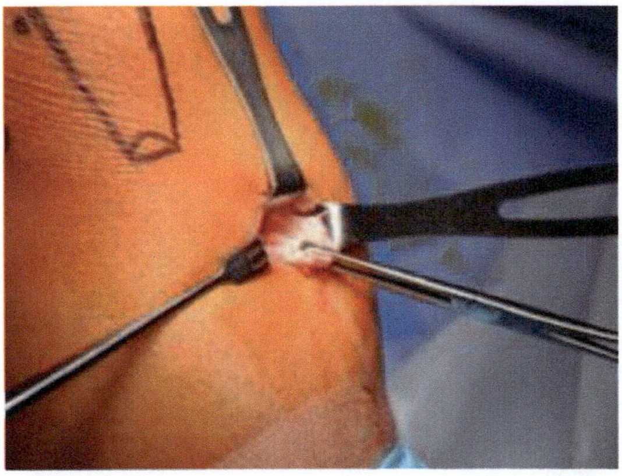

Fig. 7.13 The sartorius fascia is grasped with an Allis clamp and lifted away from the tibia

using the electrocautery and later with the scalpel (Fig. 7.12).

The sartorius fascia is grasped with an Allis clamp and lifted away from the tibia; in this moment, the protection of the underlying medial collateral ligament is very important (Fig. 7.13). The tibial insertion of the two tendons is sharply released from the crest of the tibia first with the cautery and second with the knife.

A right-angled type clamp is used to separate the two tendons from the undersurface of the sartorius fascial flap, which is preserved for later closure [21] (Figs. 7.14 and 7.15).

The gracilis tendon is sharply divided and grasped with wide Allis-Adair tissue forceps; the knife is used to free the tendon from the undersurface of the sartorius fascia [21] (Fig. 7.16).

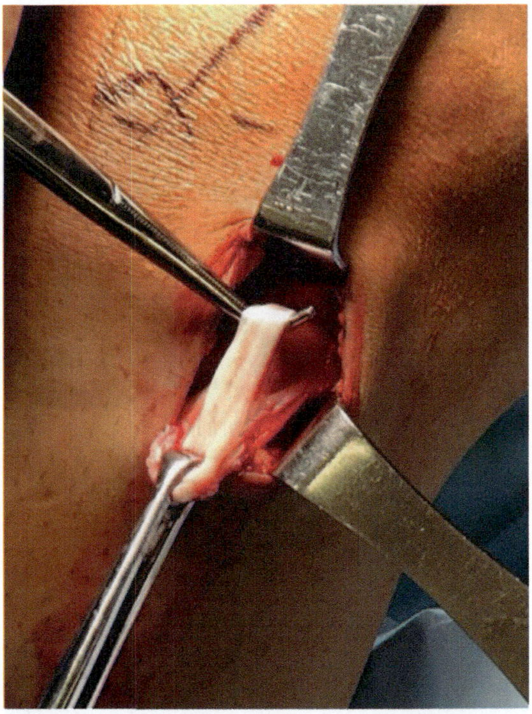

Fig. 7.14 A right-angled clamp is used to separate the two tendons

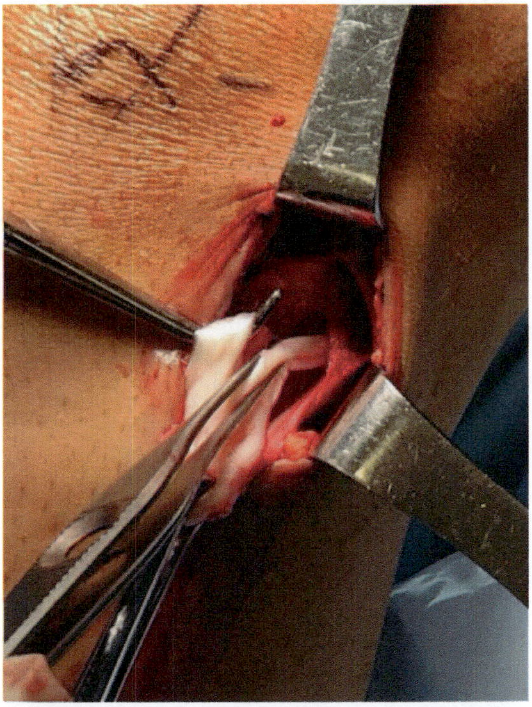

Fig. 7.15 Gracilis and semitendinosus tendons

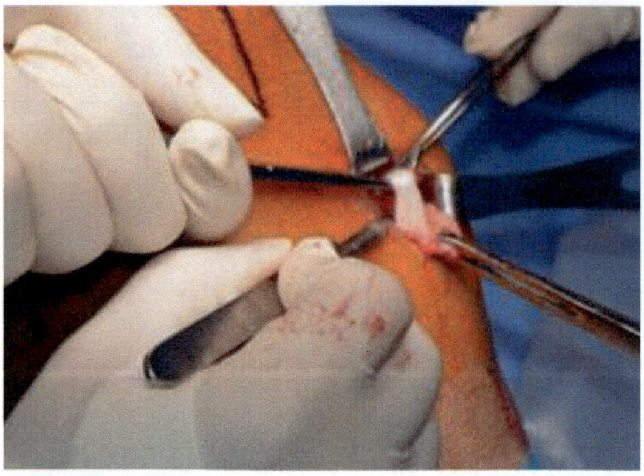

Fig. 7.16 The knife is used to free the tendon

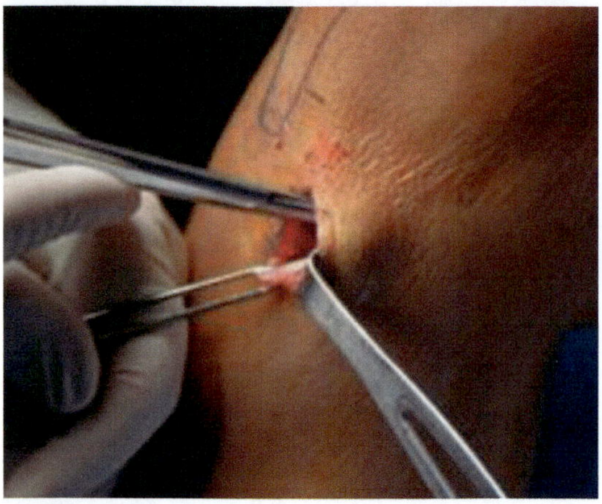

Fig. 7.17 Sharp or scissors dissection along the superior border of the gracilis

Carefully we release the interconnecting fascial bands that run between the two tendons. Sharp or scissors dissection along the superior border of the gracilis should be avoided to prevent injury to the saphenous nerve (Fig. 7.17).

Five throw sutures are placed in the free end of the gracilis tendon with a #2 non-absorbable suture (Fig. 7.18).

We pull continuously and strongly on the tendon and release with our index finger 360° around the tendon for any interconnecting fascial bands attaching to the tendon (Fig. 7.19).

We use a closed tendon stripper to harvest both tendons. The gracilis tendon is harvested by flexing the knee to 90° and advancing the tendon stripper parallel to the tendon using a slow, steady, rotating motion. The semitendinosus tendon is harvested in a similar fashion. However, there are more extensive fascial connections that extend from the inferior border of the semitendinosus tendon to the medial head of the gastrocnemius (Fig. 7.20).

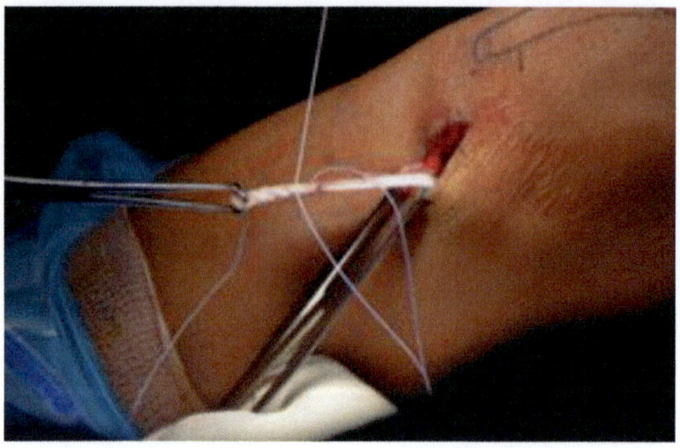

Fig. 7.18 Five throw sutures are placed in the free end of the gracilis tendon

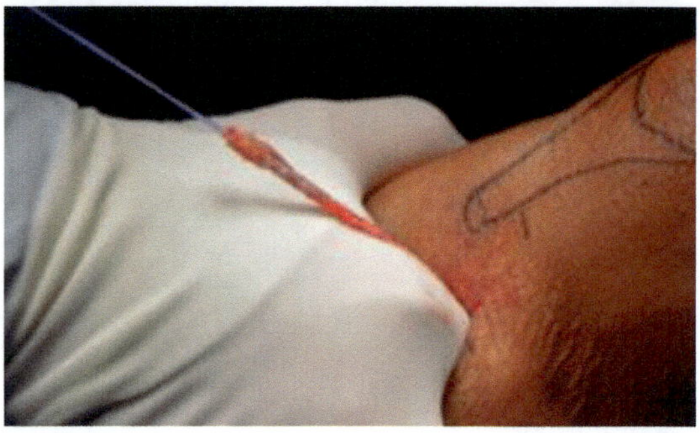

Fig. 7.19 Release with our index finger 360° around the tendon

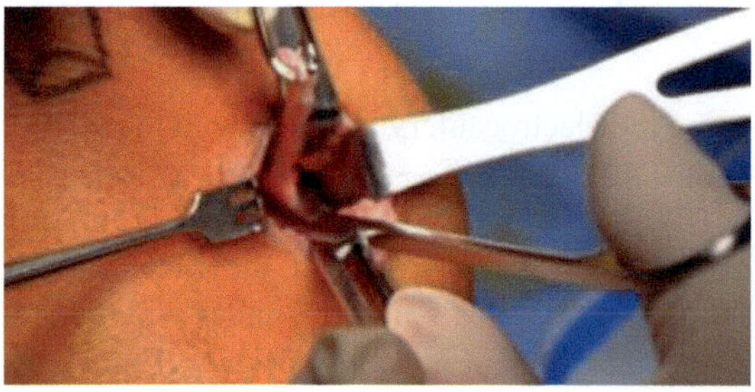

Fig. 7.20 Extensive fascial connections

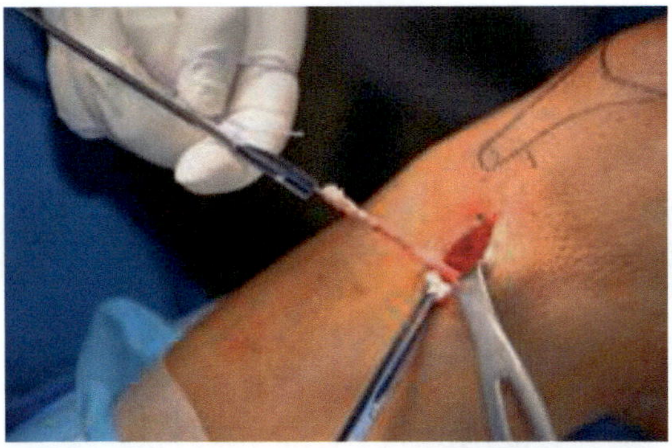

Fig. 7.21 Advancing the tendon stripper parallel to the tendon

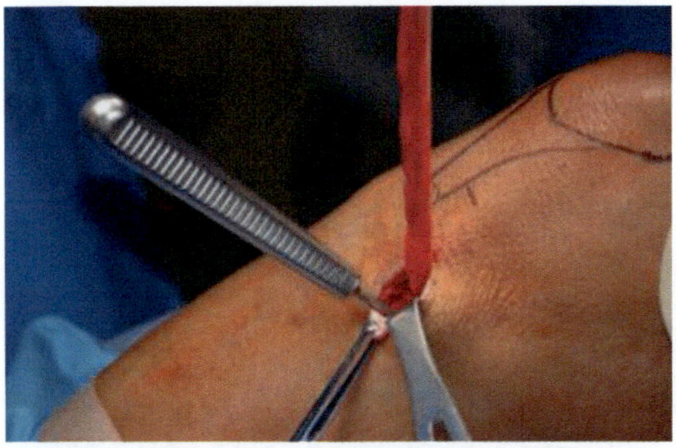

Fig. 7.22 Harvesting the tendon

These fascial connections must be released to prevent premature amputation of the semitendinosus tendon (Figs. 7.21 and 7.22).

More proximally in the thigh, the surgeon may encounter a second potential troublesome area at a band of thickened semimembranosus fascia that courses inferior and medial to the semimembranosus tendon [22]. Premature amputation of the semitendinosus tendon can occur if the tendon stripper passes outside of the tendon's normal path. If excessive resistance is encountered in the advancement of the tendon stripper, we should decrease the tension on the tendon and push the stripper harder using rotatory movements. This maneuver will often lead to success. A successful graft harvest typically results in graft lengths of 20–26 cm for the gracilis and 24–30 cm for the semitendinosus tendon.

7.5.3 Preparation of the Six Bundle Hamstring Tendon Graft

As soon as the gracilis tendon has been harvested, the assistant starts with the preparation of the tendon using the work station (Smith and Nephew endoscopy). During the preparation of the gracilis tendon, the surgeon continues with harvesting the semitendinosus tendon. The residual muscle fibers on the proximal end of both tendons should be removed using blunt dissection with a metal ruler, a large curette or one arm of a sharp scissors (Fig. 7.23).

One tendon is prepared at a time, and the proximal end of each tendon is tubularized with a continuous #2 non-absorbable suture. The sutures on each end of the tendon grafts are tensioned (Fig. 7.24).

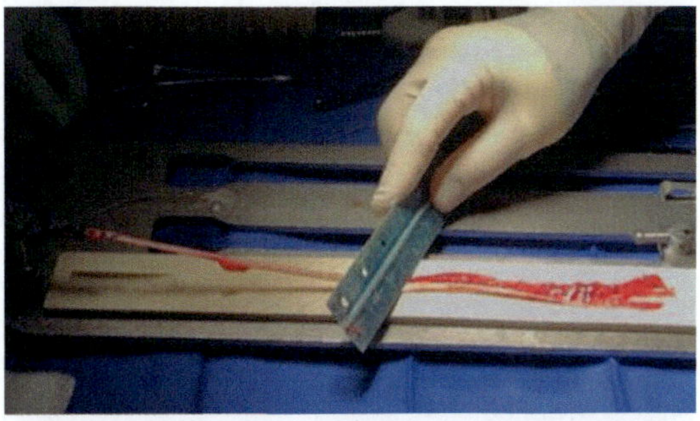

Fig. 7.23 The residual muscle fibers are removed

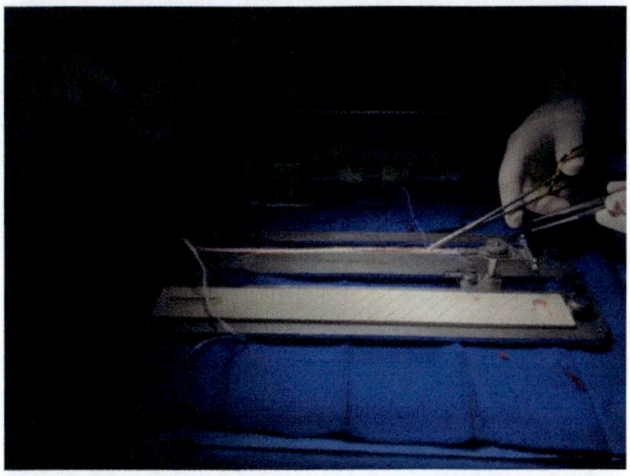

Fig. 7.24 The proximal end of each tendon is tubularized

The two tendon grafts are sutured with cottony tape at the end of each graft (Fig. 7.25).

We apply the EndoButton devise into the Smith and Nephew support, and we tie the whipstitches from the gracilis and the semitendinosus tendons to the EndoButton loop (Fig. 7.26, 7.27, and 7.28).

We pass the tip of each tendon through the EndoButton loop and then we pass the distal tip of the tendon inside the same tendon loop creating a triple strand graft, we tie the tip distally and we tension it (Figs. 7.29, 7.30, 7.31 and 7.32).

After finishing the gracilis tendon, we repeat the same process for the semitendinosus tendon (Figs. 7.33 and 7.34).

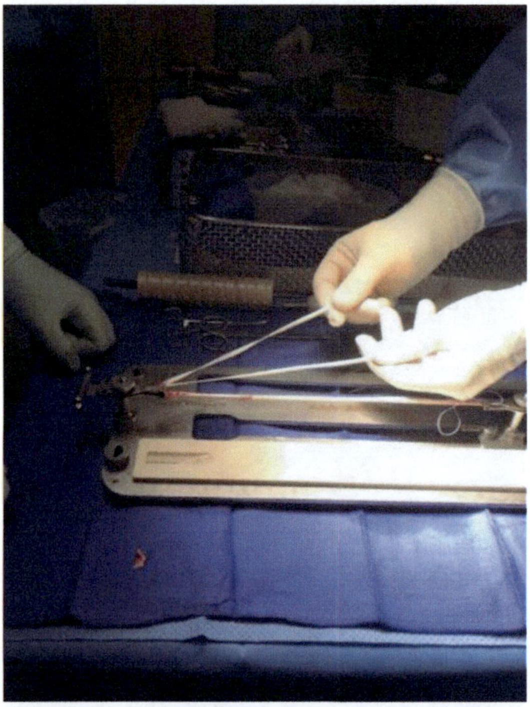

Fig. 7.25 White cottony tape

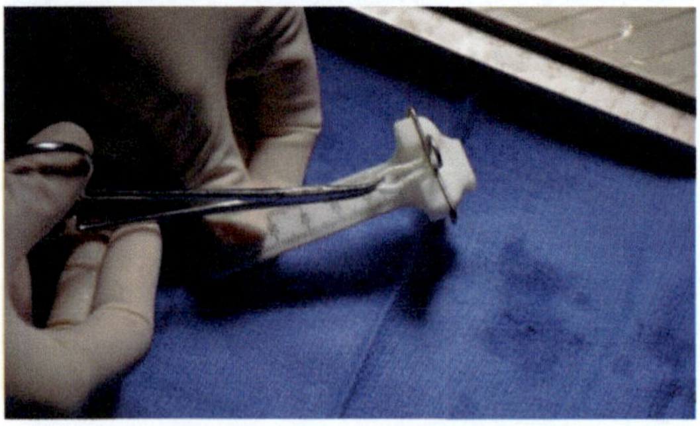

Fig. 7.26 EndoButton devise into the Smith and Nephew support

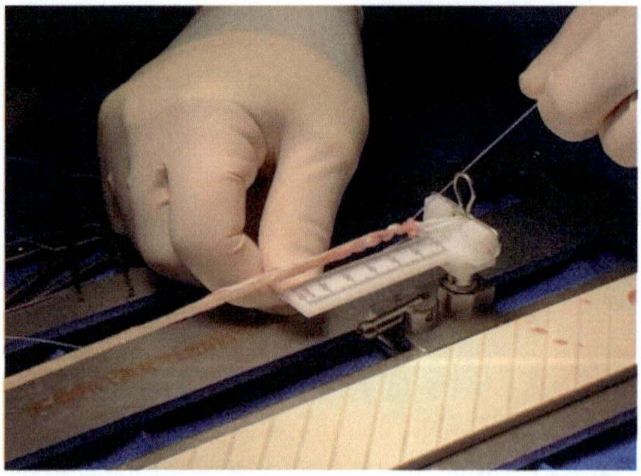

Fig. 7.27 Tying whipstitches from the gracilis and the semitendinosus tendons to the EndoButton loop

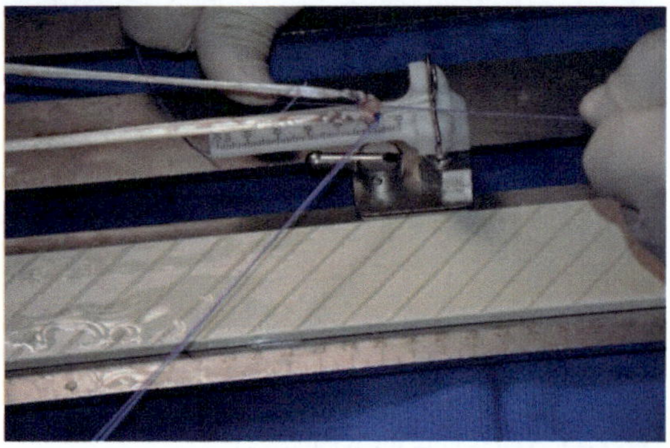

Fig. 7.28 Both tendons at EndoButton loop

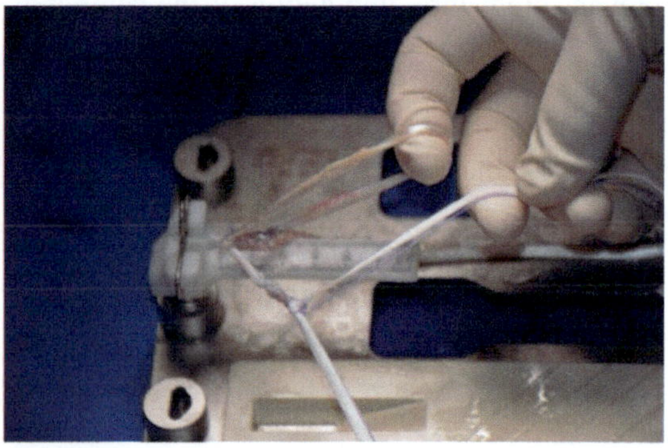

Fig. 7.29 Passing the tip of the tendon through the EndoButton loop

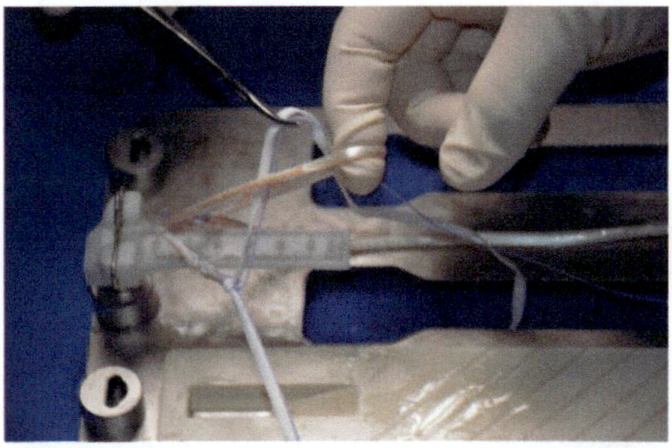

Fig. 7.30 Passing the gracilis tendon's tip through the tendon loop

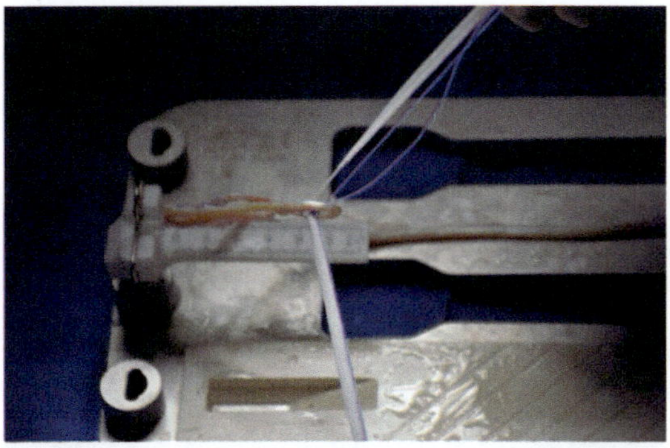

Fig. 7.31 Triple bundle gracilis tendon

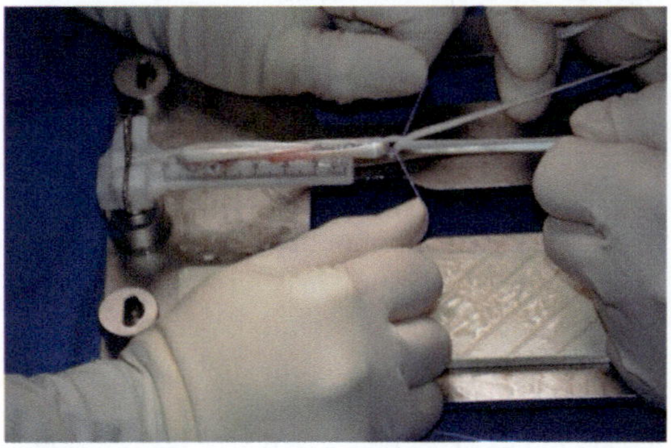

Fig. 7.32 Triple bundle gracilis tendon tied and tensioned

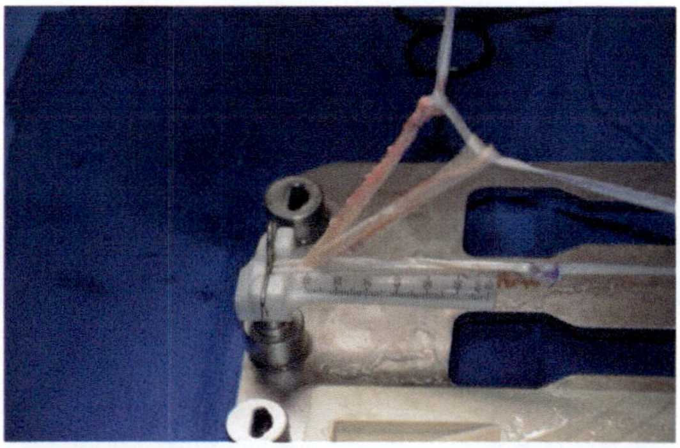

Fig. 7.33 Semitendinosus tendon above and gracilis tendon at bottom

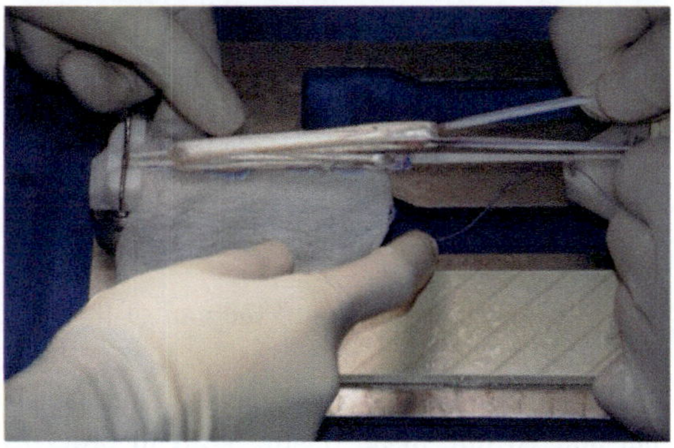

Fig. 7.34 Semitendinosus tendon above and gracilis tendon at bottom

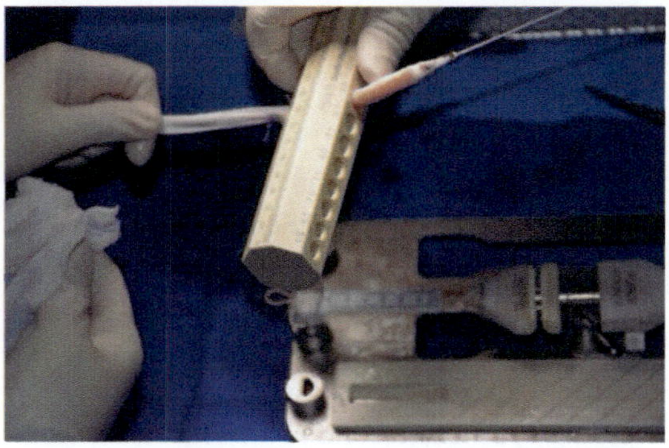

Fig. 7.35 The diameter of the TGST graft is measured

The diameter of the TGST graft is measured using Smith and Nephew measuring devise (a 0. 5-mm incremental sizing block or sizing tubes) (Fig. 7.35).

This facilitates the use of a graft-tensioning device later in the procedure. The TGST graft is covered with a wet pad containing antibiotic fluid. Graft is pre-tensioned on the work station

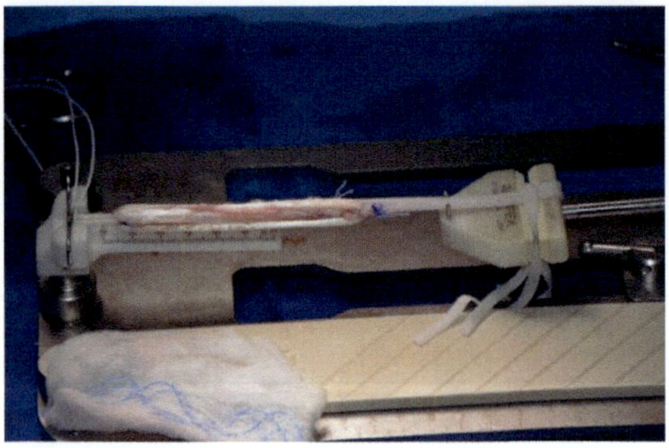

Fig. 7.36 Pretensioned on the graft

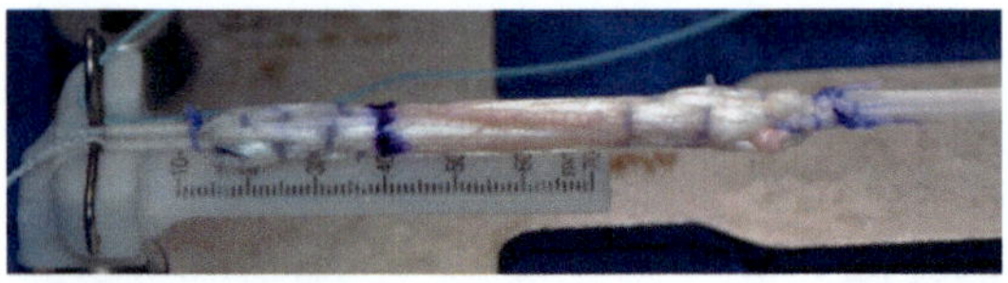

Fig. 7.37 We suture six tendon strands 30 mm proximal and 20 mm distal on the graft

applying 15–20 pounds until the end of the procedure (Fig. 7.36).

We suture the six tendon strands 30 mm proximal and 20 mm distal on the graft in order to obtain better fixation with the rigid fix pins in femoral tunnel and the interference screw in the tibial tunnel (Fig. 7.37).

7.5.4 Fibertape

There is a special situation when the hamstring six bundle graft size is smaller than 8 mm, so we add a fibertape just for protecting the graft as a seat belt in the case that the patient gets a twisted mechanism during the rehabilitation period. We do not tie the fibertape with too much stress, for avoiding over tension and avoid tear of the fibertape (Figs. 7.38 and 7.39).

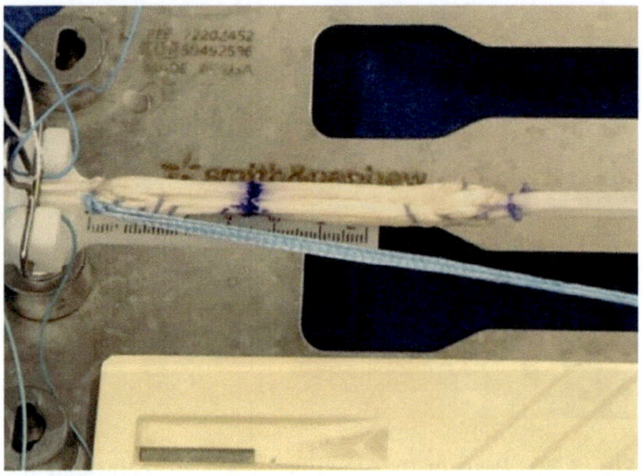

Fig. 7.38 Fibertape

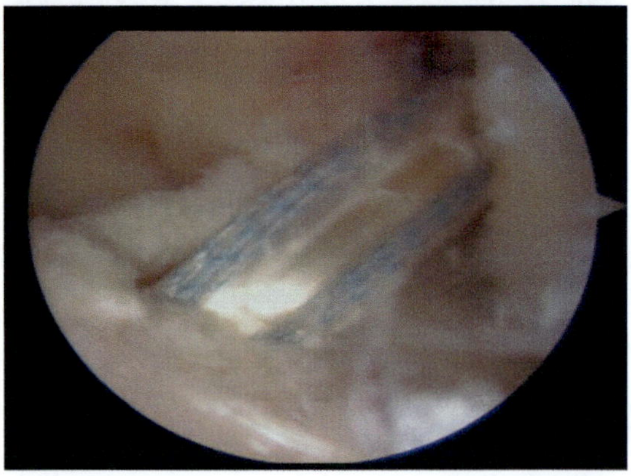

Fig. 7.39 Fibertape intraarticular

7.5.5 Arthroscopic Portal Placement

We use two portals for ACL reconstruction. A high anterolateral portal at the level of the inferior pole of the patella adjacent to the lateral border of the patellar tendon as the routine viewing portals. This portal provides an excellent view of the ACL tibial attachment site. This portal gives a frontal view of the femoral attachment site of the ACL and is more helpful in determining the clock orientation and the anatomic placement of the femoral tunnel. An anteromedial portal at the level of the inferior pole of the patella adjacent to the medial border of the patella tendon is used for instrumentation and viewing of the medial wall of the lateral femoral condyle. We usually extend distally the anteromedial portal with a scalpel for drilling the femoral tunnel but an accessory medial portal located directly inferior to the anteromedial portal at the level of the medial joint line is used for drilling of the femoral tunnel in case we do not have a good access to the correct site of femoral footprint. We change from anterolateral portal to anteromedial portal view to look for hidden injuries on the MRI for example; menisco-capsular ramp lesions, meniscal root tears, and osteochondral injuries.

7.5.6 Diagnostic Arthroscopy

Viewing the suprapatellar pouch, medial compartment, lateral compartment looking for associated meniscal and chondral injuries which must be treated before the start of the ACL reconstruction technique (Fig. 7.40). After that we proceed to the preparation of the intercondylar notch. Preparation of the intercondylar notch is necessary to allow visualization of the ACL femoral attachment site. The torn fibers of the ACL are removed from the lateral femoral condyle and the tibial attachment site by a motorized shaver, electrocautery pencil, or radiofrequency probe (Fig. 7.41). We have found that the use of a radiofrequency probe is faster, allows hemostasis to be achieved, and completely removes the soft tissue along the lateral wall of the intercondylar notch, providing better visualization of the bone anatomy. It is not really necessary to remove all the remaining fibers as they might have a biological role in revascularization. Use of the anteromedial portal technique allows the femoral tunnel to be positioned lower down to the sidewall of the lateral femoral condyle, resulting in a more horizontal orientation of the ACL graft. A more horizontal ACL graft avoids posterior cruciate ligament impingement and in most cases eliminates the need for notchplasty. However, a selective notchplasty may be required in the case of congenitally narrowed notches, more frequent in female or in chronic cases with notch stenosis due to the development of notch osteophytes (Fig. 7.42).

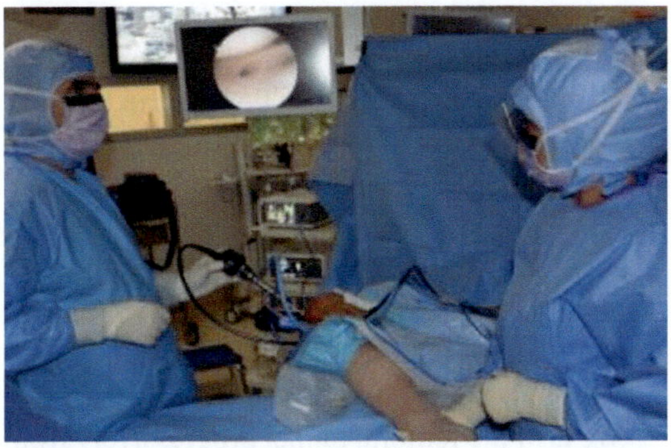

Fig. 7.40 Figure of four position to observe lateral compartment shown on screen

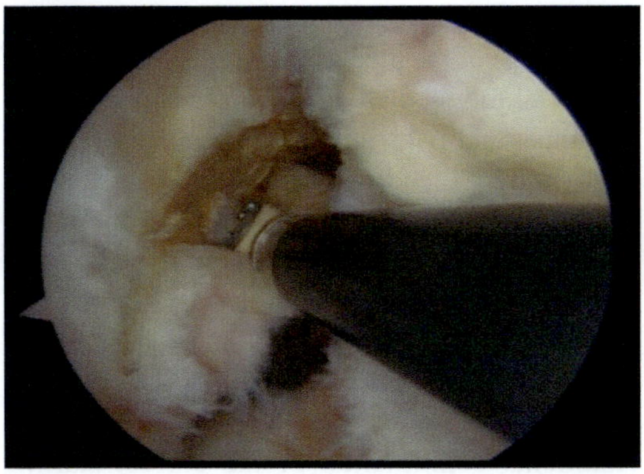

Fig. 7.41 Electrocautery pencil from anteromedial portal removing remaining ACL from lateral wall

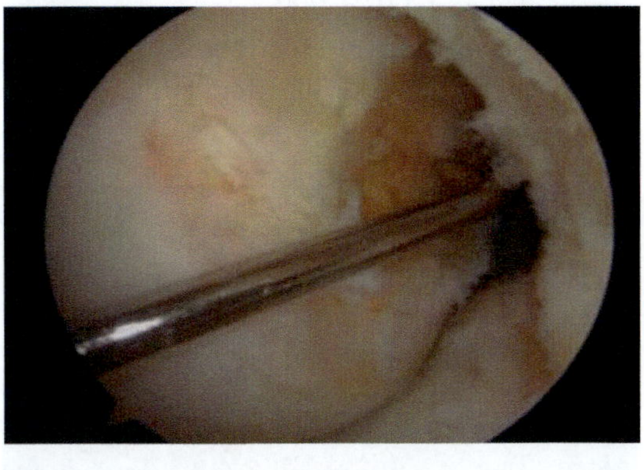

Fig. 7.42 Femoral wall after removal the ACL remnant

7.5.7 Femoral Tunnel

The best foot print position for the formal tunnel is situated between 10.00 and 11.00 clock hour in the right knee and between 1.00 and 2.00 clock hour in the left knee. The resulting longer femoral tunnel is more advantageous for femoral fixation with the EndoButton. The location for the accessory medial portal is made by an 18-gauge spinal needle. This portal is located as low as possible just above the medial joint line to avoid the damage of the medial meniscus. Placement of the portal too medially produces a shorter femoral tunnel and risks injury to the medial femoral condyle by the endoscopic drill bit during drilling of the femoral tunnel. Dilation of the portal with the blunt arthroscope obturator followed by the tips of the Metzenbaum scissors helps ease future passage of instrumentation (Fig. 7.43).

Fine tuning the awl's position is performed under arthroscopic guidance (Fig. 7.44). Additional confirmation of the correct starting point can be made by viewing the tip of the awl through the anteromedial portal. A 4- or 5-mm offset femoral aimer is passed through the accessory medial portal (Fig. 7.45).

The blade of the femoral offset aimer is placed at the center of the foot print, and the knee is slowly flexed to 120° (Fig. 7.46). A 2.7-mm drill-tipped guide pin is positioned at the site of the microfracture awl penetration mark. The 2.7-mm drill-tipped guidewire is drilled out through the soft tissues of the lateral thigh (Fig. 7.47). Inadequate knee flexion can result in the guide pin's coming to lie inferior to the intermuscular septum, placing the peroneal nerve at risk.

A 4.5-mm EndoButton drill bit (Smith and Nephew endoscopy) is used to drill the tunnel through the lateral femoral cortex (Figs. 7.48 and 7.49).

We drill the femoral tunnel progressively from 7.0 mm to the final size of the tunnel with 0.5 mm reamer guides (Fig. 7.50).

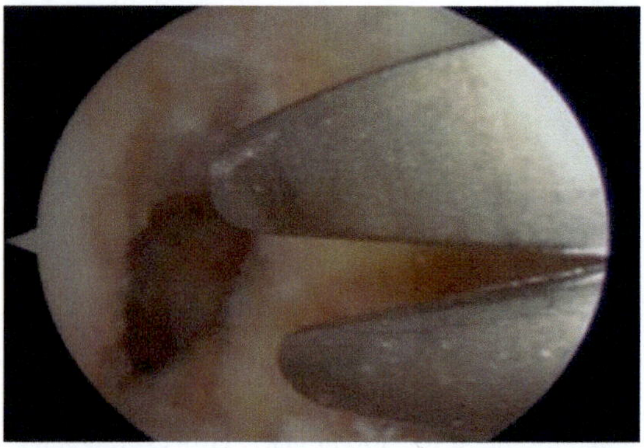

Fig. 7.43 Dilation of the anteromedial portal

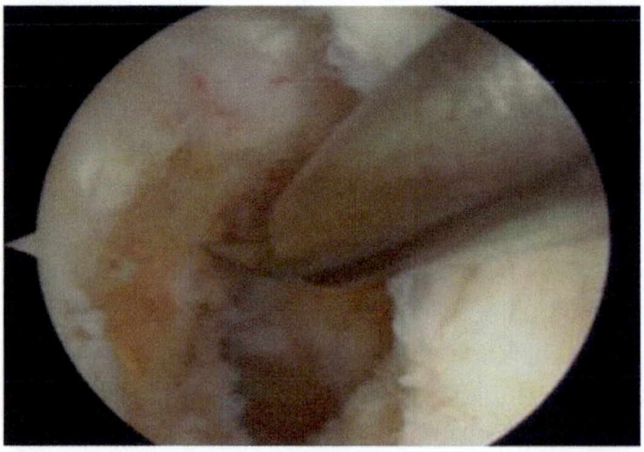

Fig. 7.44 A microfracture awl is passed through the accessory medial portal and used to mark the starting point for the femoral tunnel under arthroscopic guidance

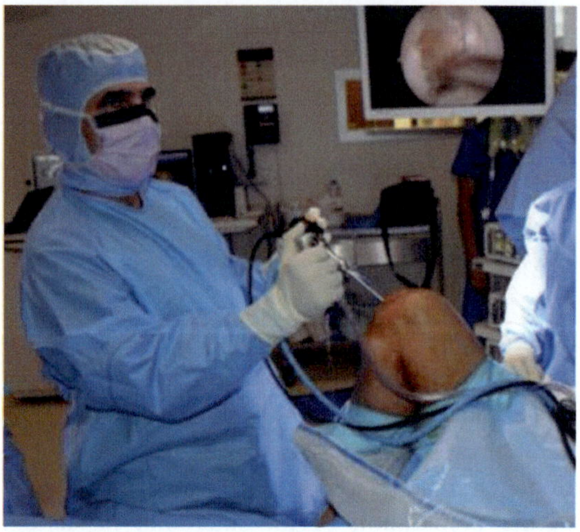

Fig. 7.45 Femoral aimer entrance and show up on screen

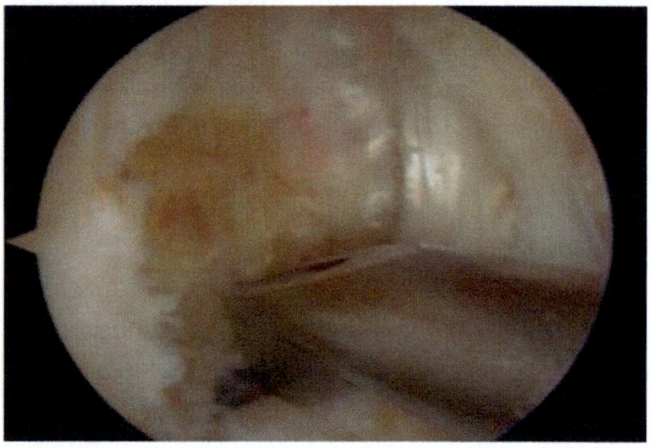

Fig. 7.46 Femoral aimer is positioned at the center of the foot print

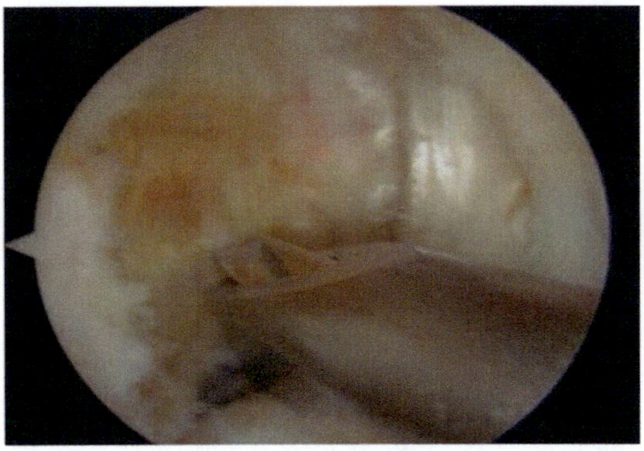

Fig. 7.47 The 2.7-mm drill-tipped guidewire is drilled out through the soft tissues of the lateral thigh

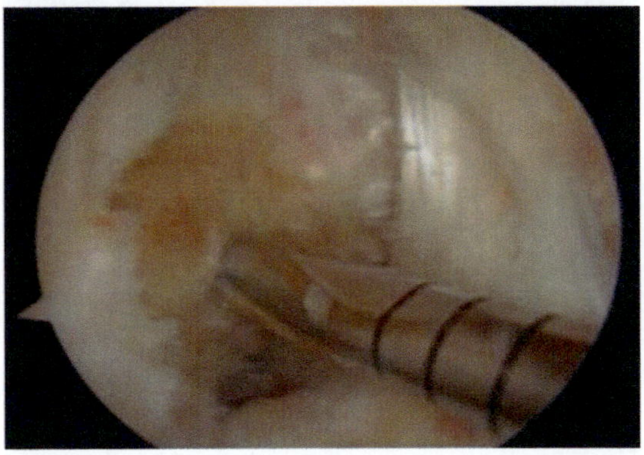

Fig. 7.48 A 4. 5-mm EndoButton drills the tunnel and the femoral cortex

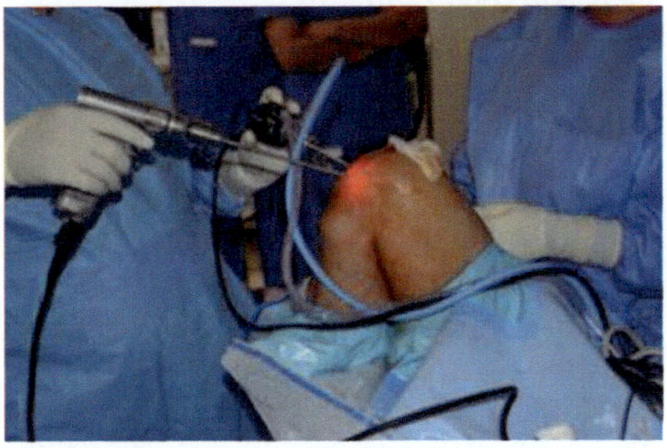

Fig. 7.49 A 4.5 reamer passing through the femoral cortex

The femoral socket depth must allow for the length of the TGST graft to be inserted into the femur (usually 25–30 mm) plus an extra 6 mm to allow the EndoButton to clear the lateral femoral

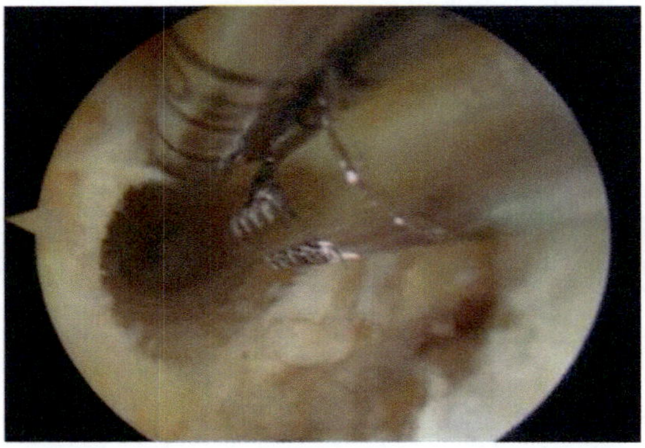

Fig. 7.50 We drill from 7.0 mm to the final size of the tunnel with 0.5 mm reamer guides

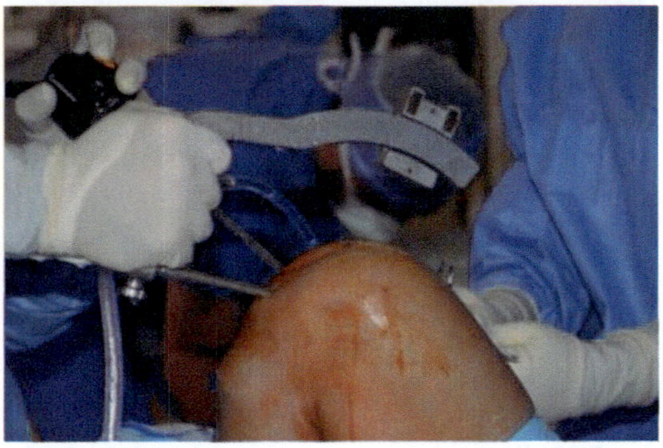

Fig. 7.51 Introduce femoral guide inside the femoral tunnel

cortex and to flip. An EndoButton depth gauge inserted through the accessory medial portal is used to measure the femoral tunnel length. The size of the tunnel should be the same than the TGST.

7.5.8 Femoral Rigidfix Curve Guide

After that, we introduce the Rigidfix U guide, and we perform the two femoral tunnels used for the cross pin Rigidfix from the medial side of the knee (Figs. 7.51, 7.52, and 7.53). We drill the tunnels, and we check from inside the tibial tunnel, the right position of the two femoral tunnels (Figs. 7.54 and 7.55). If we see fluid getting out from the cannulated pins, we know that the pins are in a right position.

A loop of #5 non-absorbable suture material is inserted into the eyelet of the passing pin, and

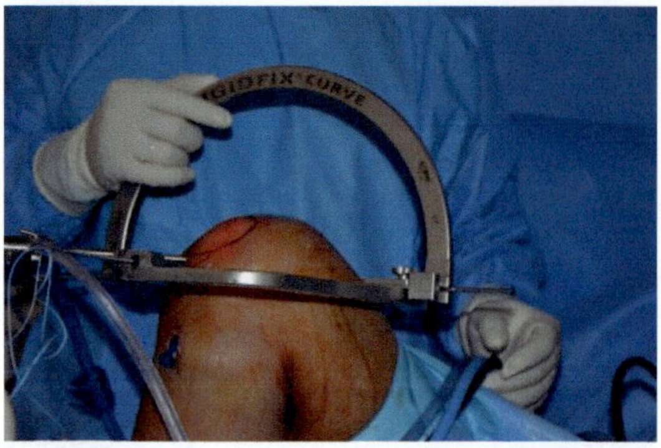

Fig. 7.52 We add the curve rigid fix to the femoral guide

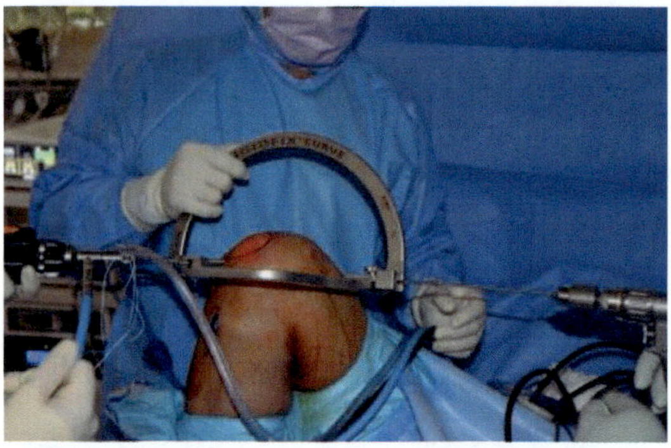

Fig. 7.53 We perform the two femoral tunnels for the cross pin device from the medial side of the knee

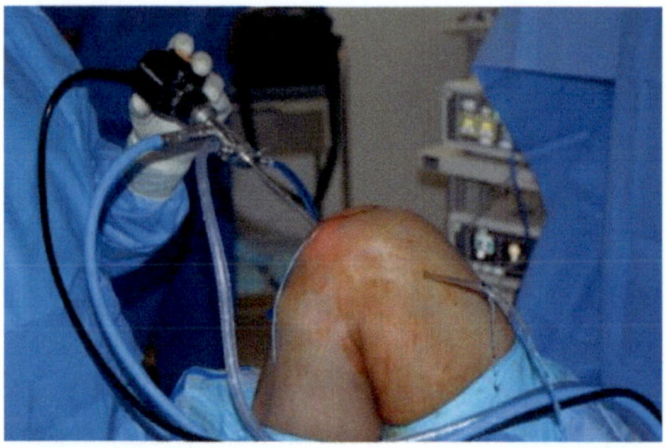

Fig. 7.54 We check by arthroscopy that the pins are inside the tunnel

the ends of the suture are passed out of the lateral thigh. The loop of suture is passed into the joint and positioned at the entrance of the femoral tunnel. This suture will be used later in the procedure to pass the hamstring graft

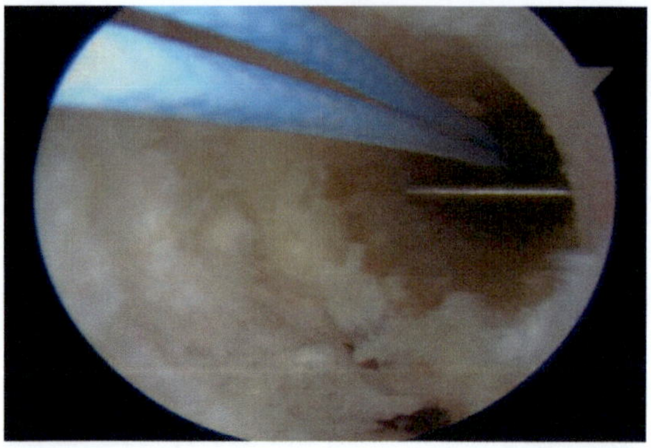

Fig. 7.55 Pins inside the tunnels

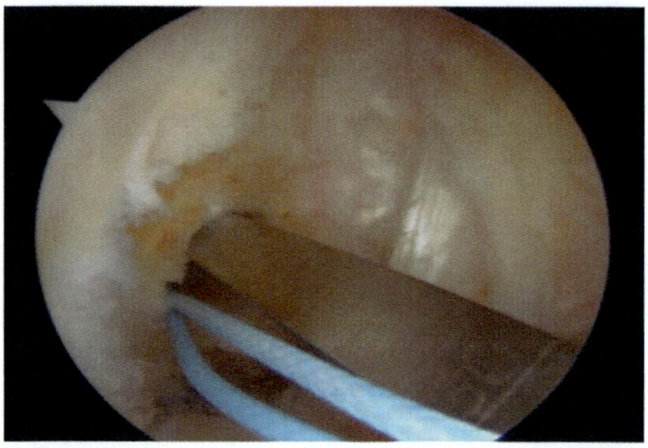

Fig. 7.56 We measure the femoral tunnel length

(Fig. 7.55). Then we measure the femoral tunnel length (Fig. 7.56).

7.5.9 Tibial Tunnel

The next step is the tibial tunnel drilling. We recommend a tibial tunnel length of 40–50 mm, with no risk that the screw will protrude into the intraarticular portion of the knee joint. Setting the adjustable tibial aimer between 50 and 55° will usually allow these tunnel lengths to be achieved (Fig. 7.57). The intraarticular position of the guide pin is situated between the anterior horn of the lateral meniscus, the medial and lateral tibial spines, and the posterior cruciate ligament (Fig. 7.58). When we adjust the tibial aimer in the right position, we place the guide pin through. The tibial tunnel is performed by drilling using

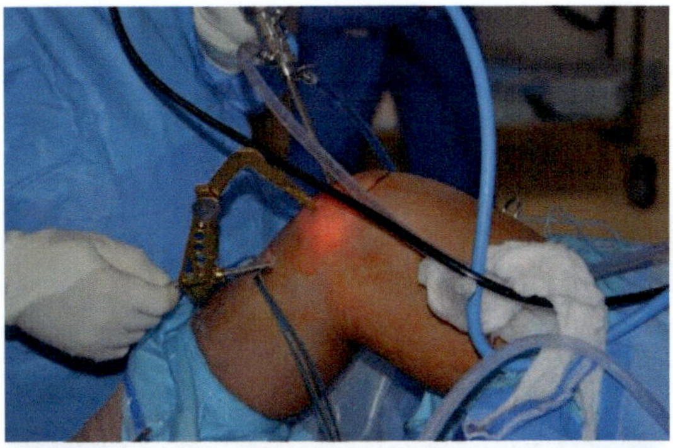

Fig. 7.57 Setting the adjustable tibial aimer

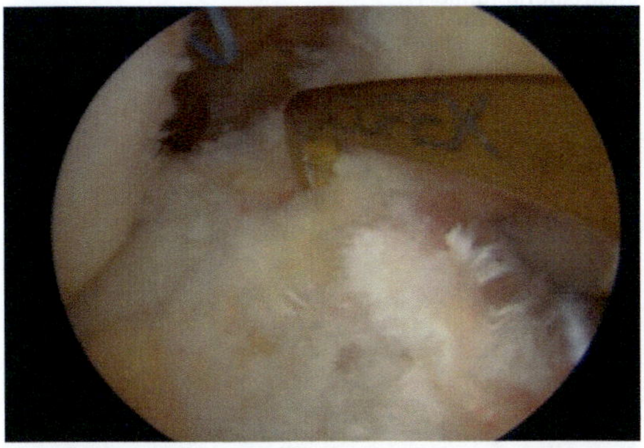

Fig. 7.58 Intraarticular position of the tibial aimer

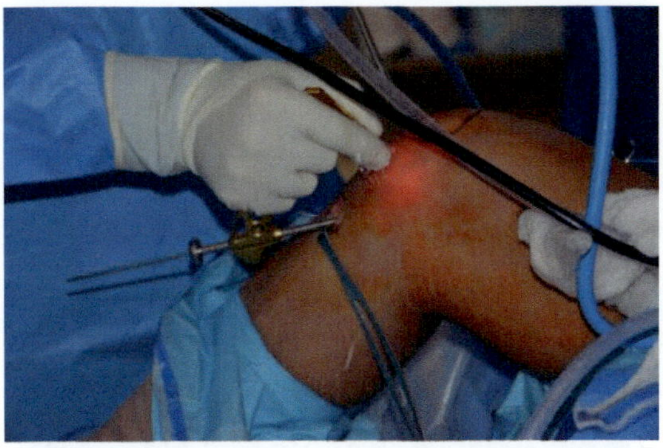

Fig. 7.59 We introduce a k-wire through the femoral aimer

5 mm, then 7.5 mm, 9.5 mm, 10 mm, and 11 mm (Figs. 7.59 and 7.60), endoscopic reamers depending on the same graft size. The articular edge of the femoral and tibial tunnel is smoothed with a rasp. Soft tissue around the external superior edge of the tibial tunnel is cleared with an electrocautery pencil and a Cobb periosteal elevator.

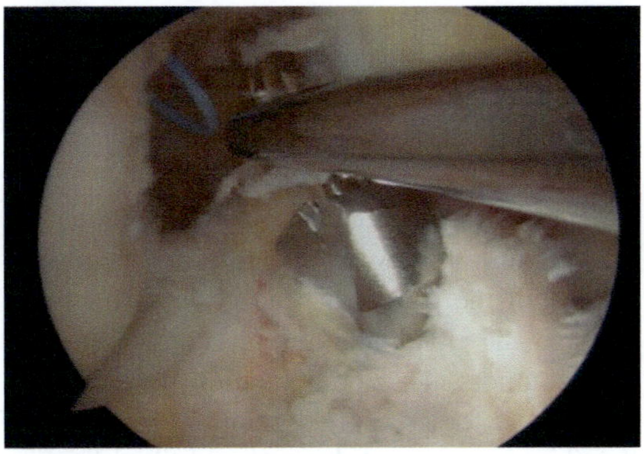

Fig. 7.60 Femoral reamer protecting the k-wire with a Kocher pincer

7.5.10 Graft Fixation

With the development of the new technology, many companies are working with enthusiasm to create new fixation devices for improving the rigidity of graft fixation and avoid slippage. Very long time is required for hamstring tendon grafts to heal to the bone, and for this reason, it is important to use graft fixations that are strong and stiff and that they resist slippage under cyclic loading in order to prevent the development of progressive laxity in the postoperative period. Attachment of rigid initial graft fixation prevents failure and minimizes elongation at the graft fixation sites during cyclic loading of the knee before healing at the graft sites has occurred. At the moment there are several fixation devises that we can use; however, in our hands, the optimal graft fixation method is the EndoButton devise plus Rigidfix cross pin in the femoral tunnel and interference screw at tibial tunnel.

There are many articles and chapters mentioning laboratory biomechanical studies that have demonstrated that the EndoButton CL and Rigidfix cross pins provide the strongest and has the fewer amount of slippage during cyclic loading. We can mention other advantages of using

EndoButton CL and Rigidfix cross pins as a femoral fixation method as tight fitting of the tendon in the bone tunnel, complete contact of the tendon against the tunnel wall, removal of the implant is not required in revision cases, and our patients do not present widening of the tunnels in the long-term follow-up. The lower bone mineral density of the proximal tibia is the main cause because the tibial fixation is controversial for many authors. The tibial fixation devices must resist shear forces applied parallel to the axis of the tibial bone tunnel. Intra-tunnel tibial fixation with interference screws seems to demonstrate high initial fixation strength and stiffness with minimal slippage under cyclic loading conditions. We prefer inter-tunnel tibial fixation with the interference of Euro bio absorbable screw, and sometimes, a double-tibial fixation adding a staple is probably safer.

7.5.11 Calculation of EndoButton CL Length and Graft Preparation

We will explain how to calculate the length on the EndoButton required and how to prepare the graft in order to introduce 30 mm inside the tunnel. Assuming the femoral tunnel length measures 48 mm, and 30 mm of TGST graft has been chosen to be inserted into the femoral tunnel, the required continuous loop length is calculated as follows: 48 mm – 30 mm = 18 mm. Because the continuous loop lengths come in 5 mm increments, a 15- or 20-mm loop comes closest to the calculated length. In general, we prefer to use the shortest possible continuous loop because this increases the stiffness of the femur-EndoButton CL-TGST graft complex. In the example mentioned before, we would choose a 15 mm length of loop. The TGST graft is pretensioned to 10 pounds on the graft preparation board. The graft is marked with a surgical marking pen at the measured femoral tunnel length (48 mm). A full-length #2 flipping suture and a #5 passing suture are passed through the end holes of the EndoButton. A second #5 suture can be inserted into the same hole as the #2 flipping suture and passed alongside the graft and out of the tibial tunnel (Figs. 7.61, 7.62 and 7.63).

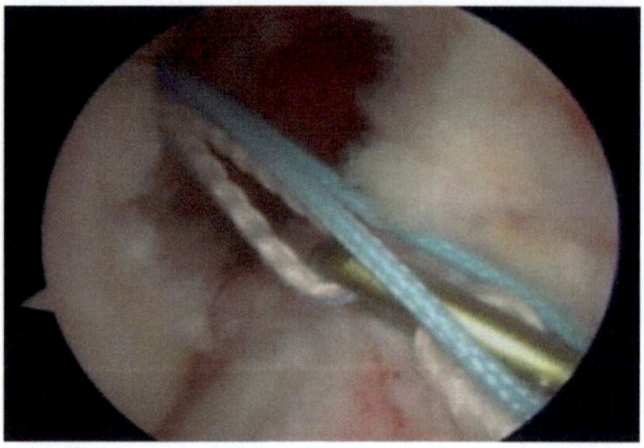

Fig. 7.61 EndoButton and graft passing through tibial and femoral tunnels

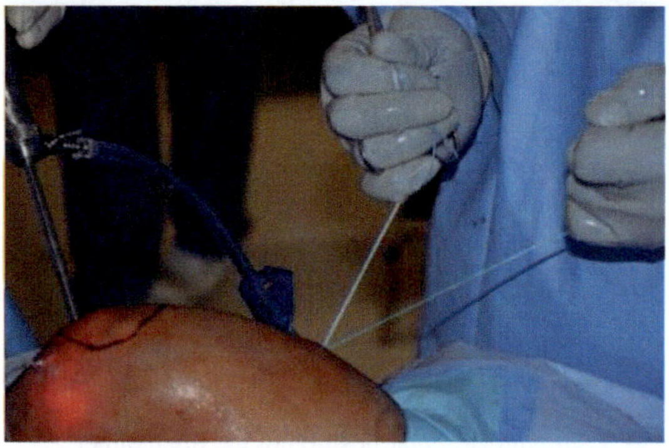

Fig. 7.62 The assistance flips the sutures

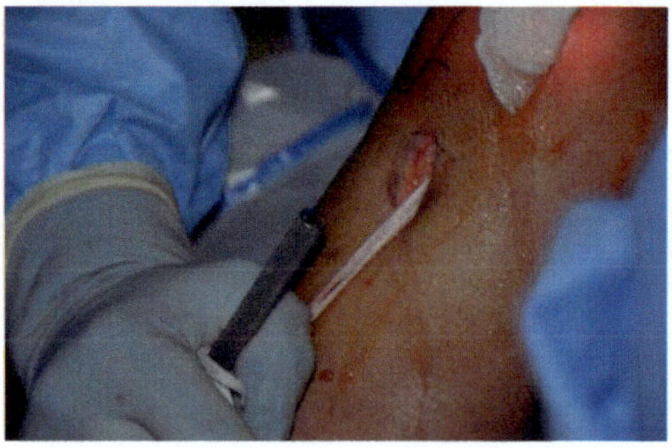

Fig. 7.63 Pulling the graft in order to be sure the femoral fixation with the EndoButton

7.5.12 Graft Passage and Femoral Fixation

The loop of #5 suture is retrieved from the femoral tunnel and pulled out of the tibial tunnel. The #2 flipping suture and #5 passing suture are passed through the loop of the #5 suture and pulled out the lateral thigh. Under arthroscopic visualization, the EndoButton and the attached hamstring tendon graft are passed across the joint and into the femoral socket using the #5 passing suture. The TGST graft must be advanced until the previously placed insertion mark is seen to pass up into the femoral socket a distance of a few millimeters. This extra distance allows the EndoButton to pass outside the lateral femoral cortex and to flip. The #2 flipping suture in a proximal direction, parallel to the femoral tunnel, and the EndoButton will be felt to flip against the lateral femoral cortex. Correct deployment can be verified by pulling on the #2 suture and feeling the EndoButton "teeter-totter" against the lateral femoral cortex. If any doubts exist about secure deployment of the EndoButton, fluoroscopy can be used to check the position of the EndoButton. Tension is applied to the hamstring tendon graft, and the previously placed mark at the insertion length will be seen to slide back down to the femoral tunnel. If the measurements are correct, this mark should lie at the entrance of the femoral tunnel. If it should become necessary to remove the graft, the #5 passing suture on the EndoButton can be pulled proximally, tipping the EndoButton away from the femoral cortex. The #5 safety suture that exited the tibial tunnel is pulled, tipping the opposite end of the EndoButton into the 4.5 mm tunnel. The EndoButton will then disengage from the femoral cortex, and the graft can be removed by applying tension to the whipstitches on the tibial end of the graft (Figs. 7.61, 7.62, 7.63 and 7.64). In this position, we proceed with the second fixation method, the Rigidfix cross pin. Keeping in tension the TGST graft from the distal tip of the tendon in 90° position knee, we introduce the two Rigidfix cross pins from the medial side of the femur (Fig. 7.65).

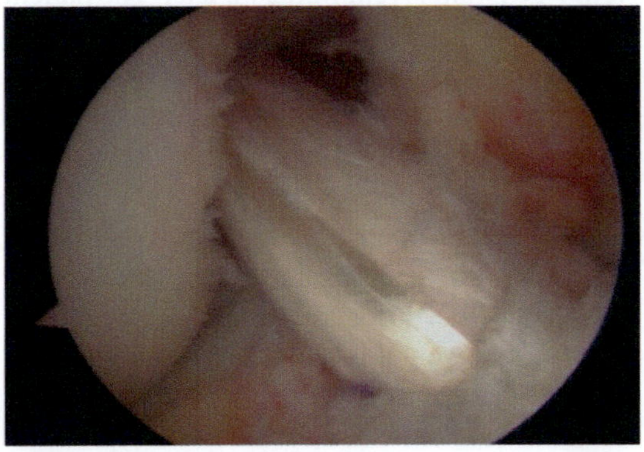

Fig. 7.64 We check tension of the graft before cross pin fixation

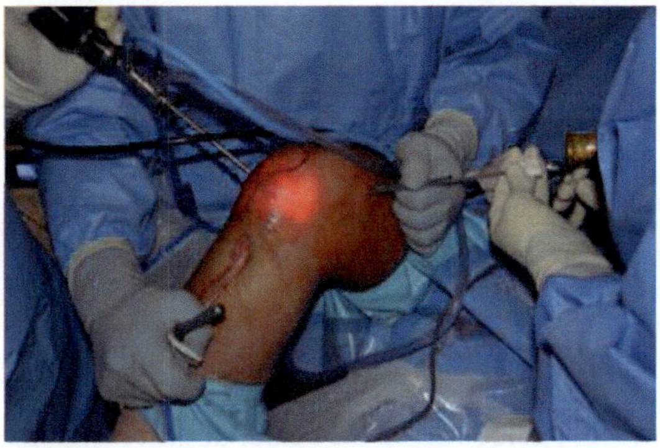

Fig. 7.65 We introduce the two Rigidfix cross pins from the medial side of the femur

7.5.13 Graft Tensioning

The opposite ends of the hamstring tendon graft are applied for tension manually at equal tension to each end of the six-stranded hamstring tendon graft allowing easier insertion of the interferential Biosure tibial screw. Application of equal tension to all six limbs of the hamstring tendon graft optimizes initial fixation strength and stiffness. The knee is cycled from 0 to 90° for a minimum of 30 cycles. Application of a preload and cycling of the knee are important steps as they allow the EndoButton CL to settle on the femoral cortex and remove creep from the polyester continuous loop, the tendon whipstitches, and the hamstring graft. At present, the optimal graft tension and knee flexion angle during tibial fixation are unknown. We tend to fix the graft with the knee positioned between 0 and 20° of flexion. The usual graft excursion pattern detected with our bone tunnel placements results in pulling the TGST graft into the tunnel (tightening) during the last 20° of terminal extension. When is a minimal graft excursion detected, we tend to fix the graft with the knee at 20° of flexion and near full flexion with greater excursions. A high graft tension force in combination with the knee flexed more than 20° may result in a permanent flexion contracture.

7.5.14 Tibial Fixation

The bioabsorbable interference screw is our fixation method of choice at the tibia. The central axis of the tibial tunnel is identified by passing a 1.1 mm guidewire up the center of the tensioning device and down the side of the six graft strands into the knee joint. An IntraFix tapered screw of 1 mm larger than the tibial tunnel diameter is chosen. For example, we use a 9 mm tapered screw for an 8 mm tibial tunnel (Fig. 7.66). Given the typical size of most TGST grafts, the 7–9 mm tapered screw is inserted into the sheath until the superior aspect of the screw head is flushed with or buried just below the tibial cortex. The best bone quality is at or next to the tibial cortex, and overly deep insertion of the screw may decrease fixation strength. Protruding or prominent areas of the polyethylene sheath are trimmed flush with the tibial cortex with a #15 blade and a small bone Rongeur. The fixation strength of any intratunnel tibial fixation device depends on the local bone mineral density. If the surgeon thinks that there was inadequate torque during the insertion of the tapered screw and the patient has soft bone, we recommend that supplemental tibial fixation be used. The stability and range of motion of the knee are checked. It is important to verify that the patient has full range of motion before leaving

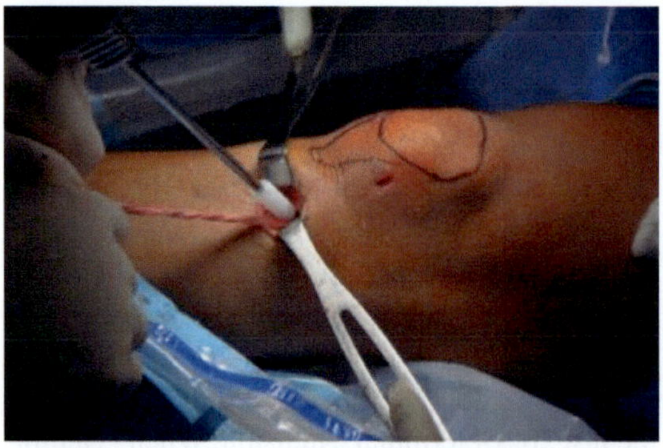

Fig. 7.66 The bioabsorbable interferential screw is our choice at tibial fixation

the operating room. The arthroscope is inserted to the knee, and graft tension and impingement are assessed. Our usual graft placement and tensioning technique result in the four strands of the TGST graft being maximally tight between 0 and 20°, with the graft tension decreasing slightly as the knee is flexed to 90°. After confirmation that the patient has a full range of motion and negative Lachman and Pivot shift test results, the passing and flipping sutures are pulled out of the lateral thigh.

7.5.15 Closure

A closed suction drain is inserted for 24 h under the sartorius fascia up into the hamstring harvest site and is helpful in preventing postoperative hematoma formation and decreasing ecchymosis along the medial side of the knee. The sartorius fascia that was preserved during the graft harvest is repaired back to the tibia with a 0 absorbable suture. The subcutaneous tissue is closed in layers with fine absorbable sutures. A running 3-0 Prolene subcuticular pullout suture produces a cosmetic suture. A second solution of 5 mg of morphine sulfate plus 20 mL of 0.25% bupivacaine with 1:100,000 epinephrine is injected into the suprapatellar pouch, and a 30 mg bolus of ketorolac is given for postoperative pain control. The continuous intravenous ketorolac infusion is continued until the patient is discharged from the day-surgery unit. A light dressing is applied over the wound, followed by a thigh-length TED antiembolism stocking, and knee immobilize. The Hemovac Drain is removed 24 h later. The patient is discharged from the day-surgery 24–48 h after surgery depending on general condition. Pain management with our protocol for the prevention of thrombosis is as follows: Clexane 40 mg daily for 3 weeks, antibiotics ciprofluoxacin 500 mg for 5 days, and pain management medication. Depending on the meniscus and osteochondral injuries repaired during the procedure, we recommend partial weight bearing, brace, and crutches and 0–90° flexion for 4 weeks in isolated ACL reconstruction or in ACL reconstruction plus partial or total meniscectomy, and avoiding weight

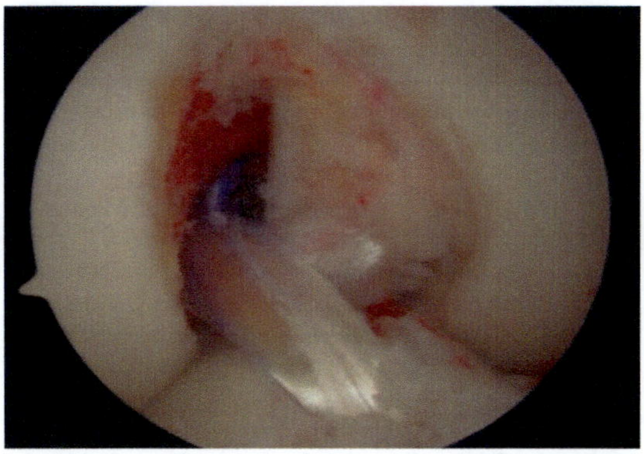

Fig. 7.67 Final assessment of the ACL graft

bearing, brace, and crutches and 0–90° flexion for 4 weeks in ACL reconstruction plus meniscal repair and/or osteochondral procedures. Final assessment of the ACL graft (Fig. 7.67).

7.5.16 Postoperative Management

7.5.16.1 Follow-Up

The patient is seen at 7–10 days for suture removal and postoperative radiographs. We follow a structured rehabilitation program.

7.5.16.2 Complications

The risks of complications such as infection, deep venous thrombosis, and loss of motion are the same as for ACL reconstructions performed with other graft sources [23]. However, we are unaware of reports of extensor mechanism rupture or patellar fracture after ACL reconstruction performed with hamstring tendon grafts. Complications unique to hamstring tendon grafts include premature amputation of the hamstring tendons [24], saphenous nerve injury [25], bleeding at the hamstring tendon harvest site, and hamstring muscle "pulls". The risk for premature amputation of the tendons can be minimized by following the recommendations outlined in the section on graft harvest. If the gracilis tendon is amputated and the semitendinosus is successfully harvested, it is possible in most cases to either triple or quadruple the semitendinosus tendon, depending on its length. In these situations, the EndoButton CL can still be used for femoral fixation; however, because of the shorter length of the graft construct, alternative tibial fixation is obtained by tying the EndoButton tape around a fixation post or an extra small non-barbed staple. If necessary, the tibial fixation can be augmented with a 25–30 mm bioabsorbable screw with a diameter 1 mm greater than that of the tibial tunnel. If the semitendinosus tendon is amputated, it will be necessary to use an alternative autograft, such as the patellar tendon or quadriceps tendon, or allograft tissue if preoperative consent has been obtained. The possibility of premature amputation of the tendons should be discussed during the informed consent process, and the patient and surgeon should agree on a course of action should this complication occur.

References

1. Blache Y, Dumas R, de Guise J, Saithna A, Sonnery-Cottet B, Thaunat M. Technical considerations in lateral extra-articular reconstruction coupled with anterior cruciate ligament reconstruction: a simulation study evaluating the influence of surgical parameters on control of knee stability. Clin Biomech (Bristol, Avon). 2018;61:136–43.
2. Lubowitz JH, Ahmad CS, Anderson K. All-inside anterior cruciate ligament graft-link technique: second-generation, no-incision anterior cruciate ligament reconstruction. Arthroscopy. 2011;27(5):717–27.
3. Desai N, Björnsson H, Musahl V, Bhandari M, Petzold M, Fu FH, Samuelsson K. Anatomic single- versus double-bundle ACL reconstruction: a meta-analysis. Knee Surg Sports Traumatol Arthrosc. 2014;22(5):1009–23.
4. Cerciello S, Batailler C, Darwich N, Neyret P. Extra-articular tenodesis in combination with anterior cruciate ligament reconstruction: an overview. Clin Sports Med. 2018;37(1):87–100.
5. Westermann RW, Duchman KR, Amendola A, Glass N, Wolf BR. All-inside versus inside-out meniscal repair with concurrentanterior cruciate ligament reconstruction: a meta-regression analysis. Am J Sports Med. 2017;45(3):719–24.
6. Lubowitz JH. All-inside anterior cruciate ligament graft link: graft preparation technique. Arthrosc Tech. 2012;1(2):e165–8.
7. Abdelkafy A. Cortical femoral suspensory fixation using screw post in anatomic single-bundle anterior cruciate ligament reconstruction: a prospective study and mid-term outcome results. Int Orthop. 2016;40(8):1741–6.
8. Abdelkafy A. Anatomic single-bundle anterior cruciate ligament reconstruction using the outside-in femo-

ral tunnel drilling technique: a prospective study and short- to mid-term results. Arch Orthop Trauma Surg. 2015;135(3):383–92.
9. Imam MA, Abdelkafy A, Dinah F, Adhikari A. Does bone debris in anterior cruciate ligament reconstruction really matter? A cohort study of a protocol for bone debris debridement. SICOT J. 2015;1:4.
10. Abdelkafy A. Protection of the medial femoral condyle articular cartilage during drilling of the femoral tunnel through the accessory medial portal in anatomic anterior cruciate ligament reconstruction. Arthrosc Tech. 2012;1(2):e149–54.
11. Harris JD, Abrams GD, Bach BR, Williams D, Heidloff D, Bush-Joseph CA, Verma NN, Forsythe B, Cole BJ. Return to sport after ACL reconstruction. Orthopedics. 2014;37(2):e103–8.
12. Yabroudi MA, Irrgang JJ. Rehabilitation and return to play after anatomic anterior cruciate ligament reconstruction. Clin Sports Med. 2013;32(1):165–75.
13. Petersen W, Fink C, Kopf S. Return to sports after ACL reconstruction: a paradigm shift from time to function. Knee Surg Sports Traumatol Arthrosc. 2017;25(5):1353–5.
14. Bastian JD, Tomagra S, Schuster AJ, Werlen S, Jakob RP, Zumstein MA. ACL reconstruction with physiological graft tension by intraoperative adjustment of the anteroposterior translation to the uninjured contralateral knee. Knee Surg Sports Traumatol Arthrosc. 2014;22(5):1055–60.
15. Weimann A, Zantop T, Herbort M, Strobel M, Petersen W. Initial fixation strength of a hybrid technique for femoral ACLgraft fixation. Knee Surg Sports Traumatol Arthrosc. 2006;14(11):1122–9.
16. Branch T, Lavoie F, Guier C, Branch E, Lording T, Stinton S, Neyret P. Single-bundle ACL reconstruction with and without extra-articular reconstruction: evaluation with robotic lower leg rotation testing and patient satisfaction scores. Knee Surg Sports Traumatol Arthrosc. 2015;23(10):2882–91.
17. Kartus J, Movin T, Karlsson J. Donor-site morbidity and anterior knee problems after anterior cruciate ligament reconstruction using autografts. Arthroscopy. 2001;17(9):971–80. Review
18. Kaeding CC, Léger-St-Jean B, Magnussen RA. Epidemiology and Diagnosis of Anterior Cruciate Ligament Injuries. Clin Sports Med. 2017;36(1):1–8.
19. Courvoisier A, Grimaldi M, Plaweski S. Good surgical outcome of transphyseal ACL reconstruction inskeletally immature patients using four-strand hamstring graft. Knee Surg Sports Traumatol Arthrosc. 2011;19(4):588–91.
20. Levy M, Prud'homme J. Anatomic variations of the pes anserinus: a cadaver study. Orthopedics. 1993;16:601–6.
21. Solomon CG, Pagani MJ. Hamstring tendon harvesting: reviewing anatomic relationships and avoiding pitfalls. Orthop Clin North Am. 2003;34:1–8.
22. Brown CH, Sklar JH, Darwich N. Endoscopic anterior cruciate ligament reconstruction using autogenous doubled gracilis and semitendinosus tendons. Tech knee SZurg. 2004;3:215–37.
23. Nadarajah V, Roach R, Ganta A, Alaia MJ, Shah MR. Primary anterior cruciate ligament reconstruction: perioperative considerations and complications. Phys Sportsmed. 2017;45(2):165–77.
24. Yasin MN, Charalambous CP, Mills SP, Phaltankar PM. Accessory bands of the hamstring tendons: a clinical anatomical study. Clin Anat. 2010;23(7):862–5.
25. Ruffilli A, De Fine M, Traina F, Pilla F, Fenga D, Faldini C. Saphenous nerve injury during hamstring tendons harvest: does the incision matter? A systematic review. Knee Surg Sports Traumatol Arthrosc. 2017;25(10):3140–5.

Arthroscopic Revision of Anterior Cruciate Ligament Reconstruction

8

Mustafa Akkaya

8.1 Introduction

There is currently an increasing incidence of anterior cruciate ligament (ACL) rupture associated with changes in lifestyle and increasing sports activities. Related to this, together with the increase in primary ACL surgeries, it has been observed that there is an increase in the number of revision operations. When this is looked at from a socioeconomic perspective, there is a need for more detailed studies of the pathology and treatment forms of ACL surgery. Although there are several surgical techniques and fixation options, it still seems to be difficult to achieve normal knee kinematics. Following primary ACL reconstruction, rates of revision because of graft failure have been reported in literature, varying from 3% to 25% [1–3].

Although it has been defined as a difficult orthopaedic procedure, knee surgery has now become routine [4–8]. However, for successful revision surgery to be able to be applied, it is first necessary to investigate the underlying cause, so that vision surgery does not have to be repeated [9, 10]. If careful surgery cannot be applied within an appropriate plan, there will be continuing laxity, an increased risk of graft failure and the formation of intra-articular meniscus and cartilage damage [8, 11].

Just as in primary ACL reconstruction, the aim of revision ACL reconstruction is to provide stabilisation of the knee joint, prevent injury to the joint cartilage and meniscus and obtain recovery of knee functions. In literature, it has been reported that the patient outcomes after revision ACL are lower compared to primary surgeries. Therefore, preoperative planning must be made appropriately, and patients must be given detailed information before the operation. In this way, an increase in success rates could be achieved [12–14].

8.2 Failure Analysis

As there are different techniques, grafts and implants in revision surgery, and patients have different accompanying pathologies, there is no evident homogeneity. Detailed physical examinations should be made of patients, and radiological images should be examined carefully. Following primary ACL surgery, re-ruptures may be seen secondary to trauma independently of the knee stability. The underlying cause can be detected after a detailed examination [15–19]. These are primarily:

I. New trauma.
II. Errors in technique (tunnel malpositioning).

M. Akkaya (✉)
Department of Orthopaedics and Traumatology, Faculty of Medicine, Ankara Yildirim Beyazit University, Ankara, Turkey

M. Bozkurt, H. İ. Açar (eds.), *Clinical Anatomy of the Knee*,
https://doi.org/10.1007/978-3-030-57578-6_8

III. Ligament pathologies.
IV. Graft selection (synthetic, allograft).
V. Biological problems (graft failure).
VI. Insufficient rehabilitation.

8.2.1 History

Careful patient evaluation and taking the history are very important for the correct planning and treatment of previously unsuccessful ACL reconstruction. The most important step in revision surgery is most probably preoperative planning. This stage is extremely important to prevent the same mistakes being made as in the primary ACL reconstruction. Detailed information should be obtained from the patient about previous operations, and the records of those should be carefully examined.

8.2.2 Clinical Symptoms

The subjective complaints of unsuccessful ACL reconstruction include instability, pain, swelling, the feeling of a gap, locking, stiffness and excessive laxity. It is important to differentiate between pain and symptoms of instability.

8.2.3 Physical Examination

In the preoperative period, a detailed physical examination must be made of the patient, and just as for primary ACL reconstruction, tests specific to all the intra-articular pathologies must be applied. Intra-articular effusion should be evaluated, the joint range of movement (ROM) should be noted, and tests of meniscus (McMurray) and knee stability (Lachman and pivot-shift) should be applied. Examination should also be made in respect of other ligament injuries that could be present [20].

8.2.4 Radiological Evaluation

Systematic radiological evaluations of the patient and archiving are very important in respect of showing the necessity for surgery and in the post-operative follow-up of healing.

The tests to be requested for evaluation are primarily:

- Standing anterior-posterior and lateral knee radiographs.
- Posterior-anterior Rosenberg radiograph taken weight-bearing in 45° flexion.
- Stress radiographs.
- Computed tomography (CT).
- Magnetic resonance imaging (MRI).

With direct radiographs, images can be obtained of the tunnel positions and the implants used in the previous surgery (Fig. 8.1). Findings of expansion in the tunnels between sclerotic edges can be determined with careful evaluation.

Varus and valgus stress radiographs can be used for the evaluation of potential damage and injuries in the medial and posterolateral corners. This avoids mistakes being made in the preoperative planning.

Evaluation with CT provides more detailed information than standard radiographs. In particular, images can be obtained of tunnel positions and expansion in the bone tunnels. 3D reconstruction can be added, and thus, potential sites for the new tunnels can be determined (Fig. 8.2).

MRI contributes in particular to the identification of other intra-articular pathologies. Problems in the cartilage, meniscus and surroundings of tissue can be seen in detail (Fig. 8.3).

8.2.5 Concomitant Pathologies

Injuries in collateral ligaments and the posterior cruciate ligament (PCL) that are not treated can cause increased stress and load distribution in the ACL graft after reconstruction. In addition, instability in the posterolateral corner (PLC) must not be overlooked, and as this is seen in 10%–15% of patients with chronic ACL damage, evaluation must be made carefully (Fig. 8.4) [21].

Treating intra-articular meniscus injuries during revision surgery requires consideration of intra-articular pathologies as a whole. With preoperative planning, it is recommended that treat-

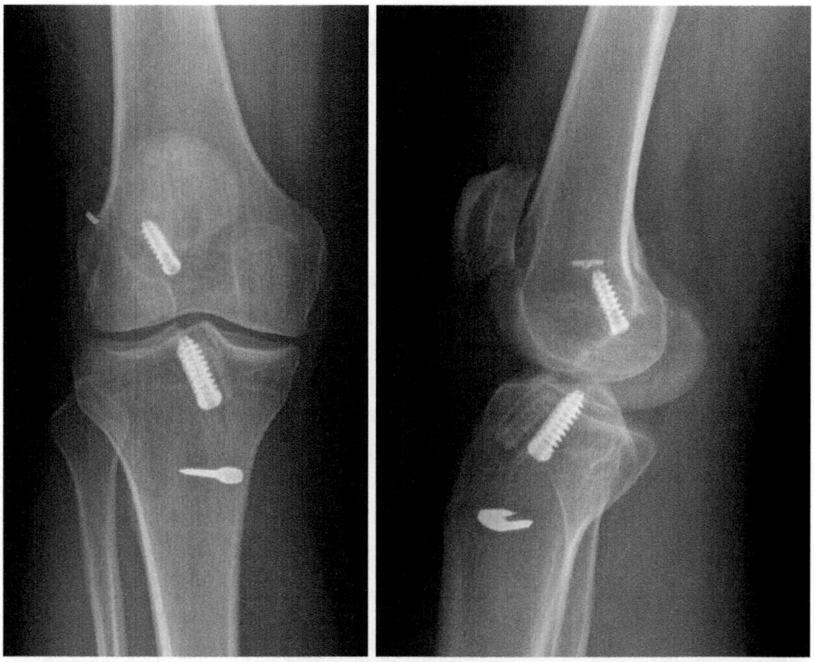

Fig. 8.1 Standing anterior-posterior and lateral knee radiographs

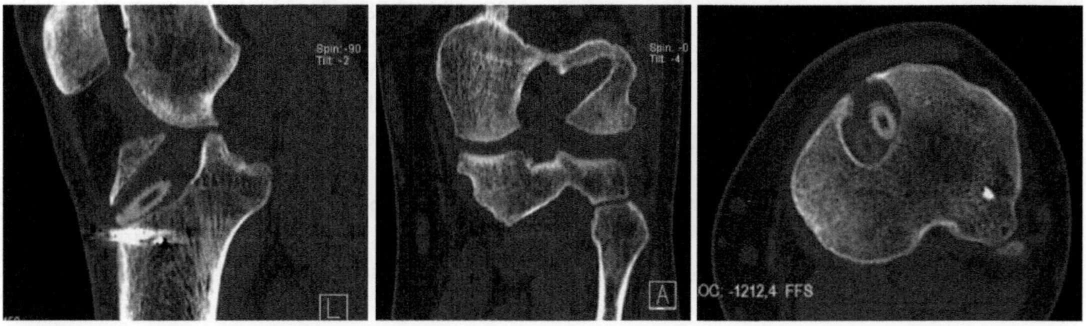

Fig. 8.2 Tunnel placement evaluation with CT

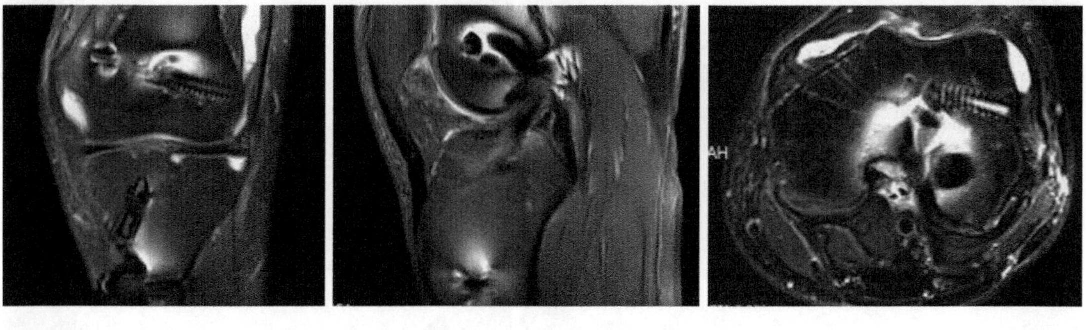

Fig. 8.3 Tunnel placement evaluation with MRI

ment is applied in the same surgical procedure to problems which could create further problems in the future (Fig. 8.5).

8.3 Surgical Steps for ACL Revision

8.3.1 The Method Used in the Old Implants

Implants assisting fixation used in the previous ACL reconstruction may endanger the creation of new tunnels and the graft fixation. Therefore, obtaining information about the previous surgery and if metal implants were used, taking two-way radiographs should be sufficient to identify their localisations. When bio-absorbable implants have been used, the old tunnels and sclerotic bones can be used as landmarks to determine the implant localisation on direct radiographs (Fig. 8.6).

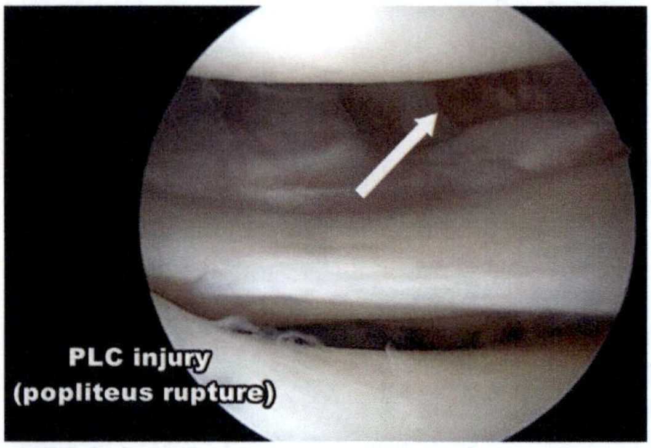

Fig. 8.4 Concomitant posterolateral corner injury

Complete removal of old implants could cause different postoperative morbidities in the bone and soft tissue or the formation of large bone defects. Therefore, old fixation implants should only be removed when there could be problems in the placement of the new tunnel or graft fixation. If old implants constitute an obstruction or partial obstruction to the formation of new tunnels, this problem can be resolved with reamerisation during tunnel dilatation. Even if biodegradable fixation materials cannot be removed during revision surgery, as they are easily fragmented with in the spongious bone during reamerisation, there is no need for complete removal. The most important stage requiring care is that debridement must be applied well to the joint after reamerisation to prevent the biodegradable implant fragments causing chondral damage, pain and local irritation in the joint. Moreover, implants which have come out from the joint and are causing widespread pain for the patient should be removed with the assistance of various implant removal devices.

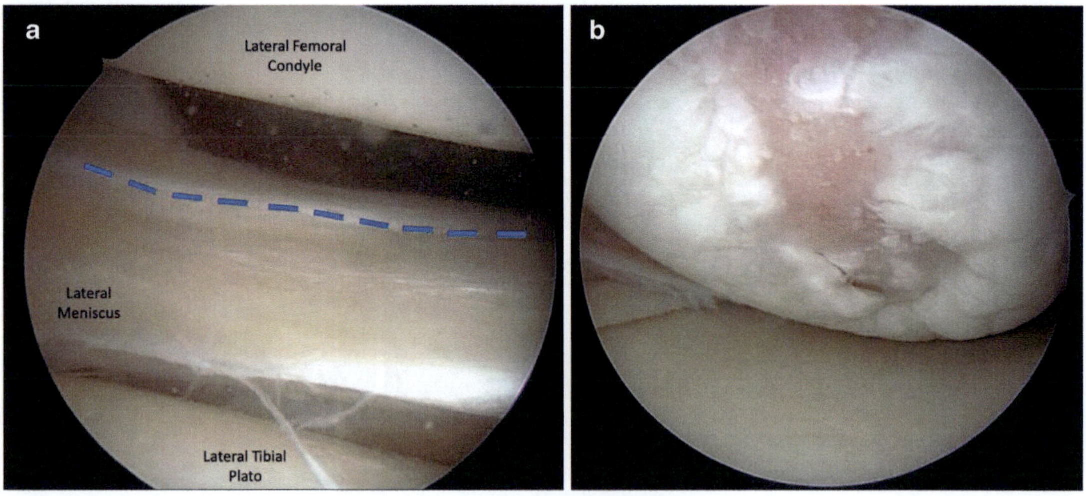

Fig. 8.5 Concomitant meniscus injuries (**a**) and chondral injuries (**b**)

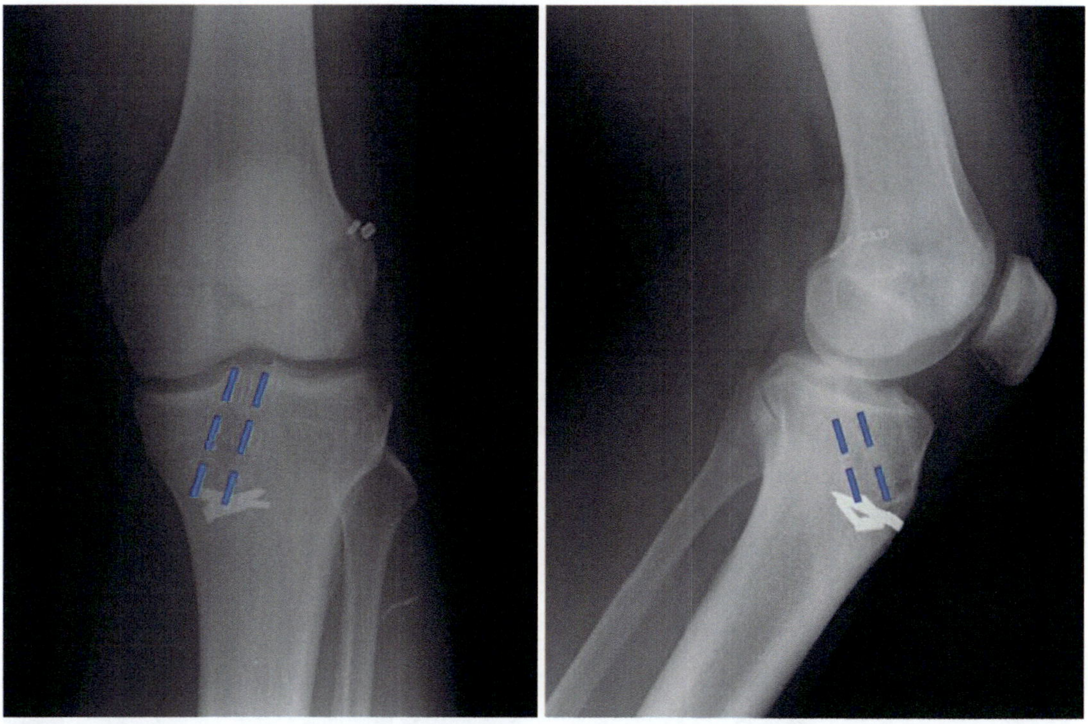

Fig. 8.6 The old tunnels and sclerotic bones

8.3.2 Tunnel Planning

Correct tunnel placement constitutes the most important step in successful ACL reconstruction. Graft impingement, elongation and graft rupture may be seen after errors made in tunnel placement [22]. According to information in current literature, anatomic landmarks should be used in femoral tunnel positioning, and the lateral femoral intercondylar ridge in particular should be taken as a guide [23]. Thus, it is possible to provide high rotational stability and decrease anterior displacement. However, in reconstructions made with transtibial techniques, while anterior translation may be reduced, rotational stability is not provided [24]. This can be confirmed with a negative Lachman test and positive Pivot-Shift test in the clinical examination.

To prevent anterior graft impingement in the knee, the intersection of the Blumensaat line and the tibial joint surface can be identified radiologically during positioning of the new tibial tunnel. Tibial tunnels that have been opened a long time ago can result in failure in the long-term follow-up as they cause impingement in extension and over-loading on the graft in flexion, and flexion loss [25].

When planning new tunnels, there are three scenarios according to the condition of the previous surgery.

8.3.2.1 Tunnels Opened in the Appropriate Position

Tunnels that have been opened in the correct position can be used again in revision surgery. After removal of the old implants, the tunnels must be debrided with a drill until a clean bone tunnel is obtained (Fig. 8.7). In cases with partial tunnel expansion or osteolysis, allograft bone plugs can be used. In cases where a bone plug cannot be applied, a double interference screw can be used in graft fixation. When tibial fixation is doubtful, the use of a bicortical screw or washer is recommended (Fig. 8.8) [26].

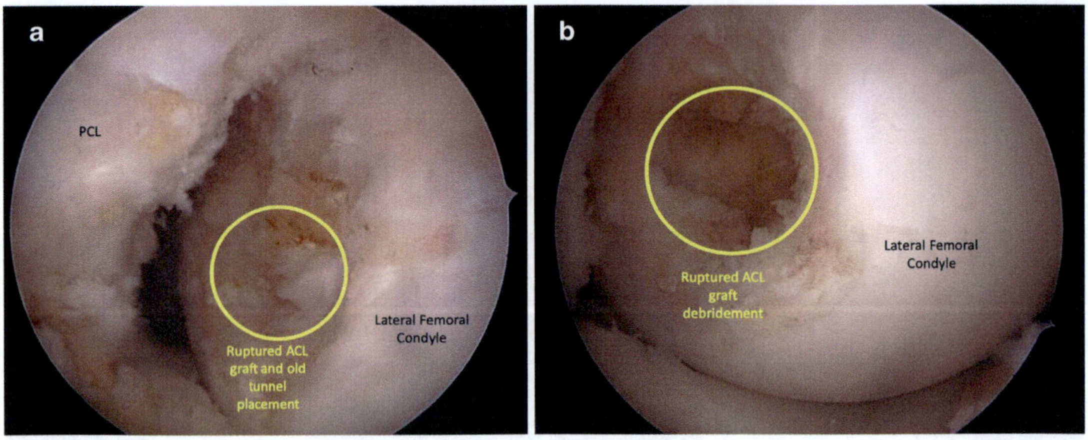

Fig. 8.7 Old tunnel placement (**a**) and ruptured ACL graft debridement (**b**)

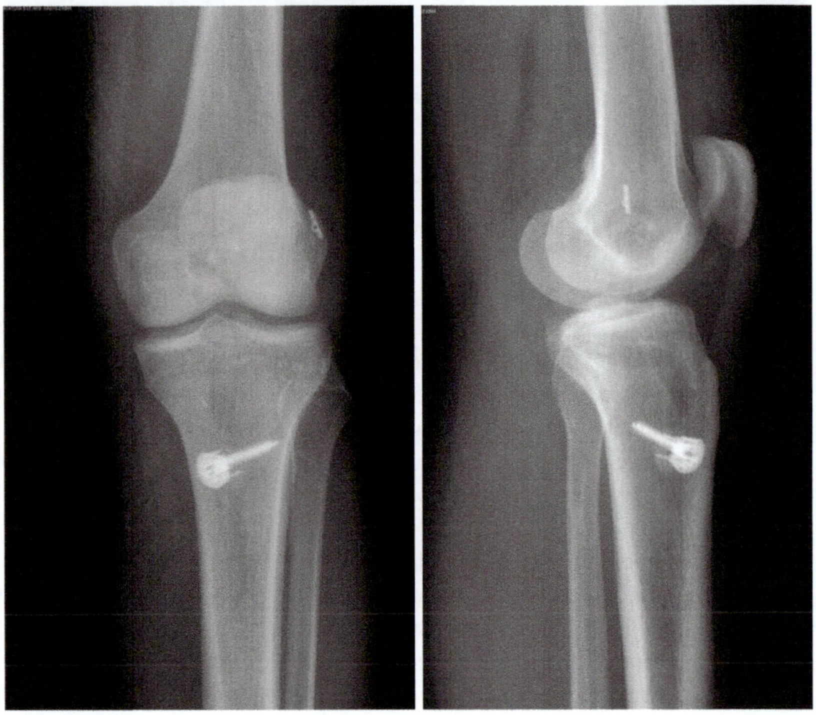

Fig. 8.8 Washer application for tibial fixation

8.3.2.2 Tunnels Opened in Partial Malposition

For all kinds of procedures to be applied to tunnels with partial malpositioning, the decision must be taken after several considerations. The most important stage of tunnel positioning is the entry site in the joint. It should be known that when the oblique position of the intra-articular entry site increases, the ovalness of the tunnel entrance will increase (Fig. 8.9).

In tibial tunnels with partial posterior placement, it is possible to open a new tunnel 2–3 mm anterior. The use of an interference screw during graft fixation will also facilitate the new positioning of the graft (anterior-posterior). The most important point to which attention must be paid is that if

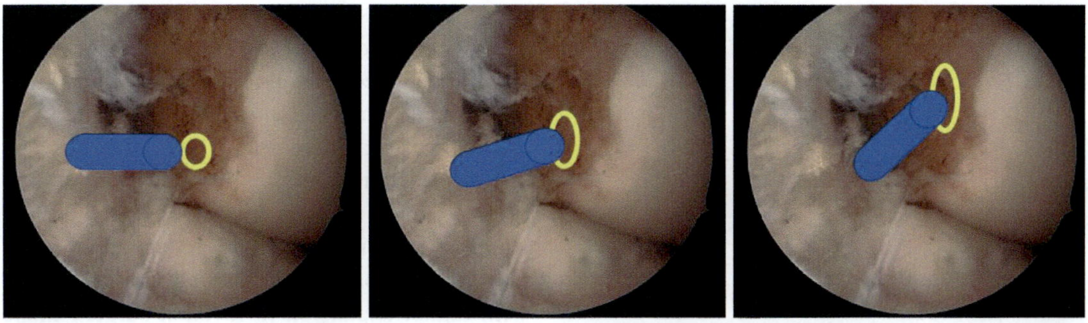

Fig. 8.9 Intra-articular tunnel positioning

the interference screw remains too much in the anterior when being placed, a cortical fracture could occur. Short anterior tunnel placement could also create secondary problems such as intra-articular extension of the fixation screw and impingement or cartilage damage. Of the problems that can be experienced in the femur, tunnels with high placement can intersect with new tunnels. In this case, composite screws should be selected. In this way, there is slow resorption in the bone, resulting in good healing with the surrounding tissue.

8.3.2.3 Tunnels Opened in Malposition

In cases with accepted malpositioning of the previous tunnel, it may be necessary to apply a different surgical approach to open a tunnel with new anatomic placement. As removal of the implant in these types of cases can cause bone and tissue damage, it may be left in place to avoid endangering the new fixation. If the new tunnel is not intersected by the old tunnel, the graft can be primary fixed. However, when the tunnels intersect, it is necessary to increase the cortical fixation. In this case, it is appropriate to transfer to hybrid fixation and use an interference screw together with a cortical button in the femur. With the hybrid fixation principle in the tibia, a bicortical screw and washer can be used [27].

8.3.3 Surgical Method

It is very important that the surgical method of the procedures applied in the first operation is known. In particular, the graft type used plays a determinant role in the tunnel positioning. As bone-patellar, tendon-bone (BPTB) grafts provide partial ossification within the tunnels, this provides a more advantageous intra-articular environment for new tunnel positioning. However, in cases where soft tissue grafts have been used, positioning of the new tunnel can be the most difficult stage.

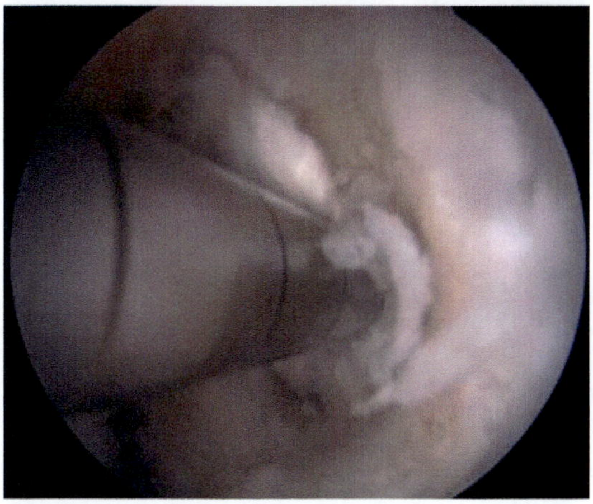

Fig. 8.10 Soft tissue remnants cleaned with a drill

If old tunnels have been opened with a diameter of ≤8 mm in the appropriate position, they can be used again depending on the desired fixation type. After advancing the guidewire into the old tunnel, soft tissue remnants can be cleaned with a drill (Fig. 8.10). Then, the procedure to freshen the tunnel until there is no sclerotic bone left is applied according to the thickness of the new graft.

In cases with tunnel malpositioning, preparations must be made to counter damage that could occur in the bone tissue depending on the graft

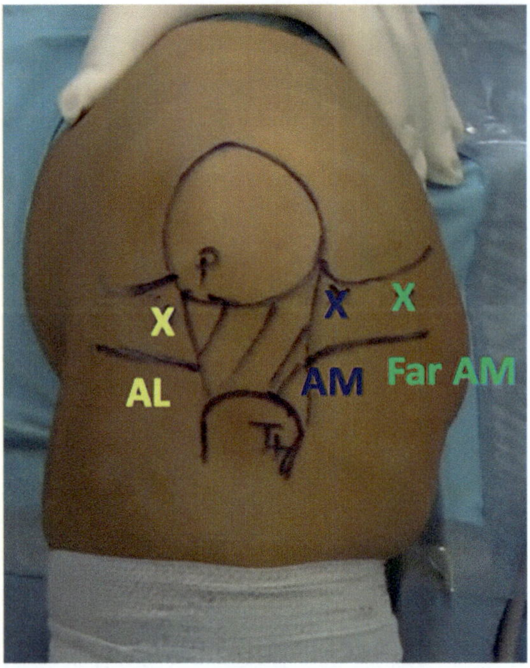

Fig. 8.11 Anteromedial portals

type. Benefit can be taken from interference screws in the re-positioning.

The correct position can be checked with fluoroscopy first, then it can be decided whether the position is correct with the guidewire. Sequential expansion and enlargement should be planned with 4–5 mm drills. In this way, the tunnels are prepared in a controlled manner [28].

The tunnel orientation should be evaluated not only according to the entry site but also in the coronal plane, and if necessary, should be examined again with fluoroscopy. The transtibial technique, which has retained its popularity, is preferred by several surgeons for primary surgery, and the routine use of the anteromedial portal technique to provide a more anatomic approach in revision surgery is recommended (Fig. 8.11) [29].

8.4 Graft Selection and Fixation

8.4.1 Graft Selection

Graft selection is of importance in revision ACL surgery. The continuously increasing rates of revision surgery have created a need for grafts at increasing rates. While autografts and allografts are still in widespread use, synthetic grafts have been abandoned because of high revision rates and chronic knee inflammation [30]. Allografts are being used at increasingly higher rates as they can be used in multiple ligament injuries and offer a choice of various sizes and thicknesses, shorten operating time and reduce donor site morbidity. In particular, bony allografts in cases with tunnel expansion in revision surgery make the surgery easier and can be used in the filling of bone defects. Compared to autografts, the most important disadvantage is delayed tunnel incorporation. Before the selection of allograft, the patient must be informed about potential complications (increased risk of infection, re-rupture), and consent must be obtained.

With the development of modern fixation techniques, although similar clinical results are seen in follow-up, there has been increased use of hamstring tendon grafts rather than bone grafts. As complications have been observed that could cause severe chronic problems such as anterior knee pain and patella fracture, there has been a significant reduction in the use of BPTB grafts [31].

The designation of autograft before revision is very important for surgical success. When autografts were used in the primary surgery, it is necessary to evaluate and plan the other ipsilateral graft options (Achilles, quadriceps, BPTB) or hamstring grafts in the contralateral knee.

8.4.2 Graft Fixation

One of the most important steps in revision ACL reconstruction is graft fixation. Graft survival will be longer with successful fixation. Hybrid fixation techniques for the femur and tibia together are at the forefront of current surgical techniques. Hybrid fixation which is often applied to the femur is also applied to the tibia. The importance of tibial fixation is especially increased when proximal tibia bone density is low, and the graft extends parallel to the tibial tunnel direction [32]. With the use of interference screws, anatomic fixation is supported and intra-tunnel movement is reduced (Fig. 8.12).

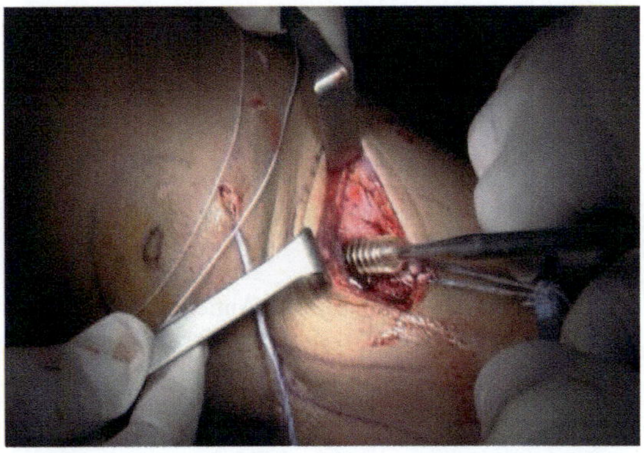

Fig. 8.12 Tibial tunnel fixation with interference screw

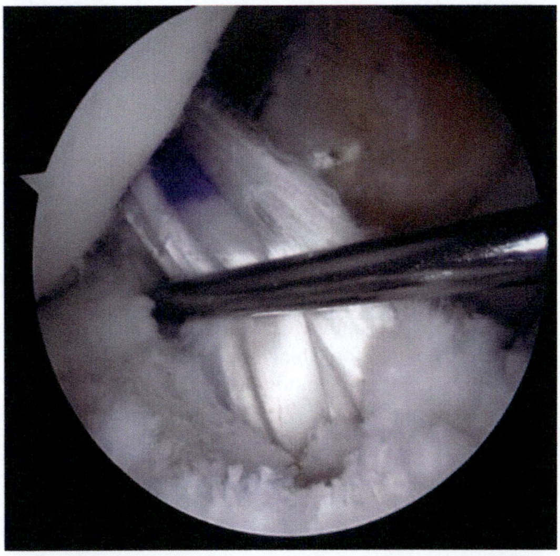

Fig. 8.13 Arthroscopic stability control

During fixation, it is important to know the graft length and the section remaining within the joint. Therefore, the tunnel length must be measured and appropriate length interference screws must be used. If biodegradable screws are to be used, fixation with a screw one size larger is recommended, taking into consideration the width of the tunnel opened. However, if composite and metal screws are to be used, the use of a screw of the same length as the tunnel can overcome tunnel problems which could be experienced. At all stages of fixation, intra-articular arthroscopic examination must be applied and then stability must be checked with fluoroscopy (Fig. 8.13).

References

1. Bach BR Jr. Revision anterior cruciate ligament surgery. Arthroscopy. 2003;19(Suppl 1):14–29.
2. Saltzman BM, et al. Economic analyses in anterior cruciate ligament reconstruction: a qualitative and systematic review. Am J Sports Med. 2016;44(5):1329–35.
3. Wolf RS, Lemak LJ. Revision anterior cruciate ligament reconstruction surgery. J South Orthop Assoc. 2002;11(1):25–32.
4. Uribe JW, et al. Revision anterior cruciate ligament surgery: experience from Miami. Clin Orthop Relat Res. 1996;325:91–9.
5. Carson EW, et al. Revision anterior cruciate ligament reconstruction: etiology of failures and clinical results. J Knee Surg. 2004;17(3):127–32.
6. Group MK, et al. Ten-year outcomes and risk factors after anterior cruciate ligament reconstruction: a MOON Longitudinal Prospective Cohort Study. Am J Sports Med. 2018;46(4):815–25.
7. Borchers JR, et al. Intra-articular findings in primary and revision anterior cruciate ligament reconstruction surgery: a comparison of the MOON and MARS study groups. Am J Sports Med. 2011;39(9):1889–93.
8. Brophy RH, et al. Association between previous meniscal surgery and the incidence of chondral lesions at revision anterior cruciate ligament reconstruction. Am J Sports Med. 2012;40(4):808–14.
9. Chen JL, et al. Differences in mechanisms of failure, intraoperative findings, and surgical characteristics between single- and multiple-revision ACL reconstructions: a MARS cohort study. Am J Sports Med. 2013;41(7):1571–8.
10. Group M, et al. Surgical predictors of clinical outcomes after revision anterior cruciate ligament reconstruction. Am J Sports Med. 2017;45(11):2586–94.
11. Group M. Meniscal and articular cartilage predictors of clinical outcome after revision anterior cruciate ligament reconstruction. Am J Sports Med. 2016;44(7):1671–9.
12. Noyes FR, Barber-Westin SD. Revision anterior cruciate ligament reconstruction: report of 11-year experience and results in 114 consecutive patients. Instr Course Lect. 2001;50:451–61.
13. Group M, et al. Subsequent surgery after revision anterior cruciate ligament reconstruction: rates and risk factors from a multicenter cohort. Am J Sports Med. 2017;45(9):2068–76.
14. Taggart TF, Kumar A, Bickerstaff DR. Revision anterior cruciate ligament reconstruction: a midterm patient assessment. Knee. 2004;11(1):29–36.
15. Cinque ME, et al. Outcomes and complication rates after primary anterior cruciate ligament reconstruction are similar in younger and older patients. Orthop J Sports Med. 2017;5(10):2325967117729659.

16. Engelman GH, et al. Comparison of allograft versus autograft anterior cruciate ligament reconstruction graft survival in an active adolescent cohort. Am J Sports Med. 2014;42(10):2311–8.
17. Eysturoy NH, et al. The influence of graft fixation methods on revision rates after primary anterior cruciate ligament reconstruction. Am J Sports Med. 2018;46(3):524–30.
18. Fauno P, Rahr-Wagner L, Lind M. Risk for revision after anterior cruciate ligament reconstruction is higher among adolescents: results from the Danish registry of knee ligament reconstruction. Orthop J Sports Med. 2014;2(10):2325967114552405.
19. Jaecker V, et al. High non-anatomic tunnel position rates in ACL reconstruction failure using both transtibial and anteromedial tunnel drilling techniques. Arch Orthop Trauma Surg. 2017;137(9):1293–9.
20. Granan LP, et al. Associations between inadequate knee function detected by KOOS and prospective graft failure in an anterior cruciate ligament-reconstructed knee. Knee Surg Sports Traumatol Arthrosc. 2015;23(4):1135–40.
21. Weiler A, et al. Primary versus single-stage revision anterior cruciate ligament reconstruction using autologous hamstring tendon grafts: a prospective matched-group analysis. Am J Sports Med. 2007;35(10):1643–52.
22. Morgan JA, et al. Femoral tunnel malposition in ACL revision reconstruction. J Knee Surg. 2012;25(5):361–8.
23. Burnham JM, et al. Anatomic femoral and Tibial tunnel placement during anterior cruciate ligament reconstruction: Anteromedial portal all-inside and outside-in techniques. Arthrosc Tech. 2017;6(2):e275–82.
24. Musahl V, et al. Varying femoral tunnels between the anatomical footprint and isometric positions: effect on kinematics of the anterior cruciate ligament-reconstructed knee. Am J Sports Med. 2005;33(5):712–8.
25. Howell SM, Taylor MA. Failure of reconstruction of the anterior cruciate ligament due to impingement by the intercondylar roof. J Bone Joint Surg Am. 1993;75(7):1044–55.
26. Cheatham SA, Johnson DL. Anticipating problems unique to revision ACL surgery. Sports Med Arthrosc Rev. 2013;21(2):129–34.
27. Group M, et al. Descriptive epidemiology of the Multicenter ACL Revision Study (MARS) cohort. Am J Sports Med. 2010;38(10):1979–86.
28. Forkel P, Petersen W. Anatomic reconstruction of the anterior cruciate ligament with the autologous quadriceps tendon. Primary and revision surgery. Oper Orthop Traumatol. 2014;26(1):30–42.
29. Mulcahey MK, et al. Transtibial versus anteromedial portal anterior cruciate ligament reconstruction using soft-tissue graft and expandable fixation. Arthroscopy. 2014;30(11):1461–7.
30. Cerulli G, et al. ACL reconstruction: choosing the graft. Joints. 2013;1(1):18–24.
31. Goldblatt JP, et al. Reconstruction of the anterior cruciate ligament: meta-analysis of patellar tendon versus hamstring tendon autograft. Arthroscopy. 2005;21(7):791–803.
32. Verioti CA, Sardelli MC, Nguyen T. Evaluation of 3 fixation devices for Tibial-sided anterior cruciate ligament graft backup fixation. Am J Orthop (Belle Mead NJ). 2015;44(7):E225–30.

Posterior Cruciate Ligament Anatomical Reconstruction

9

Ibrahim Tuncay and Vahdet Ucan

9.1 Introduction

Posterior cruciate ligament (PCL) is the primary structure which prevents posterior translation of the tibia relative to the femur. PCL origins from posterior tibial sulcus below the articular surface and attaches to medial femoral condyle. It has a broad, crescent-shaped footprint. PCL has two bundles: anterolateral (AL) and posteromedial (PM). AL bundle is tight in flexion; strongest and most important for posterior stability at 90° of flexion. PM bundle is tight in extension. Also ligaments of Wrisberg and Humphrey originate from the posterior horn of the lateral meniscus and insert into PCL.

Isolated PCL injuries are rare. The incidence of PCL injuries varied depending on the population studied. The incidence is as low as 3% in the outpatient setting and as high as 37% in the traumatic setting [1, 2]. PCL injuries can occur at low velocity or high velocity. The mechanism of PCL rupture in athletes is usually a fall on the flexed knee with a plantar-flexed foot or hyperflexion of the knee [3]. High-velocity mechanism is a direct blow to proximal tibia with a flexed knee (dashboard injury). High-velocity injuries usually include multiple ligament ruptures and dislocations.

PCL injuries are classified into three grades. In grade 1 tears (partial tear), posterior tibial translation is between 1 and 5 mm. Tibia remains anterior to the femoral condyles. In grade 2 tears (complete tear), posterior tibial translation is between 6 and 10 mm. The anterior tibia is flush with the femoral condyles. In grade 3 tears, posterior tibial translation is >10 mm. Tibia is posterior to the femoral condyles and usually indicates an associated anterior cruciate ligament (ACL) and/or posterolateral corner (PLC) injury [4].

9.2 Physical Examination

When a PCL injury is suspected, inspection, range of motion control, neurovascular examination, and specific tests such as posterior drawer test, posterior sag test, quadriceps active test, reverse pivot shift test, dial test, and posterolateral external rotation test should be performed (Chap. 4).

9.3 Imaging

Anteroposterior and supine lateral radiographs are essential. Posterior tibiofemoral subluxation or an avulsion fracture can be seen. Arthrosis may be present with chronic injuries. Also a lateral stress view is helpful for diagnosing and quantifying PCL injuries. Asymmetric posterior

I. Tuncay · V. Ucan (✉)
Department of Orthopedics and Traumatology, School of Medicine, Bezmialem Vakif University, Istanbul, Turkey

M. Bozkurt, H. İ. Açar (eds.), *Clinical Anatomy of the Knee*, https://doi.org/10.1007/978-3-030-57578-6_9

tibial displacement indicates PCL injury, and contralateral knee differences >12 mm on stress views suggest a combined PCL and PLC injury [5–7]. Magnetic resonance imaging (MRI) has become the gold standard for confirming the presence of an acute PCL tear and to diagnose associated injuries with a sensitivity of up to 100% [8]. MRI provides important information about meniscus, articular cartilage, and other ligaments in the knee that affect the course of treatment [9].

9.4 Treatment

When deciding the treatment method in PCL rupture, patient's complaints, activity level, degree of injury, and presence of additional injuries are considered. Although some studies (that included large percentages of patients with partial PCL deficiency) reported that patients did well when treated conservatively, other investigations described noteworthy symptoms and functional limitations after the injury [10–15].

Controversy regarding the treatment of PCL continues because it is not clear whether posterior laxity causes patient complaints or accelerates the development of degenerative joint disease (DJD). Furthermore, it is unclear whether reconstruction sufficiently reduces laxity to result in clinical improvement and slow the development of DJD. Only a few clinical studies have sufficient sample sizes and duration of follow-up. The biggest problem in the studies is the heterogeneity of the studies. Most studies involve multiligament injuries with PCL injury rather than isolated PCL injury. This makes comparisons difficult. Prospective randomized trials are needed to clarify this issue.

Generally, acute and isolated Grade I or II PCL tears or partial PCL tears do well with conservative treatment. The avulsion fractures, acute grade III PCL tears combined with a PLC injury, and/or other multiligamentous injuries and chronic grade II to III injuries with symptoms of instability or pain need surgery.

9.4.1 Nonoperative Treatment

Conservative treatment is recommended in patients with acute and isolated Grade I or II PCL tears [15, 16]. The aim of conservative treatment is to prevent this injury from turning into a Grade 3 injury. Therefore, it is essential to overcome the gravitational force and hamstring muscle strength which causes the tibia to shift to the posterior of the femur. Based on the literature findings, a three-phase rehabilitation protocol was proposed [17].

In Phase 1 (6 weeks after injury), partial weight bearing is recommended. Hamstring and gastrocnemius stretching and quadriceps strengthening are necessary to prevent posterior displacement of the tibia. For ligament healing, immobilization is performed with angle-adjustable brace or cylindrical leg cast [18, 19].

In Phase 2 (6–12 weeks), progressive strengthening, improving proprioception, and reestablishment of full range of motion are essential.

In Phase 3 (13–18 weeks), the patient is allowed to running and do sports-specific exercises. Return to sports usually requires a 6-month period.

9.4.2 Operative Treatment

Displaced avulsion fractures and PLC injuries should be repaired within the first 3 weeks [20]. Acute grade III PCL tears combined with a PLC injury and/or other multiligamentous injuries and chronic grade II to III injuries with symptoms of instability or pain need reconstruction. Surgical options for PCL reconstruction are transtibial and tibial inlay reconstruction techniques with single- or double-bundle reconstruction. These techniques can be performed both arthroscopically and open. However, it is not clear which is the best method for PCL reconstruction. A simple single-tunnel procedure is often effective in cases of accompanying multiligament injury. In the case of isolated posterior cruciate ligament rupture, a double-tunnel procedure can be performed.

There are various graft options for PCL reconstruction. Bone–patellar tendon–bone (BPTB), hamstring tendons, and quadriceps tendons are autologous graft options. Tibialis anterior tendon, Achilles tendon, BPTB, and quadriceps tendon can be used as allograft.

The use of allograft has the advantage that it does not cause donor site morbidity and shortens the operative time. However, the risk of infection is higher than in autologous group. Cost effectivity should also be considered.

9.4.2.1 Arthroscopic Single-Bundle Technique

Examination under anesthesia should be done before operation on both the nonoperative and the operative knees. A tourniquet is applied to the operative extremity, and the surgical leg is prepared and draped in a sterile fashion (Fig. 9.1). Tourniquet should be deflated before wound closure to ensure that there is no injury to the popliteal vessels. Use a leg holder to maintain 80–90° of knee flexion during the procedure. A padded lateral post to assist with valgus stress is necessary. The joint is thoroughly evaluated arthroscopically using standard anterolateral and anteromedial portals. If a meniscal repair is performed, the sutures should be tied after the ligament reconstruction is completed. Debride the soft tissue and residual stamp of PCL. A 70° arthroscope from the anterolateral portal or a 30° arthroscope from the posteromedial portal should be used to visualize the tibial attachment site of the PCL (Fig. 9.2). Also a transseptal portal can

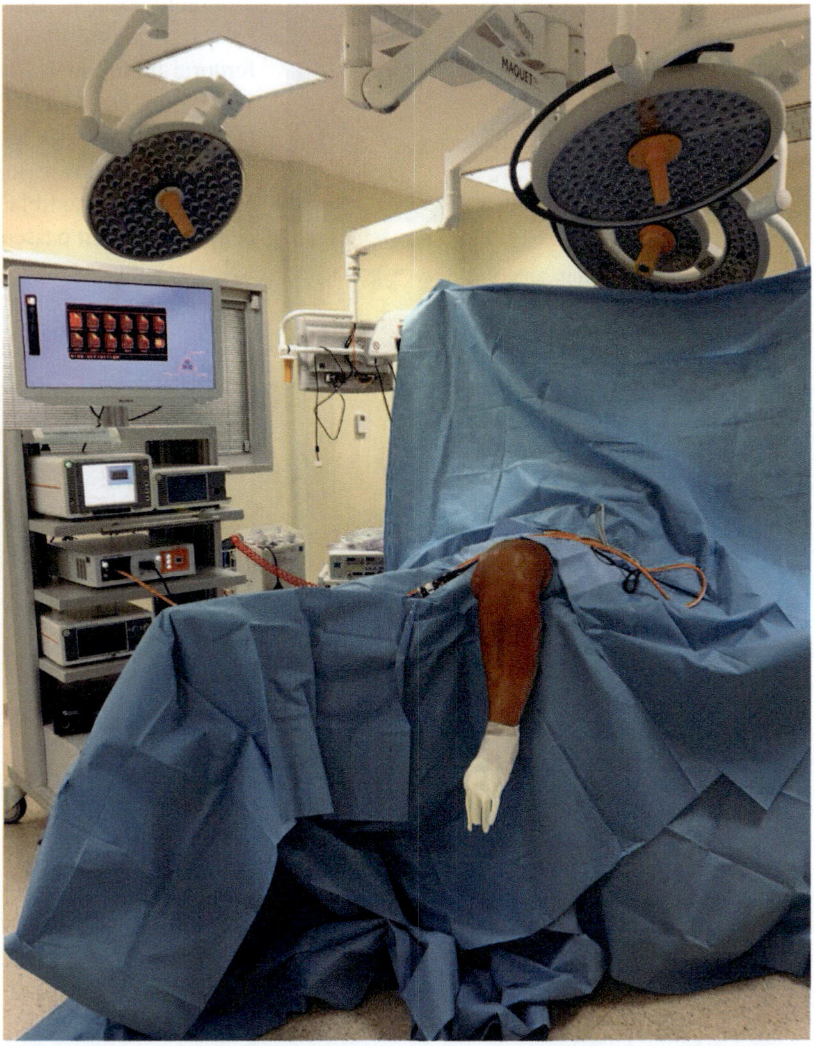

Fig. 9.1 Positioning

be created for better visualization of the tibial attachment site of the PCL [21].

For drilling the tunnel safely in the appropriate position, exposure of the tibia and using an image intensification are essential (Fig. 9.3). Elevate the soft tissue from the tibia using a curved curette/radiofrequency probe passing through the intercondylar notch or the postero-medial portal (Fig. 9.4). Then place a blunt spade-tipped guidewire 10–12 mm below the joint line in the PCL facet. Set the drill guide approximately 60° to the articular surface of the tibia, starting just medial and inferior to the tibial tuberosity (Fig. 9.5). Check the position via fluoroscopy. Before drilling the tibial tunnel, the closed curve curette may be positioned to cup the tip of the guidewire. This may help in protecting the neurovascular structures (Fig. 9.6). The tibial cortex is carefully perforated by hand reaming under arthroscopic visualization with the appropriately sized cannulated reamer (Fig. 9.7).

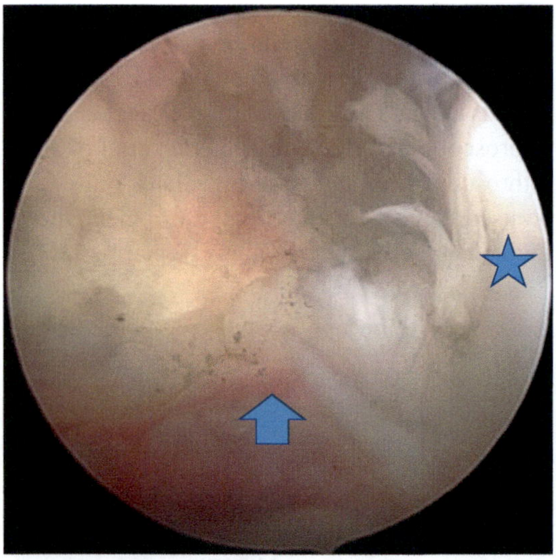

Fig. 9.2 (arrow) PCL stamp, anterior view. (star) Medial femoral condyle

To prepare the femoral tunnel, a guidewire is placed through the anterolateral portal. The starting hole is determined at 1 o'clock (right knee) or 11 (left knee) (Fig. 9.8). The femoral physiometric point is approximately 8 mm proximal to the articular cartilage. The appropriate size reamer is passed through the guidewire carefully. Then femoral tunnel is drilled. For graft passage, bent wire loop is passed through the tibial tunnel (Fig. 9.9). This wire loop is taken out of the portal, and the suture is loaded onto this ring. Graft is passed from the tibial tunnel with this suture. A beath pin is then passed through the femoral tunnel. The sutures that are at the end of the graft are loaded onto this pin and pulled into the femoral

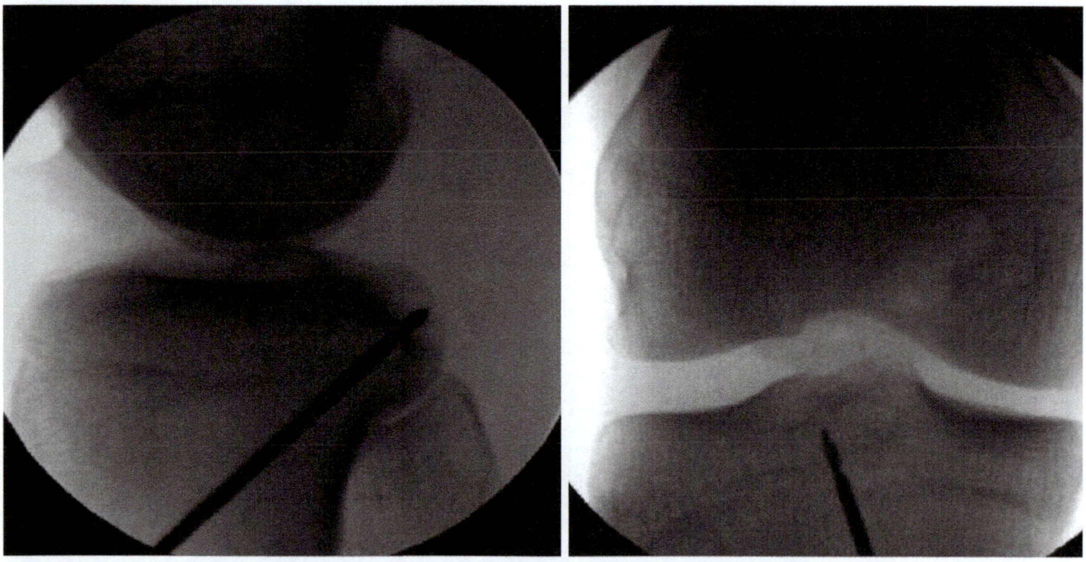

Fig. 9.3 Fluoroscopic image of transtibial tunnel guide pin placement

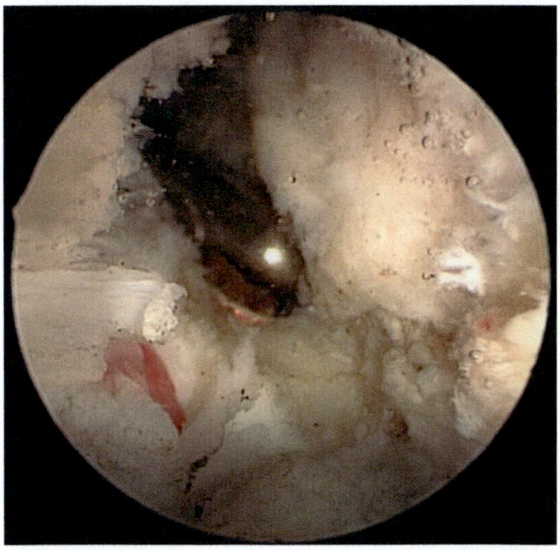

Fig. 9.4 Posteromedial view of PCL insertion and elevating the soft tissue from the tibia via radiofrequency probe

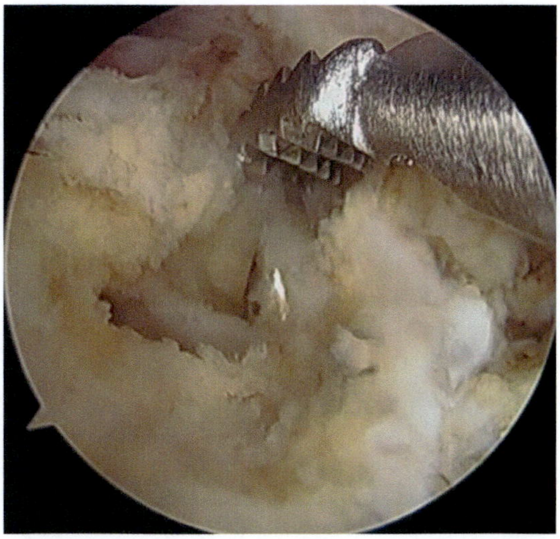

Fig. 9.6 The closed curved curette positioned to the tip of the guidewire

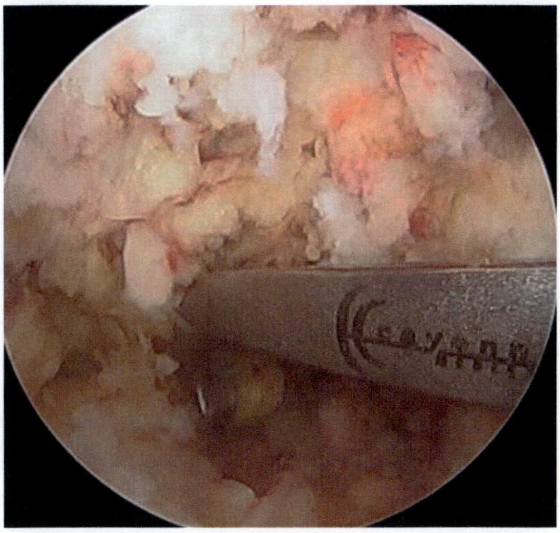

Fig. 9.5 Guidewire in the PCL facet

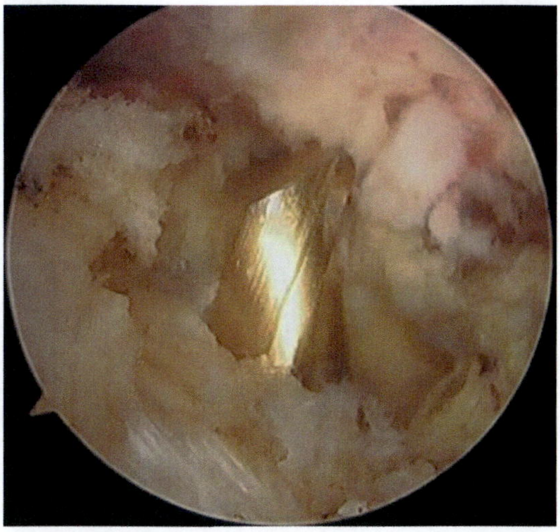

Fig. 9.7 Perforating the tibial cortex

tunnel (Fig. 9.10). According to the preferred technique (suspension system or endobutton), the graft is placed in the femoral tunnel. Maintain graft tension and put the knee through a range of motion for 20 cycles to allow stress relaxation of the graft. The tibial side is fixed at 90° flexion with a 4.5 mm cortical screw while an anterior tibial force is applied. The proper position, tension, and fixation of the graft are controlled by an arthroscope (Fig. 9.11). The incisions are irrigated and closed, and then the lower extremity is wrapped with an elastic bandage.

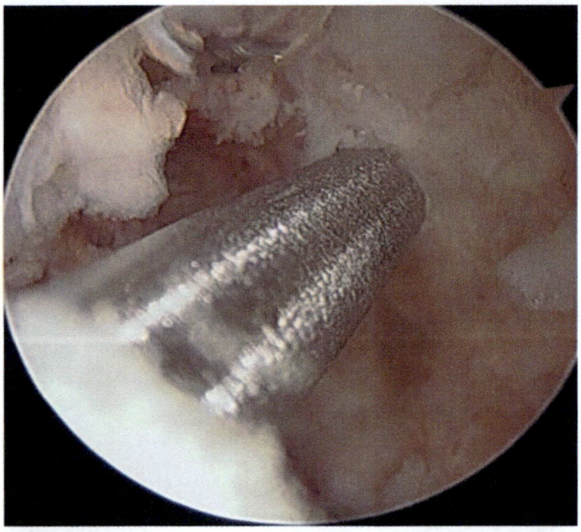

Fig. 9.8 Preparing the femoral tunnel

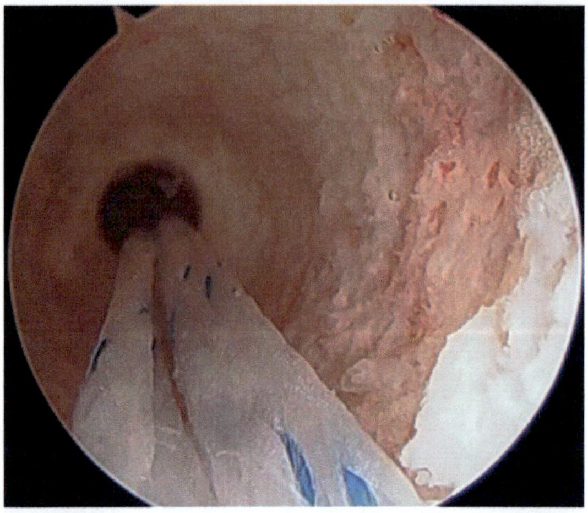

Fig. 9.10 The sutures pulled into the femoral tunnel

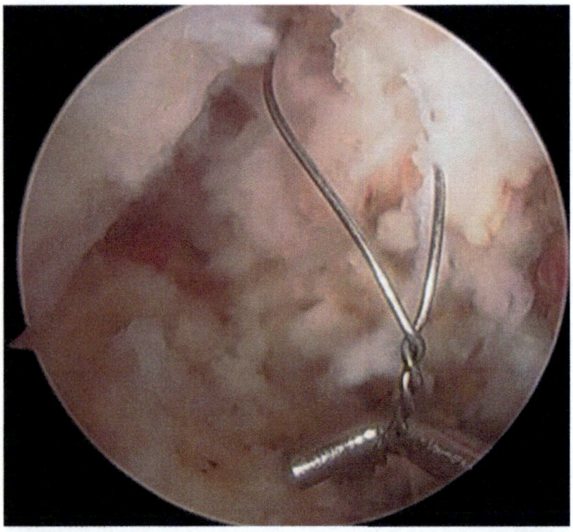

Fig. 9.9 Bent wire loop

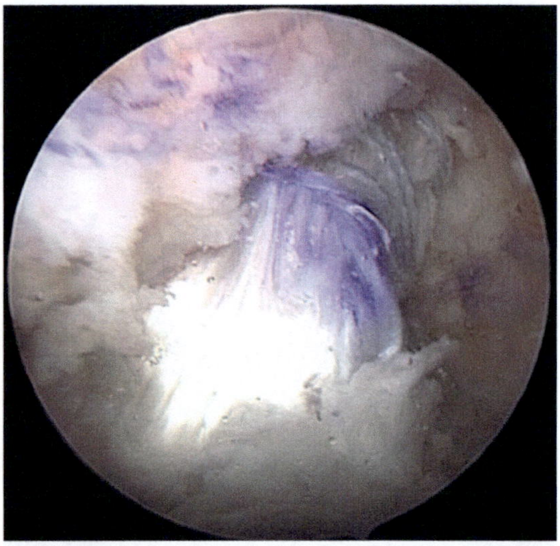

Fig. 9.11 The proper position of PCL graft

9.4.2.2 Arthroscopic Double-Bundle Technique

The steps of preparation for portal placement, arthroscopy, and drilling are the same as for single-bundle technique. In this technique, care must be taken to ensure an adequate bony bridge between the two tibial tunnels and avoid tunnel convergence. First, the AL tibial tunnel is created. It must be just distal and lateral to the PCL insertion site, same as with single-bundle reconstruction. The PM tibial guidewire enters the tibia slightly more proximal and medial than the AL guidewire. After guidewire position is checked, drill the AL tibial tunnel first with a cannulated drill. The posterior tibial cortex must be perforated by hand reaming to prevent damage to any structures. For drilling the PM tibial tunnel, same steps are repeated.

For the AL bundle's femoral tunnel, the starting hole is placed at the 1 o'clock (right knee) or 11 o'clock (left knee) position. A 4.5-mm drill is used to perforate the outer cortex of the medial femoral condyle. According to the size of the graft to be used, the tunnel is drilled to a depth of about 30 mm with a cannulated drill. The PM tunnel's starting point must be at the 3 o'clock (right knee) or 9 o'clock (left knee) position. And the tunnel must be placed parallel or slightly posterior to the AL tunnel. A 30 mm depth is enough for PM tunnel. After passing the AL graft, the PM graft is passed. Graft fixation is performed first on the femoral side. An anterior tibial force is applied to reduce the tibia before and during final tibial fixation. The AL graft is secured first at 90° flexion, and the PM bundle is then secured at 15° of flexion with screws. Finally, the proper position, tension, and fixation of the grafts are controlled by arthroscope. The incisions are irrigated and closed, and then the lower extremity is wrapped with an elastic bandage.

9.4.2.3 Single-Bundle Open Tibial Inlay Technique with Bone–Patellar Tendon–Bone (BPTB) Autograft

This technique is called *inlay* because the bone from the BPTB graft is placed into a trough in the posterior aspect of the tibia at the PCL footprint. The technique has the advantages of eliminating acute graft angle changes which is named "killer turn" and allows secure direct fixation to the posterior tibia, thus making a shorter, stiffer graft [22]. The patient can be positioned supine or in the lateral decubitus position. However, performing arthroscopy in the lateral position prevents reposition of the patient for posterior approach. At lateral decubitus position, the operative extremity can be abducted and externally rotated to facilitate the arthroscopy [23]. BPTB autograft is harvested from the ipsilateral knee in standard fashion. Graft's tibial inlay side must be prepared in a rectangular shape and femoral side in a bullet shape.

For femoral tunnel preparation, the incision begins at the medial knee anterior and superior to the medial femoral epicondyle. Dissection is carried down in line with vastus medialis to the level of the femoral condyle. The PCL guide is placed with arthroscopically at the 1 o'clock position (right knee) or 11 o'clock position (left knee), 8 mm deep in the medial femoral notch and away from the articular surface. The guide pin is drilled from outside with the use of the PCL guide while checked by arthroscopically. And the tunnel (approximately 11–12 mm in diameter, 30–35 mm in depth) is then drilled over this guidewire. A looped smooth wire is placed through the tunnel into the joint to be used later for passage of the autograft from the posterior knee into the femoral tunnel.

For distal fixation, a horizontal incision is then made in the flexion crease of the popliteal fossa (Fig. 9.12). With blunt dissection, the gastrocnemius muscle is mobilized and retracted laterally. Use Steinmann pins as a retractor (Fig. 9.13). The gastrocnemius–semimembranosus interval protects the popliteal vessels and tibial nerve. Slight knee flexion can increase the ability to laterally mobilize the medial head of the gastrocnemius and exposure of the posterior knee capsule. The popliteus muscle is commonly encountered in this interval, and the upper portion of the popliteus muscle belly can be reflected to expose the posterior cortex of the tibia. A posterior arthrotomy is made along the superior border of the popliteus. Bone trough for the inlay is prepared with an appropriate shape. Burr and osteotome can be used. The prepared BPTB graft is inlayed into the trough (Fig. 9.14). The graft is secured with two pins from a cannulated screw set, preferentially for a screw diameter of 4.5 mm. The graft is pulled into the knee joint with the previ-

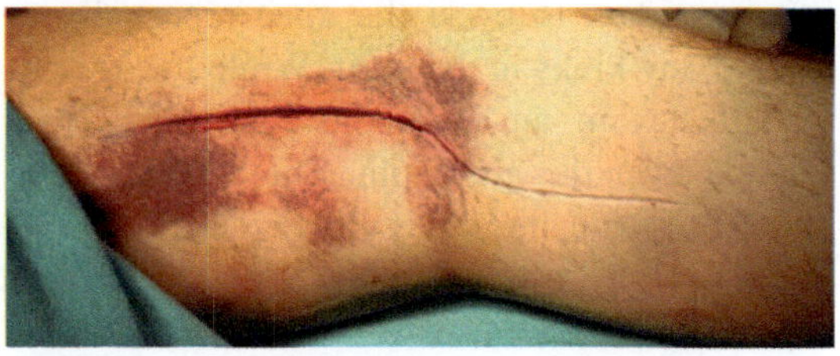

Fig. 9.12 Horizontal incision at popliteal fossa

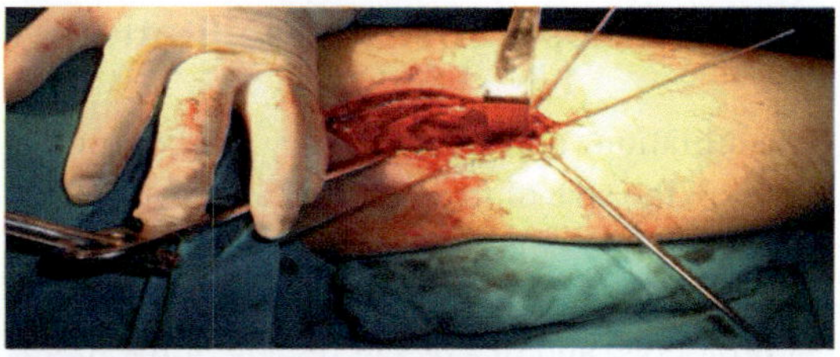

Fig. 9.13 Steinmann pins as a retractor

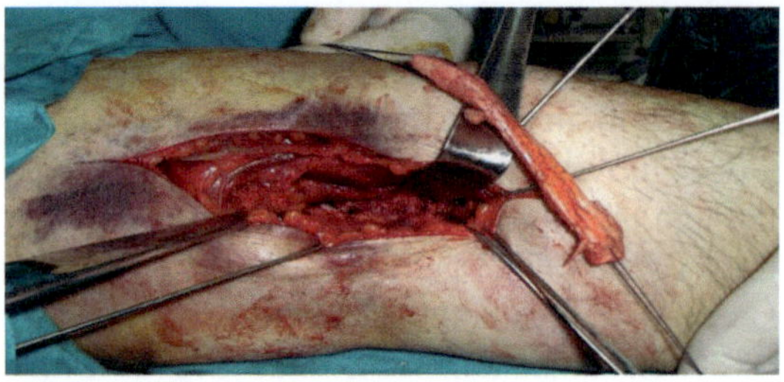

Fig. 9.14 The prepared BPTB graft is before inlaying into the trough

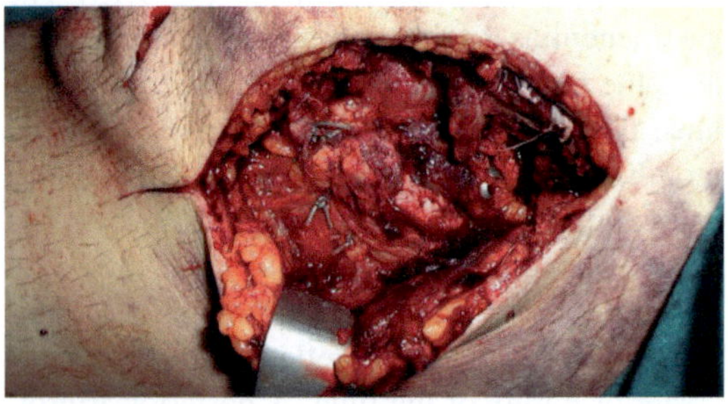

Fig. 9.15 BPTB graft after screw fixation

ously placed looped smooth wire. Screw fixation is then achieved with two bicortical screws securing the bone plug into the posterior trough (Fig. 9.15). The bullet-shaped bone plug is passed into the previously drilled and prepared femoral tunnel with the knee at 90° of flexion. While maximum manual tension is applied to the graft, it is important to cycle the knee repeatedly to remove any kinks in the graft. While the knee at 90° of flexion an interference screw is then inserted over the guide pin. The screw is then seated fully, and the graft is visualized arthroscopically [24]. The medial and posterior incisions and arthroscopy sites are also closed, and routine dressings applied.

9.5 Postoperative Rehabilitation

Rehabilitation depends on the selected graft material, the size of the patient, the expectations, and other accompanying injuries. No level I studies have been performed to compare different protocols, and in a recent review of the literature [17]. After isolated PCL reconstruction, the knee can be immobilized in a removable knee immobilizer for 4 weeks. Early range of motion and quadriceps exercises are recommended, but flexion is limited to 90° during the first 4 weeks. Patients are instructed to maintain touchdown weight bearing for 1 week. Partial weight bearing is initiated after the first postoperative visit. The brace is unlocked after 4–6 weeks and usually is discontinued after 8 weeks. Once full, pain-free ROM is achieved, strengthening is addressed. The goals for achievement of flexion are 90° at 4 weeks and 120° at 8 weeks. Hamstring strengthening is begun at 3 months. During motion and strengthening therapy, care is taken to prevent posterior tibial stress [25]. Return to sports is allowed at 9 months. Jogging on a treadmill may begin at 10–12 weeks postoperatively, but full-speed running should be avoided for 4–6 months. After 16 weeks, the patient may begin plyometrics and sports-specific activities and progress as tolerated. Return to sports, or full activity, is typically 6–12 months after PCL reconstruction after the patient has demonstrated adequate return of strength and dynamic control of the limb [26].

9.6 Complications

As with any surgery, **infection** is still an annoying complication. To reduce the risk of infection, all staff should take care to maintain the sterile technique.

Residual posterior laxity is most likely attributed to improper tensioning of the graft during graft placement or graft fixation [27].

The most feared complication during PCL reconstruction is **injury to the neurovascular structures** in the popliteal fossa. Steinmann pins placed in the posterior tibia provide sufficient, constant retraction and eliminate the risk from repetitive repositioning of retractors [28].

Starting from the femoral tunnel approximately 10 mm posterior to the articular margin helps to avoid **avascular necrosis of the medial femoral condyle** [29].

Early arthrofibrosis, which requires manipulation under anesthesia, may occur after PCL reconstruction [30].

In patients with multiligamentous injuries, extravasation of fluid during the arthroscopic portion of the case can create an **iatrogenic compartment syndrome.** The leg should be continually monitored throughout the case.

References

1. Fanelli GC, Edson CJ. Posterior cruciate ligament injuries in trauma patients: part II. Arthroscopy. 1995;11(5):526–9.
2. Miyasaka KC, Daniel DM, Stone ML. The incidence of knee ligament injuries in the general population. Am J Knee Surg. 1991;4:3–8.
3. Schulz MS, Russe K, Weiler A, Eichhorn HJ, Strobel MJ. Epidemiology of posterior cruciate ligament injuries. Arch Orthop Trauma Surg. 2003;123(4):186–91.
4. Wind WM Jr, Bergfeld JA, Parker RD. Evaluation and treatment of posterior cruciate ligament injuries: revisited. Am J Sports Med. 2004;32(7):1765–75. Review.
5. Margheritini F, Mancini L, Mauro CS, et al. Stress radiography for quantifying posterior cruciate ligament deficiency. Arthroscopy. 2003;19:706–11.
6. Schulz MS, Steenlage ES, Russe K, Strobel MJ. Distribution of posterior tibial displacement in knees with posterior cruciate ligament tears. J Bone Joint Surg Am. 2007;89(2):332–8.
7. Hewett TE, Noyes FR, Lee MD. Diagnosis of complete and partial posterior cruciate ligament ruptures. Stress radiography compared with KT-1000 arthrometer and posterior drawer testing. Am J Sports Med. 1997;25(5):648–55.
8. Esmaili Jah AA, Keyhani S, Zarei R, Moghaddam AK. Accuracy of MRI in comparison with clinical and arthroscopic findings in ligamentous and meniscal injuries of the knee. Acta Orthop Belg. 2005;71(2):189–96.
9. Munshi M, Davidson M, MacDonald PB, Froese W, Sutherland K. The efficacy of magnetic resonance imaging in acute knee injuries. Clin J Sport Med. 2000;10(1):34–9.
10. Shelbourne KD, Clark M, Gray T. Minimum 10-year follow-up of patients after an acute, isolated posterior cruciate ligament injury treated nonoperatively. Am J Sports Med. 2013;41(7):1526–33.
11. Patel DV, Allen AA, Warren RF, Wickiewicz TL, Simonian PT. The nonoperative treatment of acute, isolated (partial or complete) posterior cruciate ligament-deficient knees: an intermediate-term follow-up study. HSS J. 2007;3(2):137–46.
12. Boynton MD, Tietjens BR. Long-term followup of the untreated isolated posterior cruciate ligament-deficient knee. Am J Sports Med. 1996;24(3):306–10.
13. Torg JS, Barton TM, Pavlov H, et al. Natural history of the posterior cruciate ligament-deficient knee. Clin Orthop Relat Res. 1989;(246):208–16.
14. Fowler PJ, Messieh SS. Isolated posterior cruciate ligament injuries in athletes. Am J Sports Med. 1987;15(6):553–7.
15. Parolie JM, Bergfeld JA. Long-term results of nonoperative treatment of isolated posterior cruciate ligament injuries in the athlete. Am J Sports Med. 1986;14(1):35–8.
16. Keller PM, Shelbourne KD, McCarroll JR, Rettig AC. Nonoperatively treated isolated posterior cruciate ligament injuries. Am J Sports Med. 1993;21(1):132–6.
17. Pierce CM, O'Brien L, Griffin LW, Laprade RF. Posterior cruciate ligament tears: functional and postoperative rehabilitation. Knee Surg Sports Traumatol Arthrosc. 2013;21(5):1071–84.
18. Ittivej K, Prompaet S, Rojanasthien S. Factors influencing the treatment of posterior cruciate ligament injury. J Med Assoc Thai. 2005;88(Suppl 5):S84–8.
19. Jung YB, Tae SK, Lee YS, Jung HJ, Nam CH, Park SJ. Active non-operative treatment of acute isolated posterior cruciate ligament injury with cylinder cast immobilization. Knee Surg Sports Traumatol Arthrosc. 2008;16(8):729–33.
20. Harner CD, Waltrip RL, Bennett CH, et al. Surgical management of knee dislocations. J Bone Joint Surg Am. 2004;86A:262–73.
21. Mauro CS, Margheritini F, Mariani PP. The arthroscopic transeptal approach for pathology of the posterior joint space. Tech Knee Surg. 2005;4:120–5.
22. Papalia R, Osti L, Del Buono A, Denaro V, Maffulli N. Tibial inlay for posterior cruciate ligament reconstruction: a systematic review. Knee. 2010;17(4):264–9.
23. Gill SS, Cohen SB, Miller MD. PCL tibial inlay and posterolateral corner reconstruction. In: Miller MD, Cole BJ, editors. Textbook of arthroscopy. 1st ed. Philadelphia, PA: Saunders Elsevier; 2004. p. 717–28.
24. Cole BJ, Sekiya JK, editors. Surgical techniques of the shoulder, elbow, and knee in sports medicine. 2nd ed. p. 876.
25. Lutz GE, Palmitier RA, An KN, Chao EY. Comparison of tibiofemoral joint forces during open-kinetic-chain and closed-kinetic-chain exercises. J Bone Joint Surg Am. 1993;75(5):732–9.
26. Edson CJ, Fanelli GC, Beck JD. Postoperative rehabilitation of the posterior cruciate ligament. Sports Med Arthrosc Rev. 2010;18(4):275–9.
27. Hermans S, Corten K, Bellemans J. Long-term results of isolated anterolateral bundle reconstructions of the posterior cruciate ligament: a 6- to 12-year follow-up study. Am J Sports Med. 2009;37(8):1499–507.

28. Nemani VM, Frank RM, Reinhardt KR, Pascual-Garrido C, Yanke AB, Drakos M, Warren RF. Popliteal venotomy during posterior cruciate ligament reconstruction in the setting of a popliteal artery bypass graft. Arthroscopy. 2012;28(2):294–9.
29. Wiley WB, Owen JR, Pearson SE, Wayne JS, Goradia VK. Medial femoral condyle strength after tunnel placement for single- and double-bundle posterior cruciate ligament reconstruction. J Knee Surg. 2007;20(3):223–7.
30. Alcalá-Galiano A, Baeva M, Ismael M, Argüeso MJ. Imaging of posterior cruciate ligament (PCL) reconstruction: normal postsurgical appearance and complications. Skeletal Radiol. 2014;43(12):1659–68.

Medial Patellofemoral Ligament Reconstruction Techniques

10

Bogdan Ambrožič, Samo Novak, and Marko Nabergoj

10.1 Introduction

Patellar instability with recurrent dislocation is a common pathological condition among young and active patients. Patellar dislocation accounts for 2–3% of all knee injuries [1–3]. For acute first dislocation, conservative treatment is always a choice, but the dislocation rate ranges from 15% to 44% [4–6]. Among patients participating in high-activity sports, the rate of redislocation increases to 80% [7]. Different anatomical conditions can cause patellar instability, increased patellar height and tilt, changed tibial tuberosity-trochlear groove (TT-TG) distance, and trochlear dysplasia [8]. Over 100 surgical procedures exist for treating pattelar instability such as widely used tibial tuberosity transposition with vastus medialis plasty and lateral retinacular release [9]. Nevertheless, this procedure is associated with significant possible postoperative complications [10, 11]. Recently, more and more anatomical and biomechanical studies show that the medial patellofemoral ligament (MPFL) is the primary restraint to lateral patellar translation between 0 and 30° of knee flexion. The MPFL is injured in more than 90% of patellar dislocation cases. In adults it is mostly torn at the femoral insertion, and among children and adolescents at patellar attachment [12]. Literature has shown that patellar stabilization with MPFL reconstruction is successful treatment option for patellar instability (Fig. 10.1).

It is a mini-invasive surgical procedure associated with low postoperative complications [6, 13–24]. It also significantly improves clinical scores and allows patients to return to daily activities and even to competition sports [13–22]. The dislocation rate after surgical treatment is reported to be up to 31% [22, 25–27].

B. Ambrožič (✉) · S. Novak · M. Nabergoj
Valdoltra Orthopaedic Hospital, Ankaran, Slovenia
e-mail: bogdan.ambrozic@bisturmed.com; samo.novak@ob-valdoltra.si; marko.nabergoj@ob-valdoltra.si

10.2 Anatomy

10.2.1 Patella and Trochlea of the Femur

Patella is the largest human sesamoid bone and is a part of knee joint extensor apparatus [28]. It is shaped like an upside-down triangle with the base at the proximal part and with the apex at the distal part. Quadriceps femoral muscle is inserted at the proximal two third of the patella, while patellar ligament extends distally from apex to tibial tuberosity. On the articular side of patella, the cartilage is the thickest (about 5 mm) in human body. Trochlea of femur is cartilaginous part of the distal femur, which forms hinge joint with the patella. Inverted U-shaped area of the distal femur is concave and is asymmetrical,

M. Bozkurt, H. İ. Açar (eds.), *Clinical Anatomy of the Knee*,
https://doi.org/10.1007/978-3-030-57578-6_10

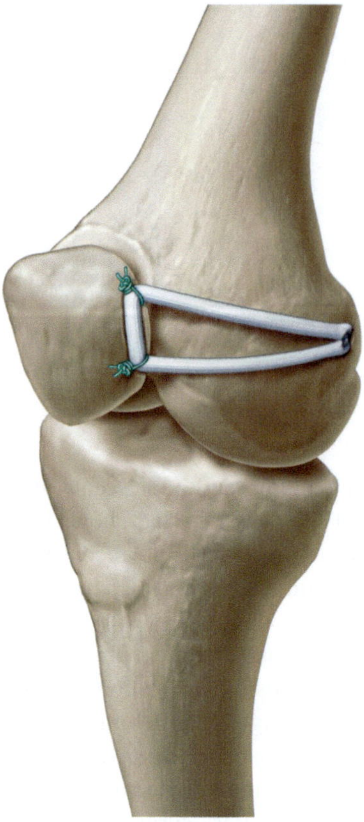

Fig. 10.1 Anatomical MPFL reconstruction (with courtesy of Dr. Arno Schmeling and Prof. Andreas Weiler, Sporthopaedicum Berlin, Germany, Operating technique: Anatomic reconstruction of the medial patellofemoral ligament with a free gracilis tendon graft)

with lateral facet being higher and extending more distally than medial. Trochlear dysplasia is one of the most important cause of patellar instability and can be evaluated on the X-ray images [29].

10.2.2 Medial Patellar Ligamentous Complex

Medial patellar ligamentous complex consists of the superficial medial retinaculum, medial patellofemoral (MPFL), medial patellomeniscal (MPML), and medial patellotibial ligament (MPTL). Waren and Marshall [30] anatomically evaluated three layers of the medial side of the knee. First being superficial retinaculum, the middle consists of the MPFL and medial collateral ligament (MCL). The deepest consists of MPTL, which is 54.6 ± 8.4 mm long and 21.8 ± 4.4 mm wide but not always present, and MPML which is 39.4 ± 3.2 mm long and 9.6 ± 1.2 mm wide.

Literature showed that medial patellofemoral ligament (MPFL) is always present in human body [31–34]. It is the most important stabilizer of the patella during initial 30° knee flexion. Forces that restraint during flexion are about 50–60% of all the strengths of the medial side [27, 35–38]. The MPFL is triangular in shape and is directed horizontally, connecting the medial part of the patella and the femur. Patellar insertion of the MPFL is mostly consistent, and it is found between the upper and medial third of the medial side of the patella [33, 34, 37, 38]. The femoral insertion described in literature is more variable. LaPrade et al. [39] showed in cadaveric study that femoral insertion is 1.9 mm anterior and 3.2 mm distal to the adductor tubercle. Nomura et al. [37] described that femoral insertion of the MPFL is 9.5 mm proximally and 5 mm posteriorly from the center of the medial femoral epicondyle. Philippot et al. [34] showed on the 23 cadaveric knees that MPFL is always present with the length 57.7 ± 5.8 mm. The medial patellofemoral ligament insertion is 12.2 ± 2.6 mm wide at femoral insertion and 24.4 ± 4.8 mm wide at patellar insertion. MPFL is also always anatomically connected to vastus medialis obliquus muscle (VMO). Contraction of the VMO muscle tenses MPFL, thus indirectly increasing stabilization of the patellofemoral joint. MPFL also consists of the nerve fibers for proprioception, neuromuscular function, and knee movement coordination [33, 40].

10.3 Biomechanics of the MPFL

Patellofemoral stability is maintained by three factors in general: bony anatomy, soft tissue restraints, and the dynamic action of muscle. The medial patellofemoral ligament is the strongest medial stabilizers of the patella and serves as the primary soft tissue restraint to lateral patellar displacement mainly between 0 and 30° of knee flexion. With higher degree of flexion, bony geometry of patellofemoral joint is becoming more significant. In the first degrees of knee flexion, the patella is mostly in contact with lateral facet of the trochlea. With increasing flexion, the center of the trochlea gradually shifts to medial and then back laterally. LaPrade et al. [41] showed on cadaveric specimen that average failure load of MPFL is 178 ± 46 N. Mountney et al. [42], Burks et al. [43], and Amis et al. [2] also showed similar tensile strength of the MPFL. On the contrary, Hinckel et al. [44] presented significant lower tensile strength of the MPFL (72 ± 32 N). The study had been done on nine knees of the donors of older age than the previous studies. However, LaPrade et al. [41] investigated all medial patellar stabilizers and emphasized the importance of all three ligaments (MPFL, MPTL, and MPML). The results and analysis showed that MPFL and MPTL have no statistically significant difference for the mean failure load. It is important to account a role of all the restructure before proceeding to reconstruction or repair. Duchman et al. [45] investigated the average lateral restraining force of native MPFL in 30° knee flexion. At 1 mm, the lateral patellar displacement force was 10.6 ± 5.7 N, at 5 mm 36.6 ± 2.7 N, and at 10 mm 69.0 ± 5.9 N. The authors also performed the same test in reconstructed MPFL group. They found that at lower lateral displacement, reconstructed MPFL acts similar to native, while at higher lateral displacement shows higher lateral restraining force. This study biomechanically proved the need for MPFL reconstruction. Studies have shown that MPFL is most isometric during knee flexion 0–90°. Steensen et al. [46] found that total changes in length ligament is only 1.1 mm. During surgical procedure, it is important to recreate origin and insertion of the native MPFL as described before.

In literature, numerous surgical techniques exist using a variety of grafts [47]. Hamstring tendons and gracilis tendon, patellar and part of the quadriceps muscle tendon, medial two-third of the adductor magnus tendon, and iliotibial band can be used. Only few studies describe techniques using allografts or synthetic grafts. For the MPFL reconstruction, it is important to use graft with similar biomechanical properties than native ligament. Nowadays, gracilis tendon graft is frequently used with its tensile strength of more than 800 N. This is much higher than native MPFL, while semitendinosus tendon graft has ultimate load of more than 1200 N. Gracilis tendon graft is preferred because of its availability and good biomechanical properties [48].

10.4 Indications for MPFL Reconstruction

In literature, isolated MPFL reconstruction is mostly indicated for the patients suffering from patellofemoral instability. It ranges from the patient with recurrent subluxation or apprehension to patellar dislocation. Recent systematic review by Yeung et al. [49] showed that in 46 of the 56 studies (82.1%), the recurrent patellofemoral instability was the most common indication for this surgical procedure. Fewer studies indicate the reconstruction of MPFL in the case of osteochondral fracture or single dislocation with persistent symptoms. Unsuccessful conservative treatment with physiotherapy and bracing was an indication only in 30.4% or 17 studies of the systematic review analyzed to date. Interestingly, in 7.1% studies, there was no clear indication for performing MPFL reconstruction. Over time, more and more clear indication is defined. Isolated MPFL reconstruction is performed to lower patellar tilt of more than 20°. Only 3% of normal population has tilt greater than this; nevertheless, it is present in 56% if there is history of patellofemoral instability. Studies found out that there is a positive correlation between greater patellar tilt and grade of trochlear dysplasia [50]. It is important to recognize risk factors for patellofemoral instability which may require additional tibial tuberosity transfer if patella alta (TT-TG greater than 20 mm) is present. Lateral retinacula contribute only 10% of the lateral patellar stability, and excessive can be released openly or arthroscopically. Numerous studies have shown that this procedure has inferior outcomes and can even increase instability of the patella. Diener et al. [51] showed that MPFL reconstruction in combination with lateral retinacula release postoperatively increases the likelihood of patellar dislocation. Moreover, Bedi et al. [52] in their study biomechanically proved these findings. Results of Fithian et al. survey have shown that expert surgeons perform this procedure in less than 2% of all cases. There is still no clear indications when to perform division of the lateral retinacula, nevertheless, it should be rarely performed alone. If there is a high-grade trochlear dysplasia, trochleoplasty can be performed. Shape of the trochlea was first described by Henry Dejour and later David Dejour [29, 53]. If there is known high patellofemoral instability, the trochlear dysplasia is presented in 96% cases. First trochleoplasty has been performed by Albee in 1915 [54] which consists of the elevation of trochlear lateral facet. In 1978 Masse [55] introduced deepening of the trochlear groove. A femoral torsional deformity at the knee level can also cause patellofemoral instability. In such cases, supracondylar femoral torsional and varisation osteotomies can be performed [56–59].

10.5 Surgical Techniques of the MPFL Reconstruction

Anatomical and biomechanical researches showed that MPFL is the most important patellar stabilizer in the first 30° of knee flexion. The MPFL is injured in more than 90% of patellar dislocation cases, mostly at the femoral insertion [12]. More than 100 different surgical techniques for patella stabilization are described in literature [60–62]. More and more studies show that performing MPFL reconstruction significantly improves clinical scores and allows patients to return to their daily and sporting activities [13–20, 22, 63, 64].

10.5.1 MPFL Reconstruction with the Gracilis or Semitendinosus Tendon

10.5.1.1 Gracilis and Hamstring Tendon Harvest

Gracilis tendon autograft is our graft of choice because of its biomechanical properties. Harvesting of the graft is done by the same way as for ACL reconstruction. Total length of the graft should be at least 180 mm. At both the sides of the graft, no. 2 nonabsorbable suture (except for the interference screw technique) is placed with a Krackow mattress technique (Fig. 10.2).

10.5.1.2 Patellar Insertion

Area of the MPFL patellar insertion is larger than the area of femoral attachment. Most authors agreed that MPFL is inserted at the upper half of the patella [33, 34, 39, 65]. There are minor differences between techniques, nevertheless most of them restore native patellar insertion. Approach for patellar insertion is made through 2-cm-long longitudinal incision over medial border of the patella, and the medial border of the patella is exposed. Fixation to the patella can be performed using different techniques.

10.5.1.3 Fixation with the Anchors

Fixation with the anchors was first described by Schöttle et al. [66]. With a bur or small raspa

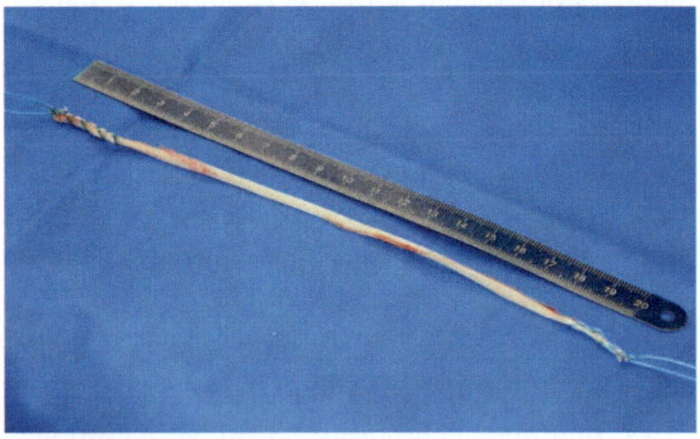

Fig. 10.2 Gracilis tendon graft—at both the sides of the graft no. 2 nonabsorbable suture

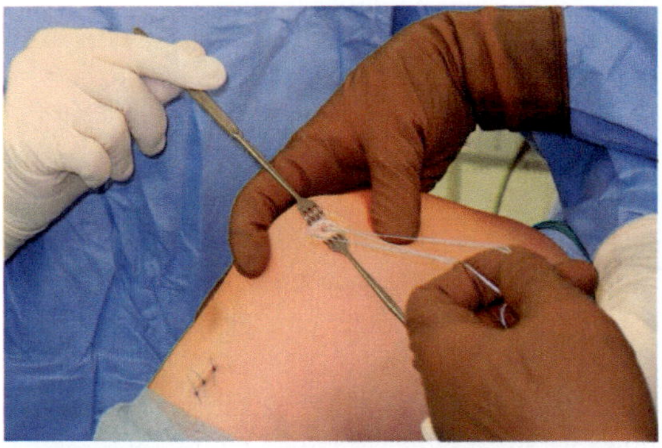

Fig. 10.3 Gracilis tendon graft is placed over medial patellar border and fixated with sliding suture knots

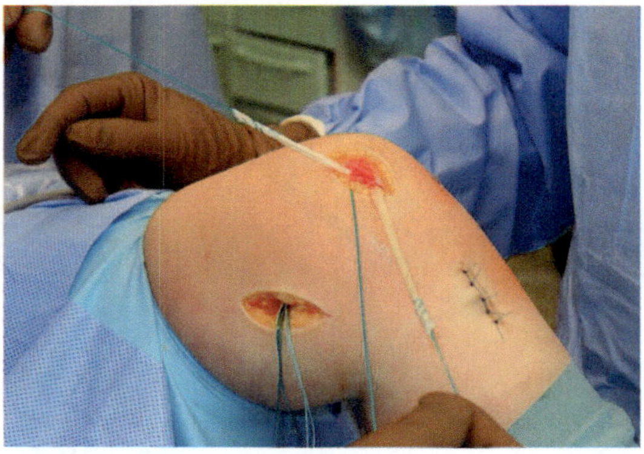

Fig. 10.4 Transosseous tunnels: The free gracilis tendon is passed through the drilled holes of the patella

shallow bony sulcus is created between the proximal and medial thirds of the patella. The authors' preferred technique is fixation of the graft with three nonabsorbable 1.4 mm single loaded suture anchors. The distance between anchors should be between 5 and 10 mm (appropriate according to the patellar size). Graft is placed longitudinal over medial border of the patella and then fixated with sliding suture knots (Fig. 10.3). Care should be taken to put on the medial border of the patella on the central part of the graft to have enough length of the free strands for femoral fixation. Benefit of this technique is to avoid placing prominent hardware into the bone. Moreover, there is also reduced risk of patella fracture intraoperatively or postoperatively.

10.5.1.4 Transosseous Tunnels

Two K wires are drilled in the convergent way in the proximal half of the patella. The distance between K wires should be 10–20 mm, depending on the patella size. Over the K wires, two 4.5–5.5 mm holes are drilled, the tunnels are connected, and the free gracilis tendon is passed through the drilled holes (Fig. 10.4). This technique was first described by Christiansen [25] and modified by Panni et al. [21] later.

10.5.1.5 Transosseous Suture Technique

This technique is similar to previous "anchor technique," but instead of anchors the transosseous nonabsorbable sutures are used. Sutures are passed across the patella from medial to lateral rim by passing small 1.6 mm guide pin. An additional lateral approach to the patella (approximately 3 mm skin incision) is made, and looped suture is withdrawn to the medial border by the same technique. This step is repeated to get two pairs of free suture ends at the medial border of patella. The graft is then secured to the patella with the suture knots. Additionally, medial retinaculum is tightened to the graft for additional stabilization [67]. Benefit of this technique is to avoid placing additional hardware into the bone. Moreover, there is also reduced risk of patella fracture intraoperatively or postoperatively.

10.5.1.6 Intraosseous Fixation with Interference Screw

Schöttle et al. [68] described double-bundle MPFL reconstruction with aperture fixation to patellar insertion. Patellar preparation is performed as described previously, and two guidewires are put into the patella. Guidewires are then overdilled with a 4–5 mm drill to a depth of 20 mm. Free sutured ends of the graft is finally fixated by two 4.75 × 19 mm bioabsorbable interference screw (Fig. 10.5).

10.5.1.7 Passing the Graft through Medial Patellar Complex

After graft fixation to the medial patellar rim, the graft should be passed to the femoral epicondyle point between the second and third layers (Fig. 10.6).

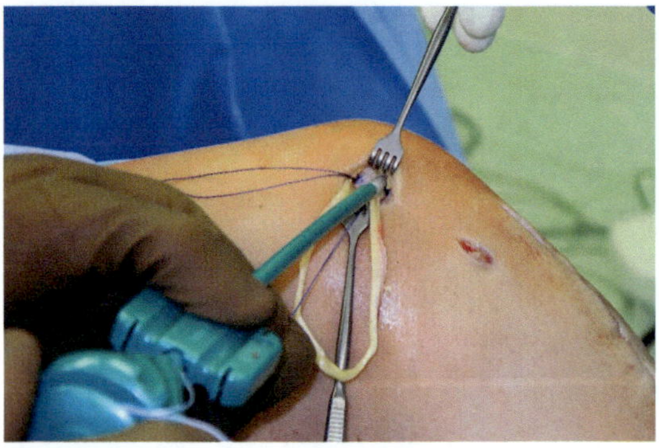

Fig. 10.5 Free sutured ends of the graft is fixated by two 4.75 × 19 mm bioabsorbable interference screw

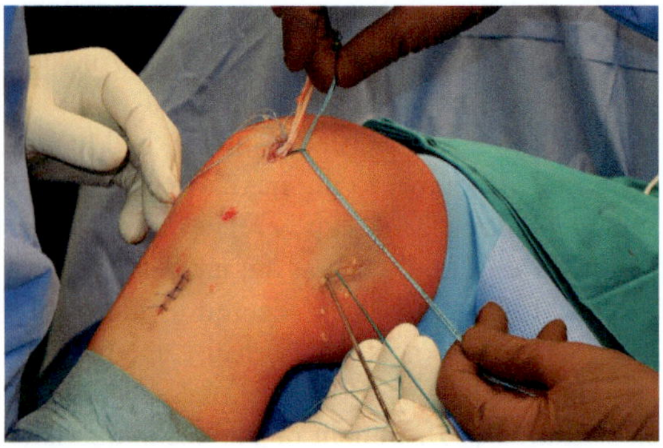

Fig. 10.6 Graft is passed between the second and third layers

The graft should not be deeper than layer 3 to avoid placing it intra-articularly (Fig. 10.7). This step can be performed also before the patellar fixation with passing the graft retrograde from the femoral part to the patellar insertion.

Femoral Insertion Site

Proper anatomical MPFL reconstruction is mandatory for achieving good clinical outcomes postoperatively. Studies showed that improper proximal-positioned or anterior- and proximal-placed femoral tunnels increase medial patellofemoral pressure. Several authors have focused on defying the correct femoral insertion site. Nomura et al. [37] described the anatomical femoral site posteriorly between the medial femoral epicondyle and adductor tubercle. Schöttle et al. in cadaveric study defined a reproducible radiographic point 1.3 mm anterior to the posterior

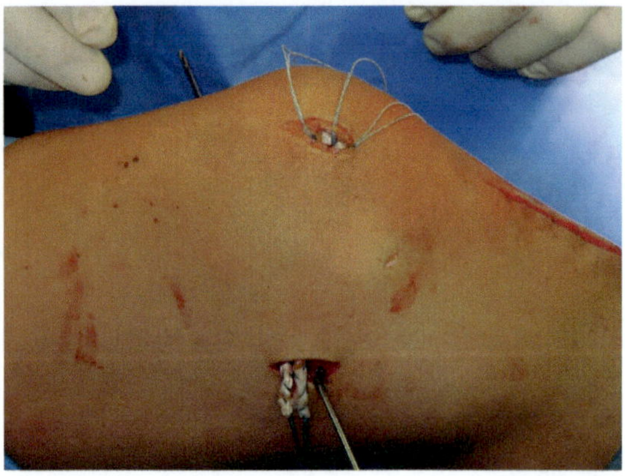

Fig. 10.7 The graft should not be deeper than layer 3 to avoid placing it intra-articularly

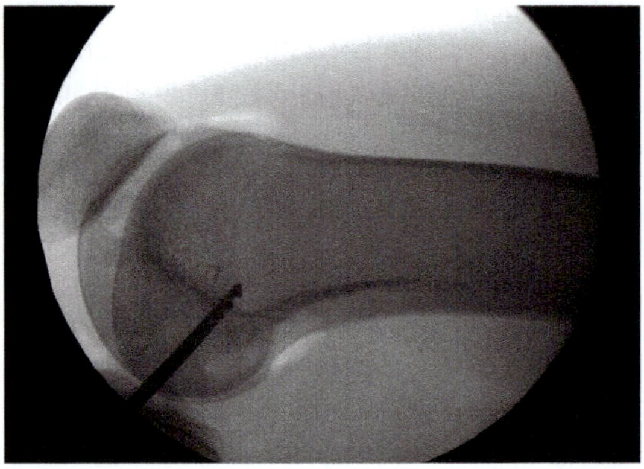

Fig. 10.8 Fluoroscopy-controlled femoral insertion point

cortex and 2.5 mm distal to the posterior origin of the medial femoral condyle and proximal to the level of the posterior point of the Blumensaat line (lateral radiographic view with overlapping posterior condyle line) (Fig. 10.8).

Servien et al. [69] analyzed on MRI femoral tunnel positioning for MPFL reconstruction in correlation with clinical results. They modified Schöttle's point to the anatomical zone to ±7 mm because of the diameter of the femoral tunnel. Therefore, the anatomical position of the tunnel was defined, if it was positioned at a distance of 7 mm from the normal position (Schöttle's point). Nevertheless, recent studies investigating anatomical position radiologically show inaccuracy and are challenging during surgery [70, 71]. Sanchis-Alfonso et al. stated that it is impossible to exactly locate the anatomic femoral tunnel placement with the Schöttle method [70].

10.5.2 Femoral Tunnel Fixation with Interference Screw

Medial femoral epicondyle is exposed through 2 cm approach. The adductor tubercle and medial femoral epicondyle are palpated. The K-wire is placed between them, slightly posteriorly. Proper anatomical femoral tunnel position is identified under fluoroscopy with K-wire. The isometry of the graft is then checked during knee flexion with K-wire position changed if needed. Finally, the femoral tunnel is drilled with a diameter of the MPFL graft (6–8 mm). After patellar fixation, the graft is secured to the femoral anatomical point with the knee in 60° flexion. The interference screw of the same diameter of the tunnel is used to fixate the graft into the tunnel. It is important to insert the screw completely into the tunnel to avoid pain and irritation on the medial part of the knee (Fig. 10.9).

Severe authors suggest different angles in which the femoral fixation of the MPFL should be performed. Thanuat and Erasmus [72] advise the fixation in full extension, Panni et al. [21] in 20°, Toritsuka et al. [73] in 45°, Nomura et al. [6] in 60°, and Schöttle et al. [68] in 30° flexion. Lorbach et al. [74] in biomechanical study analyzed patellofemoral contact pressure in different angles of knee flexion. They conclude that fixation at 60° of flexion best restores patellofemoral contact pressure compared with the intact knee. It is possible to use the arthroscope through superolateral portal to check the patella position and graft tension during knee flexion.

10.5.3 Femoral Tunnel Fixation with Extracortical Button

In this technique, the graft should be sutured around the adjustable loop cortical button fixation device. After patellar fixation, the femoral tunnel is defined and drilled with a K-wire and overdrilled with 4.5 mm drill bit. The length of the femoral tunnel is measured and then marked on the adjustable loop. The femoral tunnel is drilled with the diameter of the ligament to a depth 1 cm more than the measured length of the graft. The adjustable loop is passed, and the button is fixed on the lateral cortex. X-rays can be used to check the button position. The loop is then shortened with the knee in 60° flexion by pulling the two sutures. In this way, the graft is slowly inserted in the femoral tunnel till the desired tension is achieved (Fig. 10.10). Care should be taken not to overtight the graft because the system does not allow to undertight the fixation.

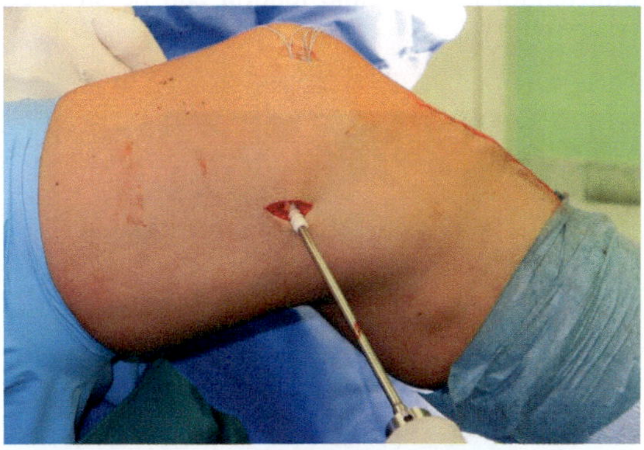

Fig. 10.9 Inserting the screw completely into the femoral tunnel

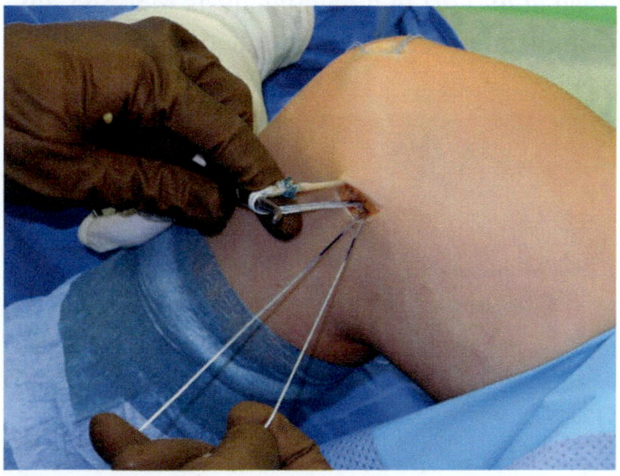

Fig. 10.10 Inserting the graft into the tunnel till the desired tension is achieved

10.5.4 MPFL Reconstruction Technique Using Quadriceps Tendon Graft

In 90° of knee flexion, a 2–3 cm transverse skin incision is placed over the superomedial pole of the patella. The prepatellar bursa is incised longitudinally, so the quadriceps tendon is exposed. The length, depth, and width of the quadriceps tendon which is then used for the MPFL graft are determined. The quadriceps tendon is harvested with a tendon stripper. The free part of the tendon is sutured with nonabsorbable sutures, and the graft is passed subperiosteally to the medial part of the patella where it is sutured with absorbable sutures (Fig. 10.11).

After patellar fixation, the graft is passed between the second and third layers to the femoral insertion side, where it is fixated with the interference screw or extracortical button as described in the previous chapter. The advantage of this technique is that it is minimally invasive with a good esthetic result; furthermore, we can also avoid implants or possible bone tunnels in the patella [75]. Thus, the technique presents a valuable alternative to common hamstring techniques for primary MPFL reconstruction as well as MPFL revision.

10.5.5 Complications

Shah et al. [14] in a meta-analysis reported 26% complications despite high rate of success of the MPFL reconstruction. The most important for successful treatment is proper indication and selection of patients. There are also technical errors and failures during the procedure. Parikh et al. [76] showed that 47% of the complications are related to technical errors. Most significant is inappropriate femoral tunnel placement. Femoral fixation point determines the kinematic behavior of the graft [77]. Excessive graft tension can lead to postoperative knee stiffness and loss of flexion. Moreover, patellofemoral contact pressure is altered in inappropriate angle of the knee flexion during fixation. Recurrence of patellofemoral dislocation reported in literature [25, 50, 78] is between 0 and 4%. Other complications are femoral pain because of the hardware placement into the bone, pain along overtensioned graft and quadriceps muscle atrophy.

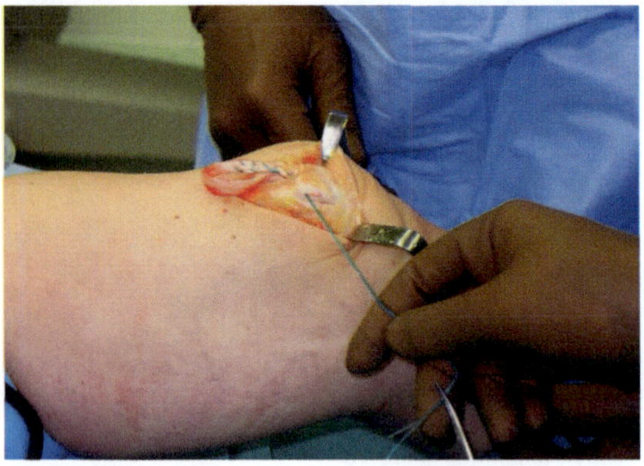

Fig. 10.11 Passing the Q-tendon graft to the medial part of the patella

10.6 MPFL Reconstruction in Skeletally Immature Patients

10.6.1 MPFL Reconstruction with the Adductor Tendon

The 5-cm-long incision is made over the distal medial part of the femur. The adductor tendon is exposed and released between hiatus and adductor tendon insertion. In the mentioned technique, the front two thirds of the large adductor tendon are relaxed in the length of 12–14 cm. In very thin tendons, the whole tendon can be stripped (Fig. 10.12).

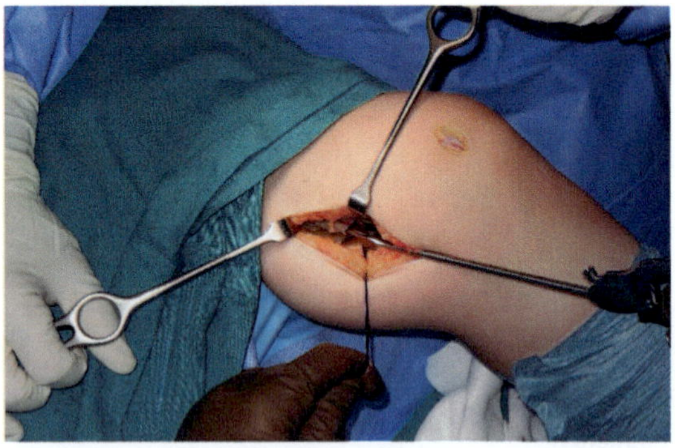

Fig. 10.12 The front two thirds of the large adductor tendon are stripped in the length of 12–14 cm

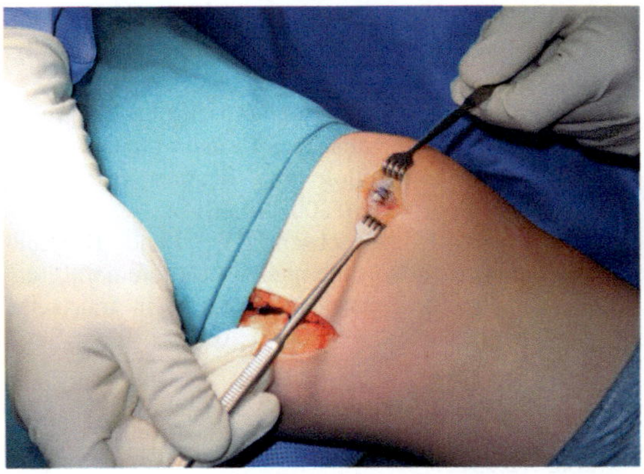

Fig. 10.13 Large adductor tendon fixated to the medial patellar border

Free tendon is then passed between the second and third layers of the medial structures of the patella to its medial edge and fixed with the suture anchor in 30° of flexion (Fig. 10.13).

With this technique, no fixation is needed to femur, and possible damage of the growth zones are avoided. The technique is suitable also in cases of higher placed patella, where the transposition of tuberositas tibiae would not be chosen. This technique was presented by Sillanpää et al. [79], modifying Avikainen technique.

10.6.2 Modified Adductor Sling Technique

Alm et al. [80] in 2017 presented modified adductor sling reconstruction technique of the MPFL. Twenty-eight children and adolescents were included in the study from 2010 to 2016 with good results. In the mentioned technique, gracilis or semitendinosus tendon is looped around adductor m agnus tendon and fixed with sutured anchors or in transosseous way in 30° of flexion to the medial patellar border (Fig. 10.14).

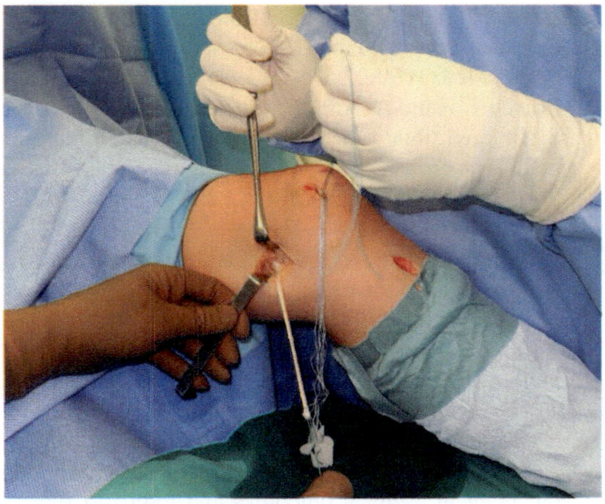

Fig. 10.14 Modified adductors linger construction technique of the MPFL

References

1. Ahmad CS, Brown GD, Stein BS. The docking technique for medial patellofemoral ligament reconstruction: surgical technique and clinical outcome. Am J Sports Med. 2009;37:2021–7.
2. Amis AA, Firer P, Mountney J, Senavongse W, Thomas NP. Anatomy and biomechanics of the medial patellofemoral ligament. Knee. 2003;10:215–20.
3. Baldwin JL. The anatomy of the medial patellofemoral ligament. Am J Sports Med. 2009;37:2355–61.
4. Cash JD, Hughston JC. Treatment of acute patellar dislocation. Am J Sports Med. 1988;16:244–9.
5. Hawkins RJ, Bell RH, Anisette G. Acute patellar dislocations. The natural history. Am J Sports Med. 1986;14:117–20.
6. Nomura E, Inoue M, Osada N. Augmented repair of avulsion-tear type medial patellofemoral ligament injury in acute patellar dislocation. Knee Surg Sports Traumatol Arthrosc. 2005;13:346–51.
7. Garth WP, Pomphrey M, Merrill K. Functional treatment of patellar dislocation in an athletic population. Am J Sports Med. 1996;24:785–91.
8. Dejour H, Walch G, Nove-Josserand L, Guier C. Factors of patellar instability: an anatomic radiographic study. Knee Surg Sport Traumatol Arthrosc. 1994;2:19–26.
9. Myers P, Williams A, Dodds R, Bülow J. The three-in-one proximal and distal soft tissue patel-

lar realignment procedure. Results, and its place in the management of patellofemoral instability. Am J Sports Med. 1999;27:575–9.
10. Pritsch T, Haim A, Arbel R, Snir N, Shasha N, Dekel S. Tailored tibial tubercle transfer for patellofemoral malalignment: analysis of clinical outcomes. Knee Surg Sports Traumatol Arthrosc. 2007;15:994–1002.
11. Tjoumakaris FP, Forsythe B, Bradley JP. Patellofemoral instability in athletes: treatment via modified Fulkerson osteotomy and lateral release. Am J Sports Med. 2010;38:992–9.
12. Nomura E, Horiuchi Y, Kihara M. Medial patellofemoral ligament restraint in lateral patellar translation and reconstruction. Knee. 2000;7:121–7.
13. Lippacher S, Dreyhaupt J, Williams SRM, Reichel H, Nelitz M. Reconstruction of the medial patellofemoral ligament: clinical outcomes and return to sports. Am J Sports Med. 2014;42:1661–8.
14. Shah JN, Howard JS, Flanigan DC, Brophy RH, Carey JL, Lattermann C. A systematic review of complications and failures associated with medial Patellofemoral ligament reconstruction for recurrent patellar dislocation. Am J Sports Med. 2012;40:1916–23.
15. Tompkins MA, Arendt EA. Patellar instability factors in isolated medial patellofemoral ligament reconstructions—what does the literature tell us?: a systematic review. Am J Sports Med. 2015;43:2318–27.
16. Kohn LM, Meidinger G, Beitzel K, Banke IJ, Hensler D, Imhoff AB, Schottle PB. Isolated and combined medial patellofemoral ligament reconstruction in revision surgery for patellofemoral instability: a prospective study. Am J Sports Med. 2013;41:2128–35.
17. Nomura E, Inoue M, Kobayashi S. Long-term follow-up and knee osteoarthritis change after medial patellofemoral ligament reconstruction for recurrent patellar dislocation. Am J Sports Med. 2007;35:1851–8.
18. Nomura E, Horiuchi Y, Kihara M. A mid-term follow-up of medial patellofemoral ligament reconstruction using an artificial ligament for recurrent patellar dislocation. Knee. 2000;7:211–5.
19. Nelitz M, Dreyhaupt J, Reichel H, Woelfle J, Lippacher S. Anatomic reconstruction of the medial patellofemoral ligament in children and adolescents with open growth plates: surgical technique and clinical outcome. Am J Sports Med. 2013;41:58–63.
20. Steiner TM, Torga-Spak R, Teitge RA. Medial patellofemoral ligament reconstruction in patients with lateral patellar instability and trochlear dysplasia. Am J Sports Med. 2006;34:1254–61.
21. Panni AS, Alam M, Cerciello S, Vasso M, Maffulli N. Medial patellofemoral ligament reconstruction with a divergent patellar transverse 2-tunnel technique. Am J Sports Med. 2011;39:2647–55.
22. Ronga M, Oliva F, Longo UG, Testa V, Capasso G, Maffulli N. Isolated medial patellofemoral ligament reconstruction for recurrent patellar dislocation. Am J Sports Med. 2009;37:1735–42.
23. Nomura E, Inoue M. Hybrid medial patellofemoral ligament reconstruction using the semitendinous tendon for recurrent patellar dislocation: minimum 3 years' follow-up. Arthrosc J Arthrosc Relat Surg. 2006;22:787–93.
24. Schöttle PB, Schmeling A, Rosenstiel N, Weiler A. Radiographic landmarks for femoral tunnel placement in medial patellofemoral ligament reconstruction. Am J Sports Med. 2007;35:801–4.
25. Christiansen SE, Jacobsen BW, Lund B, Lind M. Reconstruction of the medial patellofemoral ligament with Gracilis tendon autograft in transverse patellar drill holes. Arthrosc J Arthrosc Relat Surg. 2008;24:82–7.
26. Cossey AJ, Paterson R. A new technique for reconstructing the medial patellofemoral ligament. Knee. 2005;12:93–8.
27. Conlan T, Garth WP, Lemons JE. Evaluation of the medial soft-tissue restraints of the extensor mechanism of the knee. J Bone Joint Surg Am. 1993;75:682–93.
28. Platzer W. Anatomia umana. Atlante tascabile. Apparato locomotore. 2007.
29. Dejour H. Dysplasia of the intercondilar groove. Fr J Orthop Surg. 1990;4:113–22.
30. Warren LF, Marshall JL. The supporting structures and layers on the medial side of the knee: an anatomical analysis. J Bone Joint Surg Am. 1979;61:56–62.
31. Feller JA, Feagin JA, Garrett WE. The medial patellofemoral ligament revisited: an anatomical study. Knee Surg Sport Traumatol Arthrosc. 1993;1:184–6.
32. Smirk C, Morris H, Fulkerson JP, et al. The anatomy and reconstruction of the medial patellofemoral ligament. Knee. 2003;10:221–7.
33. Tuxøe JI, Teir M, Winge S, Nielsen PL. The medial patellofemoral ligament: a dissection study. Knee Surg Sport Traumatol Arthrosc. 2002;10:138–40.
34. Philippot R, Chouteau J, Wegrzyn J, Testa R, Fessy MH, Moyen B. Medial patellofemoral ligament anatomy: implications for its surgical reconstruction. Knee Surg Sport Traumatol Arthrosc. 2009;17:475–9.
35. Desio SM, Burks RT, Bachus KN. Soft tissue restraints to lateral patellar translation in the human knee. Am J Sports Med. 1998;26:59–65.
36. Hautamaa PV, Fithian DC, Kaufman KR, Daniel DM, Pohlmeyer AM. Medial soft tissue restraints in lateral patellar instability and repair. Clin Orthop Relat Res. 1998;349:174–82.
37. Nomura E, Inoue M, Osada N. Anatomical analysis of the medial patellofemoral ligament of the knee, especially the femoral attachment. Knee Surg Sport Traumatol Arthrosc. 2005;13:510–5.
38. Arendt EA. Medial side Patellofemoral anatomy: surgical implications in patellofemoral instability. In: Patellofemoral pain, instability, and arthritis. Berlin Heidelberg: Springer; 2010. p. 149–52.
39. LaPrade RF, Engebretsen AH, Ly TV, Johansen S, Wentorf FA, Engebretsen L. The anatomy of the medial part of the knee. J Bone Joint Surg Am. 2007;89:2000–10.
40. Larsen E, Lauridsen F. Conservative treatment of patellar dislocations. Influence of evident factors

on the tendency to redislocation and the therapeutic result. Clin Orthop Relat Res. 1982;171:131–6.
41. Laprade MD, Kallenbach SL, Aman ZS, Moatshe G, Storaci HW, Turnbull TL, Arendt EA, Chahla J, Laprade RF. Biomechanical evaluation of the medial stabilizers of the patella. Am J Sports Med. 2018;46(7):1575–82.
42. Mountney J, Senavongse W, Amis AA, Thomas NP. Tensile strength of the medial patellofemoral ligament before and after repair or reconstruction. J Bone Joint Surg Br. 2005;87:36–40.
43. Burks RT, Desio SM, Bachus KN, Tyson L, Springer K. Biomechanical evaluation of lateral patellar dislocations. Am J Knee Surg. 1998;11:24–31.
44. Demange MK, Ph D, Pereira CAM, et al. Medial patellofemoral ligament, medial patellotibial ligament, and medial patellomeniscal ligament: anatomic, histologic, radiographic, and biomechanical study. Arthrosc J Arthrosc Relat Surg. 2017;33(10):1862–73.
45. Duchman KR, DeVries NA, McCarthy MA, Kuiper JJ, Grosland NM, Bollier MJ. Biomechanical evaluation of medial patellofemoral ligament reconstruction. Iowa Orthop J. 2013;33:64–9.
46. Steensen RN, Dopirak RM, Mcdonald WG. The anatomy and isometry of themedial patellofemoral ligament. Am J Sports Med. 2004;32:1509–13.
47. Chouteau J. Surgical reconstruction of the medial patellofemoral ligament. Orthop Traumatol Surg Res. 2016;102:S189–94.
48. Amis A, Arendt EA, Deehan D, et al. The medial patellofemoral ligament. In: ESSKA Instructional Course Lectures Book. Berlin, Heidelberg: Springer; 2014. p. 113–25.
49. Yeung M, Leblanc M-C, Ayeni O, Khan M, Hiemstra L, Kerslake S, Peterson D. Indications for medial patellofemoral ligament reconstruction: a systematic review. J Knee Surg. 2015;29:543–54.
50. Schöttle PB, Fucentese SF, Romero J. Clinical and radiological outcome of medial patellofemoral ligament reconstruction with a semitendinosus autograft for patella instability. Knee Surg Sport Traumatol Arthrosc. 2005;13:516–21.
51. Dainer RD, Barrack RL, Buckley SL, Alexander AH. Arthroscopic treatment of acute patellar dislocations. Arthroscopy. 1988;4:267–71.
52. Bedi H, Marzo J. The biomechanics of medial patellofemoral ligament repair followed by lateral retinacular release. Am J Sports Med. 2010;38:1462–7.
53. Galland O, Walch G, Dejour H, Carret JP. An anatomical and radiological study of the femoropatellar articulation. Surg Radiol Anat. 1990;12:119–25.
54. Aglietti P, Buzzi R, De Biase P, Giron F. Surgical treatment of recurrent dislocation of the patella. Clin Orthop Relat Res. 1994;308:8–17.
55. Masse Y. La trochleoplastie, restauratio de la gouttierre trochleenne dans les subluxations et luxations de la rotule. Rev Chir Orthop. 1978;64:3–17.
56. Swarup I, Elattar O, Rozbruch SR. Patellar instability treated with distal femoral osteotomy. Knee. 2017;24:608–14.
57. Rg Dickschas J, Rg Harrer J, Reuter B, Schwitulla J, Strecker W. Torsional osteotomies of the femur. J Orthop Res. 2014;33:318–24. https://doi.org/10.1002/jor.22758.
58. Teitge RA. The role of limb rotational osteotomy in the treatment of patellofemoral dysfunction. In: Patellofemoral pain, instability and arthritis. Berlin Heidelberg: Springer; 2010. p. 237–44.
59. Teitge RA. Osteotomy in the treatment of patellofemoral instability. Tech Knee Surg. 2006;5:2–18.
60. Abraham E, Washington E, Huang TL. Insall proximal realignment for disorders of the patella. Clin Orthop Relat Res. 1989;248:61–5.
61. Boden B, Pearsall A, Garrett W, Feagin J. Patellofemoral instability: evaluation and management. J Am Acad Orthop Surg. 1997;5:47–57.
62. Dandy DJ, Griffiths D. Lateral release for recurrent dislocation of the patella. J Bone Joint Surg Br. 1989;71:121–5.
63. Ambrožič B, Novak S. The influence of medial patellofemoral ligament reconstruction on clinical results and sports activity level. Phys Sportsmed. 2016;44(2):133–40. https://doi.org/10.1080/00913847.2016.1148561.
64. Panni AS, Cerciello S, Vasso M, Palombi A. Medial reefing in chronic potential patellar instability. In: Patellofemoral pain, instability and arthritis. Berlin, Heidelberg: Springer; 2010. p. 159–63.
65. Nomura E, Fujikawa K, Takeda TMH. Anatomical study of the medial patellofemoral ligament (in Japanese). Bessatsu Seikeigeka. 1992;22:2–5.
66. Schöttle P, Schmeling A, Romero J, Weiler A. Anatomical reconstruction of the medial patellofemoral ligament using a free gracilis autograft. Arch Orthop Trauma Surg. 2009;129:305–9.
67. Siebold R, Chikale S, Sartory N, Hariri N, Feil S, Pässler HH. Hamstring graft fixation in MPFL reconstruction at the patella using a transosseous suture technique. Knee Surg Sport Traumatol Arthrosc. 2010;18:1542–4.
68. Schöttle PB, Hensler D, Imhoff AB. Anatomical double-bundle MPFL reconstruction with an aperture fixation. Knee Surg Sport Traumatol Arthrosc. 2010;18:147–51.
69. Servien E, Fritsch B, Lustig S, Demey G, Debarge R, Lapra C, Neyret P. In vivo positioning analysis of medial patellofemoral ligament reconstruction. Am J Sports Med. 2011;39:134–9.
70. Sanchis-Alfonso V, Ramírez-Fuentes C, Montesinos-Berry E, Elía I, Martí-Bonmatí L. Radiographic location does not ensure a precise anatomic location of the femoral fixation site in medial Patellofemoral ligament reconstruction. Orthop J Sports Med. 2017;5:2325967117739252.
71. Ziegler CG, Fulkerson JP, Edgar C. Radiographic reference points are inaccurate with and without a true lateral radiograph. Am J Sports Med. 2016;44:133–42.
72. Thaunat M, Erasmus PJ. The favourable anisometry: an original concept for medial patellofemoral ligament reconstruction. Knee. 2007;14:424–8.

73. Toritsuka Y, Amano H, Mae T, Uchida R, Hamada M, Ohzono K, Shino K. Dual tunnel medial patellofemoral ligament reconstruction for patients with patellar dislocation using a semitendinosus tendon autograft. Knee. 2011;18:214–9.
74. Lorbach O, Zumbansen N, Kieb M, Efe T, Pizanis A, Kohn D, Haupert A. Medial Patellofemoral ligament reconstruction: impact of knee flexion angle during graft fixation on dynamic Patellofemoral contact pressure—a biomechanical study. Arthrosc J Arthrosc Relat Surg. 2018;34:1072–82.
75. Fink C, Veselko M, Herbort M, Hoser C. The knee MPFL reconstruction using a quadriceps tendon graft. Part 2: Operative technique and short term clinical results. Knee. 2014;21:1175–9.
76. Parikh SN, Nathan ST, Wall EJ, Eismann EA. Complications of medial patellofemoral ligament reconstruction in young patients. Am J Sports Med. 2013;41:1030–8.
77. Sanchis-Alfonso V, Ramirez-Fuentes C, Montesinos-Berry E, Domenech J, Martí-Bonmatí L. Femoral insertion site of the graft used to replace the medial patellofemoral ligament influences the ligament dynamic changes during knee flexion and the clinical outcome. Knee Surg Sport Traumatol Arthrosc. 2017;25:2433–41.
78. Drez D, Edwards TB, Williams CS. Results of medial patellofemoral ligament reconstruction in the treatment of patellar dislocation. Arthroscopy. 2001;17:298–306.
79. Sillanpää PJ, Mäenpää HM, Mattila VM, Visuri T, Pihlajamäki H. A mini-invasive adductor magnus tendon transfer technique for medial patellofemoral ligament reconstruction: a technical note. Knee Surg Sports Traumatol Arthrosc. 2009;17:508–12.
80. Alm L, Krause M, Mull C, Frosch K, Akoto R. The knee modified adductor sling technique: a surgical therapy for patellar instability in skeletally immature patients. Knee. 2017;24:1282. https://doi.org/10.1016/j.knee.2017.08.051.

Medial Collateral Ligament Anatomical Repair and Reconstructions

11

Vlad Predescu, Ioana Enăchescu, and Bogdan Deleanu

The treatment of acute medial collateral ligament (MCL) as a solitary lesion, as well as combined with other ligamentous injuries is controversial, resulting in some cases in chronic instability. The treatment of an MCL tear is dictated by the alignment of the knee and by the associated lesions, which in many cases increase knee instability causing it to become symptomatic. Before any treatment can be considered, a full assessment of the injury must be performed, using clinical and imagistic techniques.

MCL consists of two bundles, the superficial (sMCL) and deep (dMCL) layers, and is the main stabilizer of knee valgus from 30°. From 0 to 30°, the main restrictor of valgus stress is the posterior oblique ligament (POL), which inserts immediately posterior to the MCL insertion on the femur. MCL and POL act as secondary restrictors of tibial external rotation in relation to the femur. They are usually injured together [1–4].

A frequent association is with an anterior cruciate ligament (ACL) injury. The literature regarding the surgical management in these cases is controversial. Some authors prefer to treat the MCL injury conservatively, followed by a delayed ACL reconstruction. If the medial-sided stability is not adequate after the reconstruction of the ACL, an MCL reconstruction is also performed. Others advocate for early ACL reconstruction, treating the MCL injury conservatively while some choose to treat both injuries surgically in an acute setting, i.e., ACL reconstruction and MCL repair. Insufficient medial instability causes are additional stress on the reconstructed ACL which could lead to graft failure. This controversy applies to grade III MCL injuries. For grades I and II (incomplete injuries) and isolated grade III tears, conservative treatment is the standard treatment due to the fact that the MCL has a good innate healing potential due to its vascularization and broad surface [5–7].

Indications for surgical treatment:

- Multi-ligament injuries.
- A lesion that has no healing potential with conservative treatment (extensive defect due to border separation).
- High-demand athletes.
- Avulsed bone fragment.
- Chronic rupture (i.e., a lesion that despite correct conservative treatment shows no sign of cicatrization after a minimum of 6 weeks).
- Genu valgum: if a valgus alignment of the knee is associated with an MCL rupture, then a distal femoral realignment osteotomy is performed first and only after it is healed (several

V. Predescu (✉)
Ponderas Academic Hospital, Bucharest, Romania

I. Enăchescu
Bucharest Emergency Hospital, Bucharest, Romania

B. Deleanu
University of Medicine Victor Babeş,
Timisoara, Romania

M. Bozkurt, H. İ. Açar (eds.), *Clinical Anatomy of the Knee*,
https://doi.org/10.1007/978-3-030-57578-6_11

months), if there is residual instability, ligament reconstruction is performed. If the leg alignment is not corrected, the soft tissue graft will be stretched and becomes inefficient [8].

11.1 Surgical Treatment

Before any open surgical treatment involving the MCL, it is imperative to perform a knee arthroscopy to address the associated injuries in need of repair and examine the joint to exclude lesions unapparent during clinical and MRI examination. Capsular integrity must be taken into consideration as surgical dissection might exacerbate capsular lesion and cause secondary extraarticular fluid extravasation. Recent trauma frequently presents with a damaged knee capsule. It is considered that after 5–7 days, a skilled surgeon can perform an arthroscopy, provided that the time of the intervention is maintained at a minimum.

Depending on the location of the lesion, several surgical techniques have been described. In an acute lesion, anatomical structure can be restored and should be done without delay for optimal results. In chronic tears, surgical dissection is difficult, and anatomical repair can be impossible, requiring reconstruction.

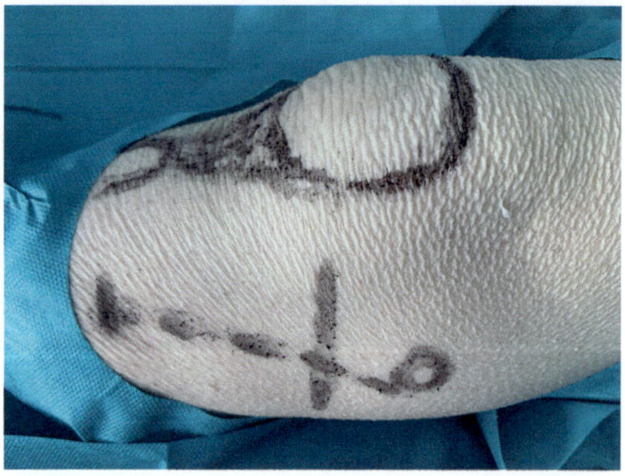

Fig. 11.1 Anatomical landmarks and planned approach to MCL

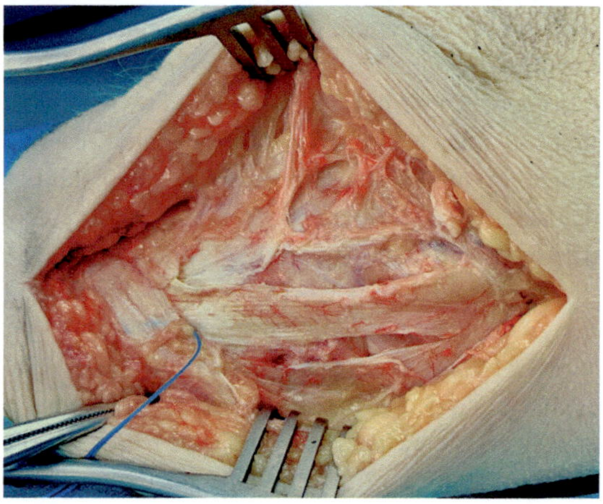

Fig. 11.2 Exposure of the MCL; pes anserinus tendons (blue suture)

11.2 Approach

The surgical approach for MCL repair is similar to the classic medial approach of the knee. The length of the incision is around 10 cm or less, according to the location and planned treatment of the injury. An incision is performed starting at 1–2 cm above the medial epicondyle, around the midline of the knee, approximately 3 cm from the patella and carried down straight in extension or anteriorly curved when the patient has been positioned with the leg at 90° flexion, toward the tibial insertion of the MCL (5 or 6 cm from the joint line) (Fig. 11.1).

The subcutaneous tissue is dissected in line with the skin incision. An oblique incision, posteriorly oriented, is carried through the sartorial fascia, to expose the MCL and also gain visualization of other medial stabilizers that might be injured—POL, the posteromedial corner (Fig. 11.2). Care should be taken to avoid the saphenous vein, running in the posterior aspect of the incision and the saphenous nerve located between the sartorius and gracilis muscles.

If an ACL reconstruction is planned at the same time, using an ipsilateral hamstring autograft, the approach for the pes anserinus tendon harvest can be extended or altered to access the distal part of the MCL. When a bone–patellar tendon–bone graft is used, the distal part of the MCL is exposed by a slight extension of the incision and medial retraction of the skin and subcutaneous tissue. This is useful only for distal MCL lesions [9, 10].

11.3 Primary Repair

A repair of the lesion is considered in cases where preoperative imaging and planning determine that anatomical restoration of the medial collateral ligament can be accomplished. The timing of the surgery should be carefully chosen, taking into account the state of the surrounding soft tissues and the associated injuries that require arthroscopic repair and a capsule that can sustain the length of the intervention.

During the arthroscopic part of the intervention, a positive "drive-through" sign is indicative of MCL lesion (exacerbated medial sided gap under valgus stress). The location of the tear is identified by performing a "liftoff" test. When applying valgus stress on the affected knee, the medial meniscus moves away from the site of the injury. If the meniscus stays on the tibia, the approach should address the proximal part of the ligament, and when the meniscus is lifted in the direction of the femoral condyle, the distal part is affected.

In proximal tears, a bone fragment (Stieda fracture) can be avulsed and should be identified with imaging (Fig. 11.3). Preoperative X-rays can differentiate an acute lesion from a chronic Pellegrini-Stieda lesion, which appear as the result of bone remodeling of a Stieda fragment or ossification at the site on a femoral-sided MCL tear, therefore aiding the planning of the intended surgery [11, 12].

The site of the superficial MCL lesion is identified—hematoma points to the site of the injury in recent tears. Adjacent structures are inspected: the deep MCL, POL, posteromedial corner/capsule, and repaired, as needed, using non-absorbable sutures or suture anchors. The superficial MCL (sMCL) is tensioned at 30° flexion, applying varus stress. The deep MCL (dMCL) and POL are repaired with the knee in extension. The repair is done starting from the deepest structures toward the most superficial. If a meniscal capsular detachment is identified, it has to be repaired.

When the MCL injury is located at an insertion site (femoral or tibial), it can be reattached using a suture anchor (Figs. 11.4 and 11.5) or a screw with a spiked washer. The sMCL has one femoral attachment (on average 3.2 mm proximal and 4.8 mm posterior to the medial epicondyle) and two tibial attachments: the proximal is a soft tissue insertion, on the anterior part of the semitendinosus (ST) tendon insertion (average 12.2 mm from the joint line); the distal is just anterior to the posteromedial tibial crest, approximately 6 cm from the joint line, mostly covered by the anserine bursae. The dMCL attaches 1 cm distal to the sMCL, to the medial meniscus and distally on average 3.2 mm from the joint line. It appears as a thickening of the joint capsule. The

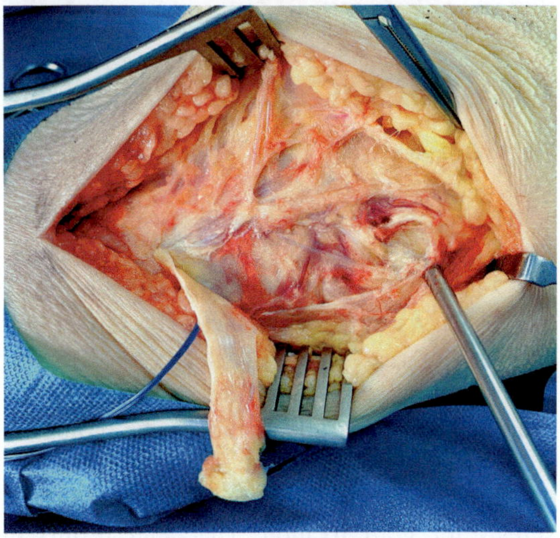

Fig. 11.3 Proximal MCL avulsion with bone fragment (Stieda)

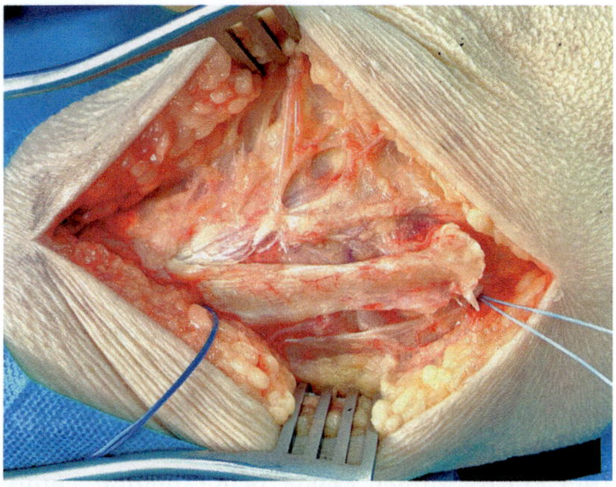

Fig. 11.4 Suture anchor placed at the origin of the proximally avulsed MCL

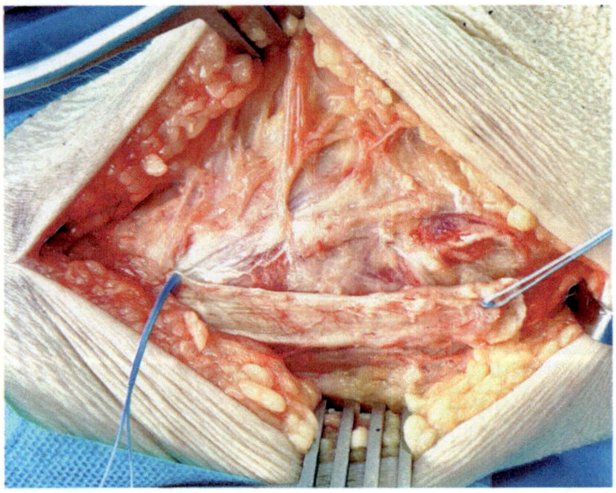

Fig. 11.5 Femoral reattachment of the avulsed MCL with a suture anchor

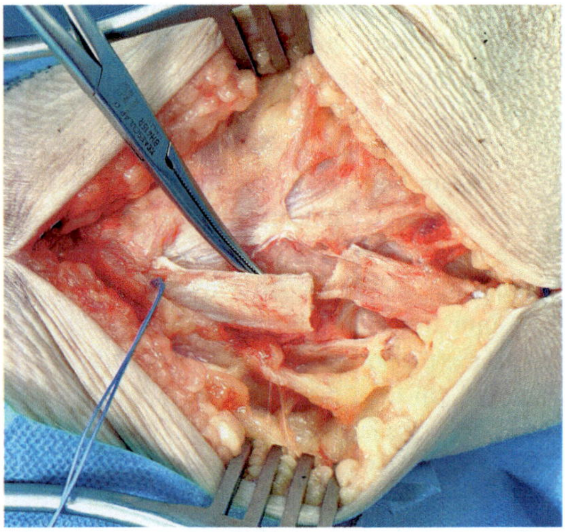

Fig. 11.6 MCL mid-substance tear

same can be said about the POL, previously described as the oblique portion of the MCL, which is rather a thickening of the posteromedial capsule that runs from its insertion slightly proximal and approximately 9.2 mm posterior from the sMCL insertion (average: 7.7 mm distal and 6 mm posterior to the adductor tubercle) toward the medial meniscus and its tibial insertion near the semimembranosus (SM) insertion (LaPrade et al., Encinas and Rodriguez) [6, 13, 14].

If there is a mid-substance MCL tear (Fig. 11.6), direct repair using tendon suture techniques is performed (Figs. 11.7, 11.8 and 11.9).

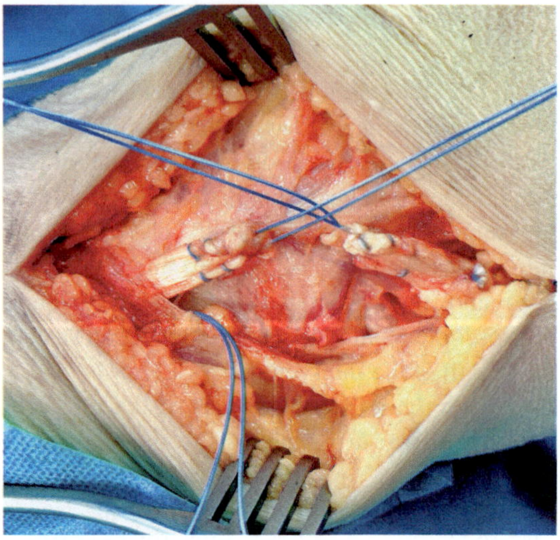

Fig. 11.7 Non-absorbable sutures placed on the torn ends of the MCL (Krackow stitch)

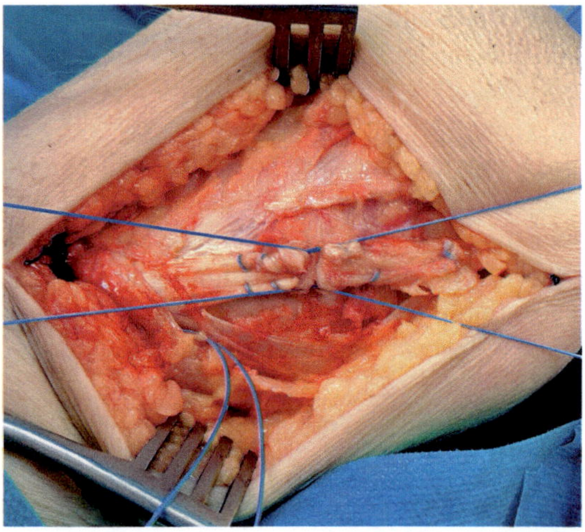

Fig. 11.8 Suture tying

The repair can be augmented using artificial biomaterials (FiberTape®) (Figs. 11.10, 11.11, 11.12 and 11.13) or tendon grafts (preferred: ST autograft). As Mackay observes, the FiberTape® strands (InternalBrace™, Arthrex) have important advantages: avoiding the graft harvest-site morbidity of autografts, they lack the biological risks of allografts, eliminating the need for the sizable tunnels made when using interference screws for graft fixation. The latter results in bone preservation, especially important in multiligament reconstruction [15].

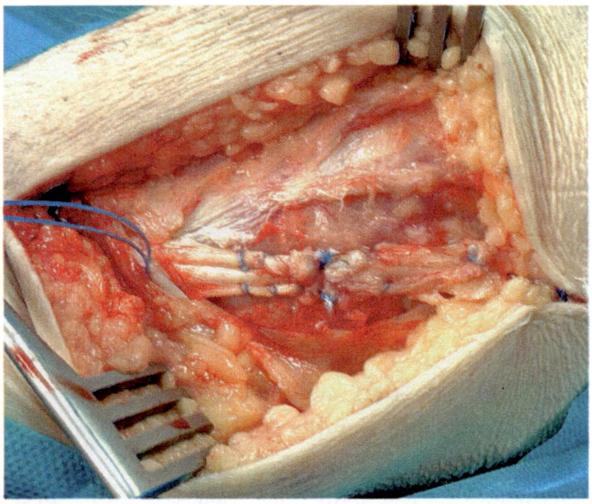

Fig. 11.9 MCL mid-substance repair

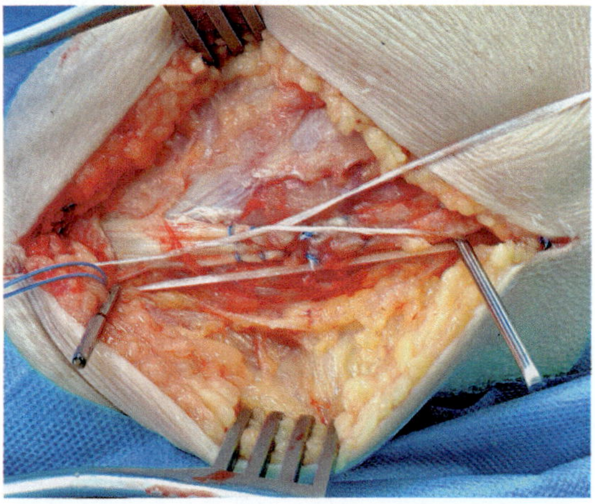

Fig. 11.10 An isometric insertion of the fibers is performed, ensuring stability

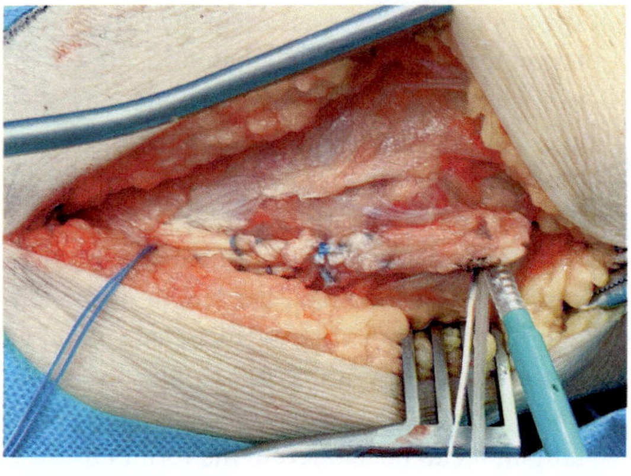

Fig. 11.11 Femoral attachment of the FiberTape® (SwiveLock® anchor)

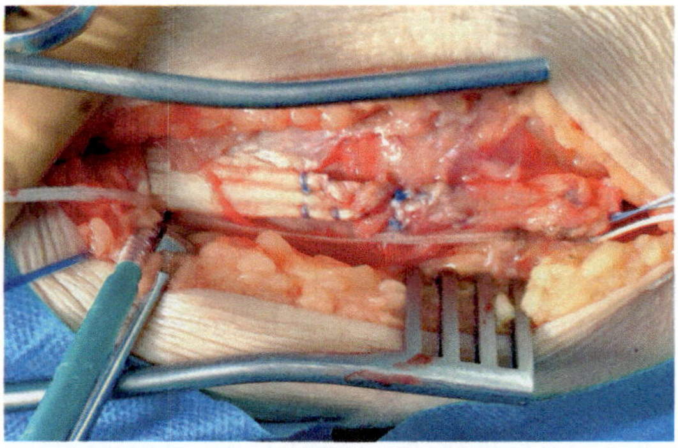

Fig. 11.12 Tibial insertion of the FiberTape® (SwiveLock® anchor)

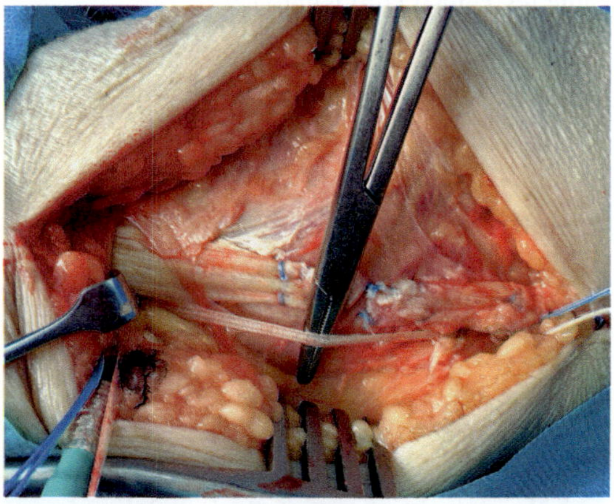

Fig. 11.13 MCL middle third repair augmented using artificial materials (FiberTape®)

Internal bracing can be performed percutaneously, using two small incisions placed over the origin and insertion of the ligament, thus medially stabilizing the knee to create optimal conditions for the healing of an acute MCL tear. The disadvantage of this technique is that it is not helpful in chronic tears.

The main complication of internal bracing is stiffness. If the FiberTape® is too tight, it can over constraint the knee, leading to a difficult rehabilitation and subsequent loss of knee motion. Proper tensioning and isometric positioning are the keys to success.

11.4 Reconstruction of the Torn MCL

The technique of reconstruction for MCL tears was initially proposed by Bosworth in which the semitendinosus tendon is transpositioned toward the femoral attachment of the medial collateral ligament [16]. This technique has undergone many modifications, most surgeons detaching only the proximal part of the tendon (Figs. 11.14 and 11.15) while some opt for the complete detachment of the ST graft in order to obtain anatomic tibial insertion of the reconstructed MCL without losing graft length in order to further stabilize the construct using the gained graft length [5].

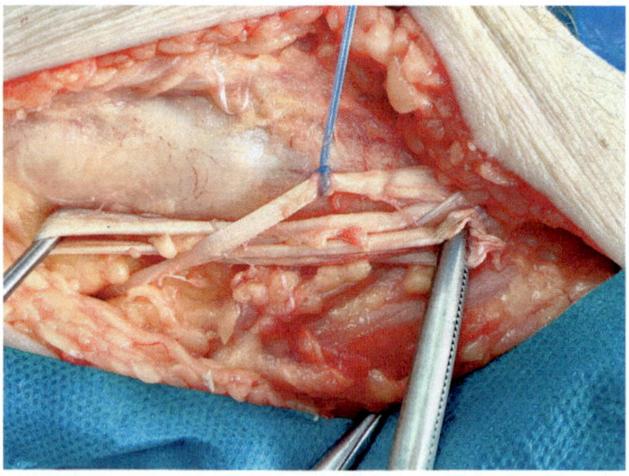

Fig. 11.14 MCL double-bundle reconstruction using ST autograft passed under the gracilis tendon (blue suture); femoral anatomic attachment posterior to the medial epicondyle (pin); tibial attachment preserved (forceps)

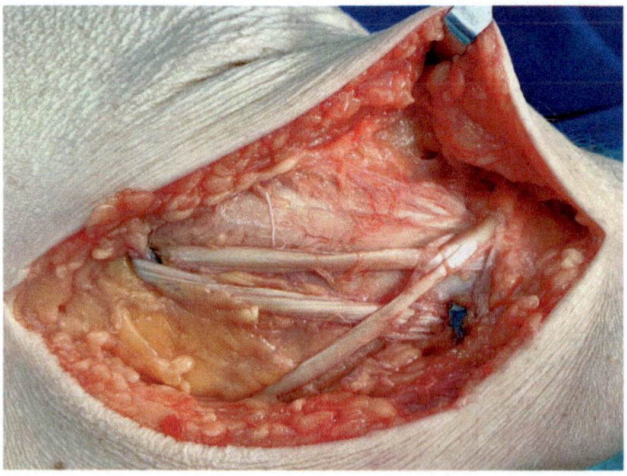

Fig. 11.15 MCL double-bundle ST autograft reconstruction: staple fixation

11.5 Modified Bosworth Technique

Through the above described approach, at the level of the pes anserinus, the ST tendon is identified and using a stripper is released proximally, leaving the tibial insertion intact. It is cleaned of any muscular attachments, and the proximal end is prepared using a running locking suture.

After this step, techniques vary. A non-anatomic double-bundle reconstruction, an anatomic double-bundle reconstruction can be performed, as well as MCL and POL reconstruction.

1. The distal end is attached just posteriorly, at the level of the tibial insertion of the sMCL using a standard 6.5 mm screw with a spiked washer. The graft comes around the screw.

   Next, the femoral insertion of the sMCL is identified, posterior to the medial epicondyle. The graft is looped around another screw, tensioned at 30° of flexion and secured with a spiked washer. Before screw insertion, isometry must be tested in full range of motion (using a K-wire). If it is not satisfactory, the insertion site is reassessed.

   The free end of the graft is then passed distally, around the first screw and the spiked washer secures the fixation, thus obtaining a double-strand reconstruction.
2. Kim et al.: The prepared graft with an intact tibial insertion is attached proximally at the site of the femoral insertion of the sMCL, posterior to the medial epicondyle using a 6.5-mm screw with a spiked washer and tensioned at 30° flexion. After passing the graft around the screw, the free end is directed obliquely distally, in the direction of the POL, attaching it to the anterior arm of the semimembranosus. When this is done, the length of the tendon plays an important factor. Care should be taken to fix the graft in full extension, avoiding flexion contracture. Depending on the length, it can be sutured to the torn POL.
3. Stannard technique is similar to Kim's technique with the difference that after the passing

of the ST graft free end under the SM, it is attached to the intact tibial insertion of the ST.

4. Alternatively, after testing the isometry, a tunnel is drilled in the femoral condyle. The graft is plicated and sutured on a length that is decided according to total length of the harvested ST tendon, without risking intercondylar notch penetration. Diameter is measured, determining the diameter of the tunnel, and appropriate interference screw is used to fix the graft in the tunnel, using a pull-through technique, tensioning it at 10° flexion. The free end is used to reconstruct the POL (Fig. 11.16), attaching it through a tunnel drilled in the tibial plateau, from posterior to anterior, slightly distally oriented, in the posteromedial corner (just proximal and medial to the superior edge of the semimembranosus groove). Adequate drill and interference screw are chosen, according to graft measurements—Lind technique.

11.6 LaPrade Technique

Using two tendon grafts (ST + gracilis or allografts), anatomic reconstruction of sMCL and POL is performed. First the grafts are attached to their correspondent tibial insertion sites using tunnels—perpendicular to the tibia at the insertion of the sMCL, just posterior to the pes anserinus, oblique in the direction of Gerdy's tubercle at the insertion of the POL, immediately anterior to the arm of the SM—and an interference screw sized to the grafts. After identifying their femoral origin and testing isometry, they are attached in a similar manner, through tunnels perpendicular to the surface of the femur, avoiding convergence. Flouroscopy can be used to aid placement. At the end, the sMCL graft is secured to its proximal tibial insertion with a staple [7, 17, 18].

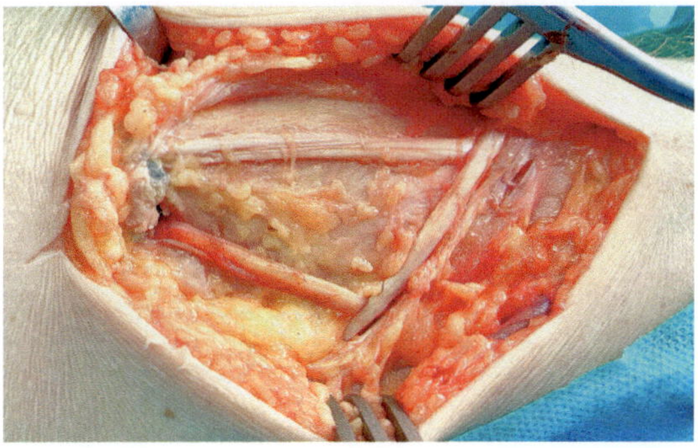

Fig. 11.16 MCL and POL reconstruction using an ST graft proximally plicated

Anatomic reconstruction of the MCL and, when required, of the POL can be performed using allografts. For this an Achilles tendon with a bone plug or a long semitendinous cadaver-harvested graft can be used, prepared using different techniques, according to the chosen bone attachment technique and graft type.

Whichever type of repair or reconstruction is performed, there are two major complications: stiffness of the knee joint and residual instability. MCL surgery is painful and a proper rehabilitation protocol is at utmost importance.

11.7 Postoperative Rehabilitation

We use a rehabilitation protocol proposed by LaPrade. He emphasizes the importance of establishing a "safe zone" for a range of motion, which is determined during the surgery, an interval of motion in which no significant tension is put on the repaired structures and communicating it to the rehabilitation specialist. A hinged long knee brace is worn, limiting flexion at 90°, non/protected touch-down weight bearing (only for reconstruction) and isometric muscle reactivation exercises for 6 weeks. Passive or passive-assisted range of motion from 0 to 90° of flexion are recommended (as tolerated with no less than 90° of flexion after the first 2 weeks) immediately after the surgery, quadriceps stetting exercises, and ankle pumps. Some authors choose to limit knee extension at 30°, but this can be detrimental for an associated ACL reconstruction. After 2 weeks, assisted active range of motion exercises are permitted, as tolerated, ideally reaching 130° of flexion at 6 weeks. Then, progressive weight bearing, discontinued brace, elliptical training, and cycling are recommended. At full weight bearing, normal gait exercises are performed. After 16 weeks, agility exercises can be initiated if the patient's leg motion, strength, and balance are restored.

References

1. Grood ES, Noyes FR, Butler DL, Suntay WJ. Ligamentous and capsular restraints preventing straight medial and lateral laxity in intact human cadaver knees. J Bone Jt Surg. 1981;63:1257–69.
2. Hughston JC, Eilers AF. The role of the posterior oblique ligament in repairs of acute medial (collateral) ligament tears of the knee. J Bone Joint Surg. 1973;55:923–40.
3. Warren LF, Marshall JL. The supporting structures and layers on the medial side of the knee: an anatomical analysis. J Bone Joint Surg Am. 1979;61:56–62.
4. Warren LF, Marshall JL, Girgis F. The prime static stabilizer of the medial side of the knee. J Bone Joint Surg. 1974;56:665–74.
5. Azar FM. Evaluation and treatment of chronic medial collateral ligament injuries of the knee. Sports Med Arthrosc Rev. 2006;14:84–90.
6. LaPrade RF, Terry GC. Injuries to the posterolateral aspect of the knee: association of anatomic injury patterns with clinical instability. Am J Sports Med. 1997;25:433–8.
7. Wijdicks CA, Griffith CJ, Johansen S, Engebretsen L, LaPrade RF. Injuries to the medial collateral ligament and associated medial structures of the knee. J Bone Jt Surg. 2010;92:1266–80.
8. Memarzadeh A, Melton JTK. Medial collateral ligament of the knee: anatomy, management and surgical techniques for reconstruction. Orthopaedics and Trauma. 2019;33(2):91–9.
9. Gwathmey FW, Miller MD. Operative techniques: knee surgery. Amsterdam: Elsevier; 2017. p. 207–15.
10. Hajnik CA, Radnay CS, Scuderi GR, Scott WN. Insall and Scott surgery of the knee, vol. 39; 2012. p. 348–54.
11. Pellegrini A. Ossificazione traumatica del ligamento collaterale tibiale dell'articolazione del ginocchio sinistro. Clin Moderna. 1905;11:433–9.
12. Stieda A. Uber eine typische verletzung am unteren femurende. Archiv klin Chir. 1908;85:815–26.
13. Encinas-Ullan CA, Rodriguez-Merchan EC. Isolated medial collateral ligament tears. EFORT Open Rev. 2018;3:398–407.
14. Saigo T, Tajima G, Kikuchi S, Yan J, Maruyama M, Sugawara A, Doita M. Morphology of the insertions of the superficial medial collateral ligament and posterior oblique ligament using 3-dimensional computed tomography: a Cadaveric Study. Arthroscopy. 2017;33(2):400–7.
15. Mackay GM, Blyth MJ, Anthony I, Hopper GP, Ribbans WJ. A review of ligament augmentation with the InternalBrace: the surgical principle is described for the lateral ankle ligament and ACL repair in particular, and a comprehensive review of other surgical applications and techniques is presented. Surg Technol Int. 2015;26:239–55.
16. Bosworth DM. Transplantation of the semitendinosus for repair of lacerations of the medial collateral ligament of the knee. J Bone Joint Surg Am. 1952;34:196.
17. LaPrade RF, Wijdicks CA. Surgical technique: development of an anatomic medial knee reconstruction. Clin Orthop Relat Res. 2012;470:806–14.
18. Lind M, Jakobsen BW, Lund B, Hansen MS, Abdallah O, Christiansen SE. Anatomical reconstruction of the medial collateral ligament and posteromedial corner of the knee in patients with chronic medial collateral ligament instability. Am J Sports Med. 2009;37:1116–22.

12 Anatomic Posterolateral Reconstruction

Bogdan Ambrožič, Marko Nabergoj, and Urban Slokar

12.1 Introduction

The posterolateral corner (PLC) of the knee is an anatomically complex unit formed by the interaction of multiple structures. PLC injury represents a complex injury pattern, with damage to important varus and external rotatory static stabilizers of the knee, which may cause significant posterolateral rotatory instability. PLC injuries account for 16% of all ligamentous knee injuries, often presenting with concomitant anterior and posterior cruciate ligament injuries and rarely occurring in isolation (1.6%) [1, 2]. Failure in detection of these injuries has been one of the principal reasons for persistent instability and unsuccessful cruciate ligament reconstructions [3]. Treatment of the PLC injuries has been challenging due to the limited knowledge of anatomy and biomechanics. However, in the past two decades, the advancement in understanding of the anatomy and biomechanics led to a development of biomechanically validated reconstruction techniques with reported good clinical outcomes. The aim of this chapter is to describe the current concepts of PLC involving surgically relevant anatomy, biomechanics, mechanism of injury, diagnostics, a guide to choose the appropriate reconstruction based upon the grade of injury, a detailed description of the most commonly used surgical techniques, and possible complications.

12.2 Anatomy of the Posterolateral Corner of the Knee

Posterolateral stability of the knee is provided by an anatomically complex and variable formation of tendons and ligaments known as the PLC. These structures can be divided into static and dynamic stabilizers. The static stabilizers are fibular collateral ligament (FCL), popliteofibular ligament (PFL), lateral capsule and arcuate ligament–fabellofibular complex. The dynamic stabilizers include biceps femoris muscle and iliotibial band (ITB). Popliteus tendon (PT) muscle is both a static and dynamic stabilizer [4]. The three main anatomically consistent, functional, and surgically relevant structures of this region are the FCL, PT, and PFL (Fig. 12.1). They prevent excessive external rotation, varus angulation, and combined posterior translation and external rotation of the tibia on the femur [5].

12.2.1 Fibular Collateral Ligament

It is a well-defined round structure with a fan-shaped insertion sites. It has an average diame-

B. Ambrožič (✉) · M. Nabergoj · U. Slokar
Valdoltra Orthopaedic Hospital, Ankaran, Slovenia
e-mail: Bogdan.Ambrozic@ob-valdoltra.si

M. Bozkurt, H. İ. Açar (eds.), *Clinical Anatomy of the Knee*,
https://doi.org/10.1007/978-3-030-57578-6_12

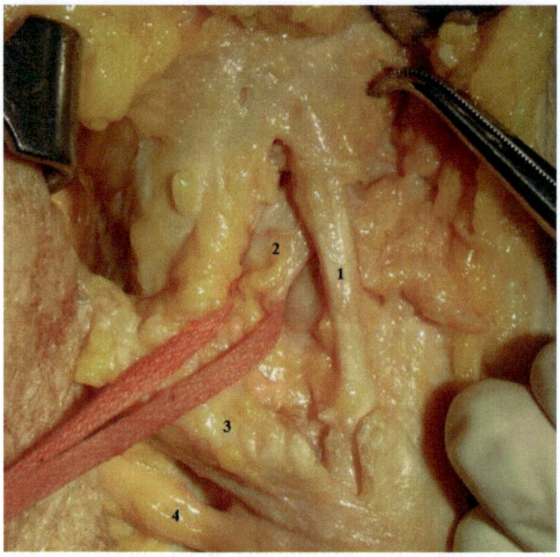

Fig. 12.1 Anatomic dissection of the posterolateral corner of the right knee. *1* LCL, *2* PT, *3* biceps tendon, *4* peroneal nerve

ter of 5.76 mm and an average length of 69.6 mm [6, 7]. Its proximal femoral attachment is typically 1.4 mm proximal and 3.1 mm posterior in relation to the lateral epicondyle. Distally, it attaches to the anterolateral aspect of the fibular head, 8.2 mm posterior to the anterior border of the fibula, and 28.4 mm distal to the tip of the styloid process with additional ligamentous extension into the fascia of peroneus longus [7].

12.2.2 Popliteus Tendon Muscle

It originates from the posteromedial aspect of the proximal tibia, and it inserts on the anterior fifth of the popliteal sulcus an average of 18.5 mm anteriorly from the femoral FCL attachment. It becomes tendinous in the lateral third of the popliteal fossa and intra-articular as it courses deep to the FCL. The average total length of the popliteus tendon from its proximal femoral attachment to its musculotendinous junction has been reported by different authors as between 36.36 and 54.5 mm [6, 7].

12.2.3 Popliteofibular Ligament

It originates from popliteus musculotendinous junction and inserts distally on the posteromedial aspect of the fibular styloid process. The mean angle of its anterodistal course is 51° relative to the longitudinal axis of the tibia. Morphological variations consist of a singular bundle (60%), a double ligament (26.7%), or an inverted Y-shaped (13.3%) structure [6].

12.3 Biomechanics

The three most essential biomechanical structures of the PLC of the knee are the FCL, PT, and PFL. They prevent varus angulation, excessive external rotation, and combined posterior translation and external rotation of the tibia on the femur [5]. Additionally, in cases of cruciate deficient knees, PLC (mainly popliteus tendon) acts as a secondary restraint to anteroposterior tibial translation at near full knee extension [8]. FCL is the primary varus stabilizer of the knee, and it limits external rotation at lower degrees of knee flexion. PT has both a static and a dynamic function in stabilizing the PLC of the knee. Together with PFL, they are the primary stabilizers of external rotation at higher degrees of knee flexion and secondary varus stabilizers. Lastly, PLC is a minor primary stabilizer in preventing internal rotation [4]. A study by LaPrade et al. has shown a small, but significant increase in internal rotation at all knee flexion angles after sectioning PT [8]. The other PLC structures act as secondary stabilizers to excessive internal rotation [4].

12.4 Mechanism of Injury

PLC injury rarely occurs in isolation and more commonly involves other ligaments in the setting of a higher energy multi-ligament injury of the knee. It is usually caused by sports injuries, falls, and vehicle accidents. It can happen in various ways, but the following examples are the most common. The injury can occur with a direct blow on the anteromedial aspect of the tibia with the knee at or near full extension. However, combined hyperextension and non-contact varus opening can also cause a PLC injury. In addition, posterior stress forces acting on a flexed knee or when the tibia is externally rotated can cause an injury to the posterolateral region of the knee. If it occurs as part of a high energy trauma in lateral knee luxation, the patient is at risk of having concomitant injuries to critical neurovascular and other ligamentous structures [4].

12.5 Diagnostics

12.5.1 Clinical Picture

The typical symptoms and signs in an isolated acute PLC injury are pain, swelling and ecchymosis on the posterolateral aspect of the knee. Patients with chronic PLC injuries complain of a broad knee pain including the medial, lateral joint line, and posterolateral region [4]. They may present with paraesthesia or numbness of the common peroneal nerve distribution. Injury of the peroneal nerve in isolated and combined PLC injuries has been reported by different authors as between 12.7 and 16.7% [9–11]. Chronic patients frequently show functional instability near full extension of the knee, such as varus thrust gait during walking or varus alignment of the knee during standing [12].

12.5.2 Clinical Examination

A thorough physical examination is required to properly identify a PLC injury. The most important clinical tests that should be performed include the dial test, varus stress testing, reverse pivot shift test, and the posterolateral drawer test. They should all be meticulously performed and compared to the uninjured contralateral knee to determine asymmetry. Additionally, it is advised to observe the lower extremity alignment when the patient is walking (varus thrust in chronic patients) or standing (varus alignment of the knee in chronic patients) [12].

Dial test is one of the most important clinical tests used to diagnose a PLC injury. It measures external rotation of the foot (tibia) relative to the femur. The patient lies in a prone position, and the thigh is in a fixed position, when the knee is either flexed to 30° or 90° and the foot is externally rotated. In the case of an isolated PLC injury, an increase of more than 10° of external rotation of the injured limb compared to the uninjured contralateral limb is observed at 30°, but not at 90° of flexion (Fig. 12.2). An increase of more than 10° at 90° of knee flexion means a combined PLC and PCL injury [13]. We have to be cautious in interpreting the results of the dial test. Forsyhte et al. showed that in case of a ruptured anterior cruciate ligament (ACL), an increase of almost 7° of tibial external rotation is found both at 30° and 90° of knee flexion [14].

Varus stress test is performed both at 30° of flexion and in full extension. Patient is in a supine position, and a varus load is applied to the tibia when the femur is stabilized. This test best isolates the FCL when it is performed at 30° of flexion. A test counts as positive when gapping of the lateral compartment happens. A positive test in full extension of the knee indicates a combined PLC and cruciate ligament injury [12].

Reverse pivot shift is conducted when the patient lies in a supine position with the knee flexed to 90°. A valgus force and an external rotation is applied to the tibia when the knee is extended. The presence of the PLC injury is indicated if a reduction of the previously subluxated lateral tibial plateau happens at around 30–40° [12].

The posterolateral drawer test is performed when the patient is supine, hip is flexed to 45°, knee is flexed to 80°, and the foot is externally rotated for 15°. A combined posterior force and external rotation is applied to the tibia. The test is positive when the tibial tubercle shows more external rotation compared to the lateral femoral condyle and is indicative of PFL and PT injury [12].

To assess the cruciate ligaments, we perform the Lachman and anterior-posterior drawer tests. Finally, it is critical to perform a neurovascular examination, especially in case of an acute PLC injury.

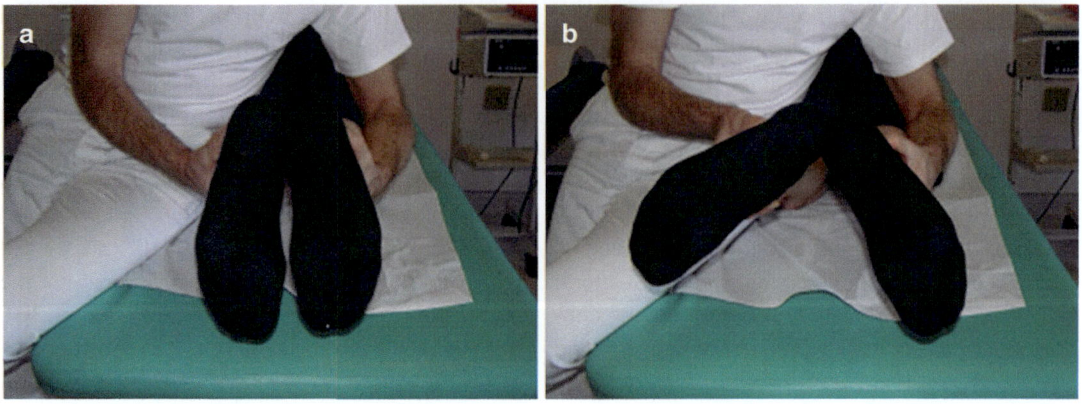

Fig. 12.2 Dial test: (**a**) Patient is lying in a prone position, the thigh is in a fixed position, and the knee is flexed to 30° with the foot neutrally rotated. (**b**) Positive dial test is observed by an increase of more than 10° of external rotation of the injured (left) limb compared to the uninjured contralateral limb at 30° of flexion

12.5.3 Imaging

If PLC injury is still in question after an adequate history is obtained and a thorough physical examination is performed, appropriate diagnostic imaging is necessary for an accurate diagnosis. Standard anteroposterior (AP), lateral, and sunrise radiographic views of the knee joint are taken to exclude other injuries such as arthrosis and avulsions. In the case of chronic injuries, a bilateral full length weight-bearing AP view X-ray of the knee in the standing position should be obtained for a potential osteotomy, which should be performed prior or during the PLC reconstruction [15].

Varus stress radiograph of the knee is critical for the diagnosis and evaluation of severity of PLC injuries. LaPrade et al. performed a bilateral varus stress radiographs at 20° of knee flexion after they sequentially sectioned PLC structures in cadaveric knees, and measured the amount of lateral compartment knee opening. They concluded that an isolated FCL tear should be suspected when the lateral compartment gapping increases by approximately 2.7 mm in relation with a clinician-applied varus stress, while a difference of approximately 4 mm indicates a grade III PLC injury [16] (Fig. 12.3).

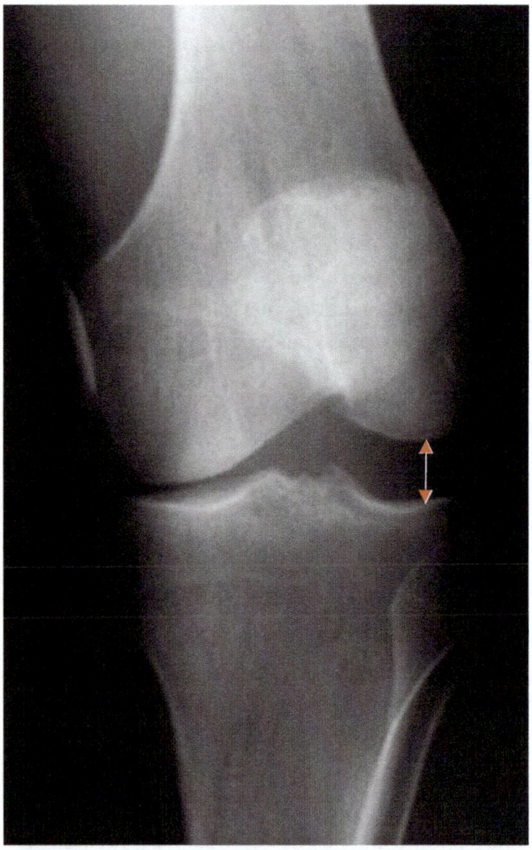

Fig. 12.3 Varus stress radiograph of the left knee, where a grade III PLC injury is recognized, characterized by opening of the lateral knee compartment by more than 4 mm

Magnetic resonance imaging (MRI) is another important diagnostic tool in identifying injured

PLC structures and concomitant meniscus or cartilage lesions. LaPrade et al. have evaluated the accuracy of MRI in identifying the injured PLC structures by using a thin-slice coronal oblique T1-weighted images through the entire fibular head which were later verified intraoperatively. They have shown a high accuracy of MRI for the identification of injury of FCL (95%) and PT origin on femur (90%). The lowest diagnostic accuracy values were for the PFL at 68%. MRI use is critical in complex cases, especially in the acutely injured painful knees, where a thorough physical examination is unobtainable [17].

Combined use of described imaging tools enhances the diagnostic accuracy of PLC injury.

12.5.4 Arthroscopy

Arthroscopy of a knee with a possible PLC injury provides us with an additional intra-articular information regarding the integrity of the PT, coronary ligament of the lateral meniscus, and posterolateral capsule. Furthermore, it enhances our surgical decision-making in choosing the proper reconstruction technique based on arthroscopic evaluation [18] (Fig. 12.4).

Lateral gutter drive-through test is performed by inserting the arthroscope in the lateral gutter through the anterolateral portal when the knee is at 30° of flexion and with neutral tibial rotation. The test is positive and indicates a posterolateral instability when the arthroscope passes into the posterolateral compartment between the PT and the lateral femoral condyle (Fig. 12.5). Feng et al. have shown in their cadaveric sectioning study that the latter gutter drive-through test was positive in two cases: when distal PT and PFL were both sectioned or when posteromedial structures (superficial, deep medial collateral ligament, and posterior oblique ligament) and anterior and posterior cruciate ligaments were all sectioned. Caution is advised in interpreting the results of this test in diagnosing the PLC injury when injury of the cruciate ligaments and posteromedial structures is suspected [19].

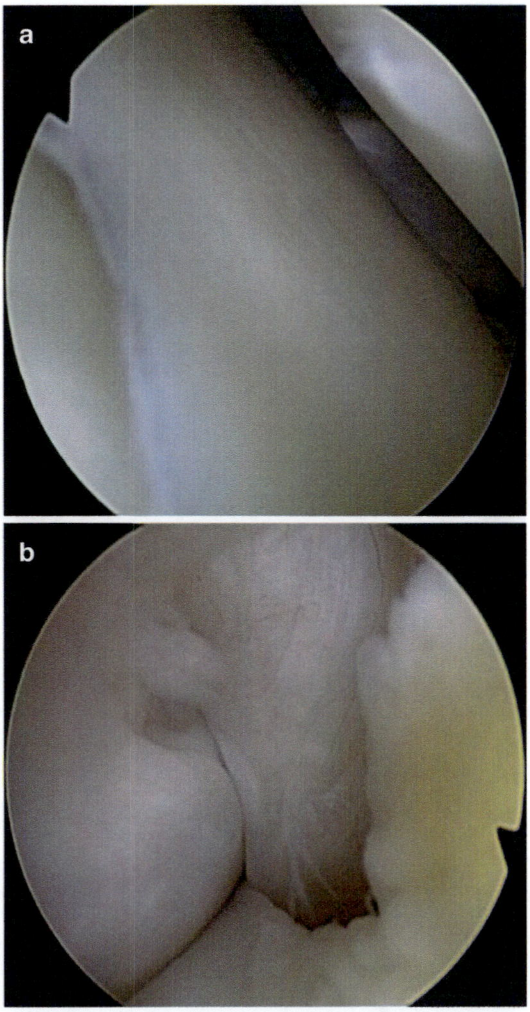

Fig. 12.4 Arthroscopy of the left knee lateral compartment: normal opening of the lateral compartment (**a**) and normal PT femoral insertion (**b**)

12.6 Classification of Posterolateral Instability

The most frequently used classification of PLC injuries by Fanelli and Larsen classifies the posterolateral instability into three types (A, B, C) based on the grade of injury (Table 12.1). Type A injury presents with increased tibial external rotation, which indicates an injury to the PFL and PT. Type B injury has increased tibial external rotation and mild 5–10 mm varus opening with a firm end point to varus load at 30° of knee flexion. It affects the PFL, PT, and weakens FCL. Type C injury presents with a severe varus instability of more than 10 mm varus gap and increased tibial external rotation. It involves the PFL, PT, complete FCL disruption, avulsion of the lateral capsule, and cruciate ligament rupture [20].

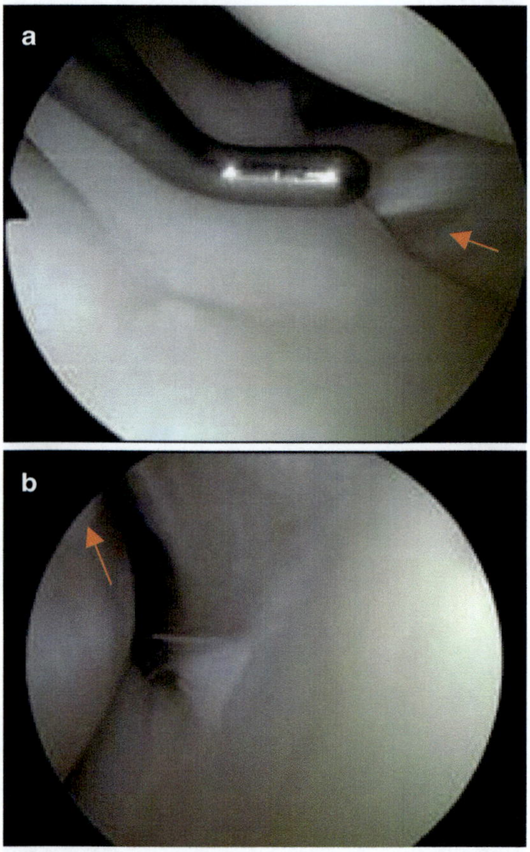

Fig. 12.5 Increased lateral opening with PT intratendinous rupture (**a**), complete popliteus femoral insertion rupture (**b**)

12.7 Treatment Options

A variety of surgical options are described, ranging from a primary repair in case of an acute injury to various reconstructions, tenodesis, and osteotomies for chronic injuries. PLC reconstructions can be further classified into anatomic and non-anatomic. Anatomic PLC reconstructions are Laprade's and Arciero's reconstruction. Non-anatomic PLC reconstructions are Larson's, Clancy's biceps tenodesis and central slip of the biceps technique [15, 20].

12.7.1 A Guide of Choosing the Appropriate Surgical Technique

In the literature, there is no clear algorithm for the selection of the appropriate surgical PLC reconstruction technique based on the grade of the PLC injury. In our surgical practice, the surgical decision-making is as follows. If the patient has a Fanelli type A or B injury, we choose Arciero's surgical technique. If the injury is

Table 12.1 Fanelli and Larsen classification of posterolateral instability

Types of posterolateral instability	Type A	Type B	Type C
Clinical signs	Increased external rotation	Increased external rotation 5–10 mm varus opening	Increased external rotation >10 mm varus opening
Injured structures	PFL, PT	PFL, PT, weakened FCL	PFL, PT, complete FCL disruption, avulsion of the lateral capsule, cruciate ligament rupture
Positive clinical tests	Dial test at 30° of knee flexion Posterolateral drawer test	Dial test at 30° of knee flexion Posterolateral drawer test Varus stress test (5–10 mm gap with a firm end point)	Dial test at 30° and 90°of knee flexion Posterolateral drawer test Varus stress test (>10 mm gap) Anterior–posterior drawer test Lachman

PFL popliteofibular ligament, *PT* popliteal tendon, *FCL* fibular collateral ligament worse and the patient has a Fanelli type C injury, we want a stronger construct and prefer Laprade's surgical technique. We also use it in revision cases after a failed PLC reconstruction. We can treat patients with an isolated PFL disruption with PCL rupture with an arthroscopic reconstruction described by Frosch. When the patient has an isolated FCL rupture, we use modified Larson's surgical technique. In case of a chronic PLC injury with a varus malalignment, we consider high tibial osteotomy as the best treatment option.

12.7.2 Surgical Approach

A surgical exposure to the PLC of the knee is performed with a patient in a supine position and the knee flexed around 70°. A lateral hockey-stick-shaped incision is made. Proximally it runs parallel to the femur, it curves distally as it crosses the proximal edge of the lateral epicondyle and finishes centered between the Gerdy's tubercle and the anterior aspect of the fibular head (Fig. 12.6). A posteriorly based fasciocutaneous flap is created by subcutaneous dissection until the superficial layer of the ITB is exposed. The three-window technique originally described by Terry and LaPrade can then be used to evaluate the deeper PLC structures [21]. Posterior window is created by fascial incision posteriorly to biceps tendon in order to visualize, release, and protect the common peroneal nerve (Fig. 12.7). Exposure of the fibular attachments of the FCL and PFL is provided by anteriorly retracting the biceps tendon and mobilized common peroneal nerve. This interval is necessary for any fibular-based reconstruction (Fig. 12.8). Middle window is created by incising between the ITB and biceps tendon. The incision starts 6–7 cm proximal to the lateral epicondyle, and it runs distally and parallel to the femur but posterior to the lateral intermuscular septum. Exposure of the FCL and the tibial attachments of the PT is achieved by dissecting through the anterior fascia of the lateral head of the gastrocnemius. This interval is necessary for tibial-based reconstructions and for passage of the graft (Fig. 12.9). Anterior window is created by fascial incision parallel to the ITB extending from Gerdy's tubercle proximally. Exposure of the femoral attachments of the FCL and PT is provided by dissecting anteriorly to the lateral head of the gastrocnemius. This interval is used for femoral-based reconstruction [22] (Fig. 12.10).

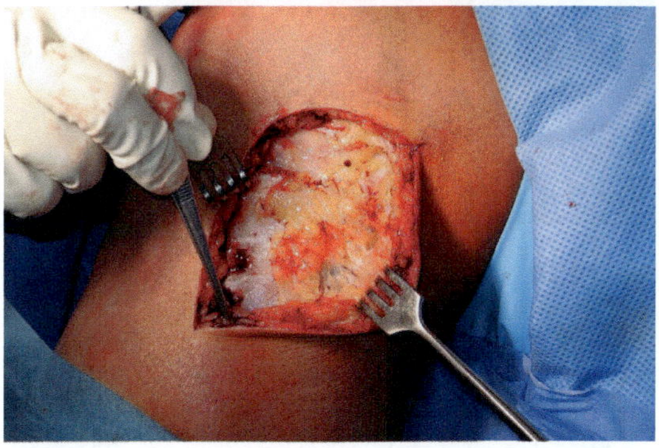

Fig. 12.6 Surgical approach to the PLC of the left knee is performed by making a hockey-stick-shaped incision on the lateral side of the knee. A posteriorly based fasciocutaneous flap is created by subcutaneous dissection

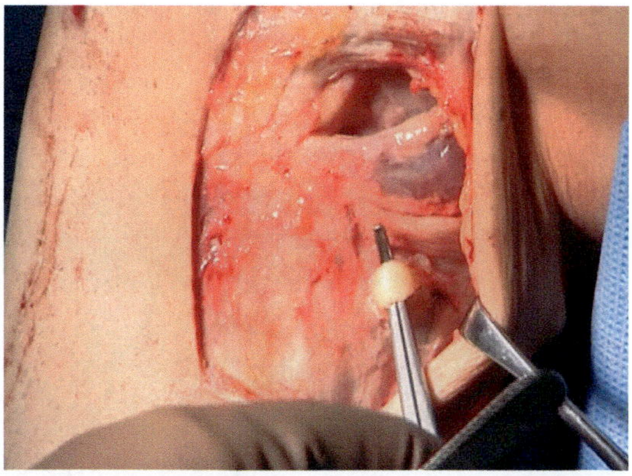

Fig. 12.7 Lateral side of the left knee. Posterior window is being created by fascial incision posteriorly to the biceps tendon. Common peroneal nerve is identified and protected

12.7.3 Preparation of Grafts

We prefer harvesting either ipsilateral or contralateral semitendinosus tendon depending on whether we are doing a simultaneous ACL/PCL reconstruction. In complex knee surgery, use of

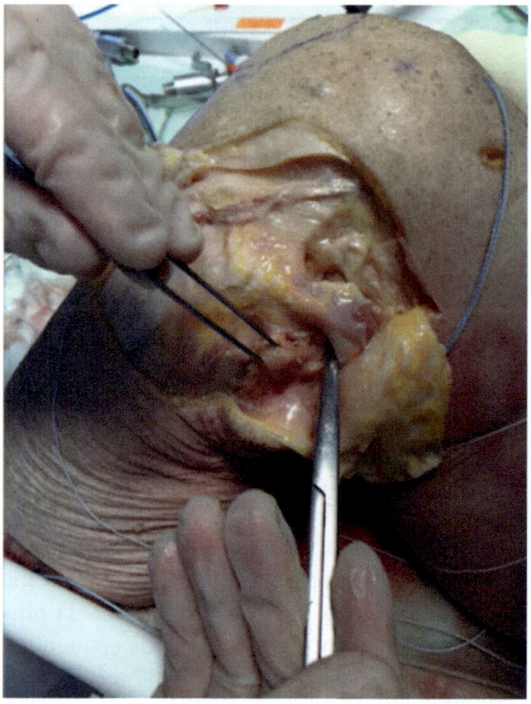

Fig. 12.8 Lateral view of the cadaver specimen of left knee. Posterior window is created by fascial incision posteriorly to the biceps tendon. The common peroneal nerve is visualized and held by the forceps. This interval is necessary for any fibular-based reconstruction

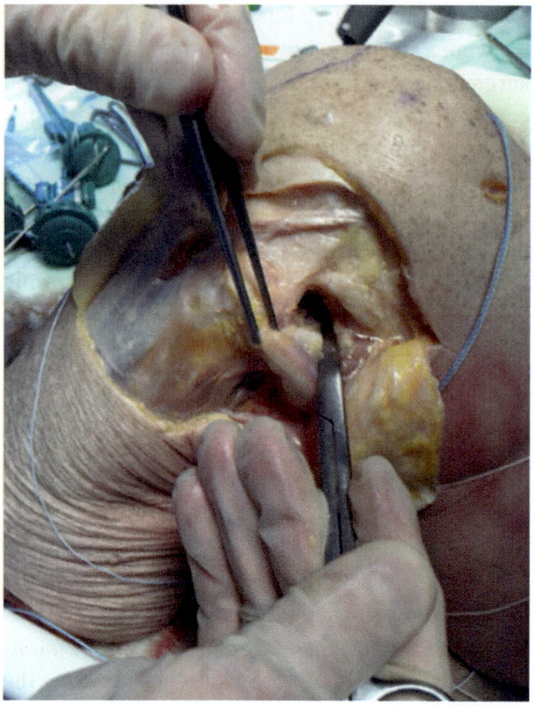

Fig. 12.9 Lateral view of the cadaver specimen of left knee. Middle window is created by incising between the ITB and the biceps tendon. The biceps tendon (**a**) and lateral head of gastrocnemius muscle (**b**) are retracted, and tibia is exposed. This interval is necessary for tibial-based reconstructions and for passage of the graft

allografts for other ligaments may be necessary. The semitendinosus tendon's overall length should be at least 22 cm. The tendon is cleaned of muscle tissue, fixed in the tendon clamps, and tagged with a stitch of approximately 2.5–3 cm in length on the free ends with a non-absorbable suture material (Ethibond) (Fig. 12.11).

12.8 Techniques

12.8.1 LaPrade's Surgical Technique

Laprade's surgical technique is the only PLC reconstruction technique that anatomically reconstructs the main three biomechanical structures of the PLC of the knee: FCL, PT, and PFL (Fig. 12.12).

The author is using personal, equivalent modification of originally described LaPrade's technique. The previously described approach

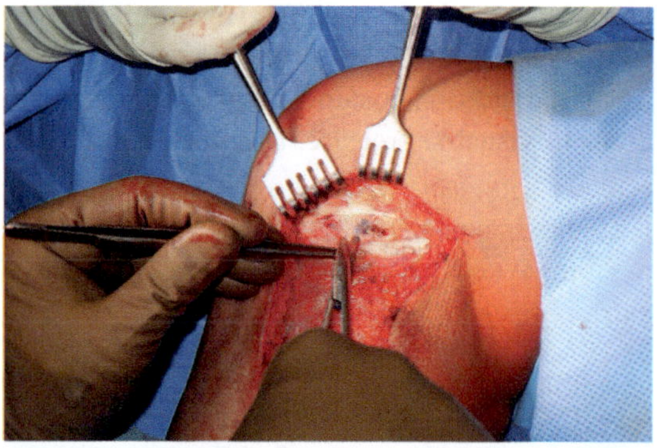

Fig. 12.10 Lateral view of the left knee. Anterior window is created by fascial incision parallel to the ITB extending from Gerdy's tubercle proximally. This interval is used for femoral-based reconstruction

and the three fascial incisions are made. Through the posterior window, the fibular attachment of the PFL and FCL are visualized, and through the middle window, the posterior tibial popliteal

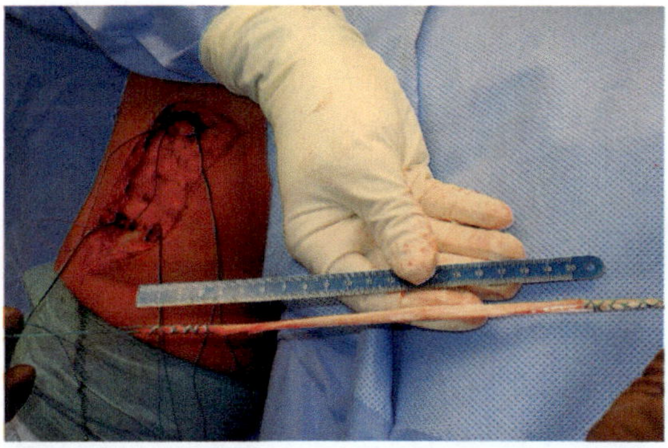

Fig. 12.11 The graft should be at least 22 cm in length. Free ends of the graft are tagged with a stitch of approximately 2.5–3 cm

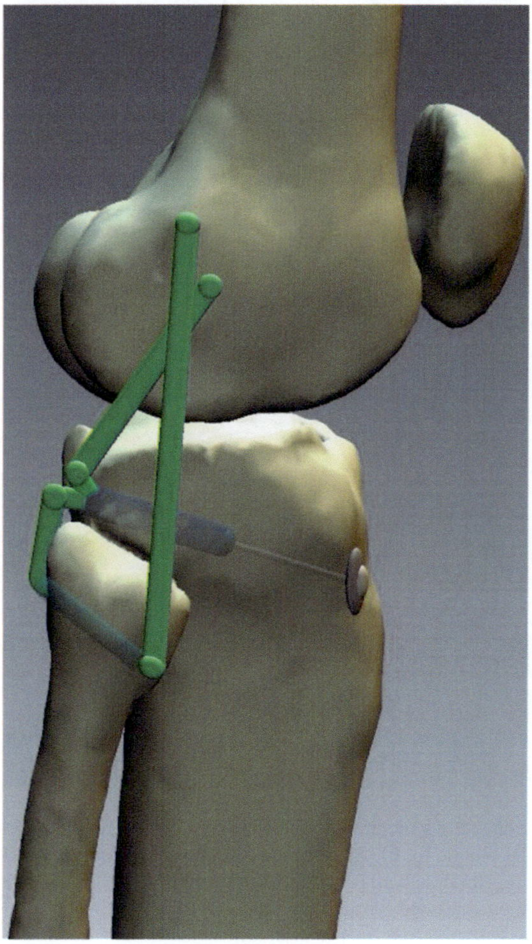

Fig. 12.12 A schematic image of lateral view of the right knee with anatomical PLC reconstruction based on LaPrade's surgical technique. Intraosseous tunnels are marked in blue color

sulcus located at the musculotendinous junction of the popliteus muscle is identified. The fibular nerve is identified and protected. The fibular tunnel is prepared by drilling a K-wire from the FCL insertion on the anterolateral aspect of the fibular head to the fibular PFL insertion site located posteromedially (Fig. 12.13). A 6-mm full tunnel is prepared by reaming over the guide pin, and a suture loop is passed through the tunnel. After that, the tibial tunnel is prepared by the second K-wire using the guide in an anteroposterior direction centered just distal and medial to Gerdy's tubercle and exited at the posterior tibial popliteal sulcus at the level of the popliteus musculotendinous junction using a finger or spoon protection for the neurovascular bundle. A 4.5-mm drill bit is reamed over the K-wire and the retro drill of 7 mm is inserted and opened at the back of the tibia. By retrograde drilling, a 7-mm tunnel socket is prepared for a length of 2–3 cm, and a suture loop is passed through the tunnel (Fig. 12.14). Through the incised anterior window, the femoral attachments of the FCL and PT are identified. Two eyelet guide pines are drilled into their attachment sites and exited proximally and anteriorly to the medial epicondyle. The tunnels are placed 15–20 mm apart depending on the anatomy of the individual (Fig. 12.15). A 7-mm tunnel to a depth of 25 mm is then created in both tunnels, and the suture loop is passed through both tunnels. The semitendinosus tendon is folded in the middle and loaded with adjustable button fixation device. The first step is the retrograde insertion (from posterior to anterior) of a folded graft loaded with fixation device into the tibial tunnel and fixation on the anterior aspect of the tibia (Fig. 12.16). After measurement of the remaining length of the two limbs of the graft, the adjustable loop is closed by pulling the sutures from the anterior part and inserting the folded semitendinosus graft fully into the tibial tunnel socket from the posterior part (Fig. 12.17). After that one limb of the graft (anterior limb) is passed directly anteriorly under the biceps tendon and iliotibial band into the anterior femoral tunnel into the insertion of popliteal tendon

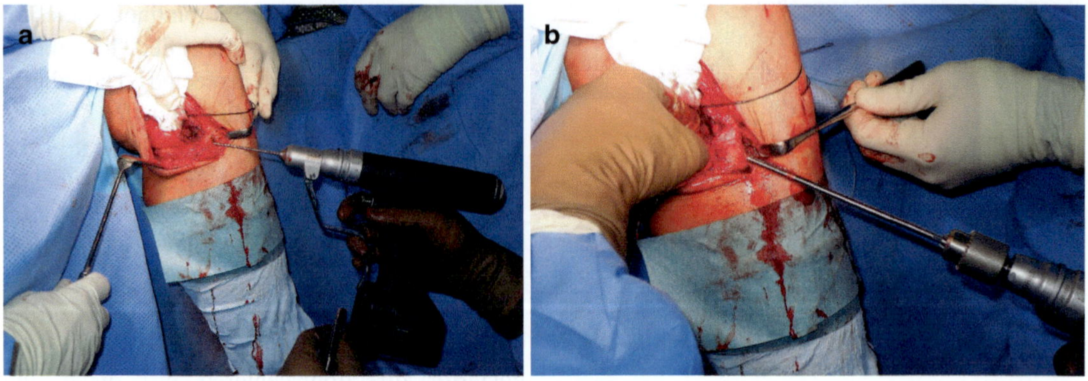

Fig. 12.13 Lateral view of the right knee. Drilling of the fibular tunnel from the FCL insertion on the anterolateral aspect of the fibular head to the fibular PFL insertion site located posteromedially. The fibular tunnel is prepared by drilling a K-wire from the FCL insertion on the anterolateral aspect of the fibular head to the fibular PFL insertion site located posteromedially (**a**). A 6-mm tunnel is prepared by reaming over the guide pin (**b**)

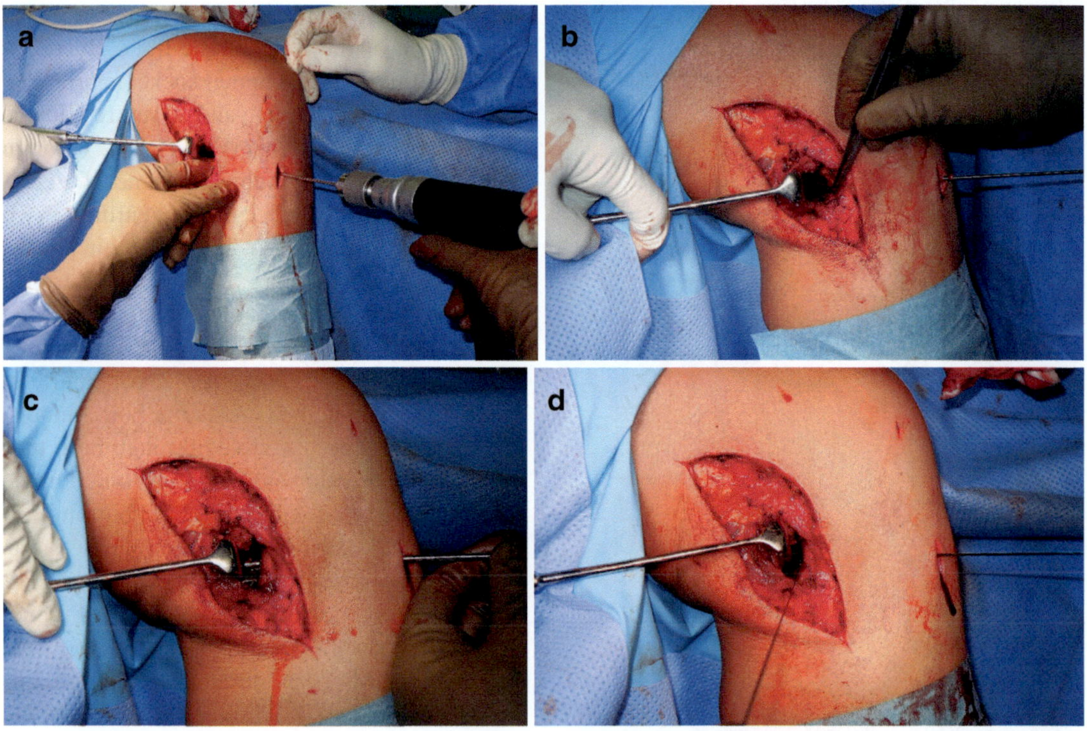

Fig. 12.14 Lateral view of the right knee. The tibial tunnel is prepared by the second K-wire using the guide in an anteroposterior direction centered just distal and medial to Gerdy's tubercle and exited at the posterior tibial popliteal sulcus at the level of the popliteus musculotendinous junction using a finger or spoon protection for the neurovascular bundle (**a, b**). Retrodrill is inserted from anteriorly and the tibial socket is drilled in retrograde way (**c**). Suture loop is passed through the tunnel (**d**)

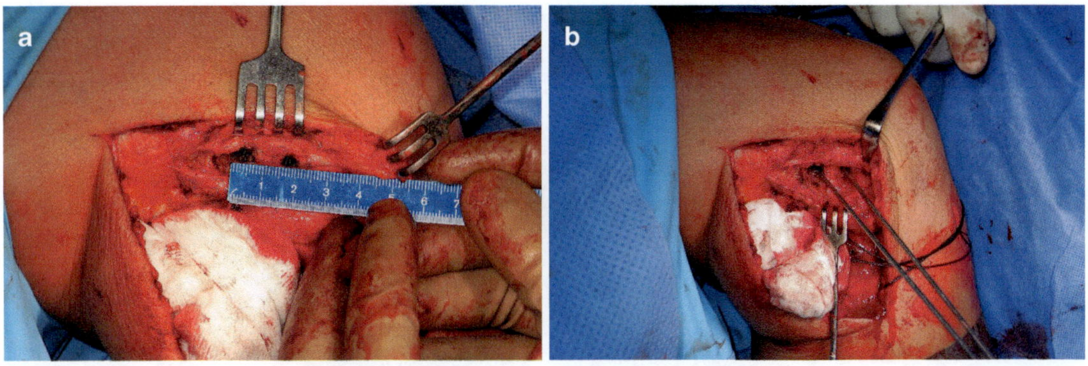

Fig. 12.15 Lateral view of the right knee. Through the incised anterior window, the femoral attachments of the FCL and PT are marked. The marks should be placed 15–20 mm apart depending on the anatomy of the individual (**a**). Two eyelet guide pines are drilled into their attachment sites and exited proximally and anteriorly to the medial epicondyle (**b**)

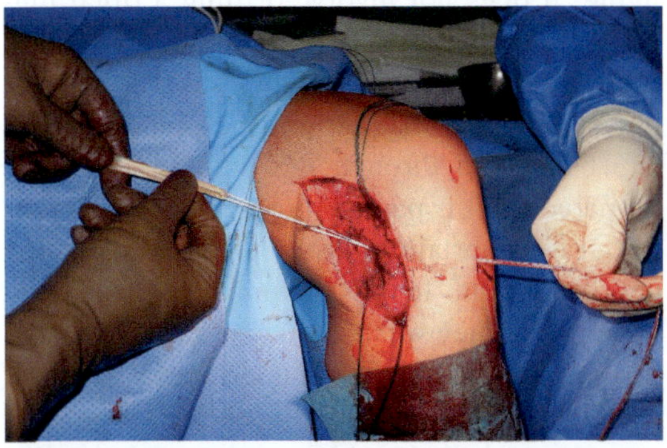

Fig. 12.16 Lateral view of the right knee. Retrograde insertion (from posterior to anterior) of a folded graft loaded with fixation device into the tibial tunnel and fixation on the anterior aspect of the tibia

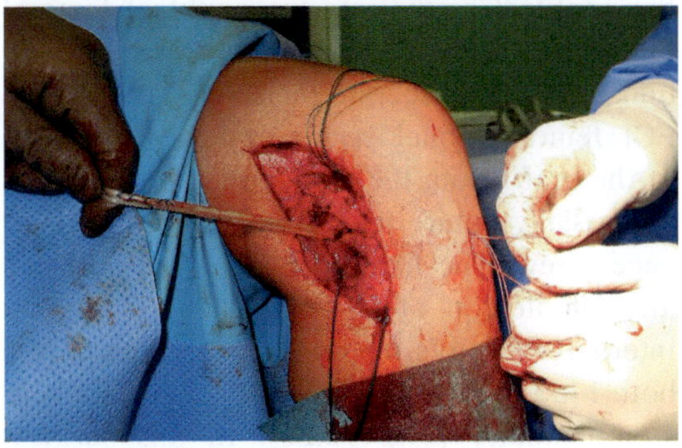

Fig. 12.17 Lateral view of the right knee. After measurement of the remaining length of the two limbs of the graft, the adjustable loop is closed by pulling the sutures from the anterior part and inserting the folded semitendinosus graft fully into the tibial tunnel socket from the posterior part

(PT). The other limb (posterior limb) of the graft is passed into the fibular tunnel from posteriorly to anteriorly and then under the iliotibial band into the posterior femoral insertion of fibular collateral ligament (FCL) (Fig. 12.18). The two limbs are crossing under the iliotibial band, and the anterior limb has to be placed under the posterior one. The two tendon grafts are both pulled by the passing sutures into the femoral tunnels. The length of the limbs can be adjusted according the femoral tunnel measurements. The portion of this graft that coursed from the fibular to the tibial tunnel represented the reconstructed PFL ligament. A fibular fixation of the graft is performed with a cannulated interference screw of 6 mm diameter and length of 23 mm, inserted from the anterior part (Fig. 12.19). The graft is tightened as the knee was cycled for 1 min through a full range of motion while the traction on the graft was applied. Both grafts residing in the fibular and tibial tunnel are pulled in anterior direction and fixed in the femoral tunnels with two cannulated interference screws of 7 mm diameter and length of 23 mm with the knee in 60° of flexion, slight valgus, and internal tibial rotation (Fig. 12.20). The graft that coursed from the femoral FCL attachment to the fibula represented reconstructed FCL, while the graft that coursed from the femoral PT attachment to the tibial tunnel represented reconstructed PT [5] (Fig. 12.21).

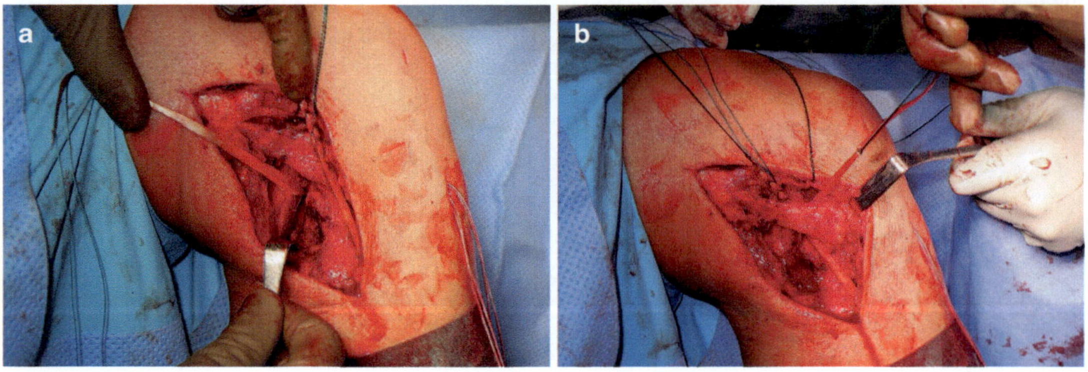

Fig. 12.18 Lateral view of the right knee. The posterior limb of the graft is already passed through the fibular head (**a**). Afterwards, the posterior limb of the semitendinosus tendon is passed under the iliotibial band, whereas the anterior limb is passed direct anteriorly under the biceps tendon, posterior limb of the graft and iliotibial band (**b**)

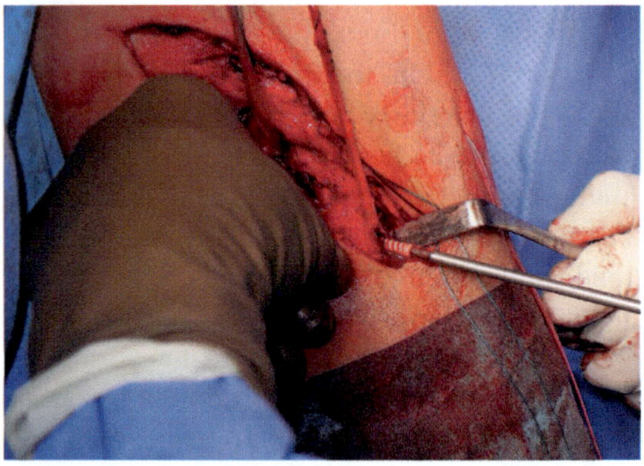

Fig. 12.19 Lateral view of the right knee. A fibular fixation of the graft is performed with a cannulated interference screw of 6 mm diameter and length of 23 mm, inserted from the anterior part

12.8.2 Modified Arciero's Surgical Technique

Arciero's Surgical Technique anatomically reconstructs ruptured FCL and PFL (Fig. 12.22). This surgical technique can be performed with classical PLC approach or with a minimally invasive surgical approach by making a double mini-open incision (Fig. 12.23). The fibular nerve is identified and protected. In comparison to LaPrade's surgical technique, Arciero's technique involves drilling of two femoral and one fibular tunnel as described in LaPrade's technique. It does not involve drilling a tibial tunnel. When the fibular and both the femoral tunnels are drilled and the passing sutures placed as described earlier, the prepared graft can be passed through the tunnels. First, the prepared graft is passed through the fibular tunnel. It is important to firmly tension the tendon and have equal length of limbs anteriorly and posteriorly, then we can proceed and fix the tendon into the fibula with an interference screw of a diameter of 6 mm (Fig. 12.24). The femoral tunnels are drilled at the insertion of the popliteus tendon and fibular collateral ligament 15–20 mm apart depending on the anatomy of the individual (Fig. 12.25). The anterior limb of the graft from fibular tunnel is passed through the popliteal hiatus into the popliteal femoral tunnel, while the posterior limb of the fibular tunnel is passed deep to the fascia lata into the FCL femoral tunnel lying over the anterior limb of the graft. Both the limbs are tunneled using the passing suture and fixed with an interference screw. The fixation is achieved with the knee in flexion of approximately 60°, slight internal tibial rotation and slight valgus, while both limbs of the graft are held in a firm tension medially (Fig. 12.26). In this surgical technique, we aim to reconstruct an FCL and a popliteus tendon-popliteofibular ligament component [23] (Fig. 12.27).

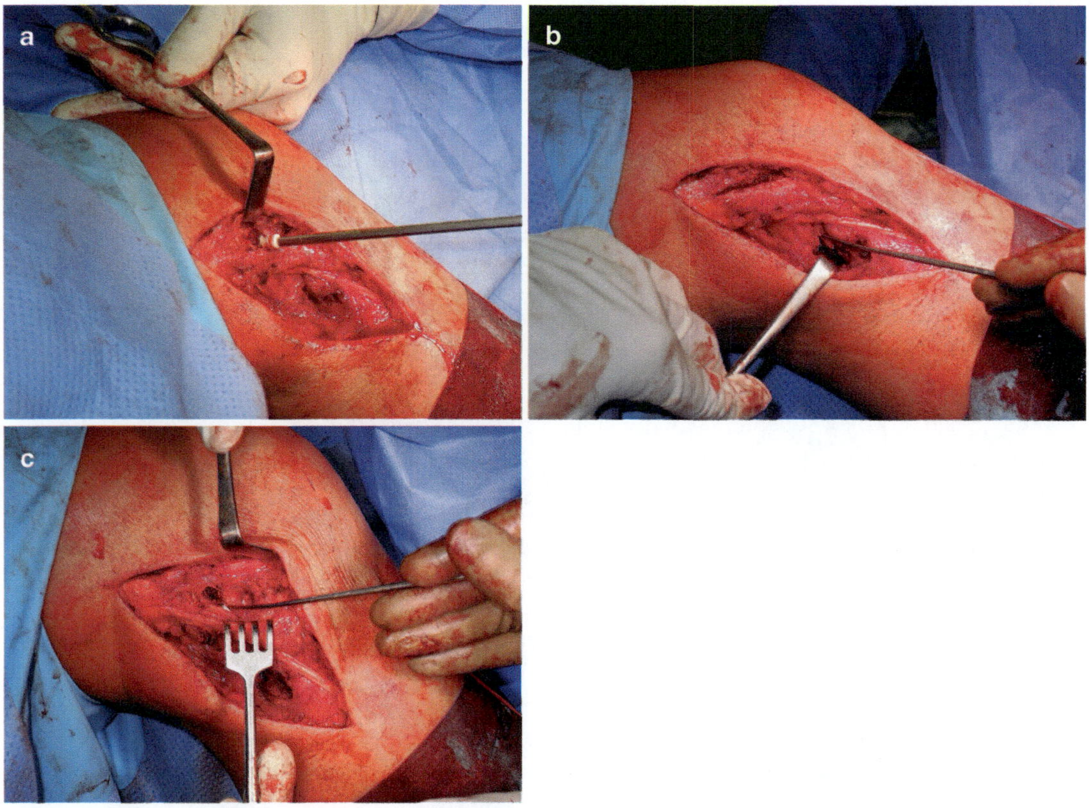

Fig. 12.20 Lateral view of the right knee. Fixation of the graft in one of the femoral tunnels with an interference screw. (**a**) Finished PLC reconstruction based on LaPrade surgical technique. The reconstructed PFL is seen through the middle window. (**b**) The femoral part of reconstructed FCL and PT is seen just above the iliotibial band (**c**)

12.8.3 Modified Larson's Surgical Technique

Modified Larson's surgical technique that we use differs from the original Larson's surgical technique by the minimally invasive approach. It involves making a double mini-open incision, which is sufficient to percutaneously reconstruct ruptured FCL and PFL [24] (Fig. 12.28).

The fibular head and the femoral FCL attachment are palpated and marked on the skin. The first skin incision is made vertically approximately 3 cm in length over the fibular head. Superficial biceps tendon is identified deep to the incision, and short longitudinal incision is made in the superficial portion of the biceps tendon posterior and its insertion on the fibula. The fibular tunnel is prepared by 1 cm skin incision on the anterior part of a fibula. A K-wire is drilled through the center of the fibular head in the posteriomedial direction while placing the small retractor through the incision on the posterior aspect of the fibular head. The K-wire is overdrilled usually by 5–6 mm drill bit, depending on the graft diameter (Fig. 12.29). A passing suture is placed in the fibular tunnel with the help of a wire loop. In this technique, it is not necessary to identify and protect the fibular nerve, unless the revision surgery is performed.

The second skin incision is made longitudinally 3 cm in length above the lateral femoral condyle. When the ITB is identified, a 3 cm incision is made just proximal to the origin of the lateral collateral ligament. A K-wire is drilled at the femoral insertion, authors preferred position is in between FCL and PT anatomical insertions (Fig. 12.30). Isometry is confirmed by placing traction on both ends of the passing suture placed

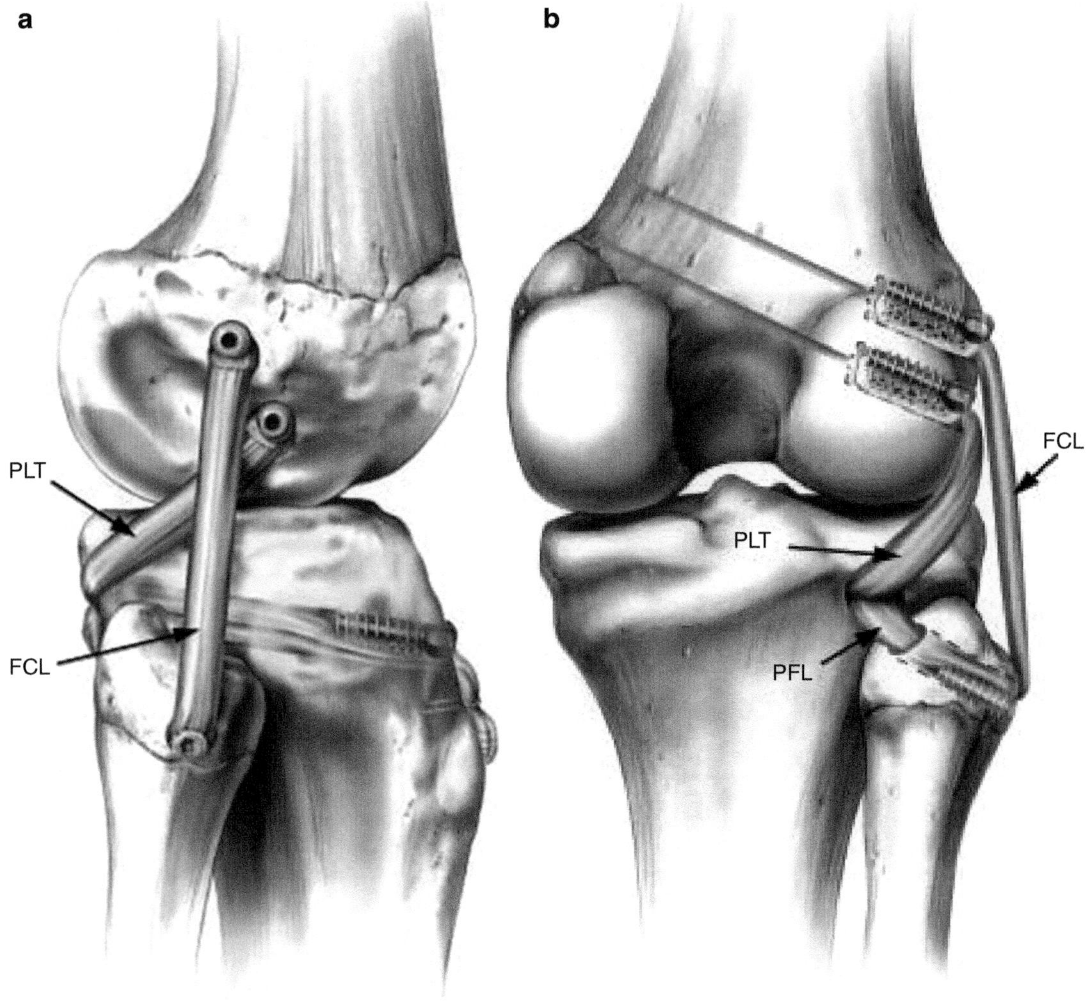

Fig. 12.21 Lateral (**a**) and posterior (**b**) view of the right knee with anatomical PLC reconstruction based on LaPrade's surgical technique. (Laprade et al. [5], Reproduced with permission)

through fibular tunnel as the knee is flexed and extended, while an anterior drawer stress is applied (Fig. 12.31). When the optimal isometric position is found, the wire is withdrawn and predrilled toward the anteromedial aspect of the femur. The wire is overdrilled with a drill of 7 mm to a depth of 25–30 mm, and a passing suture is passed through the femoral tunnel with the help of a guidewire with an eyelet.

The passage of the graft starts with the passing of the graft through fibular tunnel (Fig. 12.32). Then both the limbs are passed deep to the iliotibial tract at the femoral insertion. Both the ends of the graft are then pulled through the femoral tunnel, and the knee is cycled to tighten the graft limbs. The graft is fixed in the femoral tunnel with an interference screw (Fig. 12.33), while the knee is flexed to approximately 60°, the tibia is slightly internally rotated and the graft lead sutures are held separately under tension [25] (Fig. 12.34).

12.8.4 Arthroscopic Reconstruction by Frosch

Arthroscopic popliteus bypass reconstruction is indicated in knees classified with Fanelli A

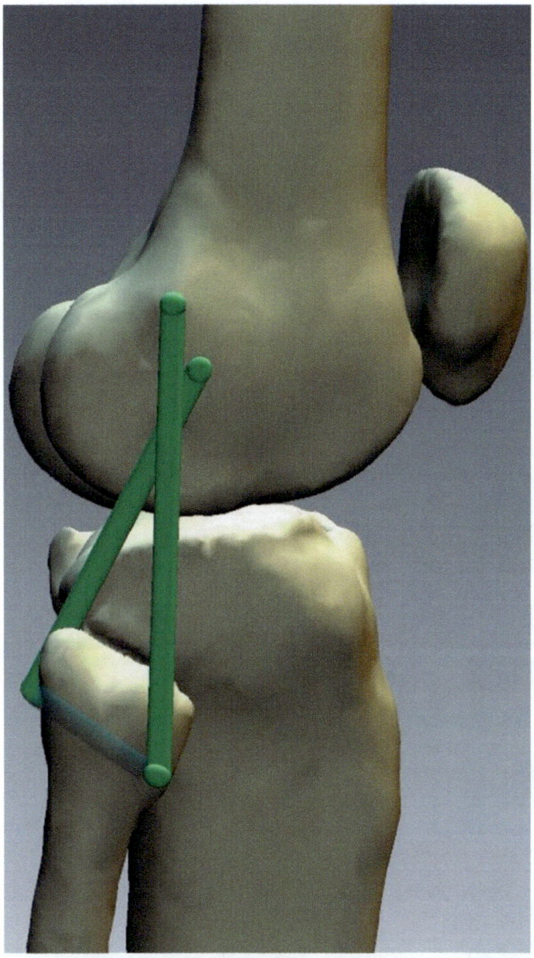

Fig. 12.22 A schematic image of lateral view of the right knee with PLC reconstruction based on Arciero's surgical technique. Intraosseous tunnels are marked in blue color

posterolateral instability with an isolated disruption of the PFL and combined PCL rupture. This kind of an injury results in a loss of a static stabilizing function of the popliteus complex, which is seen as an increase of posterior tibial translation and external rotation. For this procedure, six arthroscopic portals are needed: a high and a low anterolateral, a high anteromedial, a posteromedial, a posterolateral, and a lateral parapatellar portal. For the popliteus bypass graft, a single-stranded semitendinosus or double-stranded gracilis tendon of at least 11–12 cm length is used. A diagnostic arthroscopy is first performed. The arthroscope inserted through the high anterolateral portal is passed into the dorsomedial recessus. Dorsal septum is resected through the dorsomedial portal. Arthroscope is inserted through the dorsomedial portal where the PCL, partially dissected dorsal septum, PT, and posteromedial aspect of the lateral femoral condyle are visualized. The radiofrequency electrode is inserted through the dorsolateral portal and used to resect the lateral popliteomeniscal fibers approximately 1–2 cm in length. Then the PT is retracted by a hook through the posterolateral portal, and popliteal tunnel is visualized. First, the tibial tunnel is drilled with the help of appropriate tibial guide inserted through the high anteromedial portal. Its tip is positioned in the distal third of the popliteal sulcus, while the entering point of the drill is located between the lateral edge of the tibial tuberosity and the medial edge of the Gerdy's tubercle. The femoral popliteal attachment site is viewed arthroscopically through the high anterolateral portal with the knee at 20–30° of knee flexion and is additionally exposed by careful dissection of the capsule with a shaver inserted through the lateral parapatellar portal. The femoral tunnel is drilled percutaneously directly at the center of the femoral attachment of the PT. The popliteus bypass graft is pulled into the knee first through the tibial tunnel and then fixed in the femoral tunnel by an interference screw. It is important to note that the graft is passed below the FCL. Lastly, tibial fixation is done by an interference with the knee flexed at 90° and internally rotated for 10–20° [27].

12.9 The Role of High Tibial Osteotomy

Varus malalignment in knees with PLC injury is an indication for high tibial osteotomy (HTO), especially in the chronic setting [28]. Studies show that the use of soft tissue reconstruction techniques alone without the correction of alignment in PLC injuries associated with malalignment gives poor result [29]. HTO allows simultaneous correction of both coronal and sagittal alignment of the knee with biplanar osteotomy. It can be used alone, simultaneously with ligament reconstruction, or as a staged procedure

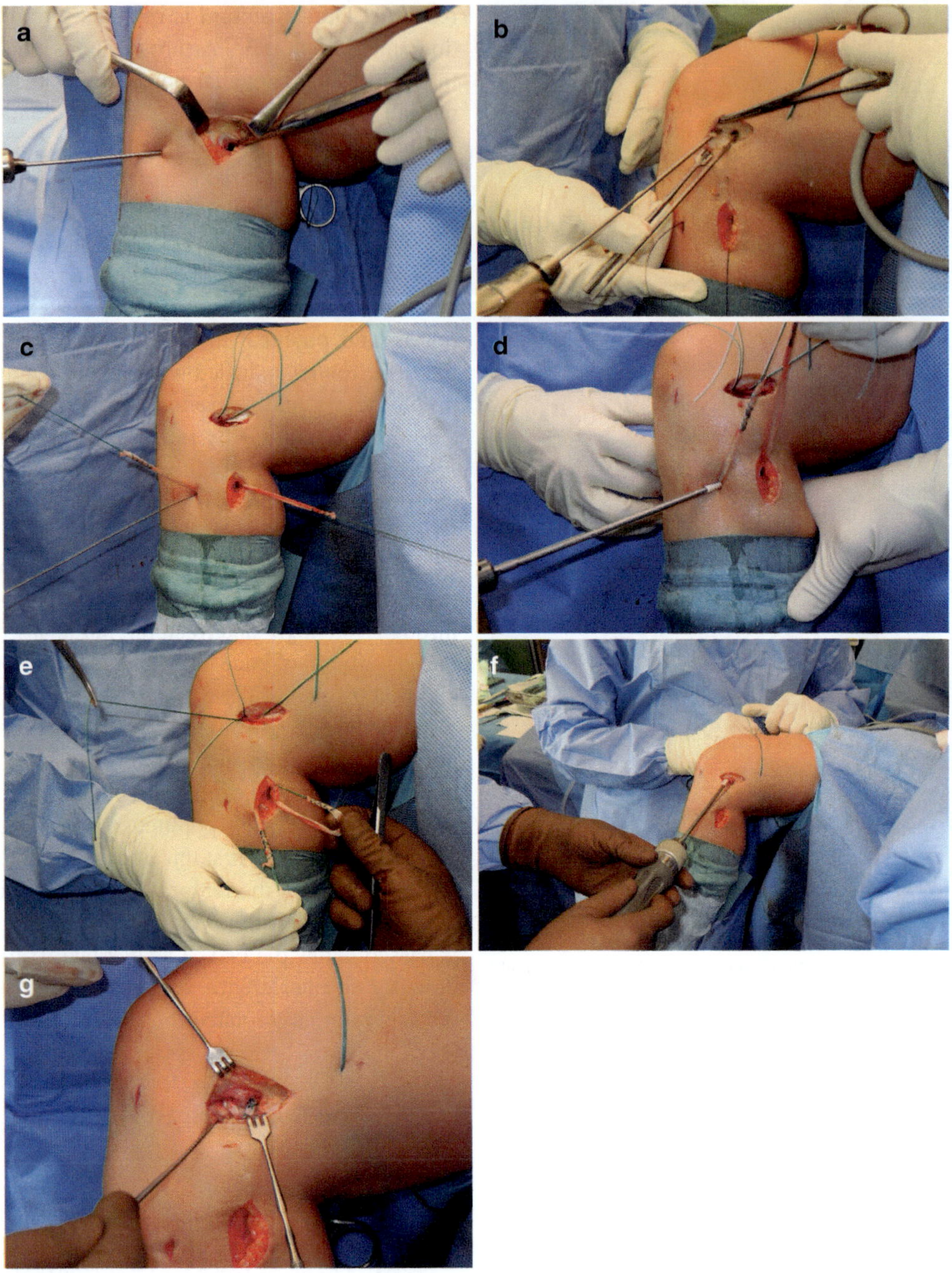

Fig. 12.23 Lateral view of the left knee. Example of minimally invasive surgical approach where a double mini-open incision is made. The fibular tunnel is prepared by drilling a K-wire from the FCL insertion on the anterolateral aspect of the fibular head to the fibular PFL insertion site located posteromedially. (**a**) K-wires are drilled in the location of the femoral insertion of the popliteus tendon and fibular collateral ligament. (**b**) The passage of the graft is started with the passing of the graft through fibular tunnel. (**c**) A fibular fixation of the graft is performed with a cannulated interference screw inserted from the anterior part. (**d**) Posterior limb of the graft is already pulled through the posterior lower incision. Afterwards, the anterior limb of the graft is passed through the popliteal hiatus into the popliteal femoral tunnel, while the posterior limb of the fibular tunnel is passed deep to the fascia lata into the FCL femoral tunnel lying over the anterior limb of the graft. (**e**) Fixation of the graft in the anterior femoral tunnel with an interference screw. (**f**) Finished PLC reconstruction with minimally invasive surgical approach based on Arciero surgical technique (**g**)

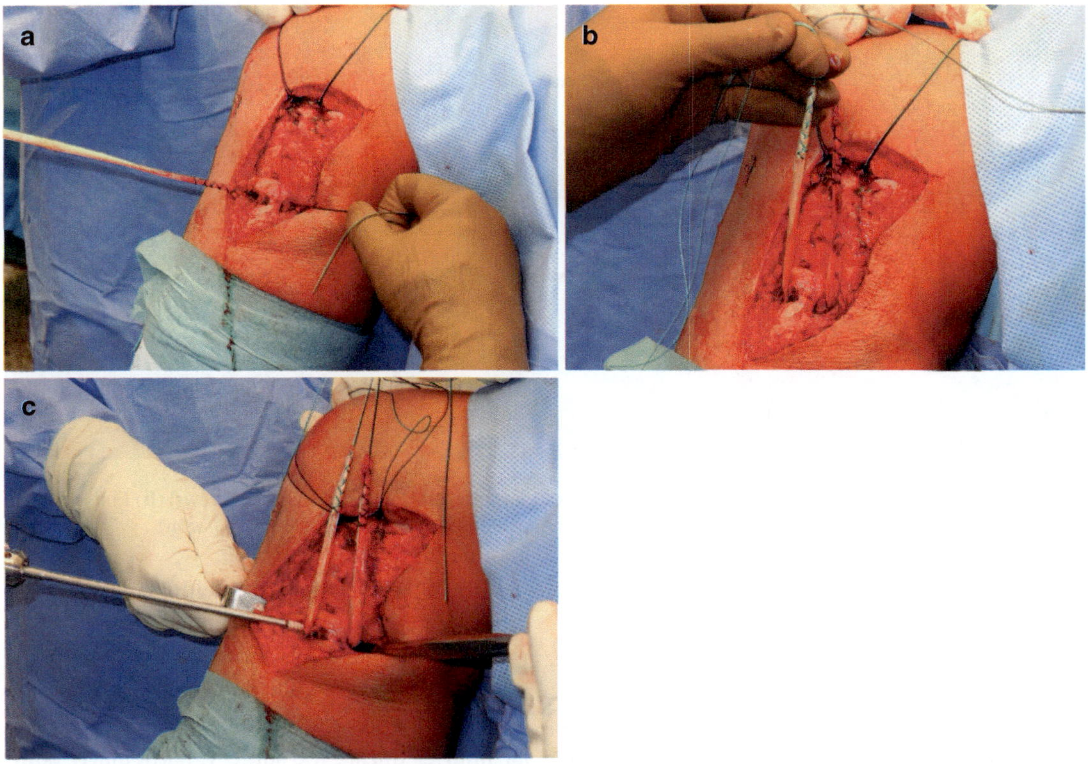

Fig. 12.24 Lateral view of the left knee. The prepared graft is passed through the fibular tunnel from anterior to posterior (**a**). It is important to firmly tension the tendon and have equal length of limbs anteriorly and posteriorly (**b**), then we can proceed and fix the tendon into the fibula with an interference screw of a diameter of 6 mm (**c**)

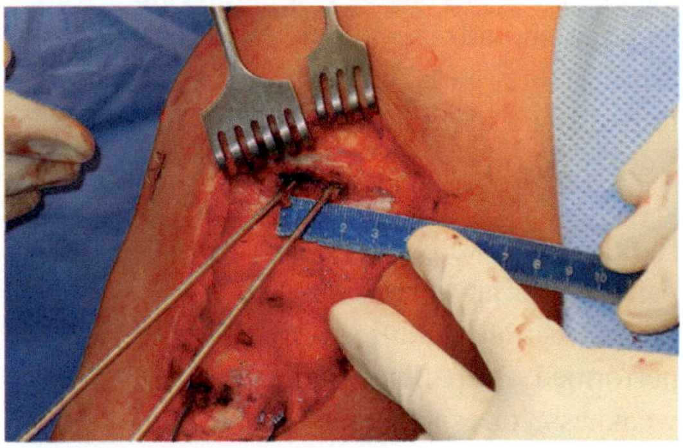

Fig. 12.25 Lateral view of the left knee with classical PLC approach. K-wires are drilled in the location of the femoral insertion of the popliteus tendon and fibular collateral ligament based on the anatomy of the individual

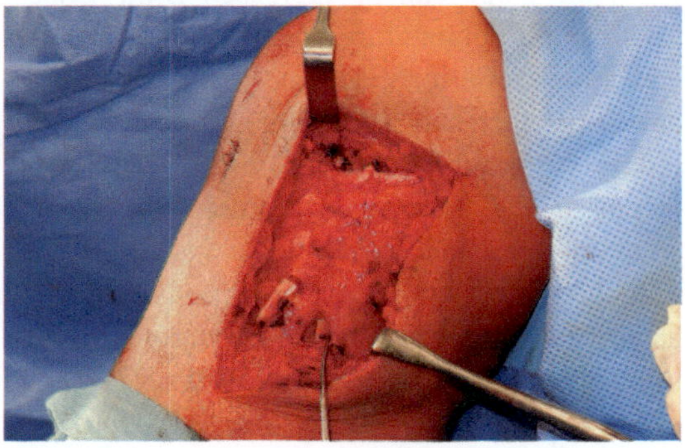

Fig. 12.26 Lateral view of the right knee. Finished PLC reconstruction based on Arciero surgical technique

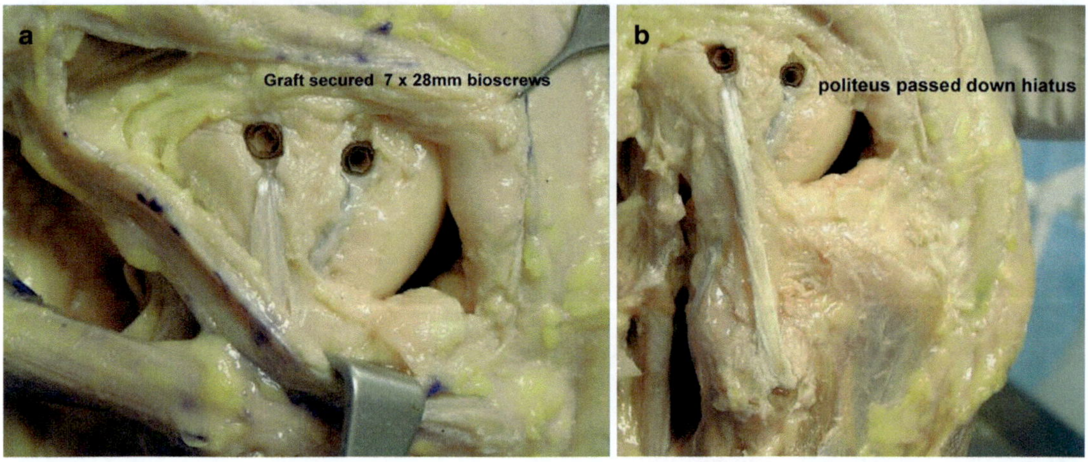

Fig. 12.27 Lateral view of the right knee. (**a**) Femoral fixation of the graft with interference screws. (**b**) Finished PLC reconstruction based on Arciero surgical technique. (Arciero et al. [23], Reproduced with permission)

where HTO is performed first followed by additional ligament reconstruction if joint instability persists and interferes with occupational or recreational activities. Different etiologies of PLC knee instability require different approaches. In chronic PLC injuries with varus malalignment the correction in coronal plane is usually necessary. With osteotomy in sagittal plane, we can change the tibial slope. In ACL-deficient knee, the slope can be decreased, and in PCL-deficient knee, the slope can be increased (Fig. 12.35). This can be obtained by opening the osteotomy gap more posteriorly (decreasing the slope) or anteriorly (increasing the slope). In general, osteotomy is usually done before any soft tissue reconstruction. A staged procedure is particularly recommended in chronic PLC instabilities or failures of previous reconstructions while in acute PLC reconstructions bone corrections are usually not a part of the procedure. HTO is an effective procedure with good reported outcomes for the treatment of PLC injuries of the knee with varus malalignment [30].

12.10 Complications

Complications after a PLC reconstruction can arise in large part due to the complex anatomy and injury pattern of this region combined with a surgeon's lack of experience with management of this uncommon and serious injury.

Complex anatomy and close proximity of important neurovascular structures in the PLC region of the knee raises the risk of neurovascular injuries during the surgery, particularly common peroneal nerve and popliteal artery. Deep vein thrombosis presents a risk after any lower limb surgery. Thus, it may develop after a PLC reconstruction procedure [31]. In high-risk patients, prophylactic low molecular weight heparin should be administered if necessary. Furthermore, early mobilization and rehabilitation also aid in preventing deep vein thrombosis from occurring [4].

Kornbluth et al. reported a case of femoral and saphenous nerve palsy after a tourniquet use in a patient after an arthroscopic PCL and open PLC reconstruction. Additional evaluation should be performed in patients with persistent muscle weakness or sensory findings after surgery involving a use of a tourniquet [32].

Possible complication after PLC reconstruction is the formation of fibrous adhesions and scar tissue which limit the knee movement. The exact incidence of arthrofibrosis after management of PLC injury is not known. Prevention consists of performing the correct surgical technique, early postoperative rehabilitation and range of motion exercises. Delay of surgery

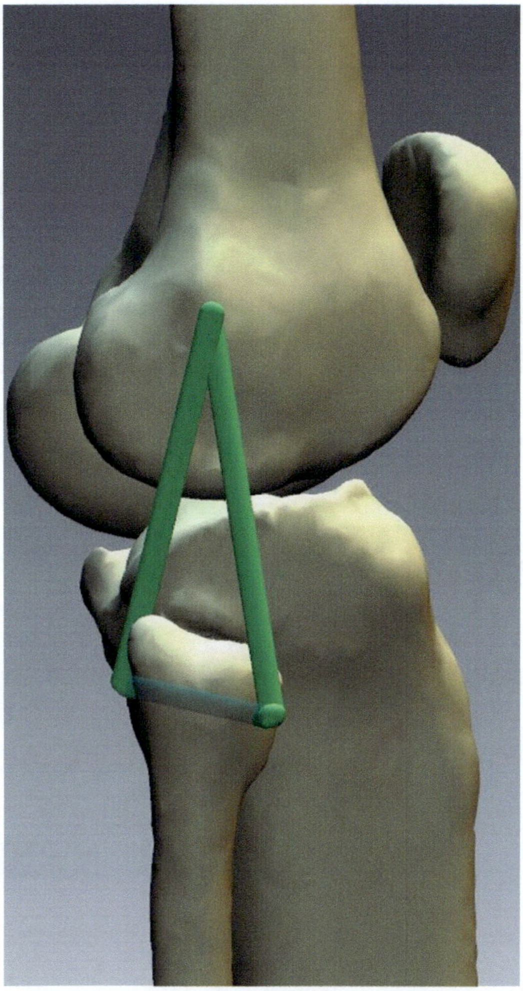

Fig. 12.28 A schematic image of lateral view of the right knee with PLC reconstruction based on modified Larson's surgical technique. Intraosseous tunnels are marked in blue color

until the resolution of acute inflammation may have a beneficial effect in occurrence of arthrofibrosis [33].

Superficial wound or deep infection is always a potential risk in knee surgery. The incidence of wound infection in open knee reconstructions ranges between 0.3% and 12.5% [34]. Additionally, large soft tissue flaps created during the PLC procedure could lead to wound dehiscence. Therefore, special caution and delicate handling with soft tissues is mandatory during surgery.

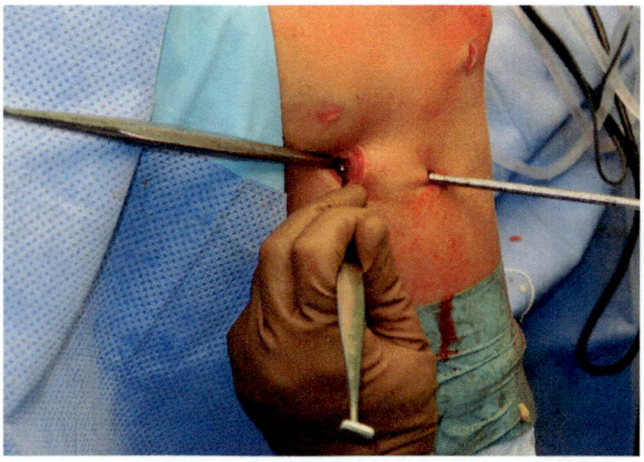

Fig. 12.29 Lateral view of the right knee. The K-wire drilled through the center of the fibular head in the antero-inferior to postero-superior direction is overdrilled usually by 5–6 mm, depending on the graft diameter

Technical flaws in PLC reconstruction can result in recurrent instability if the main anatomical and biomechanical structures are not adequately restored. Failure to do so has been shown to be an important cause of failed concomitant ACL or PCL reconstructions [33].

A biomechanical cadaveric study conducted by Nau et al. has shown that the knee can be overconstrained, when anatomic reconstruction of both the limbs of the popliteus complex is performed, which resulted in abnormal tibial internal rotation during guided movement [35].

12.11 Minimizing Technical Problems

To avoid overconstraint, a careful fixation of the PLC reconstruction is recommended in accordance with the previously detailed instructions under each surgical technique. In a multi-ligament injury, the fixation of the PLC should be done first. Then, the PCL fixation is followed in 90° of knee flexion with applied anterior drawer force. Lastly, the ACL should be fixed with the knee in full extension and the medial collateral ligament in 15° of flexion with a slight varus force [33].

Moathe et al. studied the inter-tunnel relationships of the femoral tunnels in multiple ligament reconstruction. They have shown that the least chance of convergence of the FCL and

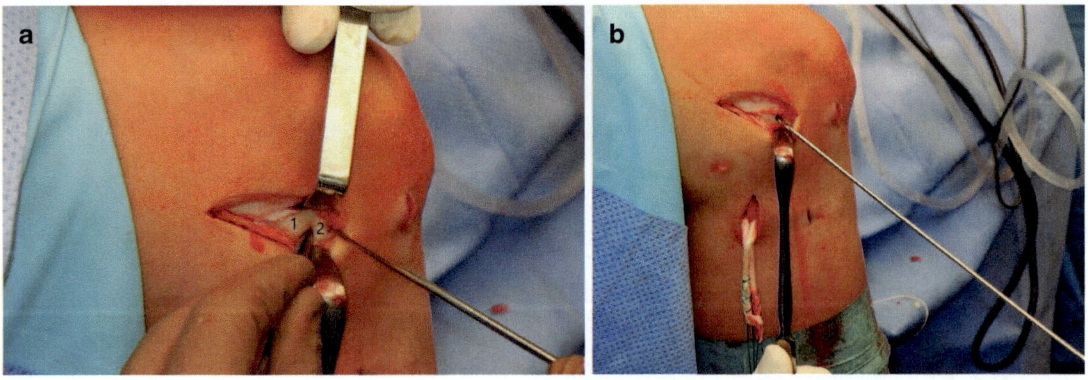

Fig. 12.30 Lateral view of the right knee. A K-wire is drilled at the femoral insertion, authors preferred position is in between FCL and PT anatomical insertions (**a**, **b**). *1* Femoral insertion of FCL, *2* Femoral insertion of PT

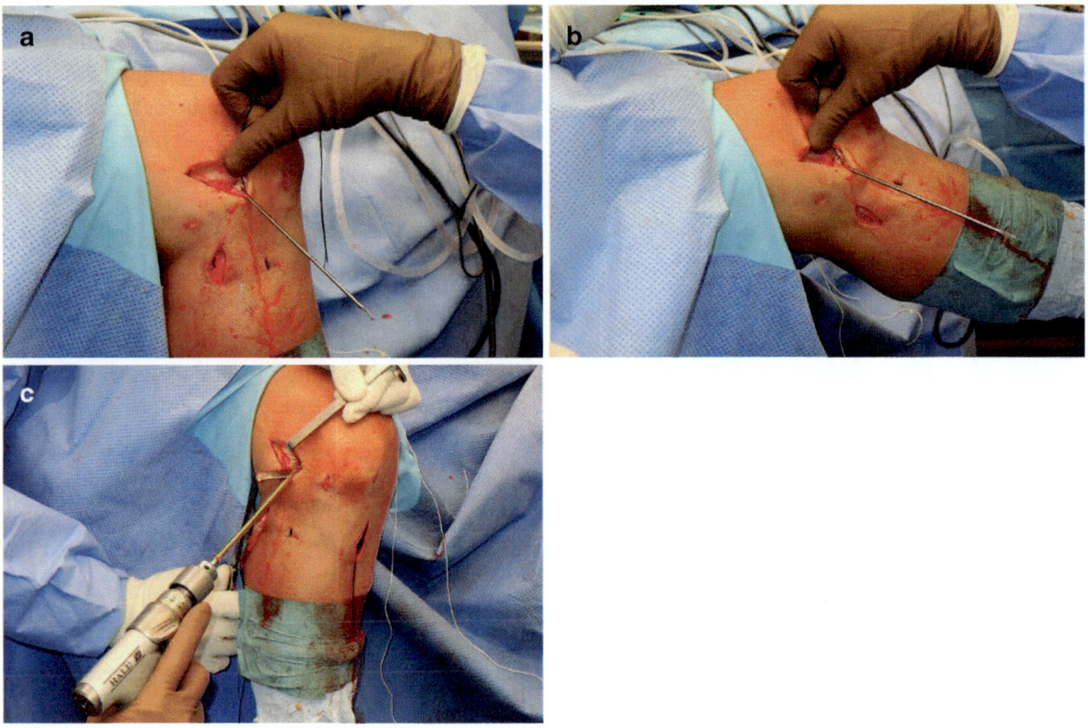

Fig. 12.31 Lateral view of the right knee. Isometry is confirmed by placing traction on both the ends of the passing suture placed through fibular tunnel as the knee is flexed (**a**) and extended (**b**), while an anterior drawer stress is applied. When the optimal isometric position is found, the wire is withdrawn and predrilled toward the anteromedial aspect of the femur. The wire is overdrilled (**c**)

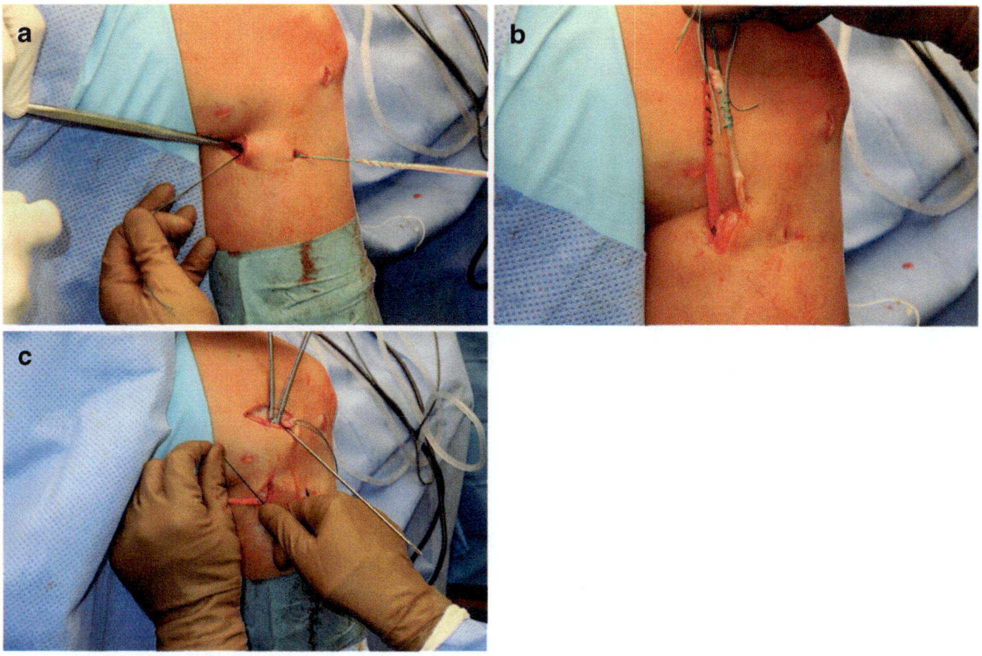

Fig. 12.32 Lateral view of the right knee. The passage of the graft is started with the passing of the graft through fibular tunnel. (**a**, **b**) Then both limbs are passed deep to the iliotibial tract at the femoral insertion. The passage of the graft is simplified with the help of pean forceps as seen on the photo (**c**)

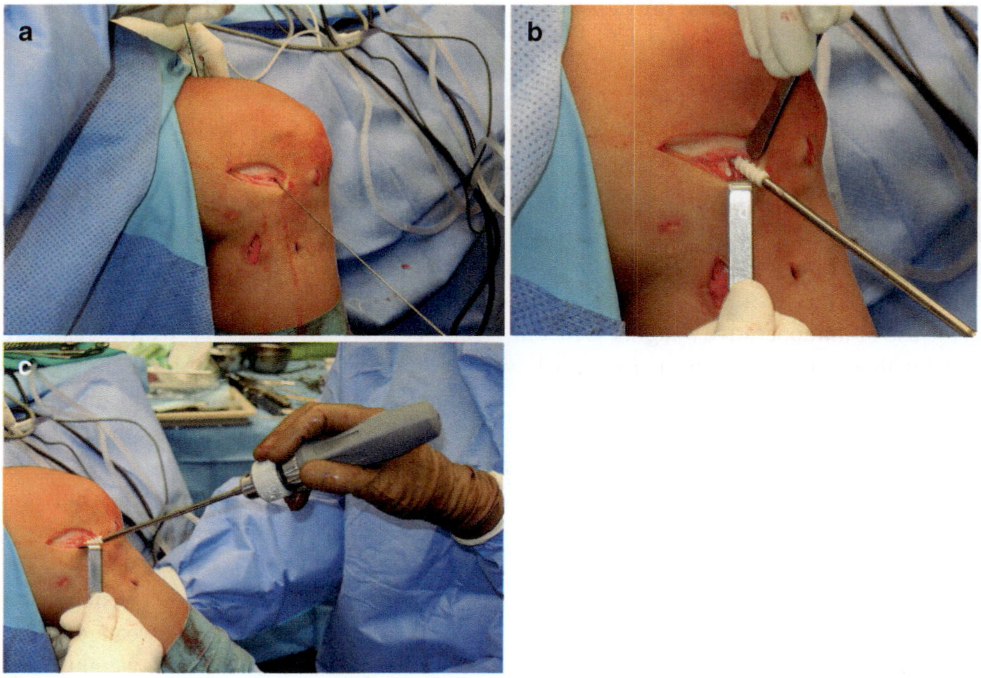

Fig. 12.33 Lateral view of the right knee. The graft is fixed in the femoral tunnel with an interference screw, while the knee is flexed to approximately 60°, the tibia is slightly internally rotated, anterior drawer stress is applied, and the graft lead sutures are held separately under tension (**a**, **b**, **c**)

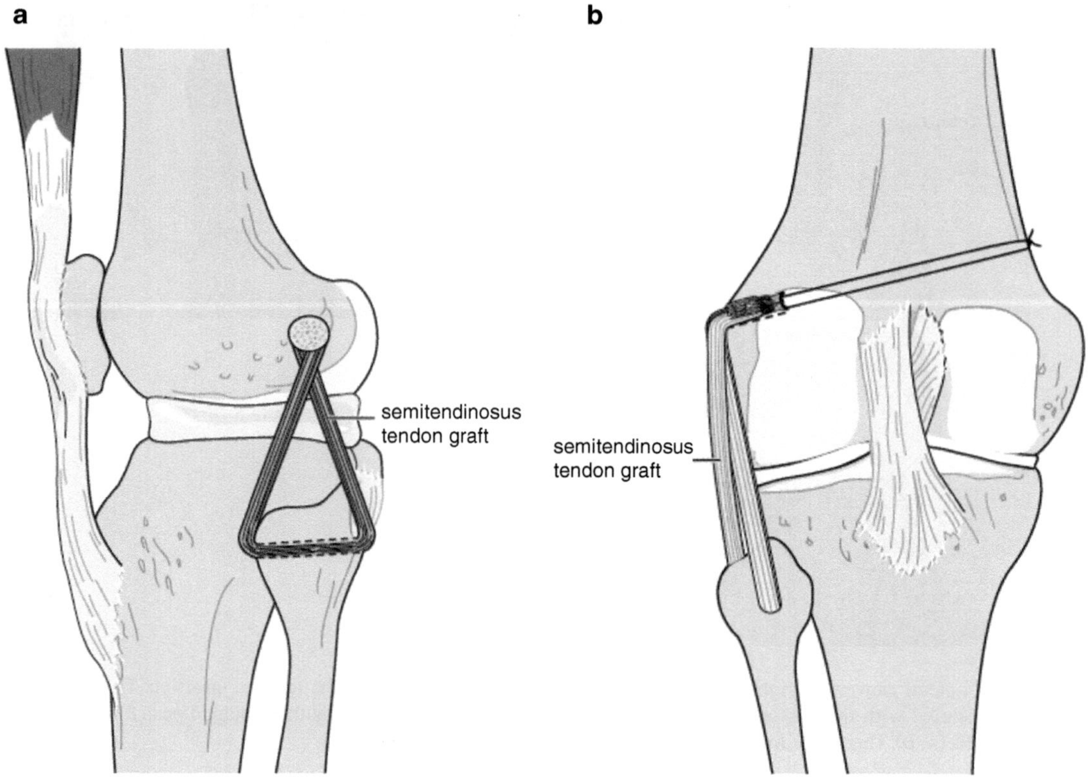

Fig. 12.34 Lateral (**a**) and posterior (**b**) view of the left knee with non-anatomical PLC reconstruction based on Larson's surgical technique. (Panzica et al. [26], Reproduced with permission)

ACL tunnel is by drilling the FCL tunnel in 35°–40° anteriorly and 0° proximally, while the PT tunnel should be aimed at 35° anteriorly in order to avoid the violation of the intercondylar notch [36].

To prevent the blow out of the tibial tunnel in LaPrade's PLC reconstruction surgical technique, the use of intraoperative imaging is recommended for those surgeons who perform a low number of these surgeries [33].

Attention to detail and technical exactness ensures a low degree of technical complications in such technically demanding procedures as PLC reconstructions.

12.12 Rehabilitation

Postoperative rehabilitation after PLC reconstruction involves patient wearing an immobilizer with the knee extended except for range of motion exercises. The patient remains non-weight bearing for 6 weeks in order the reconstruction is allowed to safely heal as varus forces on the graft encountered during ambulation are avoided. In patients with combined PCL reconstruction, the rehabilitation is following PCL rehabilitation protocol. Rehabilitation starts immediately after surgery with focus to restore tibiofemoral and patellofemoral range of motion, edema control, pain management, and restoration of quadriceps function. During the first 2 weeks, passive knee motion exercises from 0° to 60° are performed and are progressed to full range of motion, which should be achieved after 6 weeks. Afterwards, when 90° of knee flexion is reached, the patients are allowed to start using a spinning stationary bike and slowly wean off crutches. Once they can bear the full weight, they first begin with closed chain exercises in order to develop muscular endurance before advancing to muscular strength and power development. In the

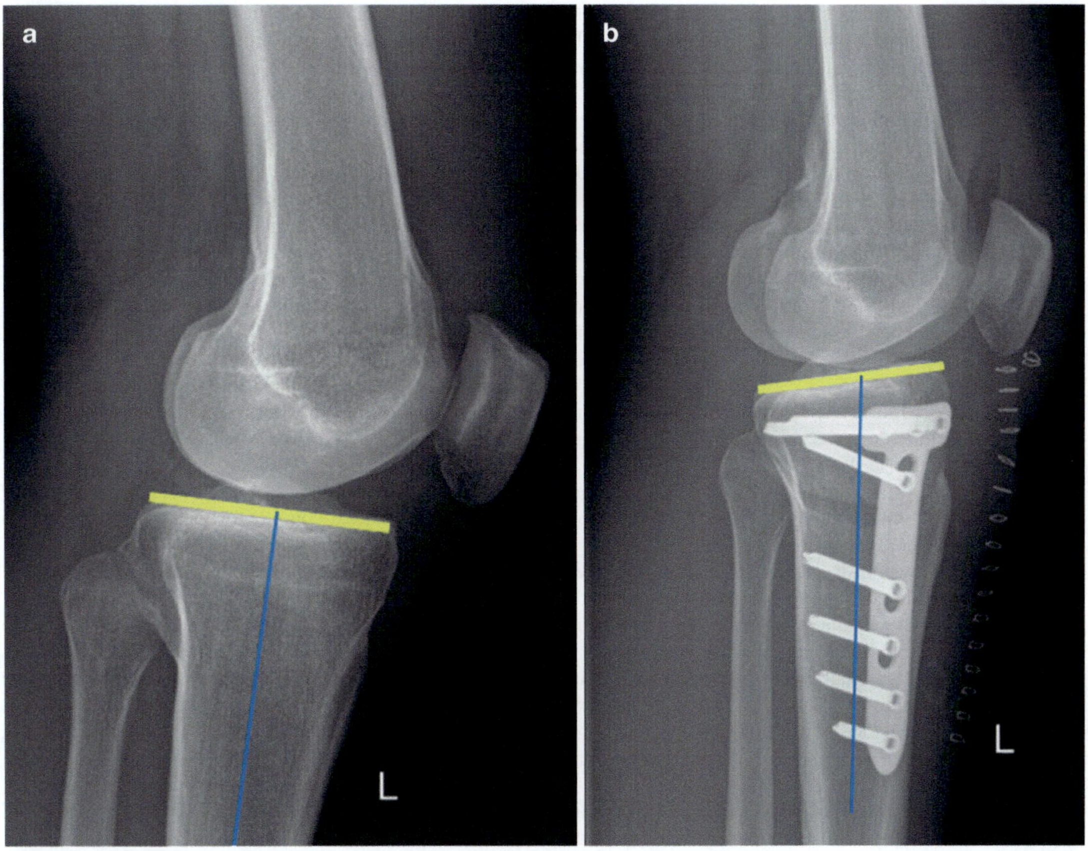

Fig. 12.35 Lateral view radiograph of the left knee. Preoperative X-ray (**a**) and postoperative X-ray with an increased tibial slope after an HTO (**b**)

first 4 months, isolated hamstring strengthening or positions in which the tibia is prone to posterior sag or external rotation are to be avoided, in order not to stress the reconstruction. Usually around 6 months after surgery, once proper strength and power have been achieved, the patients may begin with progressive jogging, along with speed and agility exercises. Return to play is permitted when adequate strength, stability, and range of motion comparable to the uninjured contralateral limb are reached [4].

References

1. LaPrade RF, Wentorf FA, Fritts H, Gundry C, Hightower CD. A prospective magnetic resonance imaging study of the incidence of posterolateral and multiple ligament injuries in acute knee injuries presenting with a hemarthrosis. Arthrosc J Arthrosc Relat Surg. 2007;23(12):1341–7.
2. Bicos J, Arciero RA. Novel approach for reconstruction of the posterolateral corner using a free tendon graft technique. Sports Med Arthrosc. 2006;14(1):28–36.
3. Pacheco RJ, Ayre CA, Bollen SR. Posterolateral corner injuries of the knee: a serious injury commonly missed. J Bone Joint Surg Br. 2011;93(2):194–7.
4. Shon O-J, Park J-W, Kim B-J. Current concepts of posterolateral corner injuries of the knee. Knee Surg Relat Res. 2017;29(4):256–68.
5. LaPrade RF, Johansen S, Wentorf FA, Engebretsen L, Esterberg JL, Tso A. An analysis of an anatomical posterolateral knee reconstruction: an in vitro biomechanical study and development of a surgical technique. Am J Sports Med. 2004;32(6):1405–14.
6. Osti M, Tschann P, Künzel KH, Benedetto KP. Posterolateral corner of the knee: microsurgical analysis of anatomy and morphometry. Orthopedics. 2013;36(9):e1114–20.
7. LaPrade RF, Ly TV, Wentorf FA, Engebretsen L. The posterolateral attachments of the knee: a qualitative

and quantitative morphologic analysis of the fibular collateral ligament, popliteus tendon, popliteofibular ligament, and lateral gastrocnemius tendon. Am J Sports Med. 2003;31(6):854–60.
8. Laprade RF, Wozniczka JK, Stellmaker MP, Wijdicks CA. Analysis of the static function of the popliteus tendon and evaluation of an anatomic reconstruction. Am J Sports Med. 2010;38(3):543–9.
9. Delee JC, Riley MB, Rockwood CA. Acute posterolateral rotatory instability of the knee. Am J Sports Med. 1983;11(4):199–207.
10. LaPrade RF, Terry GC. Injuries to the posterolateral aspect of the knee. Association of anatomic injury patterns with clinical instability. Am J Sports Med. 1997;25(4):433–8.
11. Krukhaug Y, Mølster A, Rodt A, Strand T. Lateral ligament injuries of the knee. Knee Surg Sports Traumatol Arthrosc. 1998;6(1):21–5.
12. Chahla J, Moatshe G, Dean CS, Laprade RF. Posterolateral corner of the knee: current concepts. Arch Bone Jt Surg. 2016;4(2):97–103.
13. Grood ES, Stowers SF, Noyes FR. Limits of movement in the human knee. Effect of sectioning the posterior cruciate ligament and posterolateral structures. J Bone Joint Surg Am. 1988;70(1):88–97.
14. Forsythe B, Saltzman BM, Cvetanovich GL, Collins MJ, Arns TA, Verma NN, et al. Dial test: unrecognized predictor of anterior cruciate ligament deficiency. Arthrosc J Arthrosc Relat Surg. 2017;33(7):1375–81.
15. Crespo B, James EW, Metsavaht L, Laprade RF. Erratum: injuries to posterolateral corner of the knee: a comprehensive review from anatomy to surgical treatment (Revista Brasileira de Ortopedia (2015) 50 (363-370)). Rev Bras Ortop. 2015;50(5):613.
16. LaPrade RF, Heikes C, Bakker AJ, Jakobsen RB. The reproducibility and repeatability of varus stress radiographs in the assessment of isolated fibular collateral ligament and grade-III posterolateral knee injuries. An in vitro biomechanical study. J Bone Jt Surg Ser A. 2008;90(10):2069–76.
17. LaPrade RF, Gilbert TJ, Bollom TS, Wentorf F, Chaljub G. The magnetic resonance imaging appearance of individual structures of the posterolateral knee. A prospective study of normal knees and knees with surgically verified grade III injuries. Am J Sports Med. 2000;28(2):191–9.
18. LaPrade RF. Arthroscopic evaluation of the lateral compartment of knees with grade 3 posterolateral knee complex injuries. Am J Sports Med. 1997;25(5):596–602.
19. Feng H, Song GY, Shen JW, Zhang H, Wang MY. The "lateral gutter drive-through" sign revisited: a cadaveric study exploring its real mechanism based on the individual posterolateral structure of knee joints. Arch Orthop Trauma Surg. 2014;134(12):1745–51.
20. Fanelli GC, Larson RV. Practical management of posterolateral instability of the knee. Arthroscopy. 2002;18(2):1–8.
21. Terry GC, LaPrade RF. The posterolateral aspect of the knee: anatomy and surgical approach. Am J Sports Med. 1996;24(6):732–9.
22. Dickens JF, Kilcoyne K, Kluk M, Rue J-P. The posterolateral corner: surgical approach and technique overview. J Knee Surg. 2011;24(3):151–8.
23. Arciero RA. Anatomic posterolateral corner knee reconstruction. Arthrosc J Arthrosc Relat Surg. 2005;21(9):1–5.
24. Plaweski S, Belvisi B, Moreau-Gaudry A. Reconstruction of the posterolateral corner after sequential sectioning restores knee kinematics. Orthop J Sports Med. 2015;3(2):2325967115570560.
25. Strobel M, Weiler A. In: Gagstatter F, editor. The posterior cruciate ligament: anatomy, evaluation, operative technique: Endo-Press; 2010. p. 187–208.
26. Panzica M, Janzik J, Bobrowitsch E, Krettek C, Hawi N, Hurschler C, et al. Biomechanical comparison of two surgical techniques for press-fit reconstruction of the posterolateral complex of the knee. Arch Orthop Trauma Surg. 2015;135(11):1579–88.
27. Frosch KH, Akoto R, Drenck T, Heitmann M, Pahl C, Preiss A. Arthroscopic popliteus bypass graft for posterolateral instabilities of the knee. Oper Orthop Traumatol. 2016;28(3):193–203.
28. Savarese E, Bisicchia S, Romeo R, Amendola A. Role of high tibial osteotomy in chronic injuries of posterior cruciate ligament and posterolateral corner. J Orthop Traumatol. 2011;12(1):1–17.
29. Phisitkul P, Wolf BR, Amendola A. Role of high tibial and distal femoral osteotomies in the treatment of lateral-posterolateral and medial instabilities of the knee. Sports Med Arthrosc. 2006;14(2):96–104.
30. Dean CS, Liechti DJ, Chahla J, Moatshe G, LaPrade RF. Clinical outcomes of high tibial osteotomy for knee instability: a systematic review. Orthop J Sports Med. 2016;4(3):2325967116633419.
31. Camarda L, Condello V, Madonna V, Cortese F, D'Arienzo M, Zorzi C. Results of isolated posterolateral corner reconstruction. J Orthop Traumatol. 2010;11(2):73–9.
32. Kornbluth ID, Freedman MK, Sher L, Frederick RW. Femoral, saphenous nerve palsy after tourniquet use: a case report. Arch Phys Med Rehabil. 2003;84(6):909–11.
33. MacDonald P, Vo A. Complications of posterolateral corner injuries of the knee and how to avoid them. Sports Med Arthrosc. 2015;23(1):51–4.
34. Hegyes MS, Richardson MW, Miller MD. Knee dislocation: complications of nonoperative and operative management. Clin Sports Med. 2000;19(3):519–43.
35. Nau T, Chevalier Y, Hagemeister N, Deguise JA, Duval N. Comparison of 2 surgical techniques of posterolateral corner reconstruction of the knee. Am J Sports Med. 2005;33(12):1838–45.
36. Moatshe G, Brady AW, Slette EL, Chahla J, Turnbull TL, Engebretsen L, et al. Multiple ligament reconstruction femoral tunnels. Am J Sports Med. 2017;45(3):563–9.

Anatomic Knee Joint Realignment 13

Bogdan Ambrožič, Urban Slokar, Urban Brulc, and Samo Novak

13.1 Introduction

The concept of surgical osteotomy (*osteo* = bone, *tomy* = cut) for the treatment of isolated unicompartmental osteoarthritis of the knee associated with angular deformity has been in existence for decades [1]. Originally popularized by several surgeons in mid-twentieth century, it however failed to gain popularity due to its history of unpredictable and oftentimes poor results [2–4]. The commonly associated complications included high infection rate, loss of correction, and postoperative stiffness of the joint [5, 6]. Osteotomies finally fell out of favor with orthopedic surgeons as evolution of knee prostheses in the 1980s led to subsequent success of knee arthroplasty, especially in low demand and older patients [1, 7]. Only after considerable improvement in surgical techniques, fixation devices and patient selection in the recent years, osteotomy began to regain its past reputation as a highly effective treatment option with fewer complications and reproducible functional outcomes, reaching a survival rate of more than 90% at 5 years and more than 70% at 15 years [1, 8–10].

Knee osteotomy is indicated for the treatment of malalignment with symptomatic unicompartmental cartilage disease and/or ligamentous instability of the knee [11].

The aim of the procedure is to reduce the load on the affected compartment by correcting or slightly overcorrecting the mechanical axis of the knee, so that higher percentage of the load transferred across the knee joint is shifted to the healthy compartment. This way, forces become more evenly distributed over both the medial and lateral compartment, preventing or delaying advancement of degenerative joint disease [12, 13]. Good cartilage preservation in all other compartments is therefore an important factor when considering osteotomy, as multicompartmental osteoarthritis of the knee is more often successfully treated with total knee arthroplasty [13].

Despite its fluctuating reputation historically, several recent studies have affirmed that with careful patient selection, precise preoperative planning and modern surgical fixation techniques, osteotomy around the knee is considered an effective biological treatment for degenerative disease, deformity and instability. It is progressively used on a routine basis either as a standalone therapy in patients with single compartment overload or as an additive procedure in patients

B. Ambrožič (✉) · U. Slokar · S. Novak
Valdoltra Orthopaedic Hospital, Ankaran, Slovenia
e-mail: bogdan.ambrozic@bisturmed.com; urban.slokar@ob-valdoltra.si; samo.novak@ob-valdoltra.si

U. Brulc
MD Medicina Sanatorij Ljubljana, Ljubljana, Slovenia
e-mail: urban.brulc@md-medicina.si

M. Bozkurt, H. İ. Açar (eds.), *Clinical Anatomy of the Knee*,
https://doi.org/10.1007/978-3-030-57578-6_13

where a biological reconstructive procedure (cartilage surgery, meniscus transplantation) is considered [1, 14].

13.2 Biomechanical Aspect

When deciding to perform a corrective osteotomy, many factors need to be taken into consideration, especially important being mechanical conditions of the concerned knee. As the largest and most complex load-bearing joint in our body, the knee joint is subjected to significant forces combined with a complex combination of rolling, sliding, and rotational movements, which contribute to accelerated wear of the articular cartilage [15]. Due to knee adduction moment during stance phase of the gait cycle, the peak joint forces are usually higher in the medial knee compartment, which is why it is the most common site of knee osteoarthritis [16]. When osteotomy of the knee is considered, deformity caused from the wear of the cartilage in the affected compartment should ideally cause the mechanical axis to pass through the damaged compartment. Also, the knee joint needs to have a good range of motion, especially extension should not be compromised, as this indicates progression of the arthritic process with osteophyte formation that cannot be halted by an alignment change.

13.2.1 Physiological Axes and Angles of the Leg

There are two important physiological axes of the lower extremity: anatomical and mechanical. The first corresponds to the diaphyseal midline of the femur and tibia [17]. There is a laterally opened angle of 173–175° between the anatomical axis of femur and tibia. The mechanical axis (also called Mikulicz line) extends from the center of the femoral head to the center of the ankle joint. Physiologically, this line runs approximately 4 (±2) mm medial to the center of the knee joint [18, 19]. Shifting of the mechanical axis in either medial or lateral direction indicates varus or valgus deformity of the knee, respectively. Together with the joint line (tangent to the femoral condyles), both anatomical and mechanical axes form a number of relevant physiological angles that need to be considered when planning or evaluating osteotomies (Fig. 13.1 and Table 13.1).

The tangents of the femoral condyles and tibial plateau run almost parallel, with 0° ± 2° joint line convergence angle (JLCA) (Fig. 13.2).

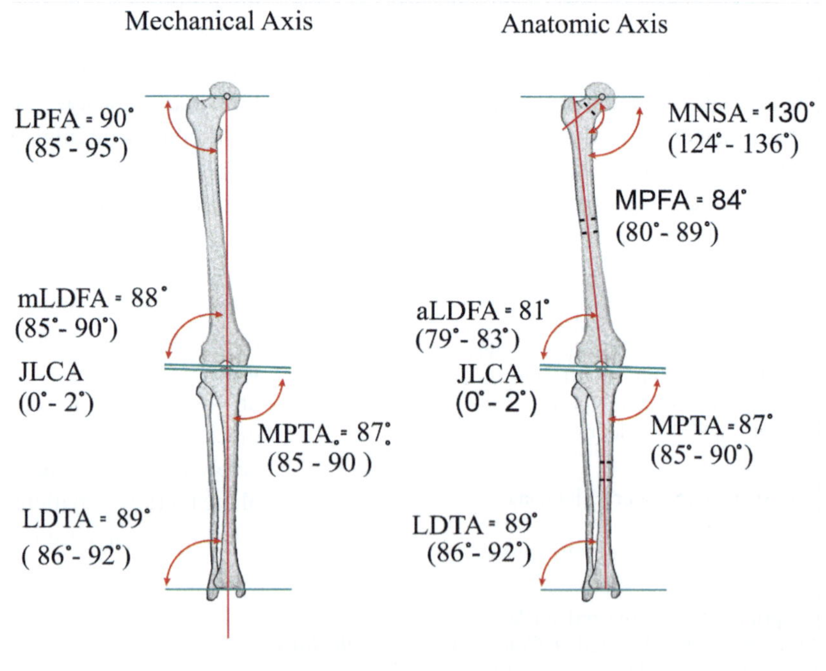

Fig. 13.1 Schematic representation of mechanical and anatomical axis of the lower extremity, with the corresponding physiological angles

For anatomic knee joint realignment, tibial slope is another important biomechanical parameter. Under physiological conditions, the tibial plateau slightly declines caudally at an angle of about 10° (9–11° medially, 6–8° laterally) [20, 21]. Every alteration of the inclination affects the kinematics of the knee joint. For this reason, the tibial slope should not be increased or decreased during osteotomy in patients with stable ligaments and normal range of motion [22, 23].

13.2.2 Leg Deformities

An essential part of preoperative preparation is defining the location, type, and amount of corrective osteotomy needed [24]. Genu varum and genu valgum are the two most common deformities of the lower extremity, which can be defined as a malalignment of a mechanical axis (medial deviation—varus deformity, lateral deviation—valgus deformity) [25]. The pathological deviation of physiological axis can occur as a result of deformation at the level of femur, tibia, both bones, or due to ligament laxity. To localize the level of the deformity, joint angles and joint line orientation must be considered [26, 27]. In the literature, numerous different preoperative planning procedures are described; however, the conventional method is performed using correctly executed long-leg weight-bearing radiographic image with the patella positioned centrally (Fig. 13.3) [28–30]. It is also important to understand that deviations of physiological axes and angles can occur in the frontal, transverse, or sagittal plane.

Table 13.1 Physiological knee joint angles

Biomechanical parameter	Value
Anatomical femorotibial angle (aFTA)	173°–175°
Anatomical lateral distal femoral angle (aLDFA)	81° ± 2°
Anatomical medial proximal tibial angle (aMPTA)	87° ± 3°
Mechanical lateral distal femoral angle (mLDFA)	87° ± 3°
Mechanical medial proximal tibial angle (mMPTA)	87° ± 3°
Joint line convergence angle (JLCA)	0° ± 2° medial convergence

13.2.3 Basic Principles of Knee Joint Realignment

The basic concept of osteotomy is to perform surgical transection of a bone to achieve realignment and subsequent transfer of weight bearing forces from damaged to healthy area of the joint surface. The main objective is correction of mechanical axis into optimal biomechanical position [1, 28]. To accomplish a successful osteotomy, precise preparation is of crucial importance. Radiographs are analyzed using the malalignment test, which determines the source of mechanical axis deviation (MAD), utilizing known normal ranges for

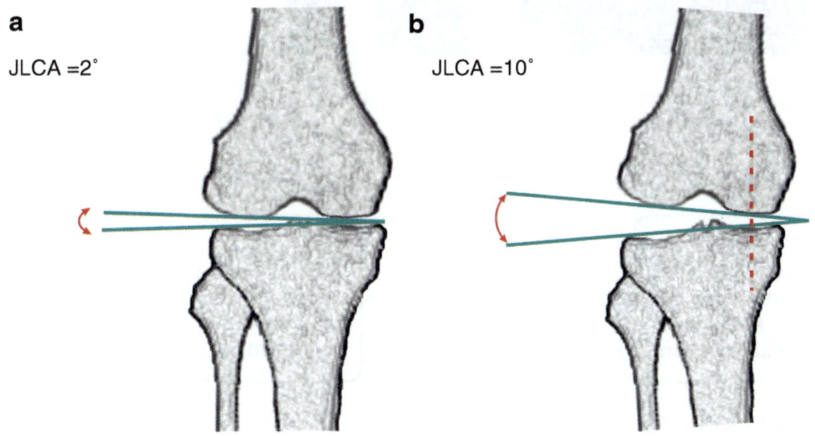

Fig. 13.2 Schematic representation of the (**a**) normal and (**b**) increased joint line convergence angle (JLCA)

the orientation of the knee joint line relative to the femoral and tibial mechanical axis. More than 15 mm of medial MAD implies a varus malalignment, whereas deviation of more than 10 mm laterally from the center of the knee joint signifies a valgus malalignment [25]. The principal concern, besides the correction of mechanical axis, is restoration of physiological horizontal joint line, which is in most cases possible to obtain with single osteotomy [28]. However, in cases where unacceptable medio-lateral slope of the knee (90° ± 3°) after a single cut is to be estimated, a double osteotomy may be considered. This is usually imperative in severe combined deformities of femur and tibia [31].

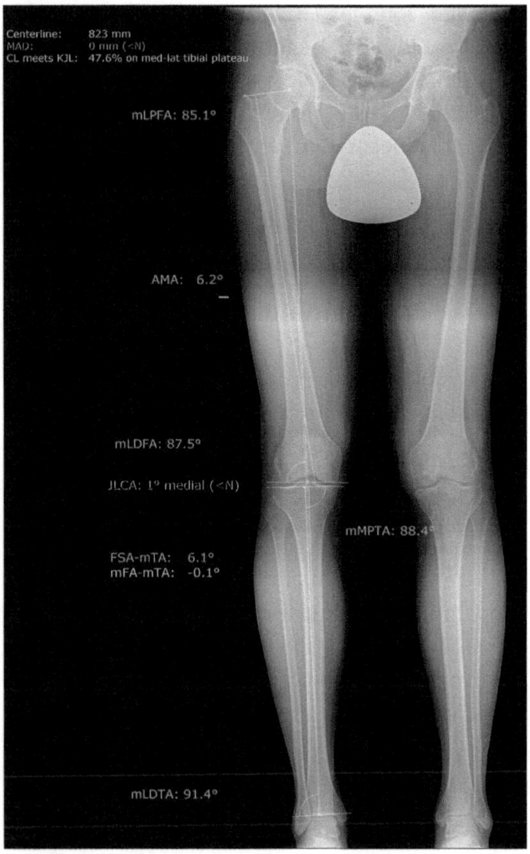

Fig. 13.3 Correctly performed long-leg weight-bearing X-ray with the physiological axes

13.3 Indications and Planning

The five-step approach will be used to demonstrate conventional preoperative planning before the osteotomy (Fig. 13.4).

13.3.1 Indications for Osteotomy and Physical Examination

Patient selection is perhaps the most important step for achieving good results with any kind of osteotomy and should be performed in a standardized fashion. Detailed anamnesis and physical examination, supported by good diagnostic imaging, are crucial for determining whether or not the patient is a suitable candidate for osteotomy. Besides previous surgeries and injuries of the knee, important aspects of patient's general history include age, pain characteristics, level of activity, and expectations. Physical examination on the other hand should assess gait, stance, range of motion, ligamentous stability, alignment of the lower extremity, leg length discrepancy, neurological status, and status of the skin and soft tissues. Unicompartmental osteoarthritis should also be verified using comprehensive physical assessment, which can confirm localized joint line tenderness, medial tibiofemoral crepitus, tenderness elicited with loading of the affected compartment, and joint space collapse during valgus/varus stress test [12].

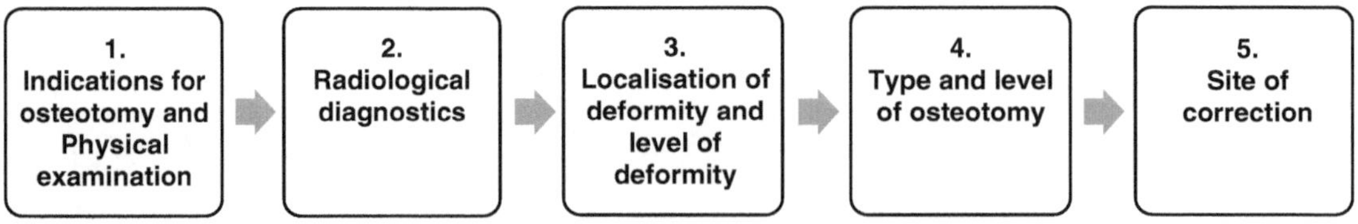

Fig. 13.4 The five-step approach to preoperative planning

Ideal candidates for osteotomy are young active patients (under 65 years old) with malalignment and unicompartmental gonarthrosis (Ahlbäck Grade II or less) [32]. Body mass index should be less than 30 kg/m^2, range of motion above 120°, and a flexion contracture below 5°. It is suggested that the patient is a non-smoker and has a certain degree of pain tolerance. In younger symptomatic patients, higher stages of osteoarthritis are also suitable for corrective osteotomy. The intervention however is not indicated in severe osteoarthritis with important bone loss or in cases of medio-lateral joint subluxation. Old age (over 70 years) and severe obesity may also be considered a contraindications for the osteotomy [12, 33–35].

13.3.2 Radiological Diagnostics

A mandatory part of preoperative preparation for osteotomy is radiographic assessment of lower extremities. Anterior-posterior (AP), lateral, and skyline views of the knee joint are necessary. Essential radiograph for pathologic malalignment evaluation is bilateral long-leg weight-bearing AP view in full extension with both patellas placed centrally (Fig. 13.5). In case of ligamentous instability, stress views should be obtained. Magnetic resonance imaging (MRI) is used to evaluate cartilage, menisci, and ligaments. Other medical imaging techniques such as scintigraphy and computer tomography (CT) may also be used in the preoperative diagnostics. CT scan is performed to obtain 3D bone reconstruction in selected patients, particularly in posttraumatic deformations and if patient-specific instruments are to be used during surgery.

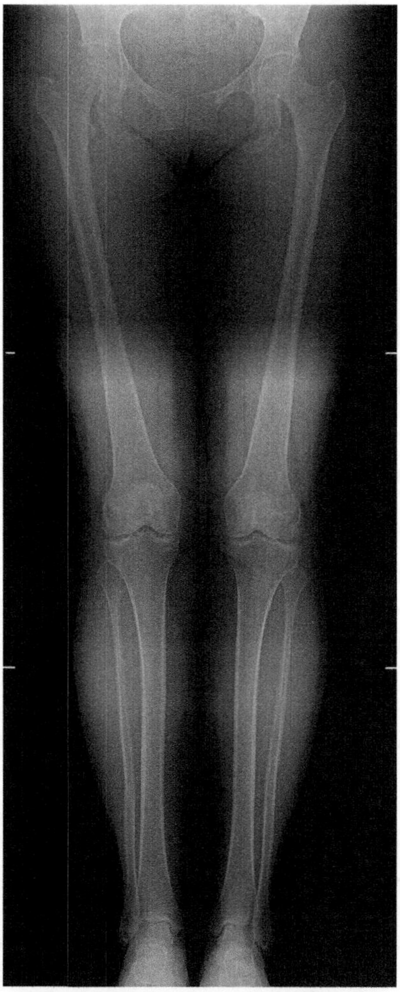

Fig. 13.5 Bilateral long-leg weight-bearing X-ray performed in anterior-posterior (AP) view. Both the legs are in full extension with patella placed centrally

13.3.3 Localization of Deformity

After careful evaluation of malalignment, the level of deformity needs to be determined. In order to correct the limb malalignment, transection of the bone should always be performed at the appropriate level [25]. If MAD is placed more than 15 mm medially, we are dealing with varus deformity which can originate from the femur (mLDFA >90°), tibia (mMPTA <85°) or both sides of the joint. If MAD is placed more than 10 mm laterally, we are dealing with valgus deformity which can also originate from the femur (mLDFA <85°), tibia (mMPTA >90°), or both sides of the joint (Fig. 13.6). It is important to analyze the joint line convergence angle (JLCA) which has normally 0° ± 2° medial convergence (Fig. 13.2). The increase of the JLCA is a consequence of lateral ligament laxity

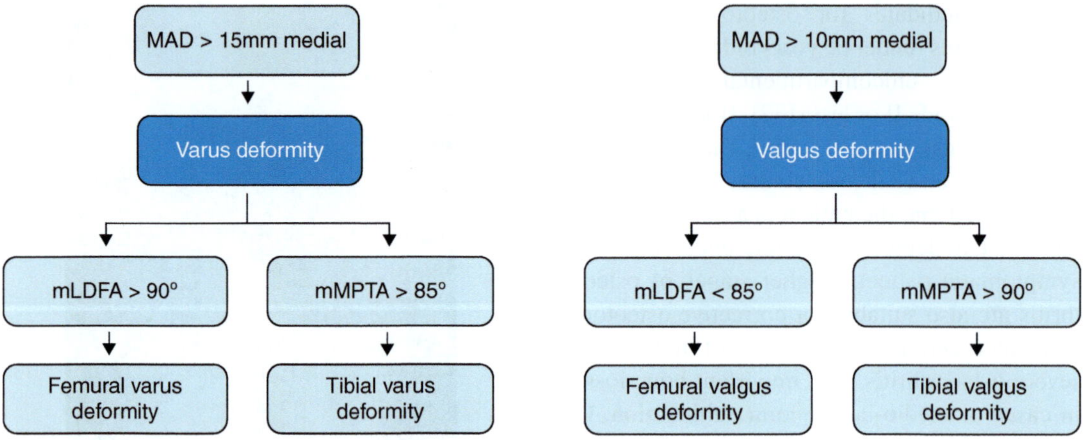

Fig. 13.6 Diagram of lower extremity deformations

or loss of the cartilage on the medial side, while decrease of the JLCA is a consequence of medial ligament laxity or loss of the cartilage on the lateral side [25].

13.3.4 Type and Level of Osteotomy

There are several different types of osteotomies around the knee joint (Fig. 13.7). The ultimate choice of the technique depends on both patient's characteristics and surgeon's preference [36]. Further steps will describe a detailed planning algorithm for opening or closing wedge high tibial osteotomy (HTO) and distal femoral osteotomy (DFO).

The evolution of anatomic knee joint realignment brought us different techniques of osteotomy. Every single technique has its own advantages and is suitable for a specific group of patients. The most commonly used types of osteotomies are listed in Table 13.2.

Generally, open-wedge osteotomies are technically less demanding and more accurate than closed-wedge osteotomies, as fine tuning of the osteotomy (adjusting the osteotomy with a spreader) is only possible with the open-wedge technique. On the other hand, closed-wedge osteotomies offer faster healing, early weight-bearing, and no need for bone grafting.

Genu varum in combination with medial unicompartmental osteoarthritis is the main indication for HTO. Advantages and pitfalls of the opening and closing wedge techniques have often been discussed controversially, so decision for either of the two techniques should be based on the accompanying anatomical features such as leg length discrepancy, patella height, tibial slope, and torsional deformities [37].Genu valgum in combination with isolated degeneration of lateral compartment is usually an indication for DFO [38]. Major valgus deformity (more than 10–12°) is associated with a joint line that slopes superolaterally, which can only be corrected with the osteotomy proximal to the knee joint. Good results are reported for DFO, despite the fact that there is no clear consensus about optimal surgical technique (opening/closing wedge) [39].

For patients with a large, combined deformity of femoral and tibial bone, double osteotomy is a valuable option [40]. Bone cuts at both the distal femur and proximal tibia enable restoration of neutral joint line (normal: 87–90°) in addition to malalignment correction [31]. Different combinations of osteotomies are possible; however, LCWDFO+MOWHTO for varus and MCWDFO+MCWHTO for valgus deformity are the two most frequently performed procedures. With a proper patient selection and accurate preoperative planning, the short- and mid-term

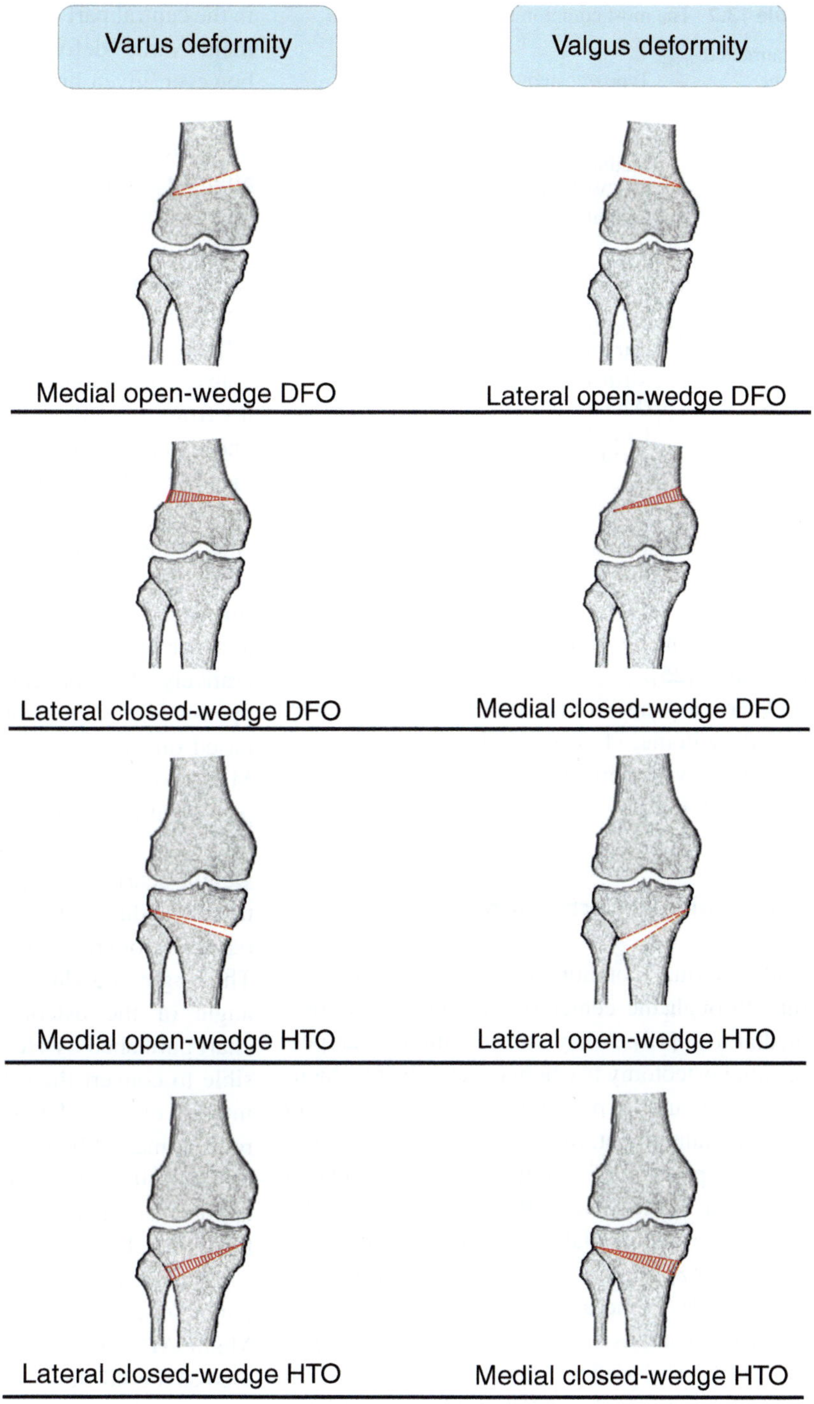

Fig. 13.7 Schematic representation of the most commonly performed osteotomies around the knee joint

Table 13.2 The most common types of osteotomies

Transected bone	Type of osteotomy
Tibia	Medial opening wedge HTO (MOWHTO) Medial closing wedge HTO (MCWHTO) Lateral closing wedge HTO (LCWHTO) Lateral opening wedge HTO (LOWHTO)
Femur	Medial closing wedge DFO (MCDFO) Medial opening wedge DFO (MOWDFO) Lateral closing wedge DFO (LCWDFO) Lateral opening wedge DFO (LOWDFO)

HTO high tibial osteotomy, *DFO* distal femoral osteotomy

results following double osteotomy are very promising [25].

There are also other types of osteotomies including dome, chevron, and rotational osteotomy. However, these surgical techniques are less commonly used.

13.3.5 Size of Correction

Under normal conditions, the mechanical axis runs through the center of the knee or slightly medial to it [18]. A key consideration when performing osteotomy is where exactly the mechanical axis should be positioned postoperatively. In varus malalignment, the majority of studies suggest transposition of weight-bearing axis beyond the center of the knee to the zone between 60% and 70% of the medial-lateral width of the tibial plateau (0%, medial edge; 100%, lateral edge) [29, 41–44]. Some authors recommend 3–6° of mechanical valgus or 8–10° of anatomical valgus [45, 46]. The definite amount of correction depends on residual cartilage in the medial knee compartment—the higher the stage of osteoarthritis, the larger the correction recommended, but the weight-bearing line after surgery should never pass the Fugisawa point [47]. The authors' recommendation is to place the mechanical axis in the central part of the knee joint in case of ligament varus deformity with no osteoarthritis; however, given higher level of osteoarthritis, the mechanical axis should project on the lateral tibial spine. Nonetheless, the joint line orientation after osteotomy should always be kept in normal ranges; hence, double-level osteotomy should sometimes be considered. In valgus malalignment, the generally accepted rule is the central position of mechanical axis after surgery [48, 49]. It is also important to analyze the joint line convergence angle (JLCA) which normally varies from 0° to 2° of medial convergence (slight knee joint varus). When the JLCA angle is increased, the difference between measured and aimed angle should be taken into account when calculating the final amount of correction.

To prolong the longevity of the native joint function, the correction angle of the osteotomy and size of the wedge must be determined preoperatively [1]. Conventional planning for medial open-wedge high tibial osteotomy is usually based on the technique, originally described by Miniaci et al. (Fig. 13.8) [30]. Lateral cortico-periosteal hinge (H) position of the tibial osteotomy is initially marked on the long-leg weight-bearing radiographic film. Afterward, the tibia is abducted until corrected mechanical axis passes through desired point in the knee joint. This angle of abduction represents the correction angle of the osteotomy. Using trigonometric chart published by Hernigou et al. [50], it is possible to convert the established correction angle into the required height of the osteotomy gap (in mm) at medial bone cortex. The same principles of determining the size of correction can also be applied for other types of osteotomies around the knee (Fig. 13.9).

Special software programs are available for computer-assisted preoperative planning. MediCAD® (Hectec GmbH, Germany), considered as a gold standard in medical planning software, enables the analysis of deformity and simulation of osteotomy (Fig. 13.10) [51].

During surgery, the amount of correction is measured by the height of the created wedge-shaped gap (opening wedge) or removed piece of wedge-shaped bone (closing wedge). In open-

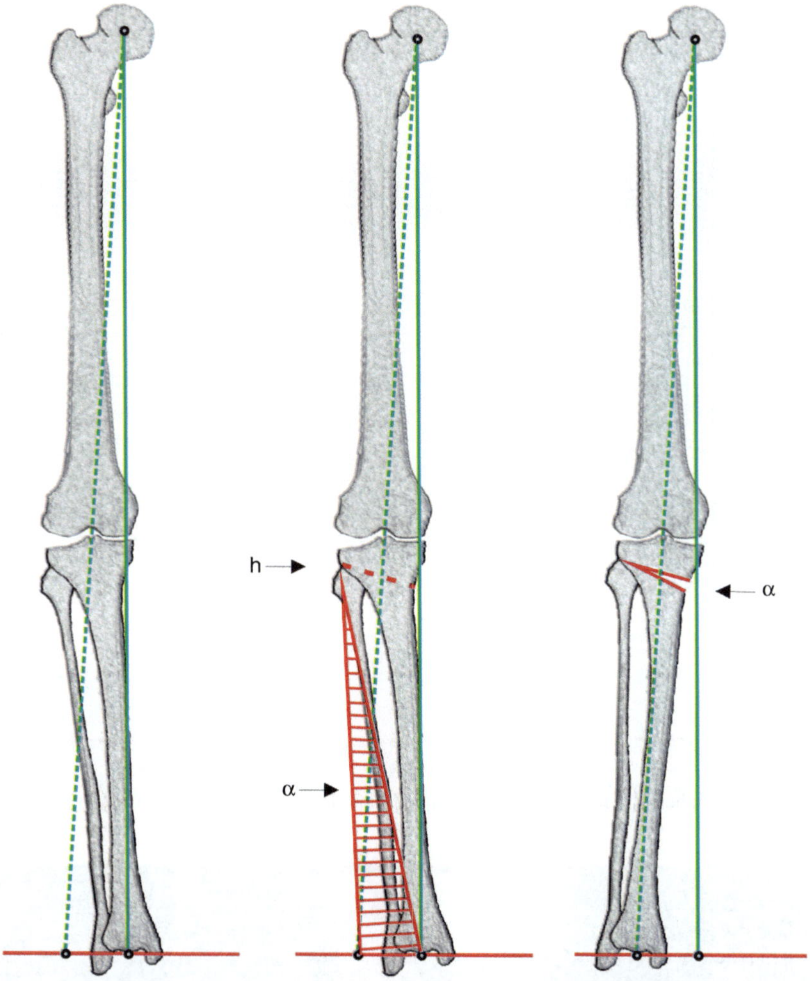

Fig. 13.8 Schematic representation of the medial open-wedge high tibial osteotomy, originally described by Miniaci et al. [30]

wedge osteotomy, the width of the saw blade should be deducted from calculation. The correction during surgery can be evaluated using the fluoroscopy and the alignment rod, centered over the femoral head proximally and the middle of the ankle joint distally. Navigation system is also used to control the correction in all three planes.

In addition to its primary objective of redistributing the forces over both medial and lateral compartment, HTO plays an important role in the treatment of ligamentous deficiency [44]. A large number of studies concluded that sagittal plane instability can be influenced by tibial slope alteration [20, 23, 52]. In patients with varus malalignment and chronic ACL insufficiency, tibial slope should be decreased to reduce anterior subluxation and ligament strain [20]. Conversely, increasing the posterior tibial slope is indicated in case of varus PCL-deficient knee [52].

13.4 Surgical Techniques: Tibial Osteotomies

13.4.1 High Tibial Osteotomy (HTO) for Varus Knee Malalignment

A widely accepted treatment of genu varum associated with medial compartment osteoarthritis is the HTO [44, 53]. The deformity has to be located in the proximal part of the tibia with mMPTA <85°. Planned corrective osteotomy should not alter the normal joint line orientation and the mMPTA after correction should not exceed 93°.

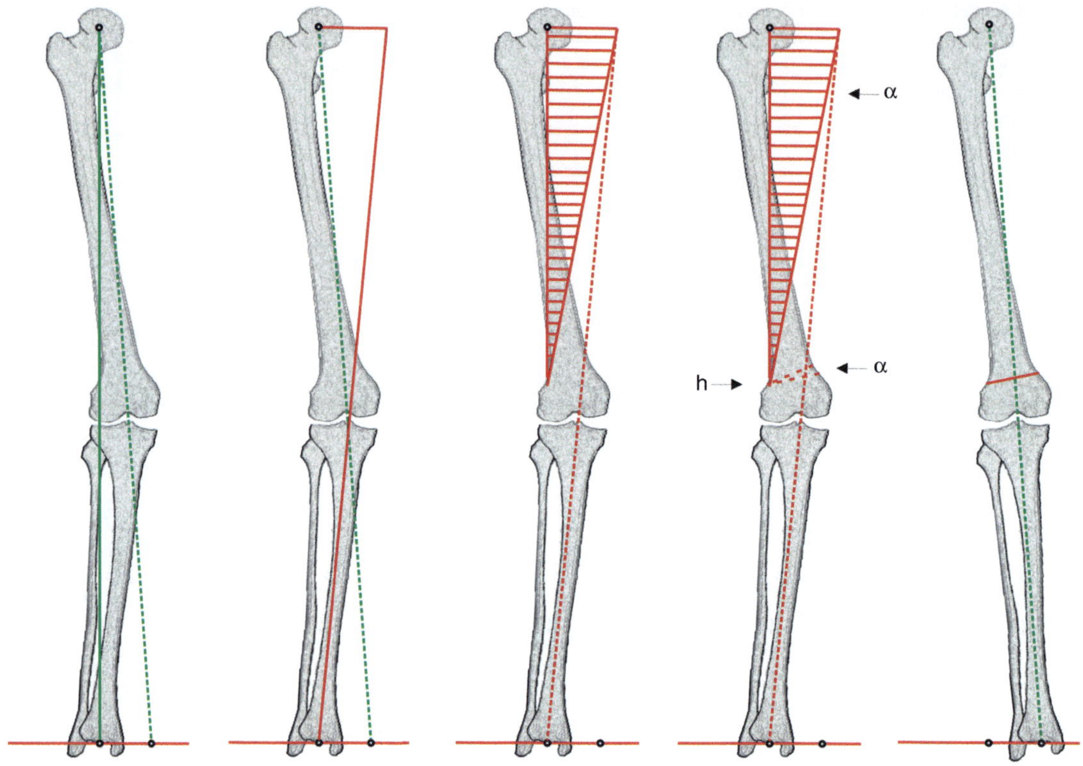

Fig. 13.9 Schematic representation of the medial closed-wedge distal femoral osteotomy

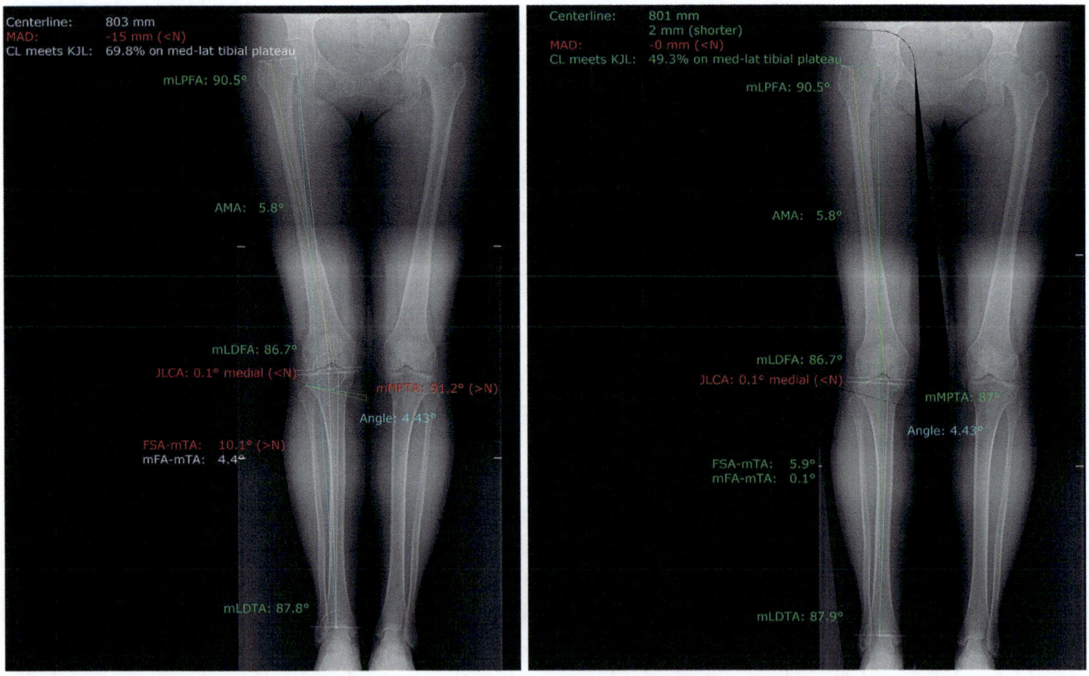

Fig. 13.10 Preoperative planning performed with MediCAD®, which enables analysis of deformity and simulation of osteotomy

For a long time lateral closing wedge (LCW) method has been considered as a gold standard as propagated by Coventry in 1965 [4]. The advantages of this technique are faster healing, early weight-bearing, and no need for bone grafting. However, the procedure requires double bone cuts and proximal tibiofibular joint disruption, which may cause peroneal nerve damage. Malalignment correction is achievable only in frontal plane and there is a possibility of the lower limb shortening [33, 53]. Nevertheless, the conventional osteotomy has demonstrated good results. In a study with 455 patients, Hui et al. [54] concluded that lateral closing wedge HTO can be effective for periods longer than 15 years, reaching survival rate of 95% at 5 years, 79% at 10 years, and 56% at 15 years.

To address the before-mentioned disadvantages of lateral closing wedge HTO, medial opening wedge (MOW) method was introduced. Advantages of the procedure included single bone cut, fibula preservation, and ability of correction in two different plains (frontal and sagittal). In addition, this surgical technique is faster, more precise, and associated with minor risk of peroneal nerve injury [33, 44, 53]. Limitations include delayed union, loss of correction, long period of weight-bearing restriction and a higher possibility of the lower limb lenghtening and a higher possibility of the lower limb lenghtening. Different survival rates of medial open-wedge HTO can be found in the literature [55–57]; however, the data is comparable to that of the lateral closing wedge HTO. Duivenvoorden et al. [58] compared clinical and radiological outcomes of both HTO procedures, and no statistically significant difference was found between the two groups. Superiority of one procedure over the other is hard to determine. Therefore, individualized surgical approach based on clinical characteristic of each patient is recommended (Table 13.3).

Table 13.3 Closed-wedge HTO vs. open-wedge HTO

Closed-wedge HTO	Open-wedge HTO
Indicated in patella Baja	Indicated in patella Alta
Longer surgery	Shorter surgery
Faster healing	Slower healing
Lower precision	Higher precision
Risk of peroneal nerve injury	No risk of peroneal nerve injury

13.4.1.1 Medial Open-Wedge High Tibial Osteotomy

The patient under general or spinal anesthesia is placed in the supine position. Manipulation of the leg from 0 to 90° of flexion has to be attainable during the procedure, and easy access to intraoperative fluoroscopy is mandatory. The procedure can be performed with or without the use of a tourniquet which is placed around the thigh in any case. It is suggested to begin the procedure with arthroscopy, which allows a thorough assessment and management of any potential intraarticular lesions. In slight flexion of the knee joint, a 5–6 cm vertical skin incision is made between the posteromedial aspect of the medial tibial condyle and the medial border of tibial tubercle. The next step involves dissection through subcutaneous tissue and exposure of MCL (medial superficial part), patellar tendon (medial border), and pes anserinus (superior border) (Fig. 13.11). The ligament is partially detached with a Cobb elevator, and a Hohmann retractor is placed behind the tibial ridge (protection of posterior neurovascular structures). Patellar tendon and pes anserinus tendons are then retracted with blunt retractors. After subperiosteal dissection of tibia, the leg is fully extended. To mark the direction of osteotomy, two guidewires are inserted under fluoroscopic control. The insertion point of the first wire is approximately 4 cm below medial joint line just in front of the posterior tibial ridge. It is directed toward the tip of the fibular head. The second guidewire is drilled 2 cm anteriorly and runs parallel to the first one. Endpoint of both wires is exactly at the lateral tibial bone cortex (Fig. 13.12). The distance from the tip of the fibular head and the tibial plateau may vary among patients and should be considered when positioning K-wires. It is important that wires are inserted parallel, with the knee in full extension. The procedure is continued with biplanar (horizontal + anterior) bone cut and distraction of the osteotomy. Using oscillating saw, the horizontal osteotomy is performed in posterior 2/3 of tibia just under guidewires to within 1 cm of the lateral

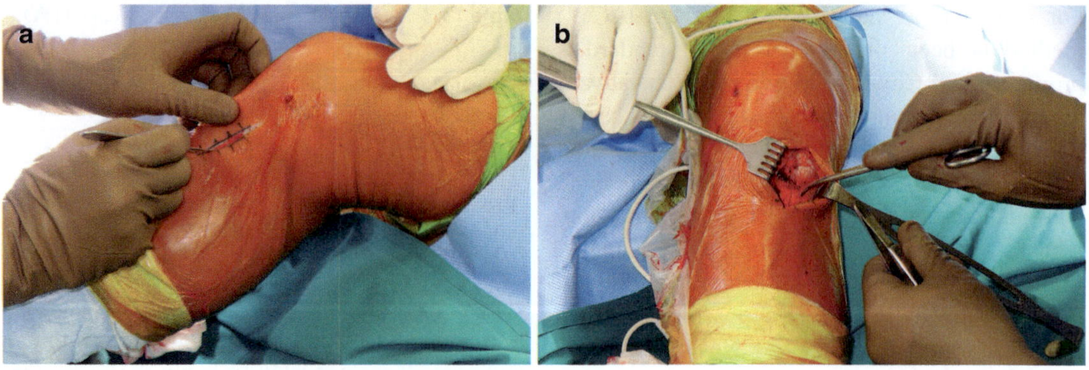

Fig. 13.11 (**a**) Vertical skin incision is made between the posteromedial aspect of the medial tibial condyle and the medial border of tibial tubercle; (**b**) Dissection through subcutaneous tissue and exposure of MCL (medial superficial part), patellar tendon (medial border) and pes anserinus (superior border)

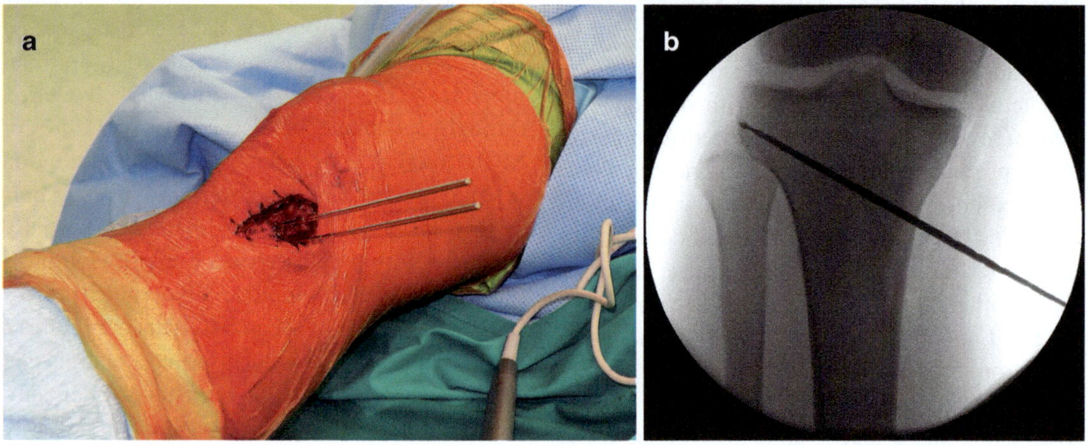

Fig. 13.12 Two parallel Kirschner wires are inserted approximately 4 cm below medial joint line as a guide for the osteotomy cut; (**a**) Intraoperative and (**b**) fluoroscopic image

cortex (Fig. 13.13a, b). Anterior ascending osteotomy is positioned behind the tibial tuberosity at an angle of 110° to the horizontal bone cut. It is important to stress out that patients with abnormally low-lying patella (patella baja) should have the osteotomy of tibial tuberosity performed distally to the patellar tendon insertion (anterior descending osteotomy) to avoid further lowering of the patella (Fig. 13.13c, d). In this case, the tibial osteotomy has to be fixed with one or two cortical screws in the frontal plane. After biplanar transection, the osteotomy is gradually opened with a spreader chisel or multiple osteotomes under continuous fluoroscopy. Special care must be taken to preserve the lateral bone hinge and posterior tibial slope inclination (Fig. 13.14). In case of normal knee stability, the tibial slope should not be altered. In ACL deficient knee the slope can be decreased, while in PCL-deficient knee, the slope can be increased. This can be obtained by opening the gap more posteriorly (decreasing the slope) or anteriorly (increasing the slope). Once the desired osteotomy gap is achieved, an image intensifier and a long measuring rod are used to control proper leg alignment in extension (Fig. 13.15a, b). Alternatively, the navigation system can be used to control the angle of correction (Fig. 13.15c, d). When the planned and desired correction is achieved, the internal fixation of the osteotomy can be performed.

The osteotomy can be fixed with different internal fixation systems. The authors' most frequently used hardware system is the TomoFix

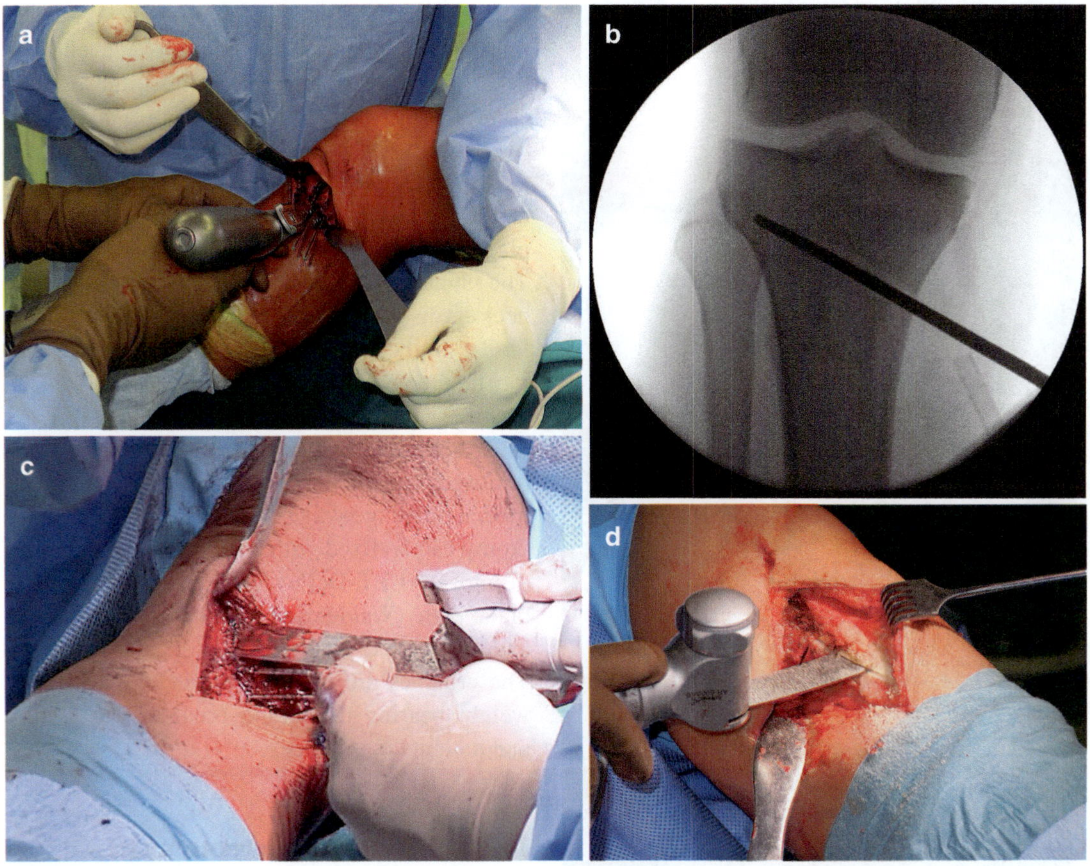

Fig. 13.13 The horizontal osteotomy is performed in posterior 2/3 of tibia just under guide wires to within 1 cm of the lateral cortex; (**a**) Intraoperative and (**b**) fluoroscopic image; (**c**) Anterior ascending osteotomy; (**d**) Anterior descending osteotomy

(DePuy Synthes) T-shaped locking compression plate (LCP). Together with eight locking screws (four proximal and four distal to the osteotomy cut), it provides rigid fixation and enables early weight-bearing. LCP is slid into subcutaneous area parallel to the tibial diaphysis, approximately 1 cm below the joint line. Subsequently, the proximal bone segment is fixed with three self-tapping monocortical screws which should not protrude beyond the lateral cortex. To apply compression on the osteotomy hinge, temporary lag screw (later replaced by a bicortical screw) is inserted into the first hole below the bone cut. From the distal to proximal end of LCP, the rest of the screws are secured (Fig. 13.16). Alternatively, other fixation systems can be used to stabilize the osteotomy (Newclip Technics® Activmotion tibial plate, Arthrex PEEKPower™ tibial plate) (Fig. 13.17). If the osteotomy gap exceeds 13 mm, the defect is filled with autologous or heterologous bone graft. The recent studies show that gap filling with bone or bone substitute when indicated, is associated with faster healing as well as decrease in postoperative pain and bleeding [59, 60]. At the end of the procedure, hardware position and alignment correction are confirmed with fluoroscopy. If necessary, combined anterior cruciate ligament reconstruction can be performed. The wound is closed in a standard fashion and compression bandage is applied. Whenever tourniquet is used during surgical procedure, the release before wound closure is recommended to control bleeding and to achieve blood clot formation in the osteotomy gap. Drainage is rarely indicated. Pain control after surgery is mandatory, and fast track reha-

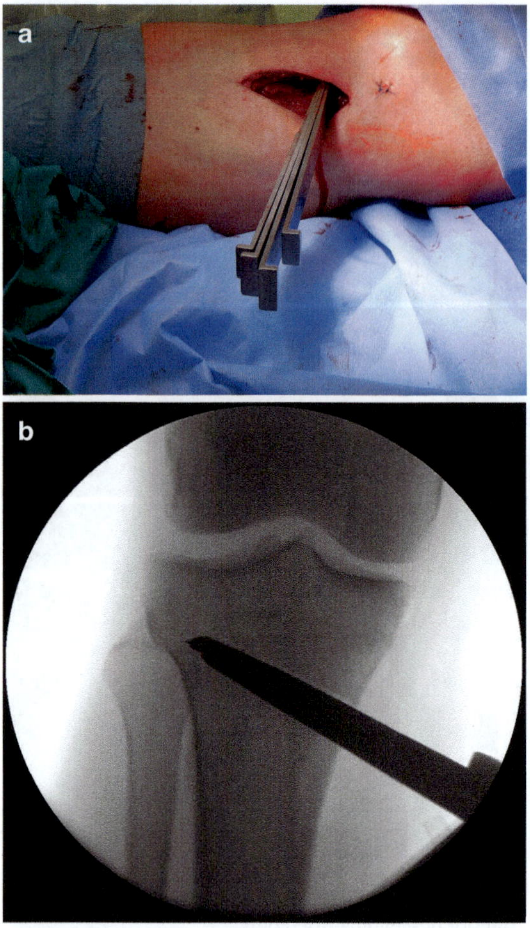

Fig. 13.14 The osteotomy is gradually opened with a spreader chisel or multiple osteotomes; (**a**) Intraoperative and (**b**) fluoroscopic image

bilitation is recommended. The crutches are used for 6 weeks with unlimited range of motion and partial weight bearing allowed, depending on the bone quality and patient compliance.

13.4.1.2 Lateral Closed-Wedge High Tibial Osteotomy

The patient under general or spinal anesthesia is placed in the supine position. The procedure is performed with the tourniquet cuff placed around the thigh. It is suggested to begin the procedure with arthroscopy, which allows a thorough assessment and management of the potential intraarticular lesions. A straight longitudinal incision is performed on the anterolateral aspect of the tibia, followed by detachment of extensor muscles from the tibia and fibula. The level of the tibial osteotomy is determined under fluoroscopic control with two K-wires placed proximally and two distally (Fig. 13.18a, b). The osteotomy is performed in two planes, leaving 5–10 mm of medial *bone bridge* intact. Proximal K-wires are placed 2 cm below and parallel to the join line. The fibula can be osteotomized at the level of the neck or distally, but alternatively proximal tibiofibular joint can be released [36]. The distance (in mm) between proximally and distally placed K-wires determines the planned correction of the osteotomy. After osteotomy, the bone wedge is removed, and the gap is carefully closed to avoid medial hinge breakage (Fig. 13.18c, d). The osteotomy can be fixed with an angular stable fixator (TomoFix® LPT, DePuy Synthes) or with Giebel® blade plate (Waldemar Link GMBH, Fig. 13.19) and the wound is closed in a standard manner. The crutches are used for 6 weeks with partial weight bearing depending on the bone quality and patient compliance.

13.4.2 High Tibial Osteotomy (HTO) for Valgus Knee Malalignment

Distal femoral osteotomy has been a widely accepted treatment of valgus knee deformity for a long time [61, 62]. When the deformity is found in the proximal part of the tibia with mMPTA >90°, the corrective osteotomy of the tibia should be considered. In severe valgus deformity with mMPTA >90° and mLDFA <85°, combined femoral end tibial osteotomy may be necessary to shift the weight bearing line in the center of the knee joint. The authors' preferred technique for the correction of valgus knee with proximal tibial deformity is biplanar medial closed-wedge high tibial osteotomy (MCWHTO). The technique of lateral open-wedge high tibial osteotomy in valgus tibia deformation has been described [63] but is rarely indicated and therefore seldom performed. The technique is technically demanding, time consuming and indicated only in selected patients.

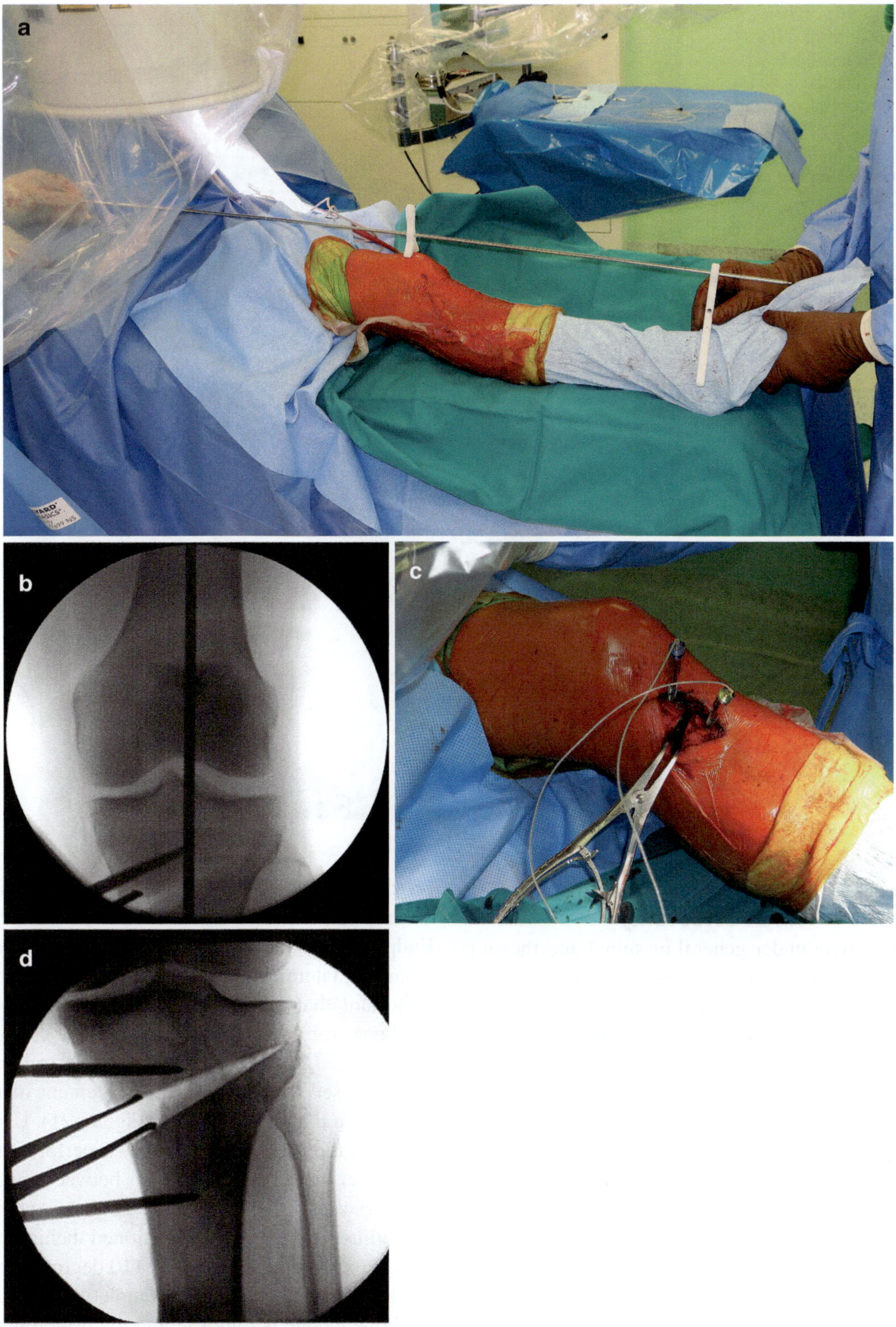

Fig. 13.15 (**a**, **b**) A long measuring rod is used to control proper leg alignment in extension; (**c**, **d**) Alternatively, navigation system can be used to control the amount of correction

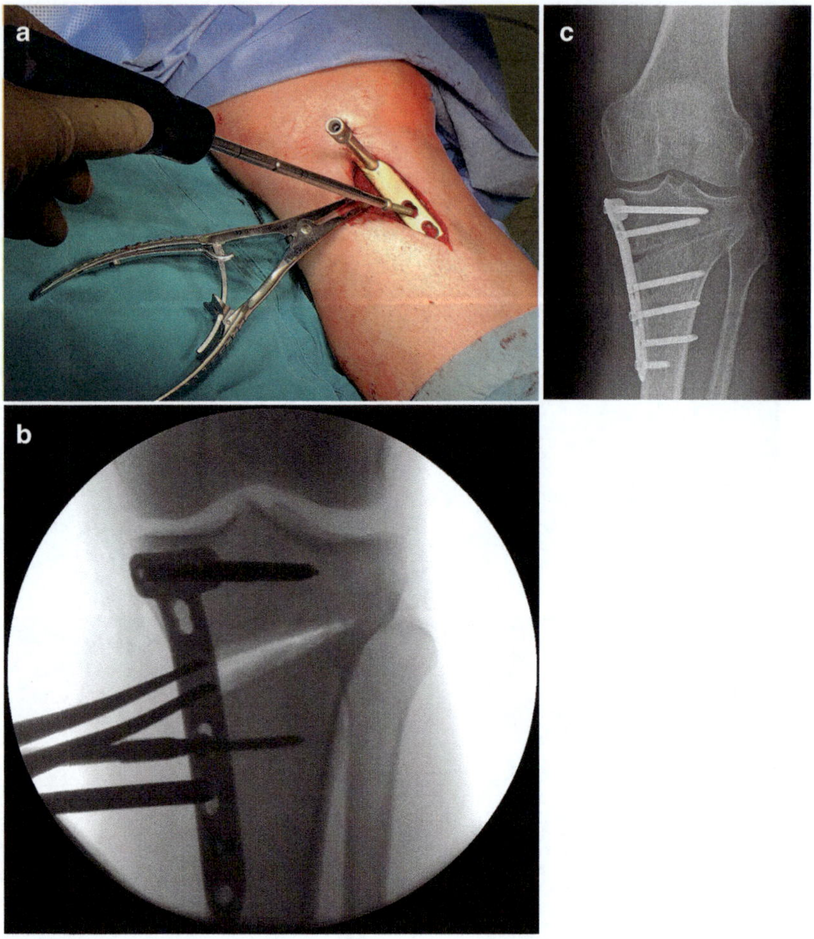

Fig. 13.16 (**a**, **b**) Temporary lag screw is inserted into the first hole below the bone cut; (**c**) Postoperative X-ray image of the TomoFix™ T-shaped locking compression plate after MOWHTO

13.4.2.1 Medial Closed-Wedge High Tibial Osteotomy

The patient under general or spinal anesthesia is placed in the supine position. The procedure is performed with or without the use of a tourniquet which is placed around the thigh in any case. The procedure starts with an arthroscopy, which allows a thorough assessment and management of any potential intraarticular lesions. A straight longitudinal incision is performed on the anteromedial aspect of the tibia. The anteromedial approach to the bone and soft tissue is the same as described in the "medial open-wedge high tibial osteotomy" technique. The level of the tibial osteotomy is determined under fluoroscopic control with two K-wires placed proximally and two distally. It is important to leave enough bone proximally to the bone cut, for a secure fixation of the tibial plate. The distance (in mm) between proximally and distally placed K-wires determines the planned correction of the osteotomy. Endpoints of all K-wires are placed 5–10 mm from the lateral bone cortex (Fig. 13.20). It is important that wires are inserted parallel, with the knee in full extension, to avoid any changes of the tibial slope. Patient-specific instruments may be used to help with the placement of the K-wires in the proximal tibia (Fig. 13.21). Using the oscillation saw, the osteotomy is performed in the posterior 2/3 of the tibia in between guidewires to within 1 cm of the lateral cortex. Anterior ascending osteotomy is positioned behind the tibial tuberosity at an angle of 110 degrees to the horizontal bone cuts. After completing the cuts, the bone wedge is removed and the osteotomy is carefully closed to avoid lateral hinge breakage. The osteotomy can be fixed with angular stable fixator (TomoFix®, DePuy Synthes) (Fig. 13.22).

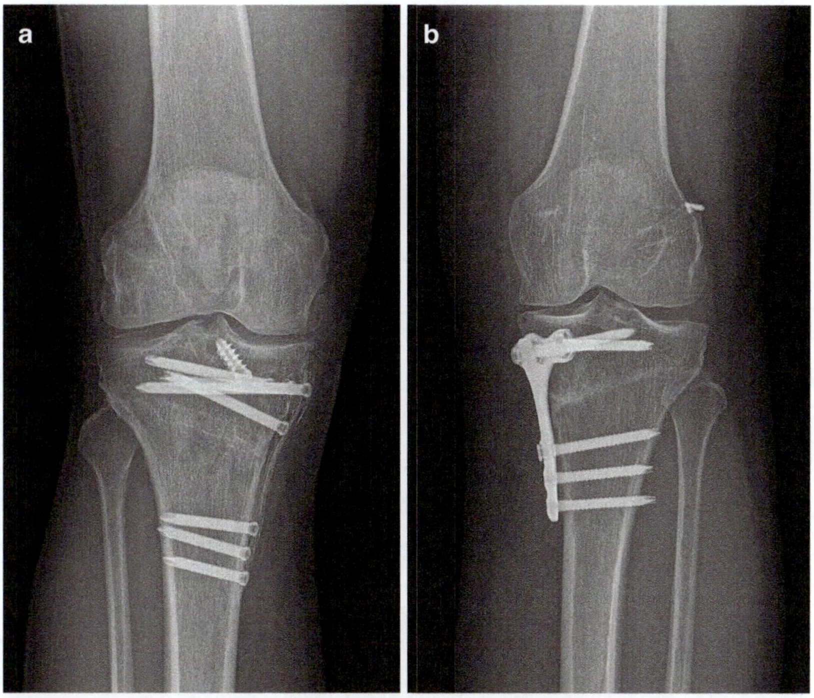

Fig. 13.17 (**a**) Arthrex PEEKPower™ HTO plate; (**b**) Newclip Technics® Activmotion HTO plate

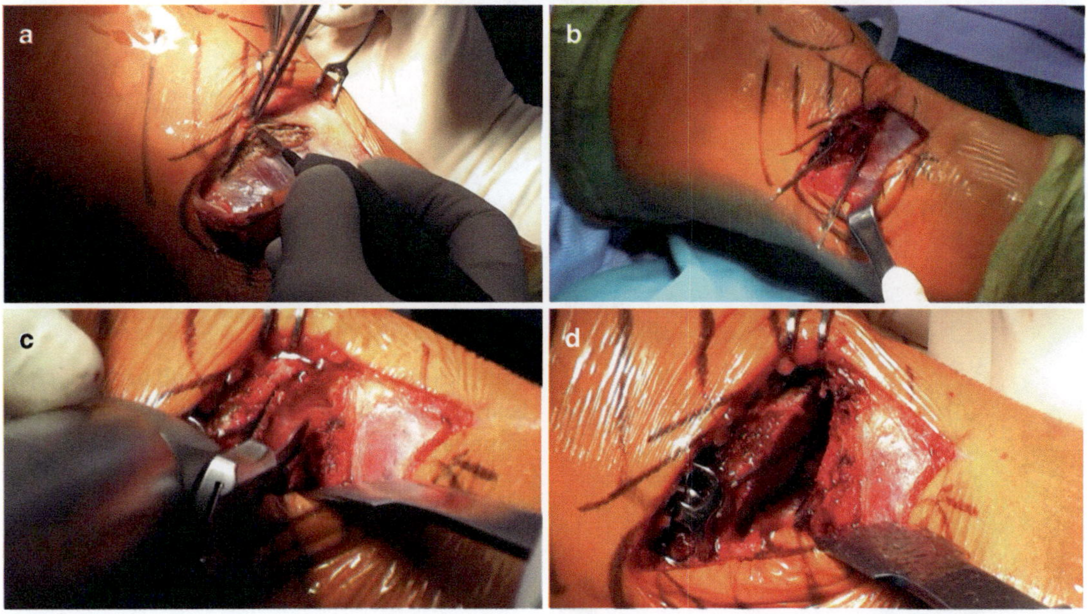

Fig. 13.18 (**a**) Skin incision on the anterolateral aspect of the tibia; (**b**) Two Kirschner wires are placed proximally and two distally; (**c**) Osteotomy is performed in two planes, leaving 5–10 mm of medial *bone bridge* intact; (**d**) Osteotomy gap after removal of the bone wedge

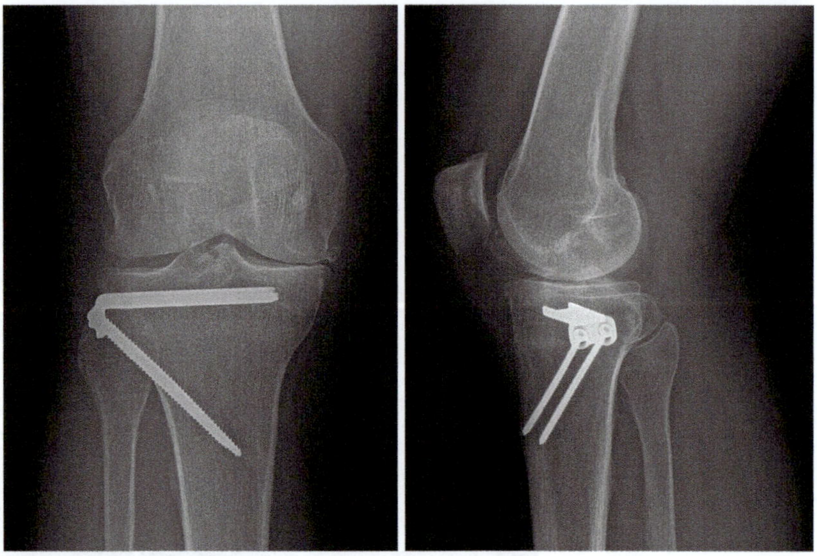

Fig. 13.19 Giebel® blade HTO plate

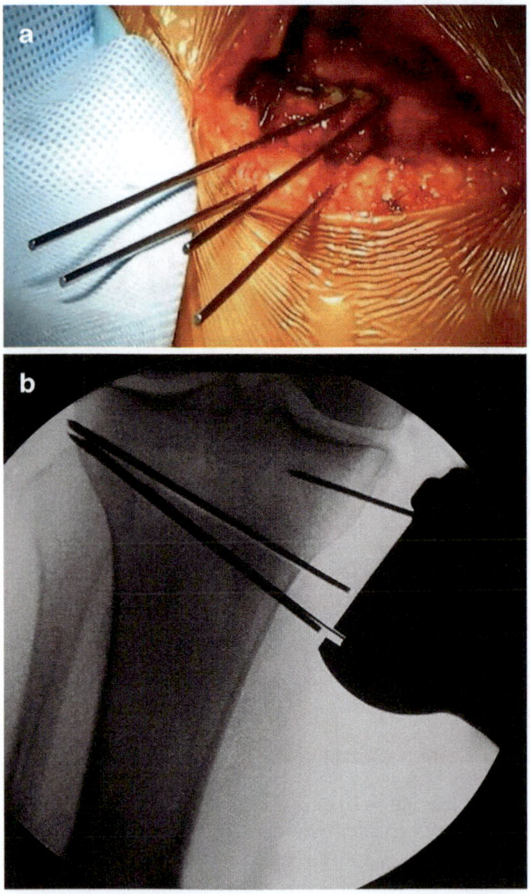

Fig. 13.20 Two parallel Kirschner wires are placed proximally and two distally. Both the pairs of wires should converge and meet 5–10 mm from the lateral cortex; (**a**) Intraoperative and (**b**) fluoroscopic image

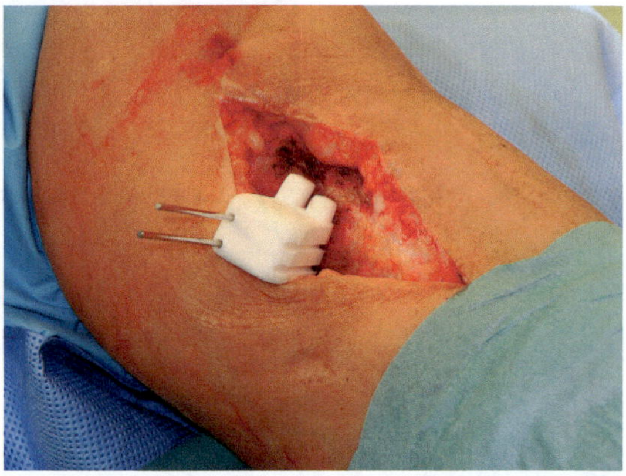

Fig. 13.21 Patient-specific instruments (PSI) may be used to help with the placement of the Kirschner wires

Crutches are used for 6 weeks with partial weight bearing depending on the bone quality and patient compliance.

13.5 Surgical Techniques: Femoral Osteotomies

13.5.1 Distal Femoral Osteotomy (DFO) for Varus Knee Malalignment

When varus deformity is located in the distal part of the femur with mLDFA >90°, corrective osteotomy of femur is indicated. In severe varus

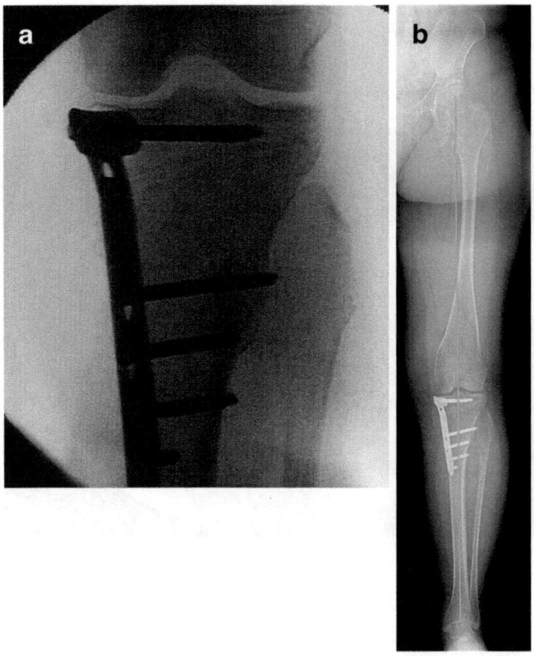

Fig. 13.22 (**a**) Intraoperative and (**b**) postoperative long-leg X-ray image of the TomoFix™ T-shaped locking compression plate after MCWHTO, with the desired position of the mechanical axis

deformity with mMPTA <85° and mLDFA >90°, combined femoral and tibial osteotomy may be necessary to shift the weight bearing line in the center of the knee joint, while maintaining the correct obliquity of the joint line. The authors' preferred technique for varus knee patients with distal femoral deformity is oblique descending biplanar lateral closed-wedge distal femoral osteotomy (LCWDFO). The technique of medial open-wedge distal femoral osteotomy in varus femur deformation has also been described [64], but is rarely performed due to high incidence of complications (delayed union or non-union). Stable osteosynthesis is of great importance in a supracondylar femoral osteotomy. It permits bone healing and functional rehabilitation. The procedure starts with an arthroscopy, which allows a thorough assessment and management of potential cartilage and meniscal damage.

13.5.1.1 Lateral Closed-Wedge Distal Femoral Osteotomy

The procedure is performed with or without the use of a tourniquet and starts with the knee in full extension. The longitudinal skin incision is placed laterally. After soft tissue division, the fascia lata is incised and vastus lateralis muscle retracted anteriorly. The perforant vessels are cauterized, and a retractor is placed on both sides of the surgical wound in order to establish a clear approach to the lateral part of the femur and lateral femoral epicondyle. Under fluoroscopic guidance and in knee extension, two K-wires are placed parallel to each other and proximal to the lateral femoral epicondyle in the direction of the predetermined hinge point, which is placed just above the medial femoral epicondyle. Another set of K-wires is then inserted proximally from the first two, while also facing the hinge point (Fig. 13.23). The distance (in mm) between proximally and distally placed K-wires determines the planned correction of the osteotomy. In this step, the aid of 3D printed patient-specific K-wire guides can be used in order to perform the correct amount of bone resection. The osteotomy is performed with oscillation saw in the posterior 2/3 of the distal femur in between two pairs of K-wires. The cuts are performed in two planes, leaving 5–10 mm of medial *bone bridge* intact. The ascending cut is then performed at an angle of 100° to the osteotomy cuts and parallel to the anterior femoral cortex, creating a few centimeters long tongue-like part of the bone, attached to the distal part of femur. Care should be taken not to break the medial hinge, so gentle sawing or drilling at the medial cortex is performed. The osteotomized piece of wedge-shaped bone is then removed, and the gap is carefully closed (Fig. 13.24). The osteotomy can be fixed with angular stable fixator (TomoFix®, DePuy Synthes). The plate is stabilized with four monocortical screws distally. The compression screw is temporarily placed into the most distal hole at the proximal part of fixation plate, which allows the remaining screws to be inserted (Fig. 13.25).

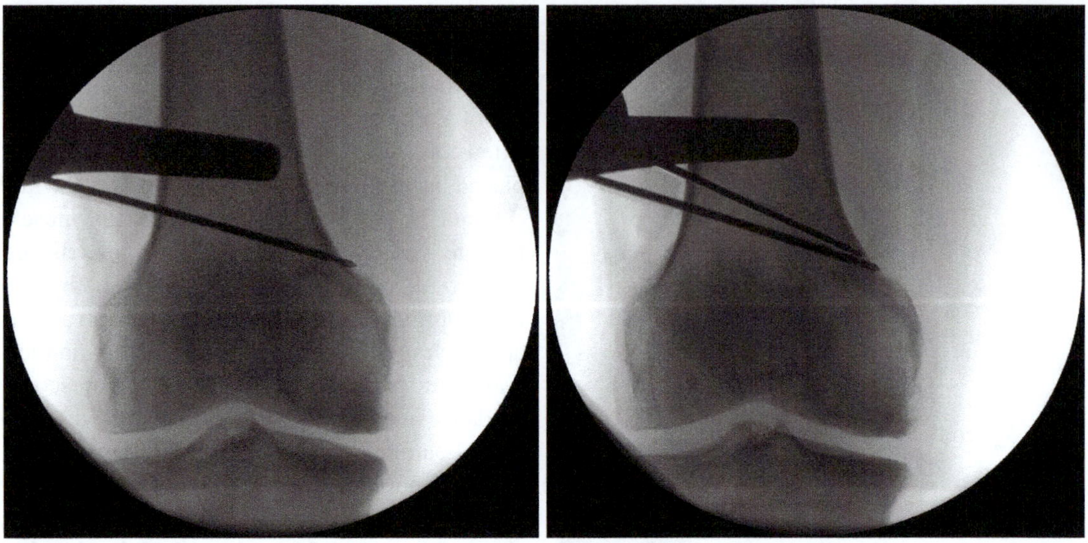

Fig. 13.23 Two parallel Kirschner wires are placed proximal to the lateral femoral epicondyle in the direction of the hinge point just above the medial femoral epicondyle. Another set of wires is placed proximally, while also facing the hinge point. Both pairs of wires should converge and meet 5–10 mm from the medial cortex

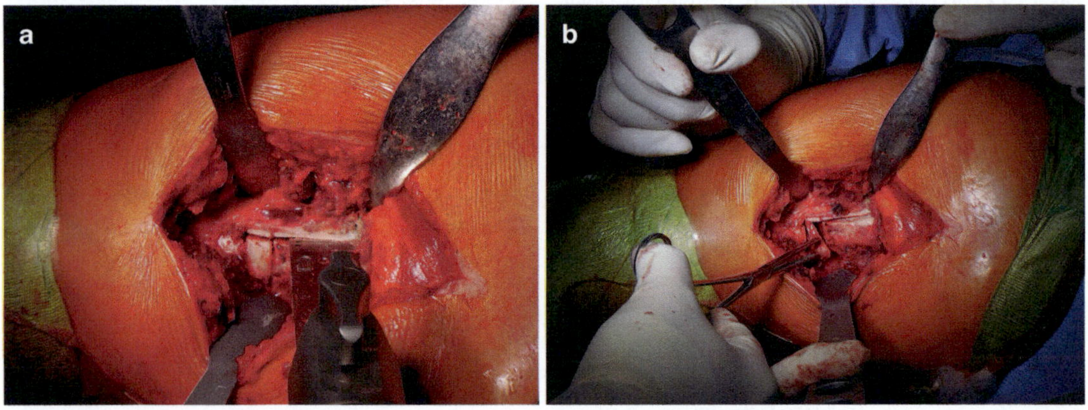

Fig. 13.24 (**a**) The ascending osteotomy creates a few centimeters long tongue-like part of the anterior femoral cortex; (**b**) Removal of the osteotomized piece of wedge-shaped bone

The crutches are used for 6 weeks with partial weight bearing depending on the bone quality and patient compliance.

13.5.2 Distal Femoral Osteotomy (DFO) for Valgus Knee Malalignment

Valgization of the distal femur is indicated in the degeneration of lateral compartment with valgus deformity or in valgus deformities subsequent to growth disorders or posttraumatic cases. If the valgus deformity is found in the distal femur with mLDFA <90°, the correction of the femur is indicated. Two different treatment options are described. The femur can be varisated with opening osteotomy from the lateral side or by closing osteotomy from the medial side of the femur. The medial closing-wedge osteotomy is more difficult to perform, but bone healing is faster. The lateral opening osteotomy is on the one hand

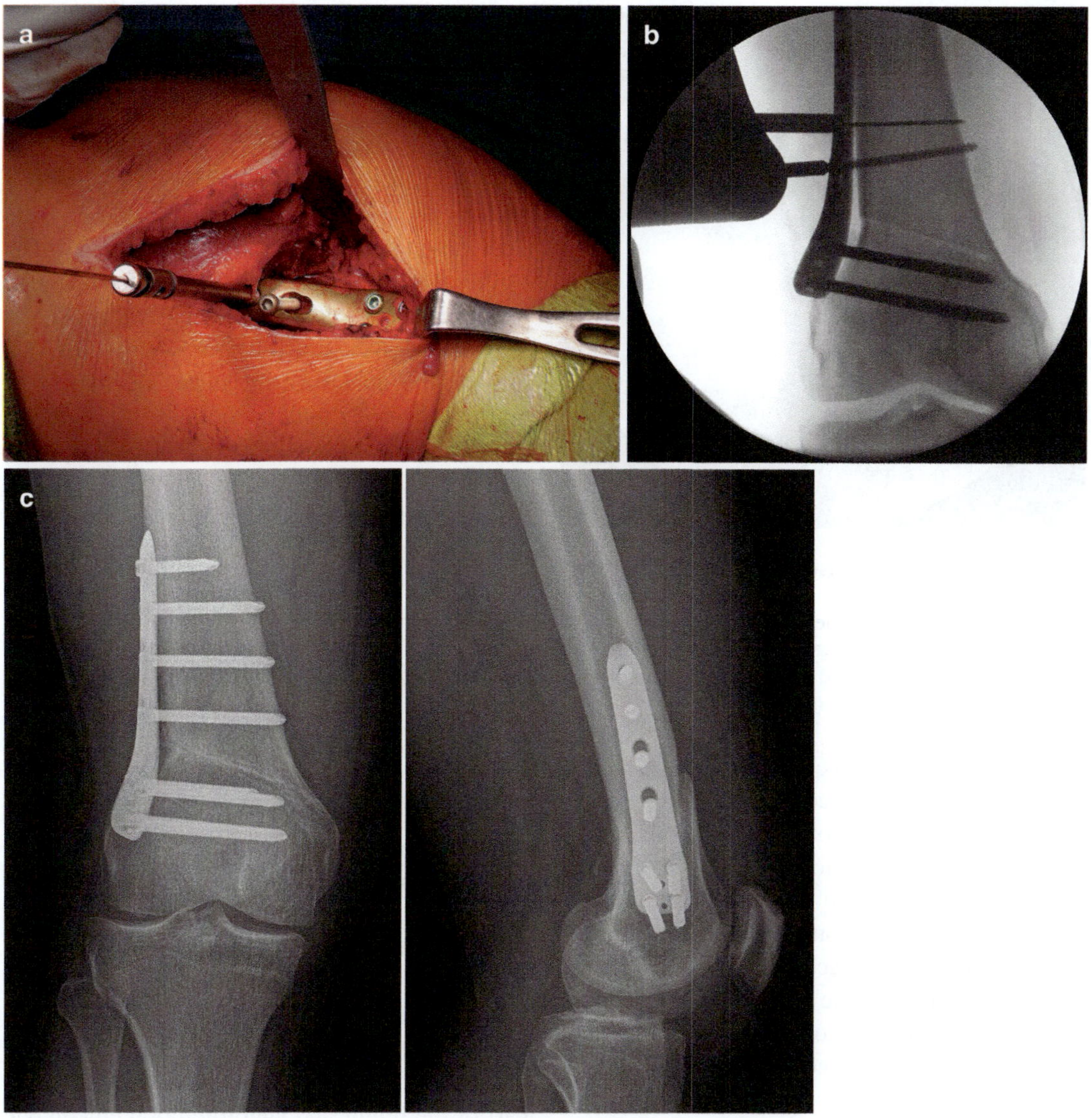

Fig. 13.25 (**a**, **b**)The compression screw is temporarily placed into the most distal hole at the proximal part of fixation plate; (**c**) Postoperative X-ray image of the TomoFix™ locking compression plate after LCWDFO

more precise and faster, but on the other hand may result in disturbed bone healing and problems with the iliotibial band sliding over the plate. In the lateral opening-wedge osteotomy, the bone (autologous or heterologous) has to be inserted in the osteotomy gap. The authors' preferred technique is biplanar medial closed-wedge DFO. The lateral open-wedge DFO is indicated only when the deformity is accompanied with shortening of the affected femur.

13.5.2.1 Medial Closed-Wedge Distal Femoral Osteotomy

The procedure is performed with or without the use of a tourniquet and starts with the knee in full extension. The longitudinal skin incision is placed medially. After soft tissue division, vastus medialis muscle and intermuscular septum are identified, followed by careful anterior retraction of the vastus medials. The perforant vessels are cauterized and retractors are placed anteriorly

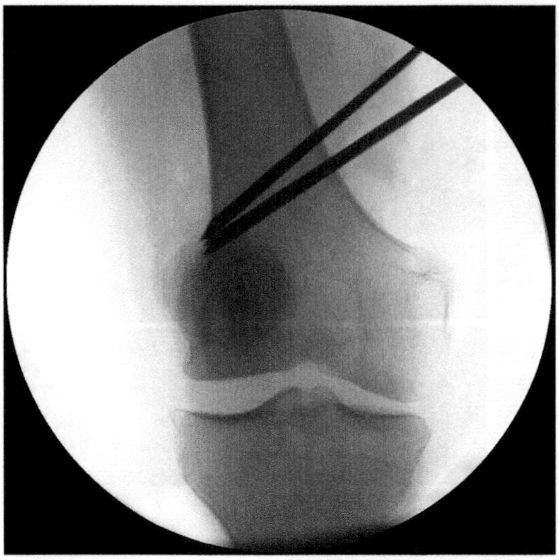

Fig. 13.26 Two parallel Kirschner wires are placed proximally to the medial femoral epicondyle in the direction of the hinge point just above the lateral femoral epicondyle. Another set of wires are inserted proximally, while also facing hinge point. Both the pairs of wires should converge and meet 5–10 mm from the lateral cortex

and posteriorly. Thus, a clear approach to medial part of the femur and medial femoral epicondyle is obtained. Under fluoroscopic guidance and with knee in full extension, two parallel K-wires are placed proximally to the medial femoral epicondyle in the direction of the predetermined hinge point which is placed just above the lateral femoral epicondyle. The wires should be placed 10 mm above the intercondylar groove. Another two K-wires are inserted proximally from the first pair and parallel to one another (Fig. 13.26). Both the pairs of K-wires should converge and meet at 5 mm from the lateral cortex. The distance (in mm) between proximally and distally placed K-wires determines the planned correction of the osteotomy. In this step, the aid of 3D printed patient-specific K-wire guides can be used in order to perform the correct amount of bone resection. The osteotomy is performed with oscillation saw in the posterior 2/3 of distal femur in between two pairs of K-wires. Both cuts should meet at 5–10 mm distance from the medial femoral cortex. The ascending cut is then performed at an angle of 100° to the osteotomy cuts and parallel to the anterior femoral cortex, creating a few centimeters long tongue-like part of the bone, attached to the distal part of femur. Care should be taken not to break the lateral hinge so gentle sawing or drilling at the medial cortex is performed. The osteotomized piece of wedge-shaped bone is then removed, and the gap is carefully closed (Fig. 13.27). The osteotomy can be fixed with angular stable fixator (TomoFix® MDF, DePuy Synthes), which has to be placed parallel to the femoral shaft (Fig. 13.28). The plate is temporarily fixated with two K-wires inserted through the guides on the plate, located proximally and distally to the osteotomy. It is then stabilized with four monocortical screws distally. Afterwards, the compression screw is temporarily placed into the most distal hole at the proximal part of fixation plate, which allows the remaining screws to be inserted. The wound is closed in a standard manner while the drainage is usually not necessary. The crutches are used for 6 weeks with partial weight bearing depending on the bone quality and patient compliance.

13.5.2.2 Lateral Open-Wedge Distal Femoral Osteotomy

The procedure is performed with or without the use of a tourniquet and starts with the knee in full extension. The longitudinal skin incision is placed laterally. After soft tissue division, the fascia lata is incised and vastus lateralis muscle retracted anteriorly. The perforant vessels are cauterized, and retractors are placed anteriorly and posteriorly. Thus, a clear approach to lateral part of the femur and lateral femoral epicondyle is achieved. Under fluoroscopic guidance and in knee extension, two parallel K-wires are placed proximal to the lateral femoral epicondyle in the direction of the predetermined hinge point which is placed just above the medial femoral epicondyle. The osteotomy is performed with oscillation saw in the posterior 2/3 of distal femur above the two K-wires. The osteotomy cut should stop around 5–10 mm from the medial cortex (Fig. 13.29). The ascending cut is then performed at an angle of 100° to the osteotomy cut and parallel to the anterior femoral cortex, creating a few centimeters long tongue-like part of the bone, attached to the distal part of femur. Under con-

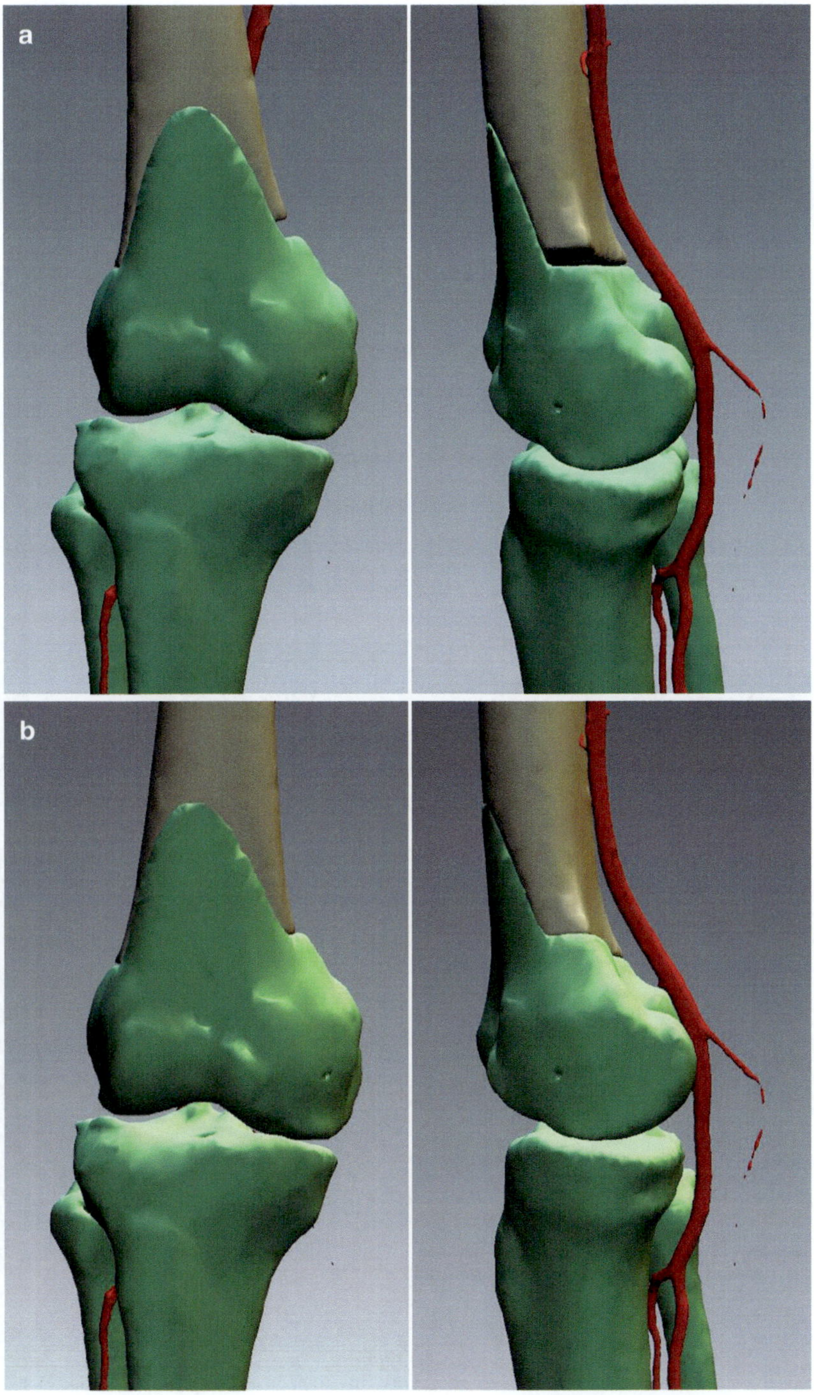

Fig. 13.27 3D simulation of the osteotomy cuts performed in MCWDFO, with (**a**) open and (**b**) closed osteotomy gap

stant fluoroscopic control, the osteotomy is gradually opened with a spreader chisel or multiple osteotomes, after which the spreader is placed in the posterior part of the gap to maintain the correction (Fig. 13.30). At this stage of surgery, great care should be taken not to break the medial hinge. The desired correction can be evaluated by measuring the opening gap in millimeters, using the alignment rod under fluoroscopy or by utilizing the navigation system. When the desired cor-

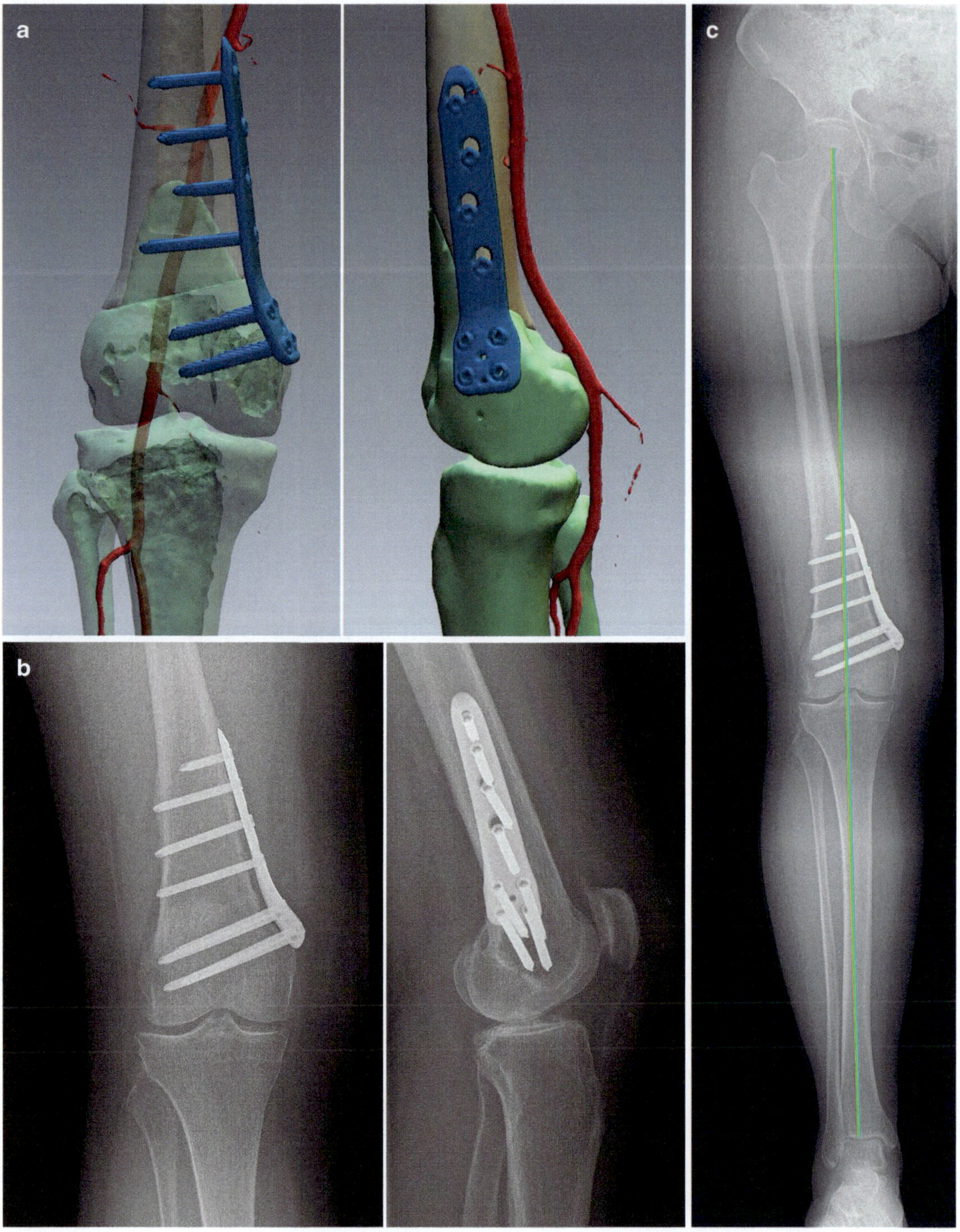

Fig. 13.28 (**a**) 3D simulation and (**b**) postoperative X-ray image of the TomoFix™ locking compression plate after MCWDFO; (**c**) Long-leg X-ray with the desired position of the mechanical axis

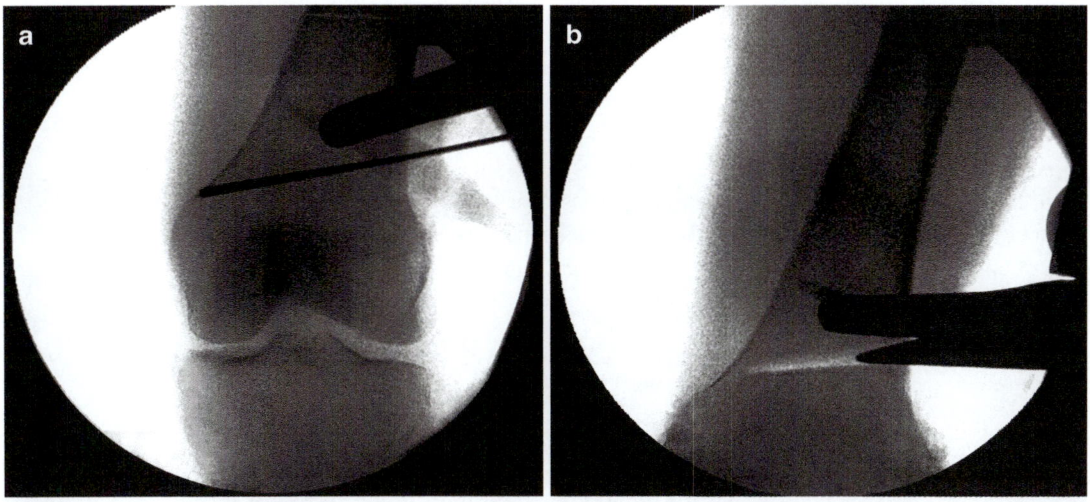

Fig. 13.29 (**a**) Two parallel Kirschner wires are placed proximal to the lateral femoral epicondyle in direction of the hinge point just above the medial femoral epicondyle; (**b**) The osteotomy cut should stop around 5–10 mm from the medial femoral cortex

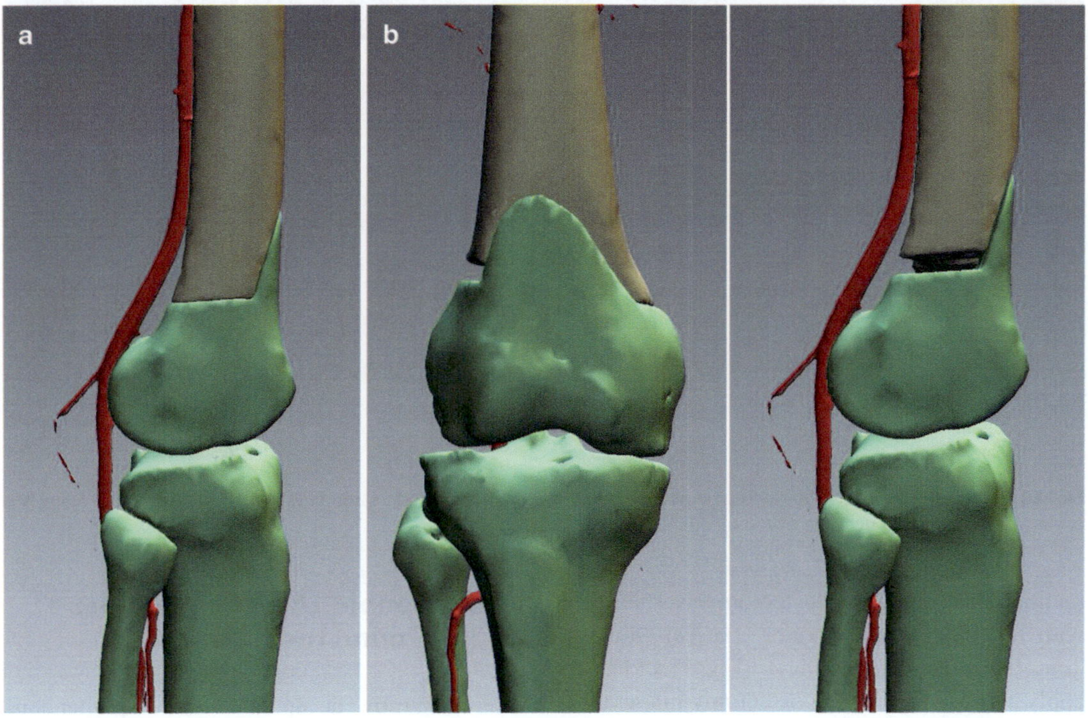

Fig. 13.30 3D simulation of the (**a**) osteotomy cuts performed in LOWDFO and (**b**) opened osteotomy gap

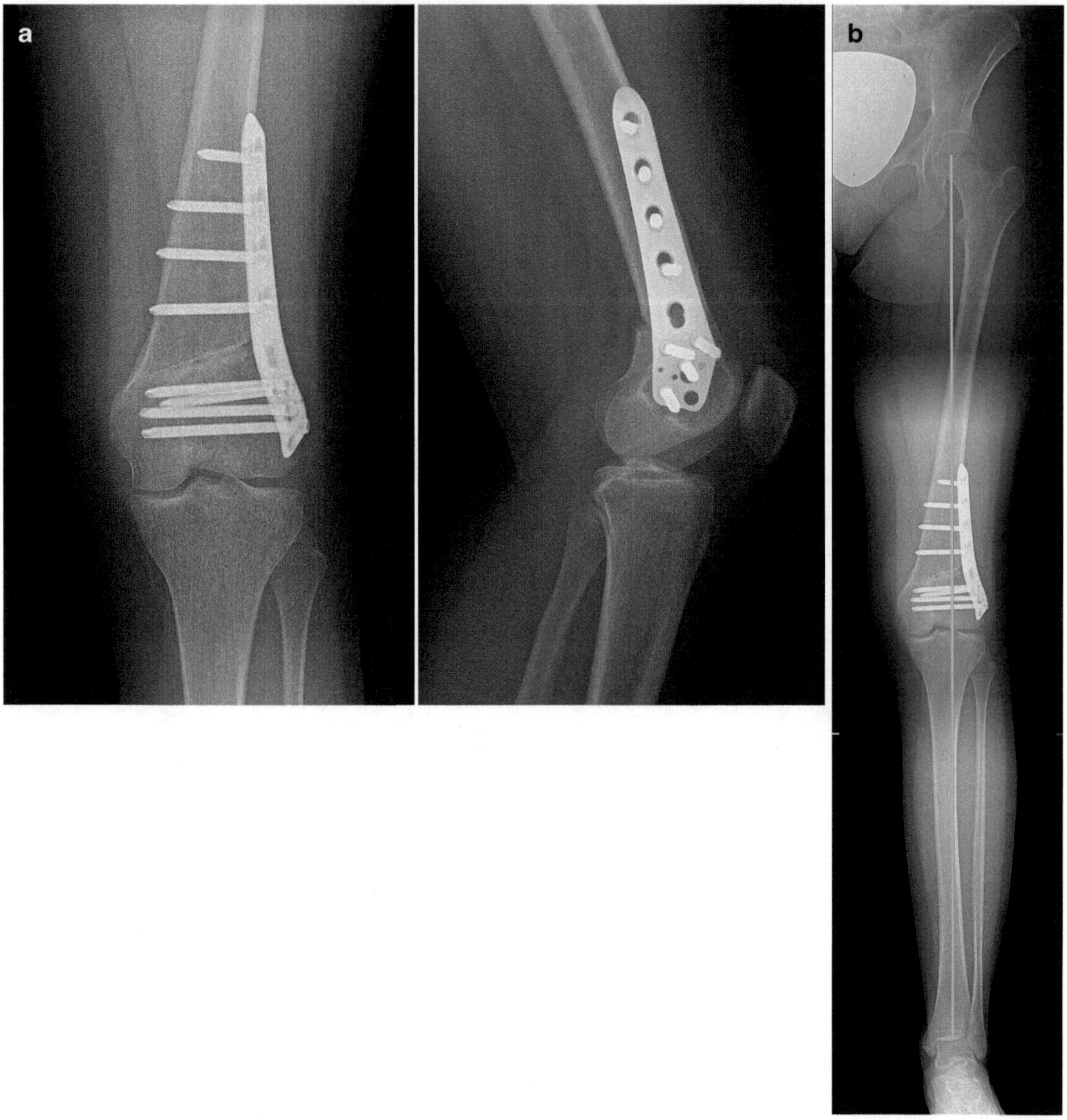

Fig. 13.31 (**a**) Postoperative X-ray image of the TomoFix™ locking compression plate after LOWDFO; (**b**) Long-leg X-ray with the desired position of the mechanical axis

rection is achieved, the osteotomy is stabilized with angular stable fixator (lateral femoral TomoFix®, DePuy Synthes) (Fig. 13.31). To apply compression on the osteotomy hinge, temporary lag screw (later replaced by a bicortical screw) is inserted into the first hole above the bone cut. With the plate in position, the bone block (autologous or heterologous) is inserted into the osteotomy gap, and the wound is closed in a standard manner. The crutches are used for 6 weeks with partial weight bearing.

13.6 Double-Level Osteotomies around the Knee

It is generally accepted that a simultaneous double-level osteotomy is indicated if a single corrective osteotomy on either femur or tibia would shift the joint line orientation beyond 90 ± 3°. This usually occurs in severe valgus or varus deformities where single-level correction would result in a straight leg axis but with the alteration of the joint line (Fig. 13.32). Precise

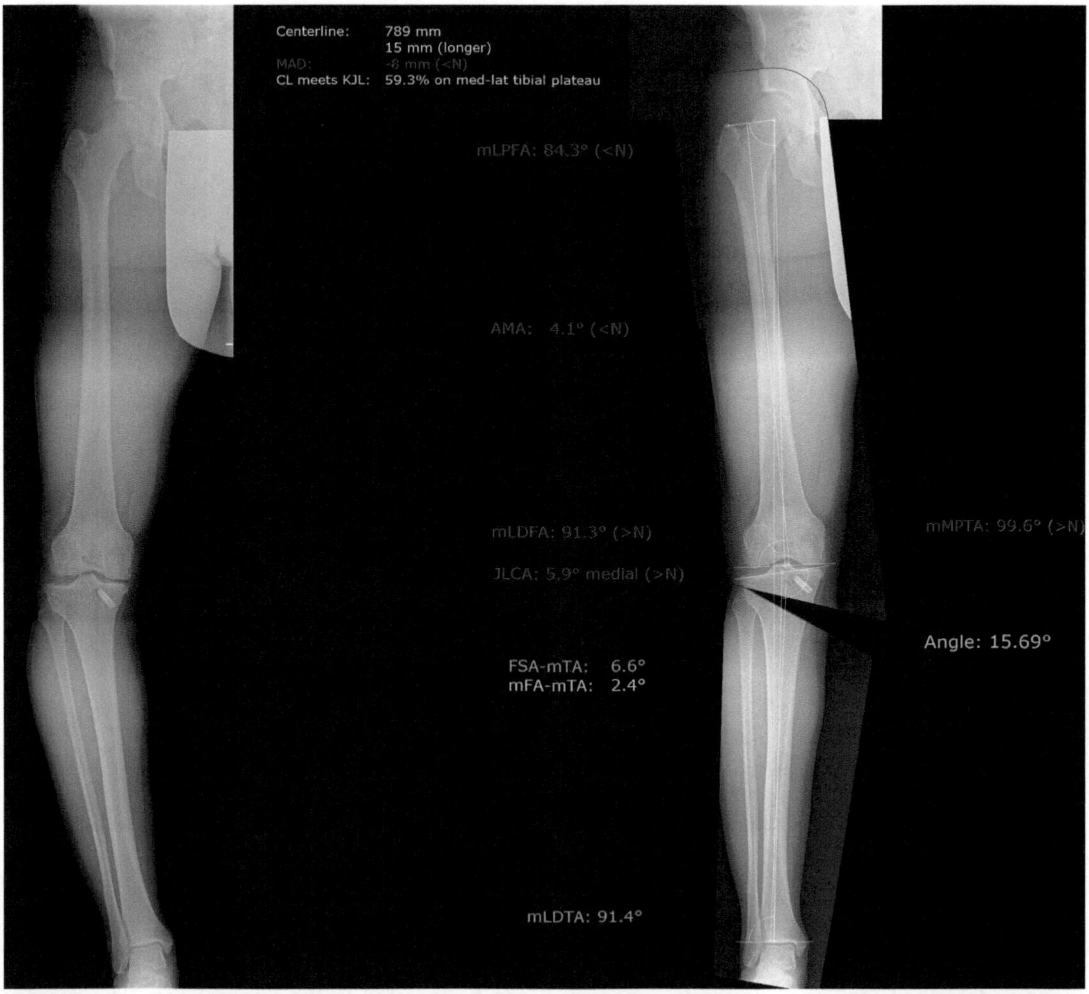

Fig. 13.32 Single-level osteotomy in severe varus/valgus deformity does not permit restoration of mechanical axis without alteration of the joint line

planning and surgical techniques are necessary to achieve desired correction and avoid possible surgical complications in this demanding procedure [25]. The use of special software programs is necessary for computer-assisted preoperative planning, which enables accurate analysis of deformity and simulation of both simultaneous osteotomies [51]. Correction osteotomies are performed medially or laterally with open- or closed-wedge technique. Different combinations of osteotomies are possible; however, LCWDFO+MOWHTO for varus and MCWDFO+MCWHTO for valgus deformity are the two most frequently performed procedures.

13.6.1 Double-Level Osteotomy in Varus Knee Malalignment

In severe varus knee malalignment, the deformity is usually found in both distal femur and proximal tibia (mLDFA >90° and mMPTA <85°). For this reason, the most commonly performed procedure is a combination of LCWDFO and MOWHTO (Fig. 13.33). The surgery starts with a distal femoral osteotomy (biplanar ascending lateral closing wedge) and continued with high tibial osteotomy (biplanar medial opening wedge) where the final tuning will shift the mechanical axis in the desired position. The sur-

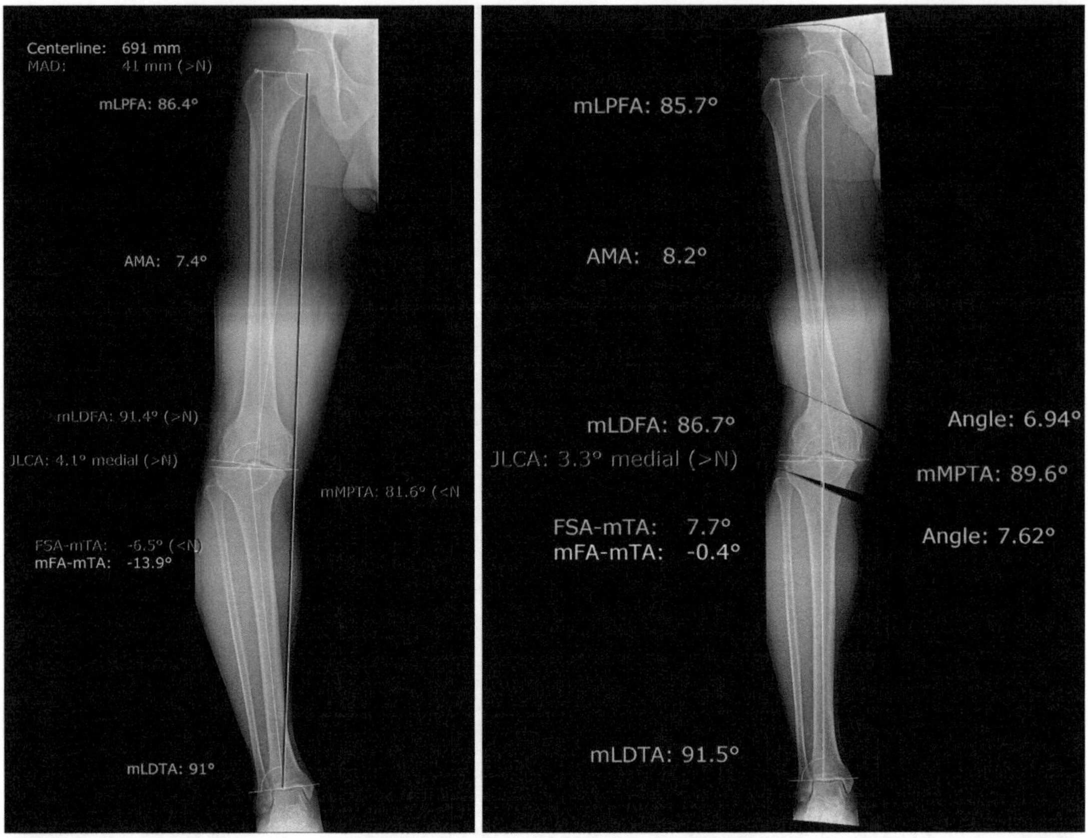

Fig. 13.33 Preoperative planning for double-level (LCWDFO and MOWHTO) osteotomy in varus knee malalignment

gery is performed according to the surgical techniques described in the previous chapters (Fig. 13.34). The bone wedge removed from the distal femur can be used to fill out the osteotomy gap created at the proximal tibia.

13.6.2 Double-Level Osteotomy in Valgus Knee Malalignment

In severe valgus knee malalignment, deformity is usually found in both distal femur and proximal tibia (mLDFA <85° and mMPTA >90°). The most commonly performed procedure is therefore the combination of MCWDFO and MCWHTO (Fig. 13.35). Lateral opening wedge osteotomy of the tibia (LOWHTO) is rarely performed. For patients with shortening of the leg, a lateral opening wedge osteotomy of the femur (LOWDFO) can also be considered, and the removed piece of bone from the tibial osteotomy can be inserted in the femoral gap. The surgery starts with a distal femoral osteotomy (biplanar ascending medial closing wedge or biplanar ascending lateral opening wedge) and continued with a high tibial osteotomy (biplanar medial closing wedge). The surgery is performed according to the surgical techniques described in the previous chapters (Fig. 13.36).

13.7 Computer-Assisted Navigation and Patient-Specific Instruments in Knee Joint Realignment Surgery

Osteotomies around the knee are technically demanding surgical procedures with a steep learning curve. Desirable long-term result of the osteotomy is achievable only when the correct

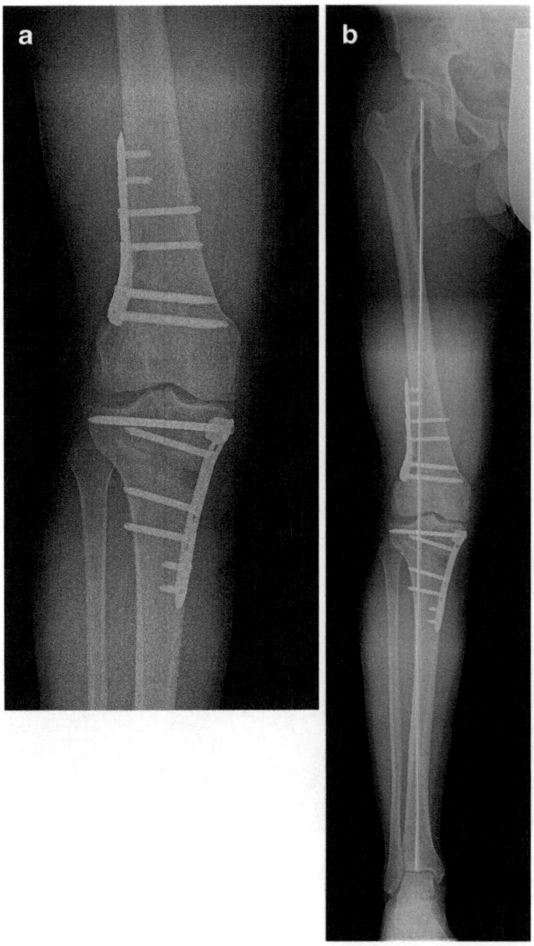

Fig. 13.34 (**a**) Postoperative X-ray image of the TomoFix™ plate system after double-level osteotomy (LCWDFO and MOWHTO); (**b**) Long-leg X-ray with the desired position of the mechanical axis

preoperative planning, together with controlled intraoperative osteotomy correction is performed. The standard preoperative planning is accomplished with an array of software tools which allow us a precise analysis of the deformity, measurement of the angles and planning of the final correction with osteotomy simulation. According to the literature, only 60–80% of patients reached satisfactory alignment after osteotomy. This can be attributed to inaccurate preoperative planning, inadequate intraoperative evaluation of the mechanical axis, or insufficient fixation with possible correction loss.

To improve the clinical results and achieve the desired correction, different computer-assisted surgical techniques are currently available. With the aid of computer assistance, the occurrence of complications related to preoperative planning and intraoperative correction can be decreased. Two different computer-assisted surgical tools are being regularly used in our daily practice when dealing with osteotomy procedures: intraoperative navigation and 3D printed patient-specific instruments.

13.7.1 Computer-Assisted Navigation

To control the osteotomy correction during surgery, we use the navigation system GUIDING STAR from Ekliptik, Ltd., with TOCOS module. The system uses electromagnetic tracking of its probes to produce precise measurements with six degrees of freedom in 3D space of the magnetic transmitter. Each probe is integrated into a threaded adapter, which can be firmly fixed to the bone. Having two bone structures with probes firmly attached, it is possible to precisely calculate and observe the difference in distances and orientations between the bone structures once they get separated, e.g., cut. By strictly following the predicted placement of the probes, the system can transform all the measurements into a form easily understood by the surgeon: varus/valgus angle, flexion angle, rotation angle, and prolonged distance.

Usually only one plane (axial) correction is necessary. Changes in other planes are controlled and usually kept close to zero. Two guiding pins are inserted in the bone, one proximally and one distally to the osteotomy (Fig. 13.37). With opening or closing of the osteotomy, the correction angles in all three planes are measured and displayed on the monitor (Fig. 13.38). When the desired correction is obtained, the osteotomy is secured in a standard manner. This technique is mainly used during open-wedge high tibial osteotomy where good control of the posterior tibial slope and the amount of correction is necessary. Other very useful indications for computer-

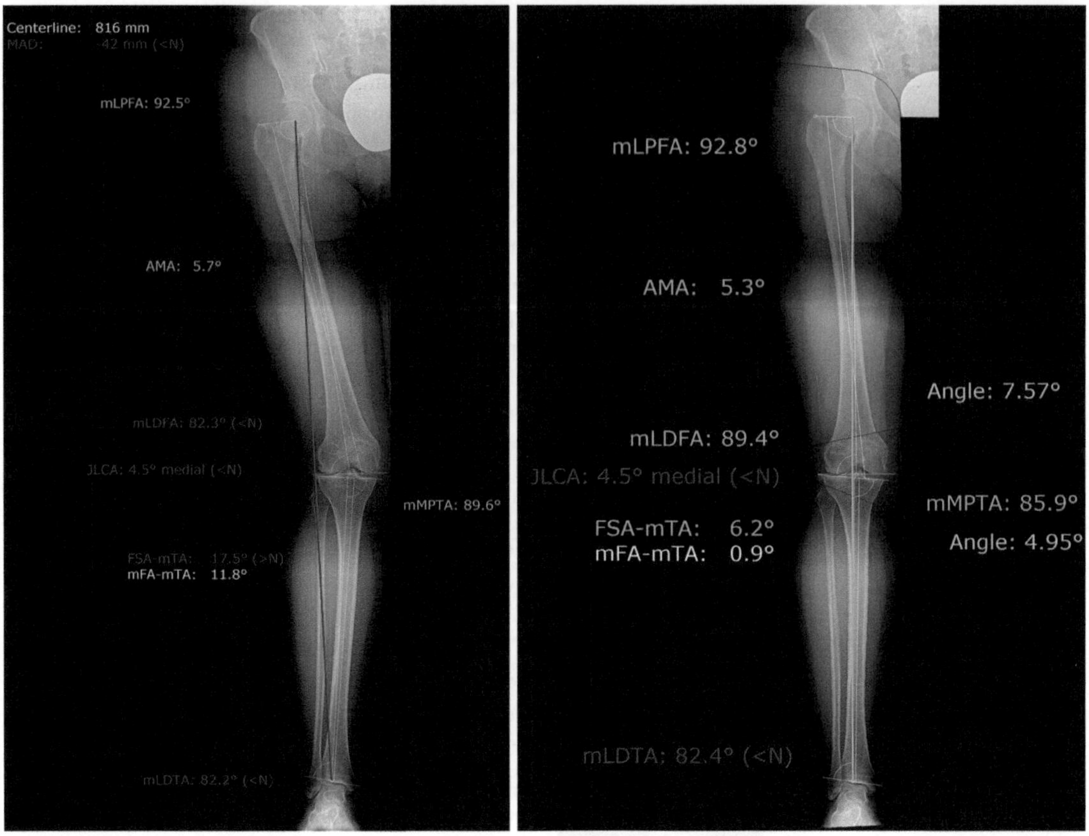

Fig. 13.35 Preoperative planning for double-level (MCWDFO and MCWHTO) osteotomy in valgus knee malalignment

assisted correction with navigation include derotational osteotomies of the femur and tibia (Fig. 13.39).

13.7.2 Patient-Specific Instruments (PSI)

In femoral osteotomies and more complex tibial osteotomies, the patient-specific instrumentation can be used. CT or MRI images are processed to produce a 3D reconstruction model of the knee. In the planning procedures, we are using EBS preoperative planning tool from Ekliptik, Ltd. (Fig. 13.40). The osteotomy cut is planned in all three planes. Based on the bone surface, angle, and direction of the cuts, an osteotomy guide is designed and printed in biocompatible plastics (Fig. 13.41). The guide (CAST—computer-assisted surgical template) is then sent into the plasma sterilization process and used during surgery as a personal template to perform osteotomy cuts. The plastic guide fits perfectly on the bone surface, allowing precise insertion of K-wires for guiding osteotomy cuts or direct orientation of the saw blade. This technique is mainly used in closing-wedge femoral and tibial osteotomies and in complex osteotomies where personalized cuts are necessary.

13.8 Complications Associated with Osteotomies around the Knee

When performing osteotomies around the knee, a surgeon must be aware of the complications that may occur and can potentially prolong the reha-

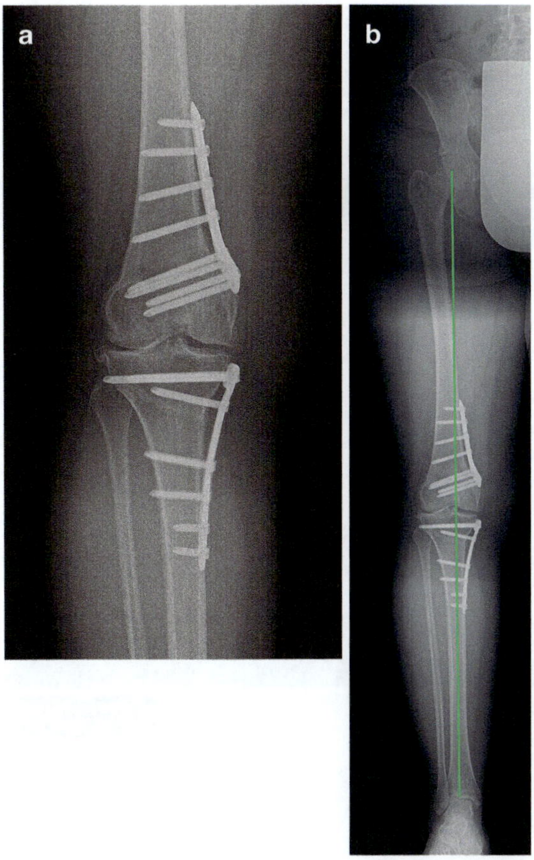

Fig. 13.36 (**a**) Postoperative X-ray image of the TomoFix™ plate system after double-level osteotomy (MCWDFO and MCWHTO); (**b**) Long-leg X-ray with the desired position of the mechanical axis

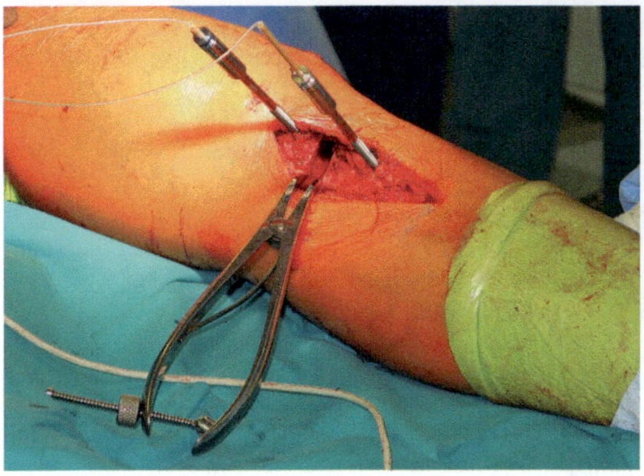

Fig. 13.37 Two guiding pins are inserted in the bone, one proximally and one distally to the osteotomy

bilitation or even lead to suboptimal clinical outcomes. At the same time, every surgeon performing osteotomies has to know how to deal with such complications properly. With modern materials and proper surgical techniques, the incidence of implant failure and loss of correction is nowadays very low (Fig. 13.42); however, other common complications may still pose a threat to the overall clinical outcome of the procedures. As with any surgical intervention, complications can be broken down into intraoperative, early postoperative, and delayed postoperative.

13.8.1 Intraoperative Complications

Intraoperative damage to the neurovascular structures in the popliteal fossa is possible, with the most dangerous being injury of the femoral artery in DFO or popliteal artery in HTO (Fig. 13.43). Situations like this usually require urgent vascular surgeon intervention. Common mechanical intraoperative complications include inadvertent propagation of the osteotomy to the tibial plateau which requires additional internal fixation as well as hinge disruption with subsequent instability of the osteotomy. Unstable hinge fracture in femoral osteotomy should be treated intraoperatively with additional contralateral osteosynthesis to avoid postoperative displacement. However, hinge fractures during open-wedge high tibial osteotomy are managed according to the Takeuchi classification [65]. In type I Takeuchi fracture (within the proximal tibiofibular joint, Fig. 13.44), the fracture is stabilized with a standard angle stable plate with no additional stabilization necessary. In type II Takeuchi fracture (distal to the proximal tibiofibular joint), there is a high risk of non-union and loss of correction, so additional fixation on a lateral side is suggested. The type III Takeuchi fracture (intraarticular fracture in the lateral tibial plateau) can be treated with long lag screws from the lateral side. Hardware malposition can be avoided by intraoperative fluoroscopic control in two planes.

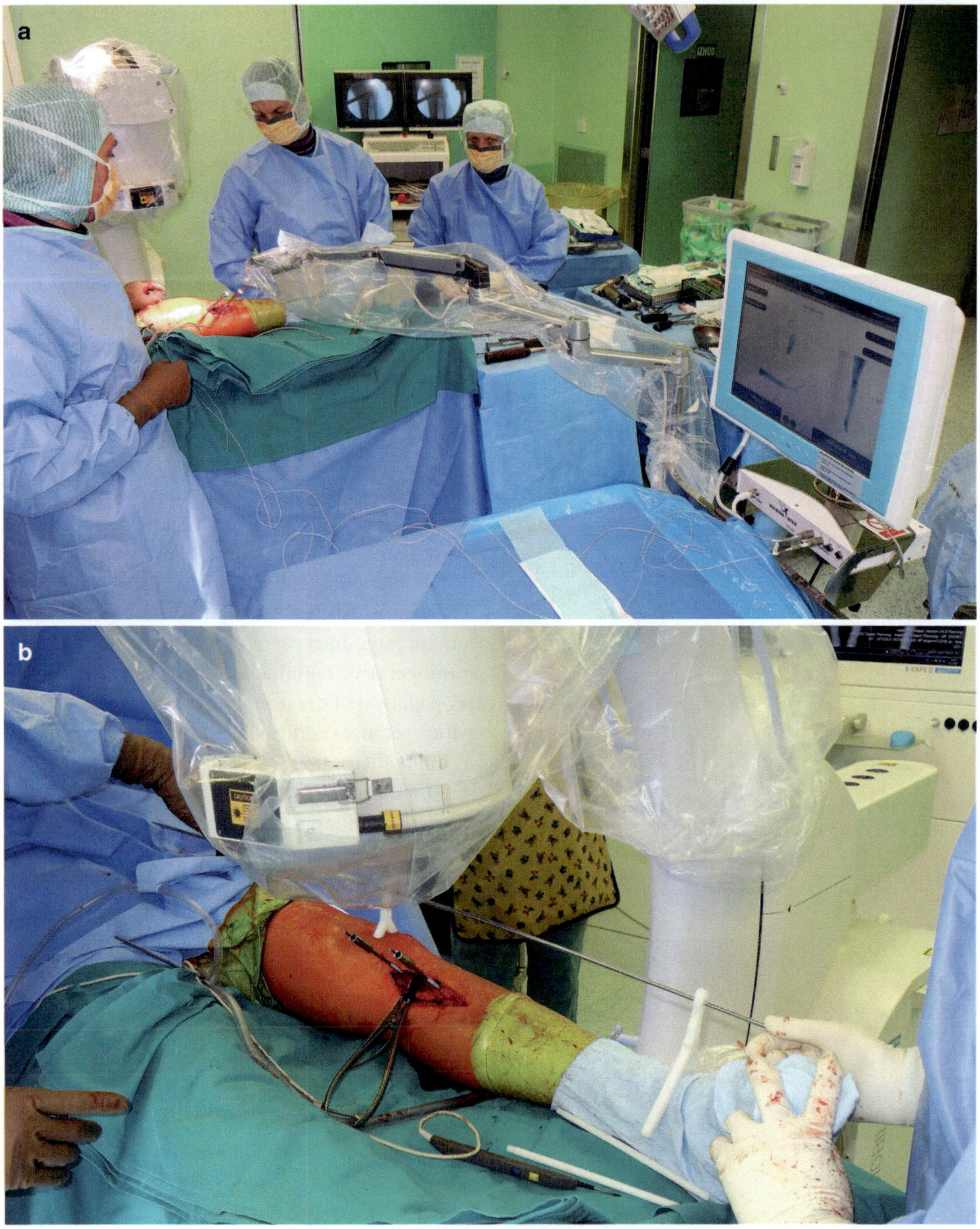

Fig. 13.38 (**a**) Navigation system setup; (**b**, **c**) Correction angles in all three planes are measured and displayed on the monitor; (**d**) The desired correction is secured in a standard manner

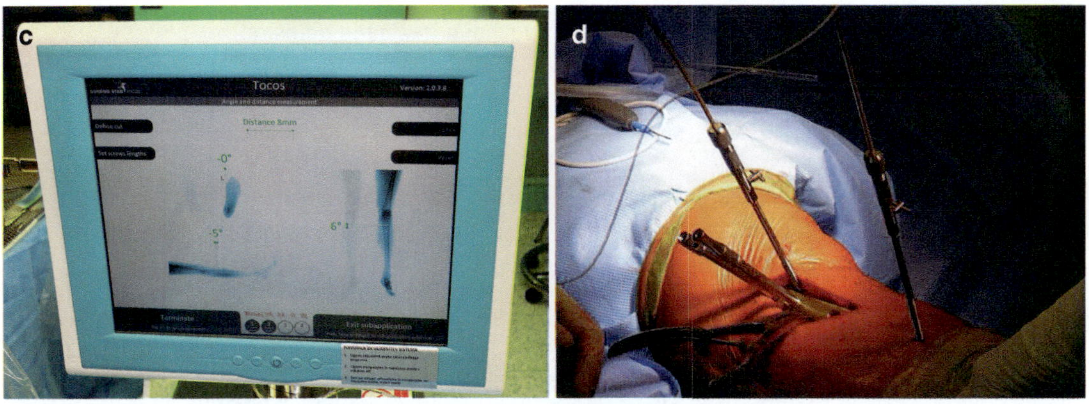

Fig. 13.38 (continued)

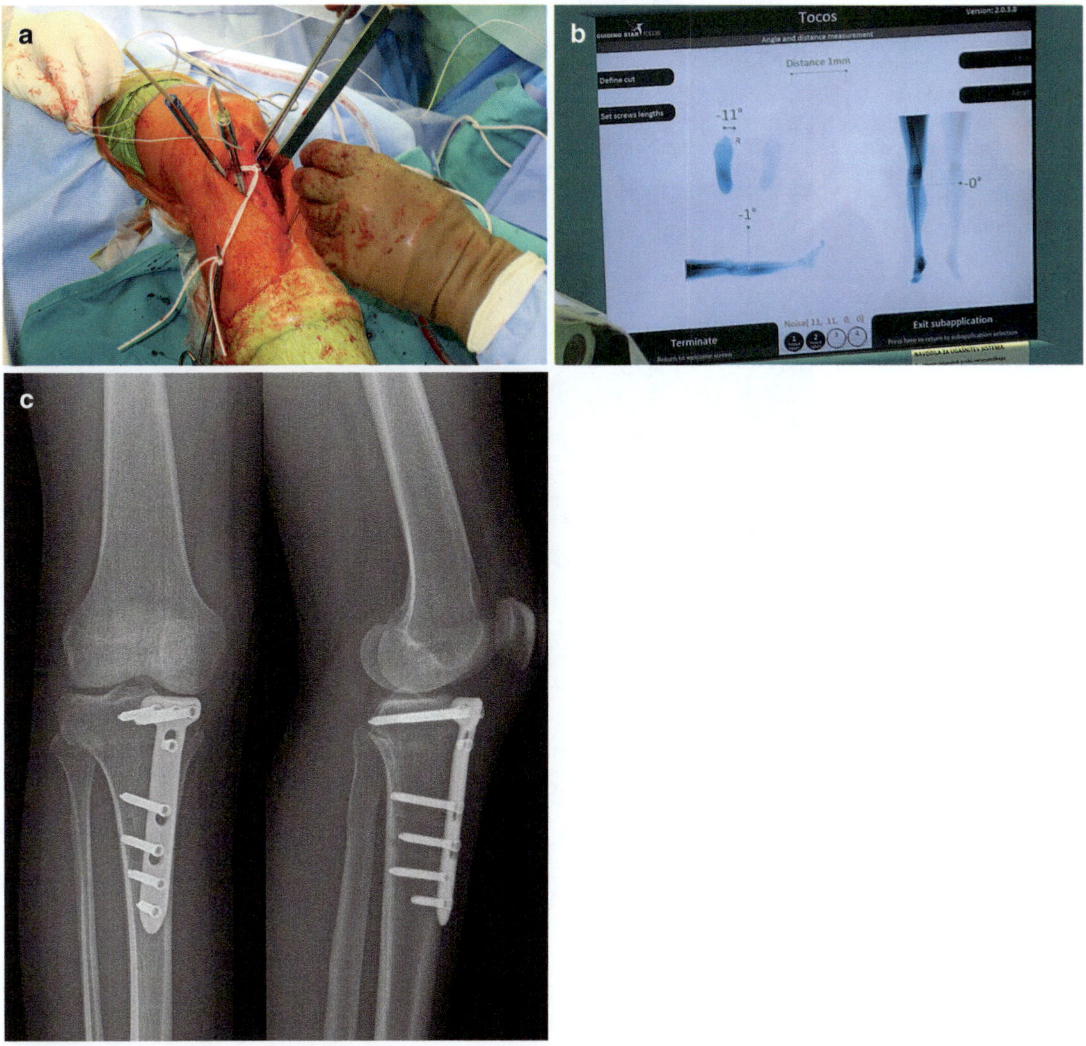

Fig. 13.39 (**a**, **b**) Derotational osteotomy of the proximal tibia performed with computer-assisted navigation system; (**c**) Postoperative X-ray image of the tibial derotational osteotomy

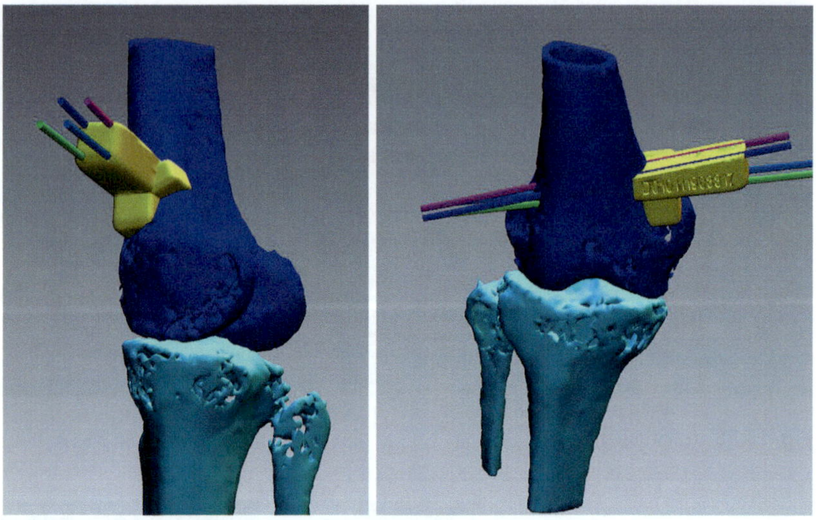

Fig. 13.40 Preoperative 3D simulation of the osteotomy using patient-specific instrumentation

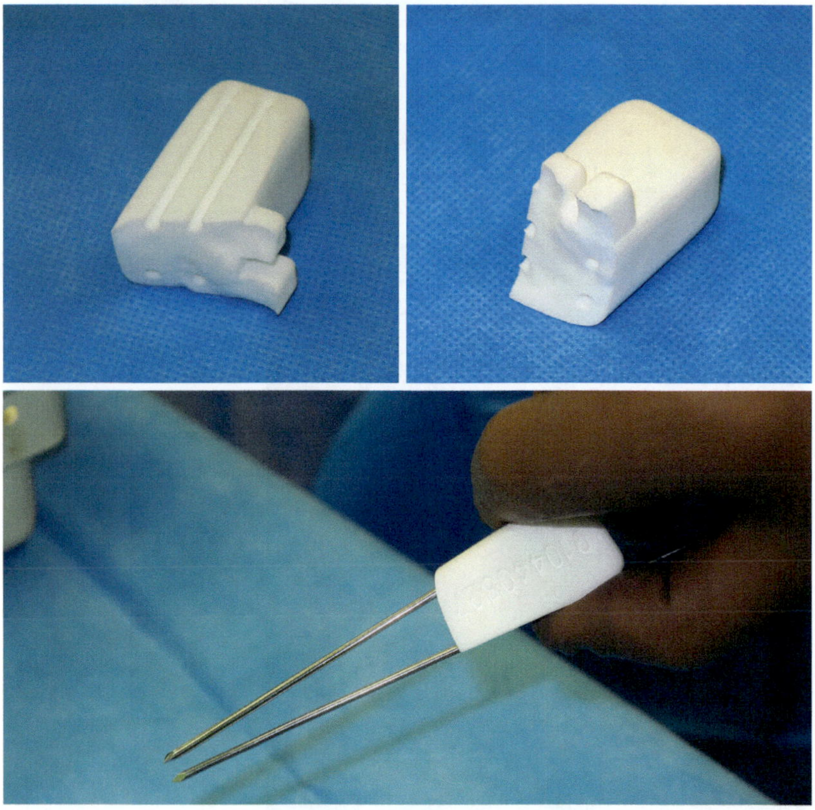

Fig. 13.41 Based on the bone surface, angle, and direction of the cuts, an osteotomy guide is designed and printed in biocompatible plastics

13.8.2 Postoperative Complications

In addition to general postoperative complications such as infection and deep vein thrombosis, other specific neurovascular complications such as compartment syndrome, bleeding from the osteotomy site, and neural palsies predominate in the early postoperative period. On the other hand,

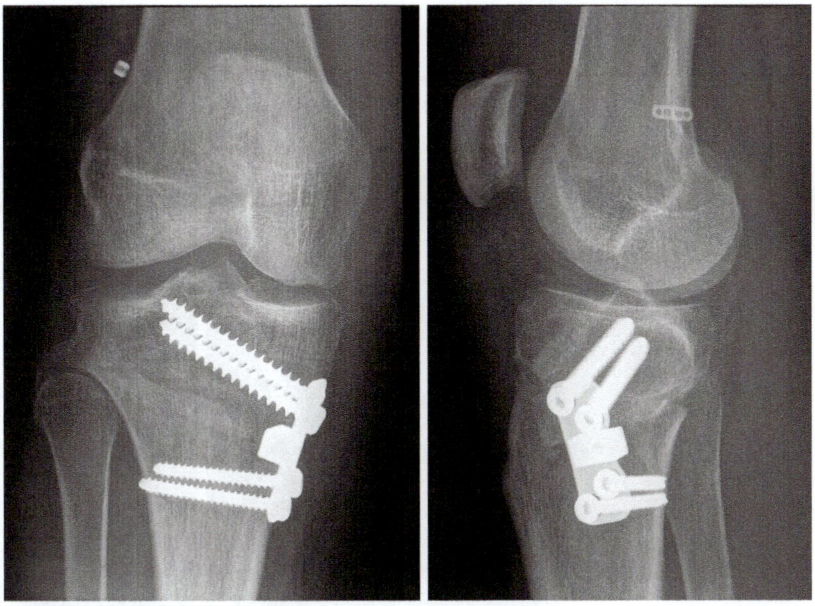

Fig. 13.42 Arthrex® Puddu plate for opening wedge high tibial osteotomy

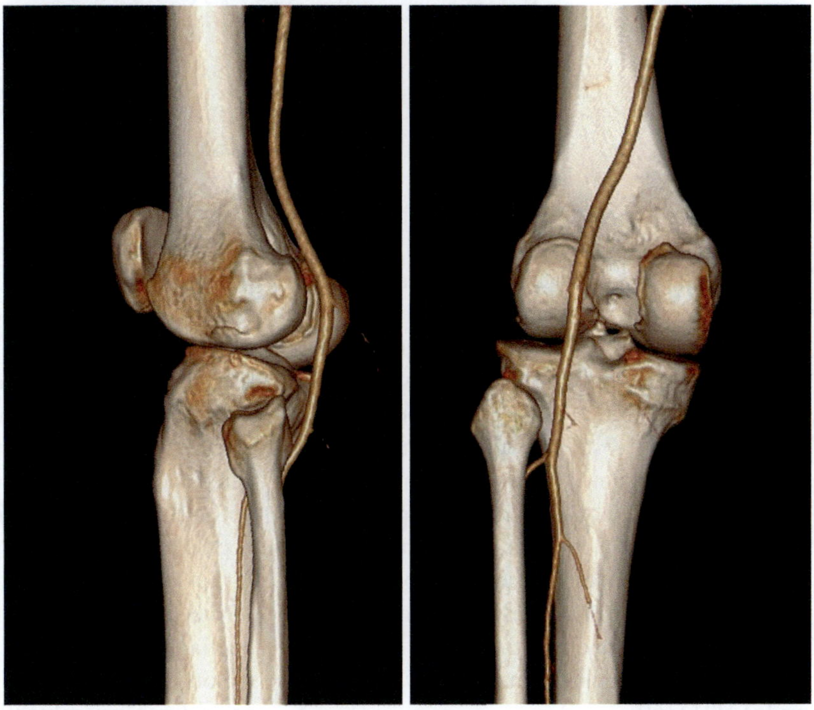

Fig. 13.43 3D CT angiography of the knee clearly demonstrates the spatial relation of the femoral and popliteal artery to the femoral and tibial posterior cortex

complications like under- and over-correction, delayed union, and hardware failure are more often encountered during delayed postoperative period [35, 66, 67].

Early infection (within the first 3 weeks after surgery) is treated with debridement, lavage, and intravenous antibiotic therapy with the plate left *in situ*. According to the author, most of these cases eventually heal completely (Fig. 13.45). In the case of late operative infection or unsuccessful treatment of early infection, the plate should be removed and an

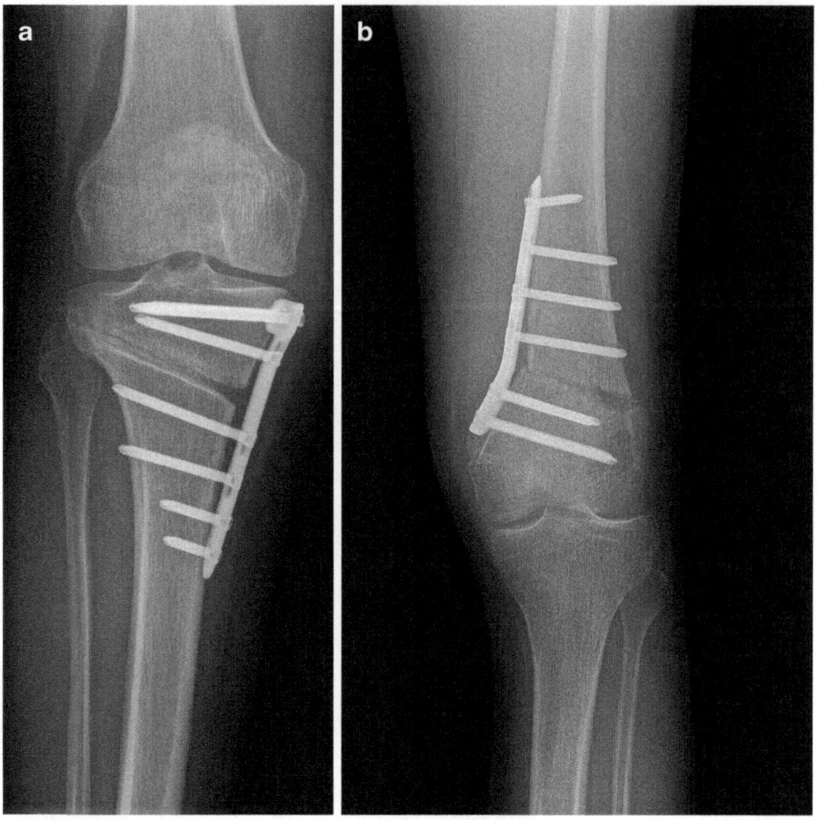

Fig. 13.44 Hinge fracture seen on postoperative X-ray image after (**a**) high tibial osteotomy (Takeuchi I) and (**b**) distal femoral osteotomy

external fixator applied to stabilize the osteotomy.

Compartment syndrome is possible in all tibial surgeries where excessive swelling of the muscle compartments develops. Frequent elevation of the leg and foot movement as well as cortisone administration can help prevent the swelling and a rise of intracompartmental pressure. If a tourniquet is used, deflation and thorough hemostasis should be achieved prior to the wound closure [68]. In addition to direct trauma, elevated compartment pressure is also thought to play a role in the pathogenesis of peroneal nerve palsy, which is one of the most commonly reported neurovascular complications in closed-wedge HTO [69].

Failure to achieve the correct alignment of the lower extremity after the procedure is getting more and more uncommon with modern surgical techniques. In case of under-correction, the weight-bearing line is shifted insufficiently, and the effect of the osteotomy does not reach its potential (Fig. 13.46). Correction can also be lost due to osteoporotic bone, which does not support the screws sufficiently. On the contrary, overcorrection leads to cosmetic problems and more importantly to overloading of the contralateral joint compartment with rapid progression of osteoarthritis. If the correction after surgery is not acceptable, the revision of the osteotomy is necessary (Fig. 13.47).

Severe pain under loading that persists up to 10 weeks after surgery may imply a delayed bone healing. In the authors' experience, this complication is rare when utilizing biplanar osteotomy procedures in combination with modern implants, correct osteotomy techniques (avoiding hinge fracture Takeuchi II and III) and careful patient selection. In case of symptomatic patients with X-ray-confirmed delayed union, a CT scan of the

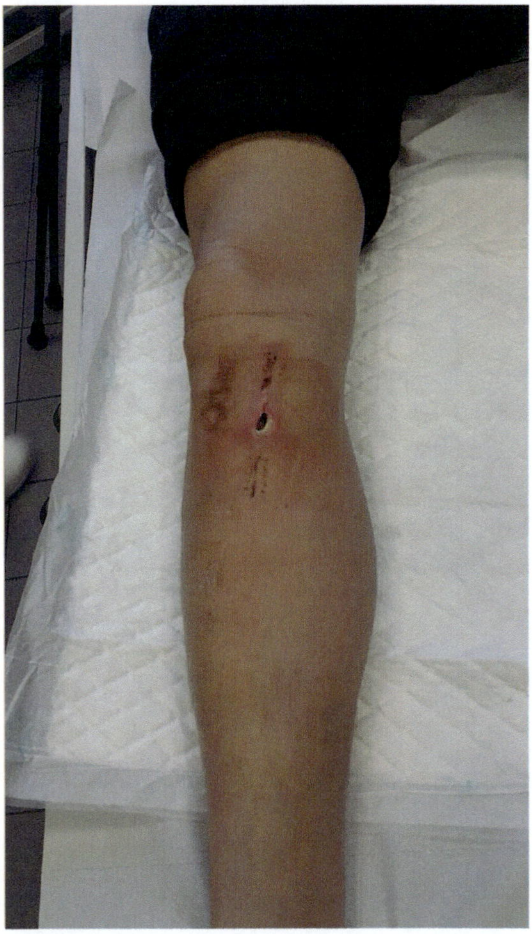

Fig. 13.45 A case of early postoperative infection of the wound, which was treated with debridement, lavage, and intravenous antibiotic therapy

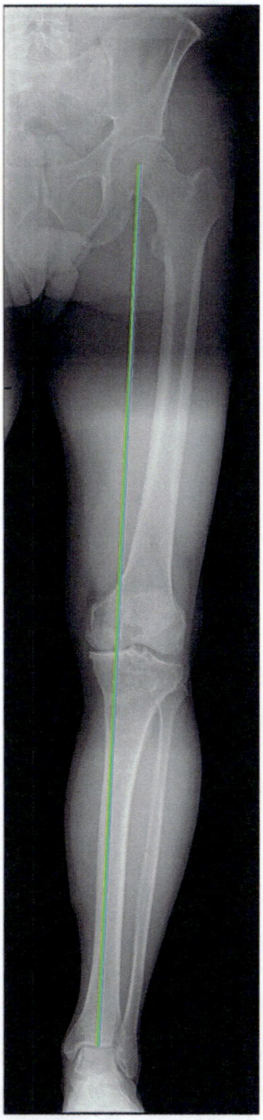

Fig. 13.46 An example of under-corrected varus deformity, where the weight-bearing line is shifted insufficiently, and the effect of the osteotomy does not reach its potential

affected area should be performed and the union regularly checked at follow-ups (Fig. 13.48). If the non-union progresses to pseudoarthrosis, the autogenous bone graft from iliac crest is used to fill out the bone defect. If delayed union is the consequence of unstable osteotomy, the revision osteosynthesis with plate or external fixator is necessary. Additionally, fixation hardware can be the cause of localized pain due to irritation of surrounding soft tissues in which case it should be removed after 9–12 months, when the osteotomy site is completely healed and the bone has remodeled.

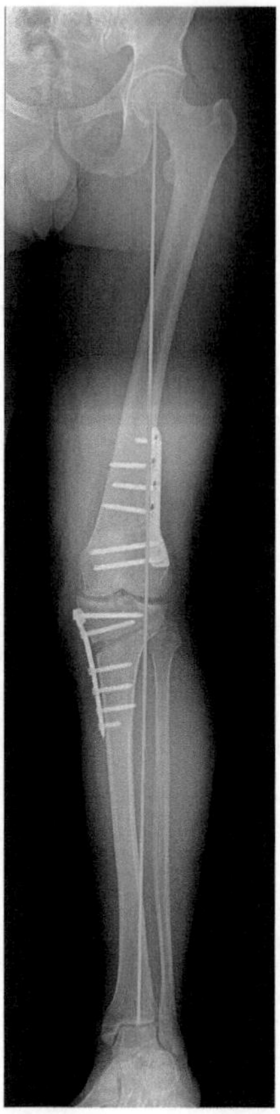

Fig. 13.47 An example of over-corrected varus deformity, where contralateral joint compartment is overloaded and rapid progression of osteoarthritis is imminent, unless the revision is performed

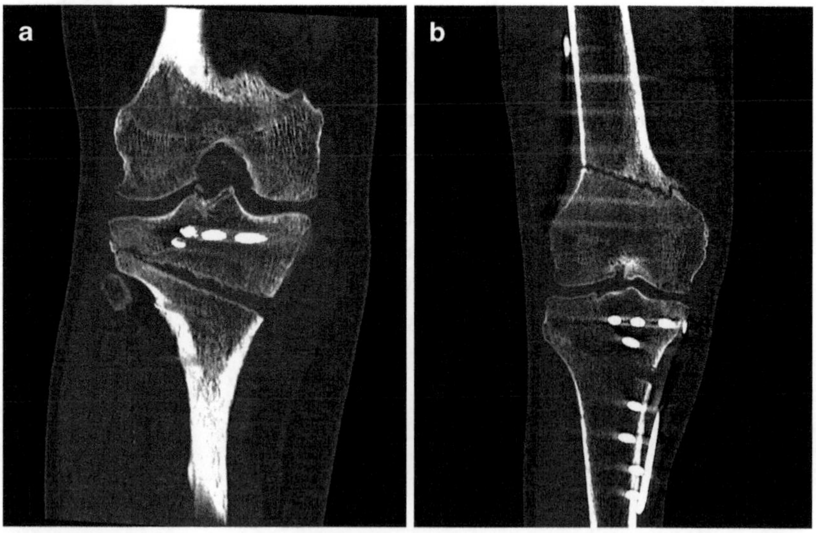

Fig. 13.48 CT scan showing delayed bone healing after (**a**) high tibial osteotomy and (**b**) distal femoral osteotomy

References

1. Smith JO, Wilson AJ, Thomas NP. Osteotomy around the knee: evolution, principles and results. Knee Surg Sports Traumatol Arthrosc. 2013;21(1):3–22.
2. Jackson JP, Waugh W. Tibial osteotomy for osteoarthritis of the knee. Proc R Soc Med. 1960;53(10):888.
3. Wardle EN. Osteotomy of the tibia and fibula. Surg Gynecol Obstet. 1962;115:61–4.
4. Coventry MB. Osteotomy of the upper portion of the tibia for degenerative arthritis of the knee. A preliminary report. J Bone Joint Surg Am. 1965;47:984–90.
5. Barwell R. Clinical lecture on antiseptic osteotomy for Ankylosis and deformity. Br Med J. 1878;1(907):705–7.
6. Macewen W. Clinical lecture on antiseptic osteotomy. Br Med J. 1879;1(957):656–8.
7. Fu D, Li G, Chen K, Zhao Y, Hua Y, Cai Z. Comparison of high tibial osteotomy and unicompartmental knee arthroplasty in the treatment of unicompartmental osteoarthritis: a meta-analysis. J Arthroplasty. 2013;28(5):759–65.
8. Lobenhoffer P, Agneskirchner JD. Improvements in surgical technique of valgus high tibial osteotomy. Knee Surg Sports Traumatol Arthrosc. 2003;11(3):132–8.
9. Niemeyer P, Schmal H, Hauschild O, von Heyden J, Südkamp NP, Köstler W. Open-wedge osteotomy using an internal plate fixator in patients with medial-compartment gonarthritis and varus malalignment: 3-year results with regard to preoperative arthroscopic and radiographic findings. Arthrosc J Arthrosc Relat Surg. 2010;26(12):1607–16.
10. Staubli AE, De Simoni C, Babst R, Lobenhoffer P. TomoFix: a new LCP-concept for open wedge osteotomy of the medial proximal tibia—early results in 92 cases. Injury. 2003;34(Suppl 2):B55–62.
11. Morrey BF. Upper tibial osteotomy for secondary osteoarthritis of the knee. J Bone Joint Surg Br. 1989;71(4):554–9.
12. Wolcott M, Traub S, Efird C. High tibial osteotomies in the young active patient. Int Orthop. 2010;34(2):161–6.
13. Gomoll AH. High tibial osteotomy for the treatment of unicompartmental knee osteoarthritis: a review of the literature, indications, and technique. Phys Sportsmed. 2011;39(3):45–54.
14. Seil R, van Heerwaarden R, Lobenhoffer P, Kohn D. The rapid evolution of knee osteotomies. Knee Surg Sports Traumatol Arthrosc. 2013;21(1):1–2.
15. Flandry F, Hommel G. Normal anatomy and biomechanics of the knee. Sports Med Arthrosc Rev. 2011;19(2):82–92.
16. Shelburne KB, Torry MR, Pandy MG. Muscle, ligament, and joint-contact forces at the knee during walking. Med Sci Sports Exerc. 2005;37(11):1948–56.
17. Luo C-F. Reference axes for reconstruction of the knee. Knee. 2004;11(4):251–7.
18. Cherian JJ, Kapadia BH, Banerjee S, Jauregui JJ, Issa K, Mont MA. Mechanical, anatomical, and kinematic axis in TKA: concepts and practical applications. Curr Rev Musculoskelet Med. 2014;7(2):89–95.
19. Paley D, Pfeil J. Principles of deformity correction around the knee. Orthopade. 2000;29(1):18–38.
20. Herman BV, Giffin JR. High tibial osteotomy in the ACL-deficient knee with medial compartment osteoarthritis. J Orthop Traumatol. 2016;17(3):277–85.
21. Genin P, Weill G, Julliard R. The tibial slope. Proposal for a measurement method. J Radiol. 1993;74(1):27–33.
22. Hohmann E, Bryant A. Closing or opening wedge high Tibial osteotomy: watch out for the slope. Oper Tech Orthop. 2007;17(1):38–45.
23. Robin JG, Neyret P. High tibial osteotomy in knee laxities: concepts review and results. EFORT Open Rev. 2016;1(1):3–11.
24. Pape D, Rupp S. Preoperative planning for high Tibial osteotomies. Oper Tech Orthop. 2007;17(1):2–11.
25. Lobenhoffer P, van Heerwaarden RJ, Staubli AE. Osteotomies around the knee: indications-planning-surgical techniques using plate fixators [internet]. AO. 2011. https://books.google.si/books?id=wc_Bo3XDeegC.
26. Paley D, Herzenberg JE, Tetsworth K, McKie J, Bhave A. Deformity planning for frontal and sagittal plane corrective osteotomies. Orthop Clin North Am. 1994;25(3):425–65.
27. Keppler P, Suger G, Kinzl L, Strecker W. Osteotomies in malalignments of the lower extremities. Chir Z Alle Geb Oper Medizen. 2002;73(10):982–9.
28. Amis AA. Biomechanics of high tibial osteotomy. Knee Surg Sports Traumatol Arthrosc. 2013;21(1):197–205.
29. Dugdale TW, Noyes FR, Styer D. Preoperative planning for high tibial osteotomy. The effect of lateral tibiofemoral separation and tibiofemoral length. Clin Orthop. 1992;274:248–64.
30. Miniaci A, Ballmer FT, Ballmer PM, Jakob RP. Proximal tibial osteotomy. A new fixation device. Clin Orthop. 1989;246:250–9.
31. Babis GC, An K-N, Chao EYS, Rand JA, Sim FH. Double level osteotomy of the knee: a method to retain joint-line obliquity. Clinical results. J Bone Joint Surg Am. 2002;84–A(8):1380–8.
32. Dettoni F, Bonasia DE, Castoldi F, Bruzzone M, Blonna D, Rossi R. High tibial osteotomy versus unicompartmental knee arthroplasty for medial compartment arthrosis of the knee: a review of the literature. Iowa Orthop J. 2010;30:131–40.
33. Bonasia DE, Governale G, Spolaore S, Rossi R, Amendola A. High tibial osteotomy. Curr Rev Musculoskelet Med. 2014;7(4):292–301.
34. Akizuki S, Shibakawa A, Takizawa T, Yamazaki I, Horiuchi H. The long-term outcome of high tibial osteotomy: a ten- to 20-year follow-up. J Bone Joint Surg Br. 2008;90(5):592–6.

35. Koyonos L, Slenker N, Cohen S. Complications in brief: osteotomy for lower extremity malalignment. Clin Orthop. 2012;470(12):3630–6.
36. Wu L, Lin J, Jin Z, Cai X, Gao W. Comparison of clinical and radiological outcomes between opening-wedge and closing-wedge high tibial osteotomy: a comprehensive meta-analysis. PLoS One. 2017;12(2). https://www.ncbi.nlm.nih.gov/pmc/articles/PMC5300239/. Accessed 29 Nov 2017.
37. Ferner F, Lutter C, Dickschas J, Strecker W. Medial open wedge vs. lateral closed wedge high tibial osteotomy - Indications based on the findings of patellar height, leg length, torsional correction and clinical outcome in one hundred cases. Int Orthop. 2018.
38. Rosso F, Margheritini F. Distal femoral osteotomy. Curr Rev Musculoskelet Med. 2014;7(4):302–11.
39. Backstein D, Morag G, Hanna S, Safir O, Gross A. Long-term follow-up of distal femoral varus osteotomy of the knee. J Arthroplasty. 2007;22(4 Suppl 1):2–6.
40. Saragaglia D, Sigwalt L, Rubens-Duval B, Chedal-Bornu B, Pailhe R. Concept of combined femoral and Tibial osteotomies. J Knee Surg. 2017;30(08):756–63.
41. Fujisawa Y, Masuhara K, Shiomi S. The effect of high tibial osteotomy on osteoarthritis of the knee. An arthroscopic study of 54 knee joints. Orthop Clin North Am. 1979;10(3):585–608.
42. Odenbring S, Egund N, Hagstedt B, Larsson J, Lindstrand A, Toksvig-Larsen S. Ten-year results of tibial osteotomy for medial gonarthrosis. The influence of overcorrection. Arch Orthop Trauma Surg. 1991;110(2):103–8.
43. Hernigou P, Medevielle D, Debeyre J, Goutallier D. Proximal tibial osteotomy for osteoarthritis with varus deformity. A ten to thirteen-year follow-up study. J Bone Joint Surg Am. 1987;69(3):332–54.
44. Lee DC, Byun SJ. High tibial osteotomy. Knee Surg Relat Res. 2012;24(2):61–9.
45. Coventry MB. Upper tibial osteotomy for osteoarthritis. JBJS. 1985;67(7):1136.
46. Ivarsson I, Myrnerts R, Gillquist J. High tibial osteotomy for medial osteoarthritis of the knee. A 5 to 7 and 11 year follow-up. J Bone Joint Surg Br. 1990;72(2):238–44.
47. Jakob RP, Jacobi M. Closing wedge osteotomy of the tibial head in treatment of single compartment arthrosis. Orthopade. 2004;33(2):143–52.
48. Sherman SL, Thompson SF, Clohisy JCF. Distal femoral Varus osteotomy for the management of Valgus deformity of the knee. J Am Acad Orthop Surg. 2018;26(9):313–24.
49. Sabbag OD, Woodmass JM, Wu IT, Krych AJ, Stuart MJ. Medial closing-wedge distal femoral osteotomy with medial patellofemoral ligament imbrication for genu Valgum with lateral patellar instability. Arthrosc Tech. 2017;6(6):e2085–91.
50. Hernigou P. Open wedge tibial osteotomy: combined coronal and sagittal correction. Knee. 2002;9(1):15–20.
51. Schröter S, Ihle C, Mueller J, Lobenhoffer P, Stöckle U, van Heerwaarden R. Digital planning of high tibial osteotomy. Interrater reliability by using two different software. Knee Surg Sports Traumatol Arthrosc. 2013;21(1):189–96.
52. Savarese E, Bisicchia S, Romeo R, Amendola A. Role of high tibial osteotomy in chronic injuries of posterior cruciate ligament and posterolateral corner. J Orthop Traumatol. 2011;12(1):1–17.
53. Sabzevari S, Ebrahimpour A, Roudi MK, Kachooei AR. High Tibial osteotomy: a systematic review and current concept. Arch Bone Jt Surg. 2016;4(3):204–12.
54. Hui C, Salmon LJ, Kok A, Williams HA, Hockers N, van der Tempel WM, et al. Long-term survival of high tibial osteotomy for medial compartment osteoarthritis of the knee. Am J Sports Med. 2011;39(1):64–70.
55. Bode G, von Heyden J, Pestka J, Schmal H, Salzmann G, Südkamp N, et al. Prospective 5-year survival rate data following open-wedge valgus high tibial osteotomy. Knee Surg Sports Traumatol Arthrosc. 2015;23(7):1949–55.
56. Laprade RF, Spiridonov SI, Nystrom LM, Jansson KS. Prospective outcomes of young and middle-aged adults with medial compartment osteoarthritis treated with a proximal tibial opening wedge osteotomy. Arthrosc J Arthrosc Relat Surg. 2012;28(3):354–64.
57. Niinimäki TT, Eskelinen A, Mann BS, Junnila M, Ohtonen P, Leppilahti J. Survivorship of high tibial osteotomy in the treatment of osteoarthritis of the knee: Finnish registry-based study of 3195 knees. J Bone Joint Surg Br. 2012;94(11):1517–21.
58. Duivenvoorden T, Brouwer RW, Baan A, Bos PK, Reijman M, Bierma-Zeinstra SMA, et al. Comparison of closing-wedge and opening-wedge high tibial osteotomy for medial compartment osteoarthritis of the knee: a randomized controlled trial with a six-year follow-up. J Bone Joint Surg Am. 2014;96(17):1425–32.
59. Akiyama T, Okazaki K, Mawatari T, Ikemura S, Nakamura S. Autologous osteophyte grafting for open-wedge high tibial osteotomy. Arthrosc Tech. 2016;5(5):e989–95.
60. Lash NJ, Feller JA, Batty LM, Wasiak J, Richmond AK. Bone grafts and bone substitutes for opening-wedge osteotomies of the knee: a systematic review. Arthrosc J Arthrosc Relat Surg. 2015;31(4):720–30.
61. McDermott AG, Finklestein JA, Farine I, Boynton EL, MacIntosh DL, Gross A. Distal femoral varus osteotomy for valgus deformity of the knee. J Bone Joint Surg Am. 1988;70(1):110–6.
62. Terry GC, Cimino PM. Distal femoral osteotomy for valgus deformity of the knee. Orthopedics. 1992;15(11):1283–9; discussion 1289–1290.
63. Collins B, Getgood A, Alomar AZ, Giffin JR, Willits K, Fowler PJ, et al. A case series of lateral opening wedge high tibial osteotomy for valgus malalignment. Knee Surg Sports Traumatol Arthrosc. 2013;21(1):152–60.
64. Freiling D, van Heerwaarden R, Staubli A, Lobenhoffer P. The medial closed-wedge osteotomy

of the distal femur for the treatment of unicompartmental lateral osteoarthritis of the knee. Oper Orthop Traumatol. 2010;22(3):317–34.
65. Takeuchi R, Ishikawa H, Kumagai K, Yamaguchi Y, Chiba N, Akamatsu Y, et al. Fractures around the lateral cortical hinge after a medial opening-wedge high tibial osteotomy: a new classification of lateral hinge fracture. Arthrosc J Arthrosc Relat Surg. 2012;28(1):85–94.
66. Willey M, Wolf BR, Kocaglu B, Amendola A. Complications associated with realignment osteotomy of the knee performed simultaneously with additional reconstructive procedures. Iowa Orthop J. 2010;30:55–60.
67. Miller BS, Downie B, McDonough EB, Wojtys EM. Complications after medial opening wedge high tibial osteotomy. Arthrosc J Arthrosc Relat Surg. 2009;25(6):639–46.
68. Gibson MJ, Barnes MR, Allen MJ, Chan RN. Weakness of foot dorsiflexion and changes in compartment pressures after tibial osteotomy. J Bone Joint Surg Br. 1986;68(3):471–5.
69. Bauer T, Hardy P, Lemoine J, Finlayson DF, Tranier S, Lortat-Jacob A. Drop foot after high tibial osteotomy: a prospective study of aetiological factors. Knee Surg Sports Traumatol Arthrosc. 2005;13(1):23–33.

Meniscal Implants and Transplantations

14

Mustafa Akkaya and Murat Bozkurt

14.1 Introduction

Problems related to the meniscus, which has an extremely important place in lower extremity functions, are being seen at increasing rates with the current rates of sporting activities and intense working tempo. The leading problem is meniscus tears, and these create loss of workforce and a serious impairment to quality of life. Several different treatment protocols can be applied to meniscus tears, including conservative observation, repair, and open or arthroscopic meniscectomy. Which treatment should be applied in which circumstances varies according to the patient and the form of the meniscus tear. Currently, repair of a torn meniscus with newly developed surgical techniques and preservation of the meniscus should always be the first treatment plan [1–3].

Despite all these developments, many surgeons prefer meniscectomy rather than repair of meniscus tears because of the technical simplicity and shorter operating time [4]. However, in patients with impaired protection of the cartilage and shock-absorbing mechanisms in the knee joint following meniscectomy or untreated meniscus tears, a degenerative process starts in the knee. This degenerative process causes loss of joint cartilage and subsequent rapid development of early stage osteoarthritis findings [5].

In young adult patients with wide meniscus tears that cannot be repaired or who develop widespread meniscal damage related to a previous meniscectomy, it is necessary to protect joint functions and prevent the onset of osteoarthritis in the early stage. To achieve this, meniscus horns are used as meniscus implants in healthy cases, while in cases with wide meniscectomy and horn damage, meniscus transplantation can be applied.

14.2 Meniscus Implants

Animal experimental models and clinical studies aimed at protecting the intra-articular structures and regaining biomechanical alignment following the complaints of pain and increasing intra-articular problems in patients who have undergone meniscectomy or have a degenerative meniscus tear that has been left untreated have led to the development of meniscus implants, and the first implants produced were of porous polymer. No foreign tissue reaction developed in the joint in these implants, and healing was provided with an overlying fibrous-cartilage-type tissue. With continued research, collagen meniscus implants were then produced, and later, with increasing numbers of patients and technological advances, new synthetic, biocompatible polyurethane meniscus implants were produced

M. Akkaya (✉) · M. Bozkurt
Department of Orthopaedics and Traumatology, Faculty of Medicine, Ankara Yildirim Beyazit University, Ankara, Turkey

M. Bozkurt, H. İ. Açar (eds.), *Clinical Anatomy of the Knee*,
https://doi.org/10.1007/978-3-030-57578-6_14

(Actifit™, Orteq Bioengineering, 2008, UK). The main aim of both the types of implants is to re-shape the intra-articular load distribution in the area of the meniscus defect and to protect the intra-articular structures.

14.3 Types of Meniscus Implants

14.3.1 Collagen Meniscus Implants (CMI)

These are GAG-rich (group antigen) implants of porcine Achilles tendon origin, which is type 1 collagen rich. In a series of 300 patients, Rodkey et al. reported that at 1 year after CMI, new meniscus tissue coverage was determined arthroscopically, and the results were good [6]. In another study, MRI at 2.5 years after CMI showed normal meniscus tissue, but in the long-term, there was seen to be a partial loss of volume [7]. In patients treated with CMI, follow-up results from 6 months to 10 years have shown decreased pain and increased levels of activity [8].

14.4 Synthetic Biocompatible Polyurethane Meniscus Implants

These implants increase cell attachment and vascularization from the vascular region due to the highly porous synthetic matrix content. In addition, the aliphatic polyurethane structure provides optimal mechanical strength, biocompatibility, and safe biodegradability. There are two types of implant: medial and lateral (Fig. 14.1).

The polymer structure is formed of two types of biomaterials. These are polyester forming the soft part and polyurethane forming the hard part. The biodegradable polyester structure of the 80% soft part of the synthetic implant is formed of poly(ε-caprolactone). The remaining 20% hard part is formed of repeated chains of two 1.4-butanediisocyanate and one 1.4-butanediol. This section is a partially biodegradable structure

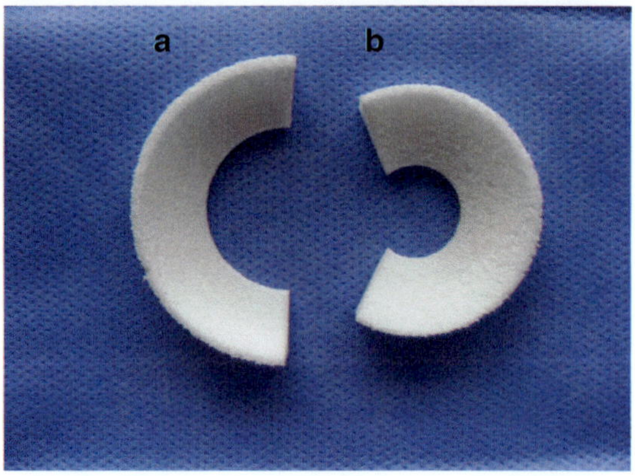

Fig. 14.1 (**a**) Actifit™ medial meniscus scaffold, (**b**) Actifit™ lateral meniscus scaffold

and is formed of structures 5000-fold smaller than human cells.

The biodegradability of the poly(ε-caprolactone) ester bridges in the implant is based on the water and hydrolysis mechanism. The hard part of the implant is a more resistant structure and completes the biodegradation process by obtaining phagocytes from macrophages. This process progresses very slowly, and while biodegradation is completed in mean 4–6 years, the implant becomes biocompatible. Previous studies in an in vitro environment, which have provided basal body temperature and pH values, have shown that after mean 1.5 years implant weight decreased 50%, but the original molecular weight did not change. This condition is supported by cellular migration.

14.5 Surgery Indications and Contra-Indications

In the patient group in which synthetic implant could be used, there must be preserved peripheral meniscus tissue and anterior-posterior horns. Preferred patients should have a stable knee joint or knee stability provided with the surgical procedure to be applied. In addition, a BMI value <35 kg/m^2 increases the chance of treatment success. These implants are not recommended for patients aged >50 years as the meniscus healing potential is low. Just as in other surgical procedures, attention must be paid to systemic diseases and infection. Intra-articular osteochondral damage should not be greater than ICRS grade 3. There have not been sufficient studies as yet about the protective effect of synthetic meniscus implants on cartilage tissue in the long term [9, 10]. However, no clear contra-indication for the use of implants has been reported in literature.

14.6 Surgical Procedure

Synthetic meniscal implants can be applied under general or spinal anesthesia, using manual instruments or as a standard arthroscopic procedure with the application of a tourniquet to the lower extremity of the patient. While stabilizing the implant, the appropriate position can be provided by taking the knee to varus or valgus.

Following diagnostic arthroscopy applied to the patient, the section where the implantation is to be made is cleaned with manual instruments to increase vascularization and attachment. It must be ensured that the area where the meniscus implant is to be placed is a "red-red" or "red-white" area. Then to measure the meniscus defect and determine the size of the implant to be used, the meniscus defect area should be measured with an arthroscopic ruler with a specifically designed tip which is inclined within the joint. To be able to provide implantation at the optimal level, the implant must be prepared 3 mm larger if the defect area is <3 cm, and 5 mm larger if the defect area is >3 cm.

Following the determination of the size of the implant with the arthroscopic ruler from the concave outer surface which will form the meniscocapsular part, an arthroscopic pen is used for marking according to the defect orientation. Using a lancet, the marked area must be cut sharply with a slope of 30°–45°. Both the ends must be marked to be able to determine the intra-articular orientation and surface of the implant. The anteromedial or anterolateral portal that is to be used according to the surgeon's preference should be widened with a lancet to allow comfortable entry of the implant into the joint. Then, according to the side and orientation of the meniscus damage, holding the far end with a clamp, the implant should be gently placed within the joint.

After placement of the implant in the joint appropriate to the defect, it must be fixed to the surrounding natural meniscus tissue using intra-articular fixation methods (all inside, inside-out, outside-in) (Fig. 14.2). When fixing the implant, the intervals between the sutures should be a maximum of 0.5 cm, and in each suture of the implant with a height of 8 mm, care must be taken that at least 4 mm remains in the suture line. The suture material used for fixation of the implant should be strengthened monofilament polyester or polypropylene. After fixation of the implant, stability must be carefully checked with an arthroscopic probe during 0°–90° movements of the knee.

14.7 Meniscus Transplantation

14.7.1 Surgery Indications and Contra-Indications

The development of intra-articular meniscus defects is caused by a reduced area of contact and increased intra-articular contact pressure, resulting in cartilage damage. At later stages, this leads to joint degeneration and development of osteoarthritis.

Therefore, with the meniscus allograft transplantation to be applied, it is attempted to correct the intra-articular anatomy and regain the natural intra-articular functions [11, 12].

14.7.1.1 Indications

The indications for appropriate surgery of meniscus allograft transplantation can be listed as:

1. Pain in the compartment where the meniscus defect is located.
2. A stable knee joint.
3. No malalignment.
4. Joint cartilage degeneration ≤ ICRS grade 3 (International Cartilage Repair Society).

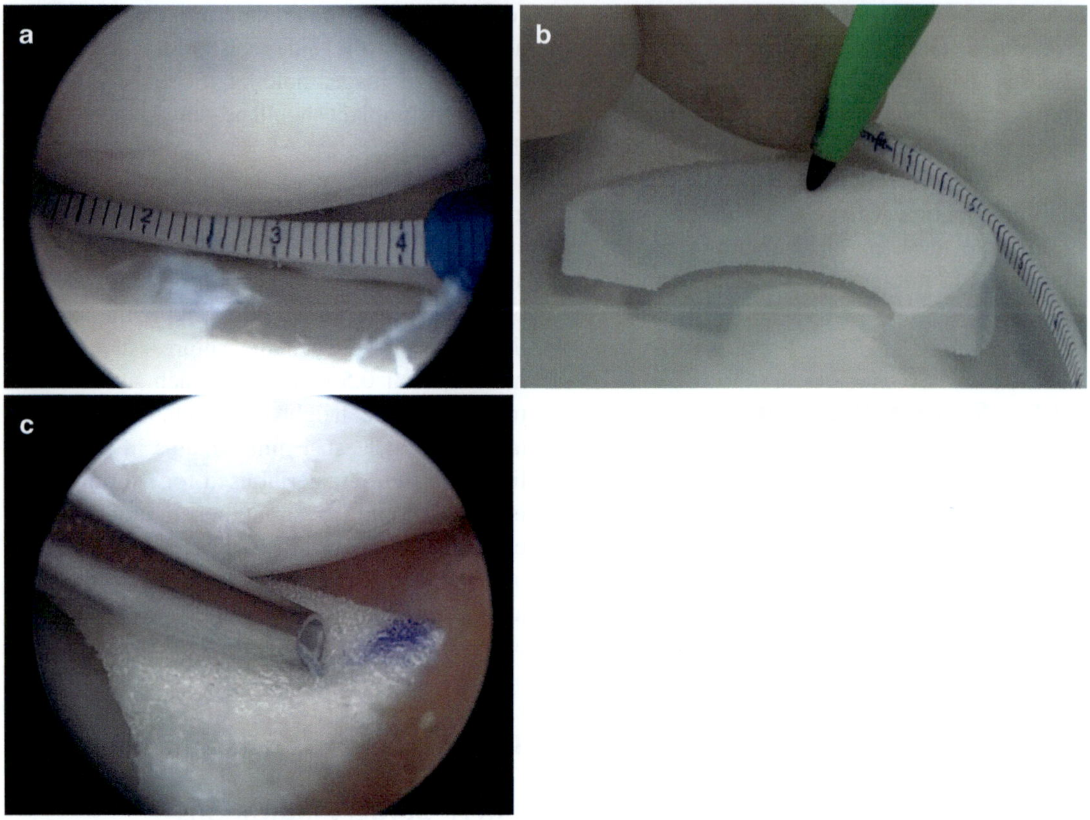

Fig. 14.2 Arthroscopic treatment steps of irreparable meniscus tears with meniscal implants. (**a**) Arthroscopic measurement of damaged meniscal area. (**b**) Sizing of meniscal scaffold comparing to damaged area. (**c**) Fixing the meniscus scaffold sutures on the joint capsule

In patients with concomitant ACL rupture, reconstruction of the ACL in the same session can provide a stable knee joint and increase the stability of medial meniscus allograft, thus ensuring a more functional allograft transplantation [13].

To prevent early joint degeneration following total meniscectomy in young and athletic patients, the transplantation can be applied at the beginning of the symptomatic period. However, successful functional results may not be obtained in a return to high-intensity sport [14].

14.7.1.2 Contra-Indications

The contra-indications for meniscus allograft transplantation can be listed as follows:

1. Advanced stage cartilage damage in the joint (ICRS grade ≥ 3).
2. Changes that impair the femoral condyle morphology and the development of osteophytes.
3. Axial malalignment.
4. Other contra-indications can be listed as obesity, skeletal immaturity, untreated knee instability, synovial diseases, inflammatory arthritis, and previous intra-articular infections [15].

Previous studies have shown cartilage degeneration to be a risk factor in respect of being able to obtain successful results in meniscus allograft transplantation. However, cartilage degeneration in the early stage and <ICRS grade 3 should not be seen as a risk factor. Moreover, concomitant localized cartilage defects can be treated simultaneously with transplantation [16].

Transplantation is not usually appropriate for patients aged >50 years with cartilage damage.

However, successful results have been reported from transplantations applied after corrective osteotomies [17].

14.7.2 Surgical Procedure

Currently, arthroscopic meniscus transplantation is significantly superior to open surgery techniques in respect of surgical morbidity and the ease of meniscus repair techniques. In a 2013 study that compared open and arthroscopic meniscus allograft transplantation methods, it was shown that radial displacement of the allograft was greater following open surgery [18]. When applying the arthroscopic procedure, all the intra-articular compartments must be evaluated in detail by first performing routine diagnostic arthroscopy. In this way, intra-articular problems can be examined and recorded. Previous studies in literature have shown that when applying allograft fixation, applying bone and soft tissue fixation together rather than isolated soft tissue fixation is more appropriate in respect of intra-articular load distribution [19].

Generally, meniscal allografts can be transplanted using the two techniques of "bone bridge" or "bone tunnel" (Fig. 14.3). In both the techniques, it is aimed to make safe transplantation of the anterior and posterior horns. With the bone bridge technique, transplantation can be made protecting the anterior and posterior horns of the meniscus to be transplanted. This technique can be applied in both medial and lateral meniscus transplantations. In the bone tunnel technique, it is difficult to adjust the distance of the tunnels to be opened. There is also a risk of the tunnels intersecting because of the short distance between the anterior-posterior horns of the lateral meniscus. Therefore, the bone tunnel technique is more appropriate for medial meniscus transplantation.

During intra-articular reduction of the allograft, an anteromedial arthrotomy approximately 4 cm from the tibial tubercle and adjacent to the patellar tendon and a 3 cm vertical posteromedial incision are generally sufficient [20].

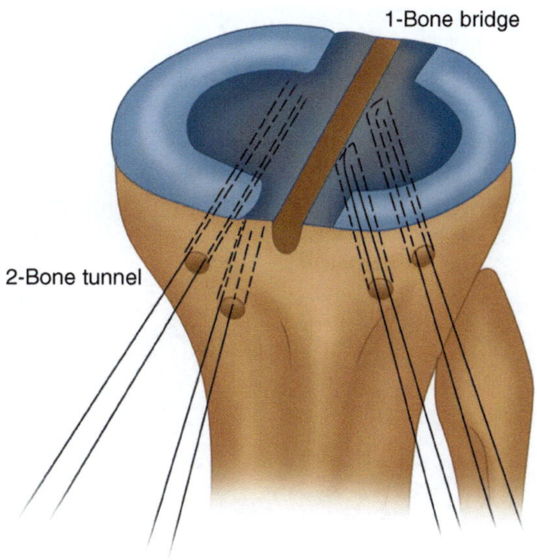

Fig. 14.3 Bone bridge and bone tunnel technique illustration

14.7.2.1 Bone Tunnel

Two tunnels are opened when the anatomic localization is determined in the tibia for the anterior and posterior horns of the meniscus allograft. In this process, guides and manual instruments that are used to open ACL tunnels can be of use. Angulation of 60° should be applied when opening the anterior tunnel over the tibial guide, and 45° for the posterior tunnel. The anterior-posterior horns of the graft should be marked using a marker pen before forming bone plugs over the allograft, and marks should be placed to determine the upper and lower surfaces of the graft. The correct position of the graft after reduction can be checked with these marks.

Then the allograft anterior and posterior horn plugs are prepared. With the exception of the root sections from the tibial plateau, the freed anterior and posterior horns of the meniscus allograft are cut with a motorized cutter, and manual instruments to form plugs compatible with the tunnels are opened in the tibial plateau. Then, suspended sutures are passed through these plugs. During graft reduction, these suspended sutures facilitate fixation by passing within the tunnels opened in the tibia. In addition, two different traction

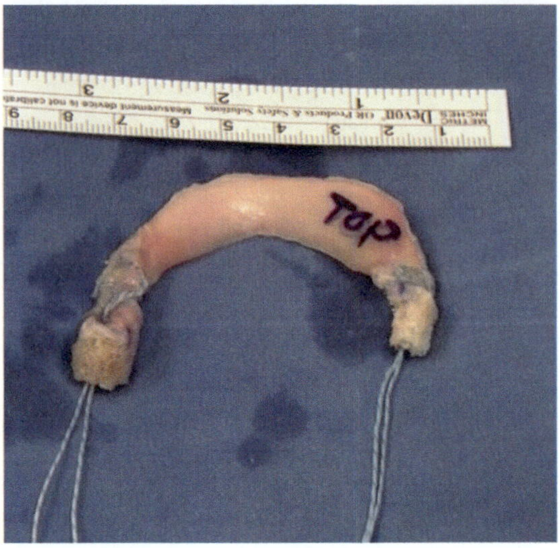

Fig. 14.4 Allograft prepared according to bone tunnel technique

sutures pass over the allograft, the first of which traverses the anterior-mid segment of the allograft and the second passes at the level of the mid-posterior segment. These traction sutures provide seating within the joint at the medial level of the meniscus and on the correct surface (Fig. 14.4). It is necessary to open three portals when applying this technique: medial, lateral parapatellar, and posteromedial portals. After intra-articular reduction of the graft, the procedure is completed using meniscus suture techniques (all inside, inside-out, outside-in).

14.7.2.2 Bone Bridge

To be able to place the lateral meniscus allograft over the tibia, a rectangular slot is opened with a vibrating saw, at a width of 8 mm and depth of 10 mm, adjacent to the ACL attachment point and extending as far as the tibia posterior wall. The allograft can be easily entered into the joint when opening this slot, and an anterolateral arthrotomy should be made with the vibrating saw 3 cm lateral to the patellar tendon.

The mean length of the slot to be opened within the joint, which varies according to the tibial morphology, is 35–40 mm. During all these procedures, there should be a ruler which is used arthroscopically.

Following the preparation of the graft to be placed within the slot, traction and suspended sutures are passed sequentially through the center and meniscus horns (Fig. 14.5). The traction suture is passed from a mini-incision opened with arthroscopy support in the lateral of the joint. This suture allows the prepared allograft to be pulled into the correct position within the joint. The suspended sutures are opened from the posterior and anterior with the tibial tunnel guide which is used during ACL reconstruction. The tunnels are opened with a 4.5 mm endobutton drill to be 60° for the anterior tunnel and 45° for the posterior tunnel. Knotting the suspended sutures together in the endobutton drill slot entry point provides an additional contribution to fixation. The traction suture is removed following intra-articular reduction of the allograft. The procedure is then completed using meniscus suture techniques (all inside, inside-out, outside-in).

14.8 Conclusion

Meniscus defects in young adult patients cause the rapid development of several intra-articular degenerative changes. This leads to several problems such as pain in the joint, disrupted quality of life, and workforce loss. Appropriate treatment of meniscus defects has increasing importance for the reconstruction of biomechanical alignment and prevention of the intra-articular degenerative process. Therefore, depending on the preoperative patient evaluation, meniscus implants in partial defects, and meniscus allograft transplantation in meniscus defects that are extensive and have horn lesions, present pleasing treatment results.

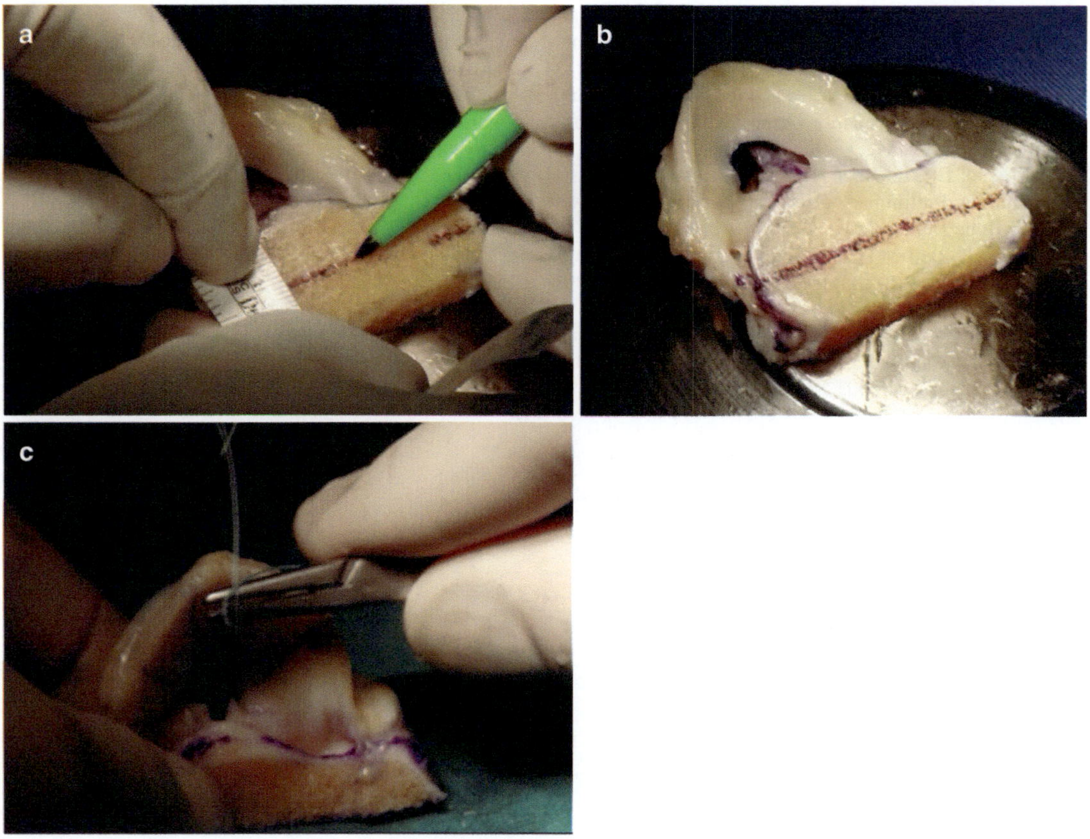

Fig. 14.5 Allograft prepared according to bone bridge technique

References

1. Lubowitz JH, Poehling GG. Save the meniscus. Arthroscopy. 2011;27(3):301–2.
2. Yoon KH, Park KH. Meniscal repair. Knee Surg Relat Res. 2014;26(2):68–76.
3. Stein T, et al. Long-term outcome after arthroscopic meniscal repair versus arthroscopic partial meniscectomy for traumatic meniscal tears. Am J Sports Med. 2010;38(8):1542–8.
4. Montgomery SR, et al. Cross-sectional analysis of trends in meniscectomy and meniscus repair. Orthopedics. 2013;36(8):e1007–13.
5. Papalia R, et al. Meniscectomy as a risk factor for knee osteoarthritis: a systematic review. Br Med Bull. 2011;99:89–106.
6. Rodkey WG, et al. Comparison of the collagen meniscus implant with partial meniscectomy. A prospective randomized trial. J Bone Joint Surg Am. 2008;90(7):1413–26.
7. Bulgheroni P, et al. Follow-up of collagen meniscus implant patients: clinical, radiological, and magnetic resonance imaging results at 5 years. Knee. 2010;17(3):224–9.
8. Grassi A, et al. Clinical outcomes and complications of a collagen meniscus implant: a systematic review. Int Orthop. 2014;38(9):1945–53.
9. Verdonk P, et al. Successful treatment of painful irreparable partial meniscal defects with a polyurethane scaffold: two-year safety and clinical outcomes. Am J Sports Med. 2012;40(4):844–53.
10. De Coninck T, et al. Two-year follow-up study on clinical and radiological outcomes of polyurethane meniscal scaffolds. Am J Sports Med. 2013;41(1):64–72.
11. Krause WR, et al. Mechanical changes in the knee after meniscectomy. J Bone Joint Surg Am. 1976;58(5):599–604.

12. Levy IM, Torzilli PA, Warren RF. The effect of medial meniscectomy on anterior-posterior motion of the knee. J Bone Joint Surg Am. 1982;64(6):883–8.
13. Alford W, Cole BJ. The indications and technique for meniscal transplant. Orthop Clin North Am. 2005;36(4):469–84.
14. van Arkel ER, de Boer HH. Human meniscal transplantation. Preliminary results at 2 to 5-year follow-up. J Bone Joint Surg Br. 1995;77(4):589–95.
15. Noyes FR, Barber-Westin SD. Meniscus transplantation: indications, techniques, clinical outcomes. Instr Course Lect. 2005;54:341–53.
16. Cole BJ, Carter TR, Rodeo SA. Allograft meniscal transplantation: background, techniques, and results. Instr Course Lect. 2003;52:383–96.
17. Noyes FR, Barber-Westin SD. Repair of complex and avascular meniscal tears and meniscal transplantation. J Bone Joint Surg Am. 2010;92(4):1012–29.
18. De Coninck T, et al. Open versus arthroscopic meniscus allograft transplantation: magnetic resonance imaging study of meniscal radial displacement. Arthroscopy. 2013;29(3):514–21.
19. Wang H, et al. Bone plug versus suture-only fixation of meniscal grafts: effect on joint contact mechanics during simulated gait. Am J Sports Med. 2014;42(7):1682–9.
20. Peters G, Wirth CJ. The current state of meniscal allograft transplantation and replacement. Knee. 2003;10(1):19–31.

15 Cartilage Treatment Techniques

Safa Gursoy and Murat Bozkurt

15.1 Introduction

Joint cartilage lesions are seen as a major musculoskeletal system problem that have a negative effect on quality of life and are observed at increasing rates [1]. In a prospective examination of 1000 knee arthroscopies by Hijelle et al., chondral and osteochondral lesions were determined incidentally at the rate of 61%. In a similar series, grade III–IV lesions at least 1 cm^2 in size were determined in 5.3% of patients aged ≤40 years [2]. According to this, it can be said that focal symptomatic cartilage lesions are frequently seen in young patients.

Depending on the age of the cartilage, 65%–80% of the weight is water. In respect of the viscoelasticity of weight-bearing joints and allowing changes under weight-bearing, water is important. Macrofibrils constitute up to 10%–20% of the basic component of the structure, and the majority are type II collagen, providing resistance to tensile forces. Another 10%–20% is formed of proteoglycans. These protein-polysaccharide molecules expressed by chondrocytes are the primary structures providing tissue resistance to compressive forces. Chondrocytes comprise only a 1%–5% part. They are responsible for the synthesis of extracellular matrix components and the regulation of metabolic activity.

Cartilage tissue covering the joint surface has no neurological, vascular, or lymphatic organization. Therefore, the intrinsic healing potential is limited [3]. In an article presented to the Royal Medical Society by the Scottish scientist William Hunter, in 1743, it was stated that, "if all the surgical knowledge is considered from the time of Hippocrates to today, it can be concluded that an ulcerated cartilage will create a highly problematic disease in the future, and when damaged, will never heal" [4].

Symptoms of cartilage lesions are pain, swelling, locking in the joint, stiffness, and sounds from the joint [5, 6]. Just as these symptoms can prevent work and sporting activity, they can also restrict daily activities [7].

Despite current advances in technology and the accumulated knowledge of basic sciences, still problems are remaining in the treatment of cartilage injuries.

15.2 Diagnosis

In patients with joint cartilage damage, the clinical presentation is generally nonspecific. Patients usually have complaints of pain and mechanical symptoms such as lacking and restricted move-

S. Gursoy (✉) · M. Bozkurt
Department of Orthopaedics and Traumatology, Ankara Yildirim Beyazit University, Ankara, Turkey

M. Bozkurt, H. İ. Açar (eds.), *Clinical Anatomy of the Knee*,
https://doi.org/10.1007/978-3-030-57578-6_15

ment. In the physical examination, effusion and sensitivity in the joint line can be determined. Following a detailed examination, radiological evaluations should be made. The radiological techniques often used in the diagnosis of cartilage lesions can be listed as follows:

X-ray: Especially in cases with no subchondral component, the use of X-ray provides extremely limited information. The standard series to be used in the diagnosis of cartilage damage is: standing anteroposterior (AP), lateral, 35°–45° tangential patella, and especially if there is suspicion of OCD, notch (tunnel) radiographs [8]. AP, lateral, Rosenberg, and patella tangential standing radiographs are important in respect of showing early degenerative arthritis findings [9].

Computed tomography arthrography: Spiral 3D tomography is extremely useful in respect of determining the extent of osteochondral damage. By adding arthrography to this technique, it is possible to evaluate fissures and cracks in the upper layer of cartilage [10].

Single-photon emission computed tomography (SPECT): This technique can be used to determine hemostasis and the physiology of subchondral tissue adjacent to treated and untreated cartilage defects [11].

Magnetic resonance imaging (MRI): This is extremely useful in respect of showing the location, size, depth, and extension of the cartilage defect. The most sensitive imaging method in showing cartilage lesions is T1-weighted, fat-suppressed, 3D gradient echo sequences [12]. Previous studies of MR evaluation of cartilage lesions have found sensitivity to be 81%–93%, specificity 91%–97%, and accuracy 91%–97% [13–15].

Scanners with high magnetic field power (1.5 or 3 Tesla) should be used to show cartilage defects. Scanners of 0.2 Tesla are not sufficiently reliable in the visualization of cartilage. Methods such as biochemical MR techniques T2 mapping and dGEMRIC are used in the determination of early stage cartilage damage [16]. Nevertheless, arthroscopy is still the gold standard in the diagnosis and classification of cartilage lesions (Fig. 15.1).

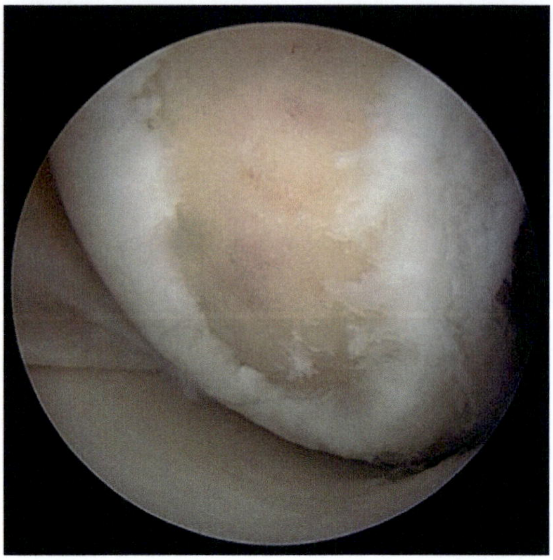

Fig. 15.1 Arthroscopic view of a large chondral lesion

Table 15.1 Outerbridge classification

Grade 0	Cartilage of normal character
Grade 1	Edema and softening
Grade 2	Fibrillation and fragmentation. Fissure formation <½ in.
Grade 3	Fragmentation. Fissure formation >½ in.
Grade 4	Cartilage lesion descending as far as the subchondral bone

Table 15.2 International Cartilage Repair Society (ICRS) classification

Grade 0	Cartilage of normal character
Grade 1	Softening (a) or superficial fissures (b)
Grade 2	Defect of <50% of cartilage thickness
Grade 3	Defect of >50% of cartilage thickness
Grade 4	Cartilage lesion descending as far as the subchondral bone

15.3 Classification

In 1961, Outerbridge developed a classification system according to cartilage abnormalities observed during knee arthrotomy (Table 15.1). With the interest that developed in cartilage damage and repair, the International Cartilage Repair Society (ICRS) developed its own classification system, which is more widely used (Table 15.2).

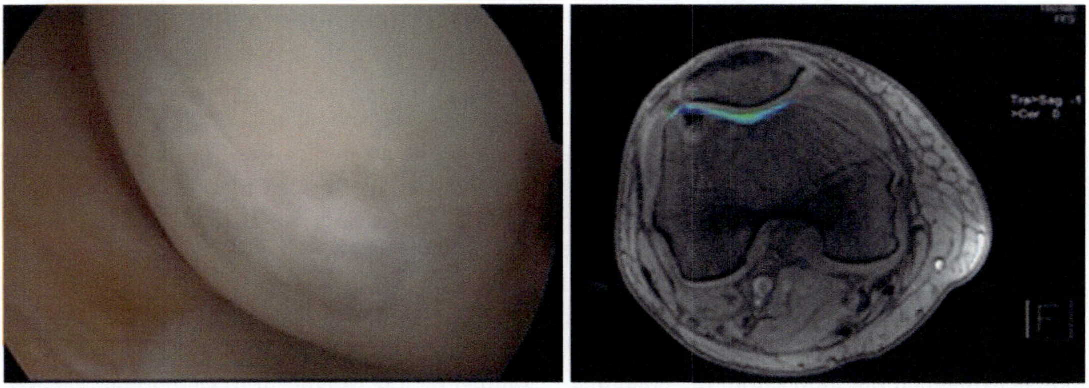

Fig. 15.2 On the left, under control arthroscopy following microfracture, the defect can be seen to have been filled by fibrous cartilage. On the right, with the T2-mapping technique on MRI, donor site morbidity can be seen in the lateral femoral condyle following mosaicplasty

15.4 Treatment

Joint cartilage has poor regeneration potential because of its avascular nature and that the extracellular matrix (ECM) is formed of collagen and proteoglycans. The absence of blood flow limits the healing process by inhibiting inflammatory mediators to be directed to the defect [8, 10]. In addition, the ECM does not allow cellular migration [6]. Consequently, even if chondrocytes try to form a response to tissue damage and fill the defect, complete repair of the damaged cartilage cannot be provided [17]. Over time, these lesions become symptomatic and in the long term, cause cartilage erosion and osteoarthritis [16, 18]. Therefore, when cartilage damage is determined, treatment in the early stage of damage is important in respect of eliminating the symptoms, regaining joint functions and preventing the development of osteoarthritis.

In the treatment of a symptomatic patient with full-thickness cartilage damage, the cartilage formed as a result of the ideal surgical treatment method should have the following properties: there must be hyaline in the cartilage structure, the defect area must be completely filled, there must be full integration of the cartilage tissue with the surroundings, and it must have a long life and normal mechanical properties.

As a result of acquired experience, various treatment algorithms have been developed, taking into consideration parameters such as the age and expectations of the patient and the size of the defect.

Although there is an ongoing debate about which surgical treatment method can be used on which patients, the current treatment options are abrasion chondroplasty, bone marrow stimulation technique (microfracture and drilling), mosaicplasty—osteochondral autograft transfer system (OATS), osteochondral allografts, and cell-based treatment methods such as autologous chondrocyte implantation (ACI) and matrix-induced autologous chondrocyte implantation (MACI). Although these techniques have their own specific problems (Fig. 15.2), there is evidence showing that they are effective in restoring knee functions [19–31].

15.4.1 Bone Marrow Stimulating Technique

The bone marrow stimulating technique was first described by Kenneth Pridie in 1959 [32]. However, unwanted results have been shown to emerge from the drilling used in this technique and the abrasion methods described below [33].

Currently, the most frequently used bone marrow stimulation technique is the microfracture method described by Steadman, who reported 10-year successful results from a group applied with the same technique [34]. In this technique, which can be applied arthroscopically, following debridement of the cartilage area until healthy cartilage tissue is reached, it is aimed to reach the bone marrow by opening holes of 3–4 mm with an awl of 2 mm diameter angled to the subchondral bone.

This is based on the principle of pluripotent mesenchymal cell migration from the bone marrow to the damaged area with the penetration of the subchondral bone plate. The tissue expected to be formed at the end of the treatment is type 1 collagen-weighted fibrous cartilage (Fig. 15.3a–e). In the long term, this becomes a poor, worn tissue which worsens clinical results [34]. Previous studies have shown that the microfracture technique gives better results in relatively small lesions [34], in patients aged <40 years and in those with higher ICRS and Cincinnati scores and better MRI data [35].

15.4.2 Mosaicplasty-Osteochondral Autograft Transfer System (OATS)

Although the OATS technique was first described in 1993, the mosaicplasty method in which small osteochondral grafts are aligned within the defect, was defined in 1997 by Hangody et al. [21], and the same team then popularized its use [36].

The method is based on the placement in the defect area of osteochondral plugs taken from the weight-bearing area of the knee. In this technique, which can be applied arthroscopically or with mini-open techniques, the diameter of the autograft to be taken is determined by the shape and diameter of the cartilage defect. The advantages of mosaicplasty are that it is a single-stage treatment and the cartilage formed is hyaline cartilage. Poor integration of the graft with surrounding tissue and donor site morbidity are the main disadvantages [37] (Fig. 15.4a–f).

15.4.3 Osteochondral Allografts

Just as in the OATS technique, bone loss accompanying wide and deep osteochondral cartilage damage is an indication for treatment. The advantages of the OATS technique of osteochondral allografts are that there is no donor site morbidity, and it can be used in very wide defects.

The clinical use of this technique, which was developed for the treatment of large defects, is limited by low suitable graft support and the risk of contagious disease [38].

When current literature is examined, there are many studies that have shown that extremely successful results can be obtained from the application of the allograft technique in primary lesions or in cases where previous treatments have been unsuccessful or when there are concomitant pathologies [39–41].

15.4.4 Cell-Based Treatments

Cell-based treatments can be evaluated as three generations. First-generation treatment is autologous chondrocyte implantation (ACI) in which cultured autologous cells are placed in the defect area, and closure is with autologous periosteal cover [7, 42, 43]. Despite many reports about ACI, because of difficulties in the surgical technique, that it is extremely invasive, costly, and complications related to periosteal coverage in particular, second-generation treatments which use collagen coverage were introduced as the collagen-covered autologous chondrocyte implantation technique (CACI) [6, 44, 45]. These second-generation treatments provide similar results to those of first-generation treatments, but hypertrophy complications continue.

With recent innovations and advances in basic sciences, third-generation treatments using biodegradable covers were developed to resolve these problems. This technique is the matrix-induced autologous chondrocyte implantation technique (MACI).

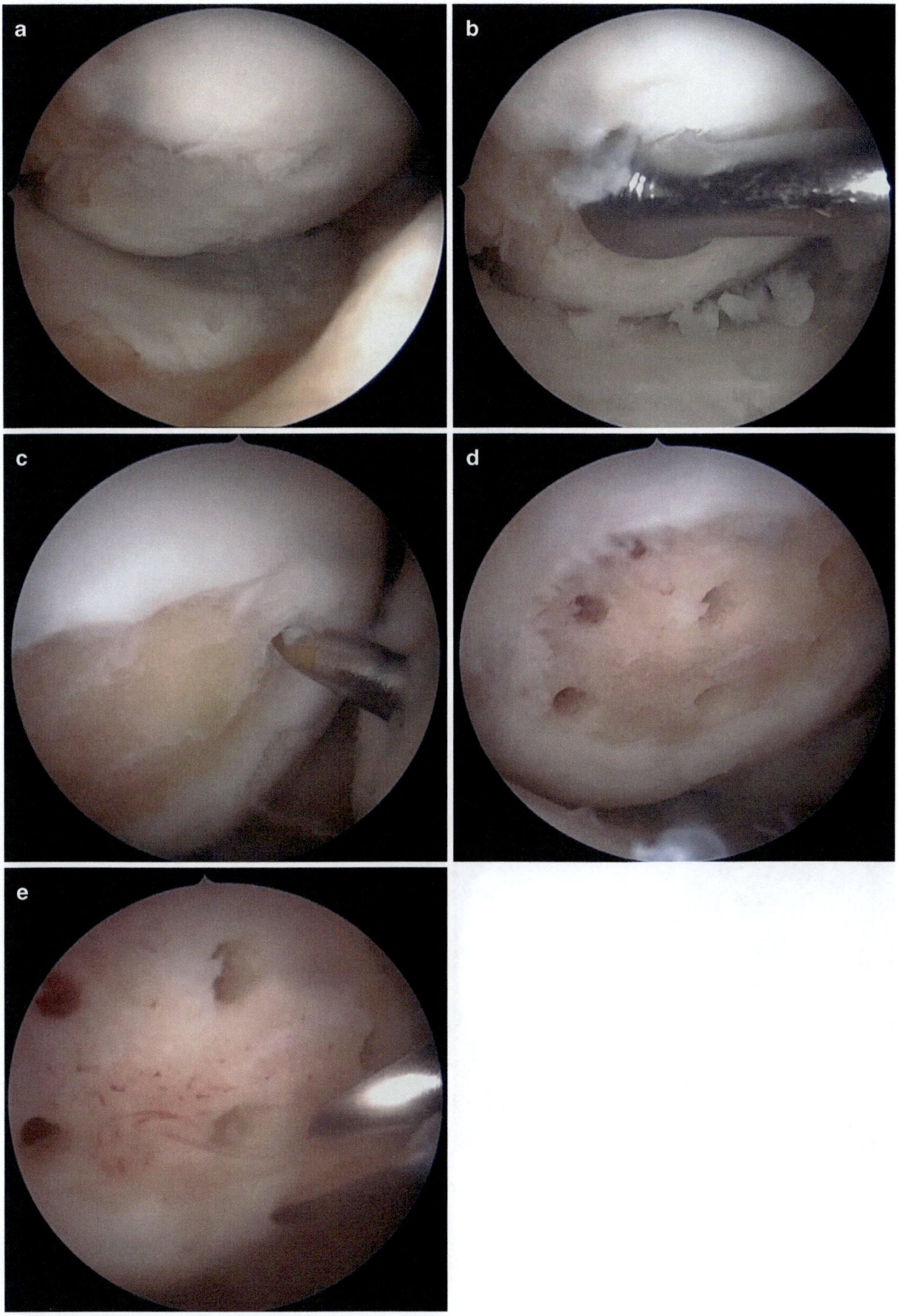

Fig. 15.3 (**a**–**e**) Step-by-step surgical technique of an arthroscopic microfracture technique

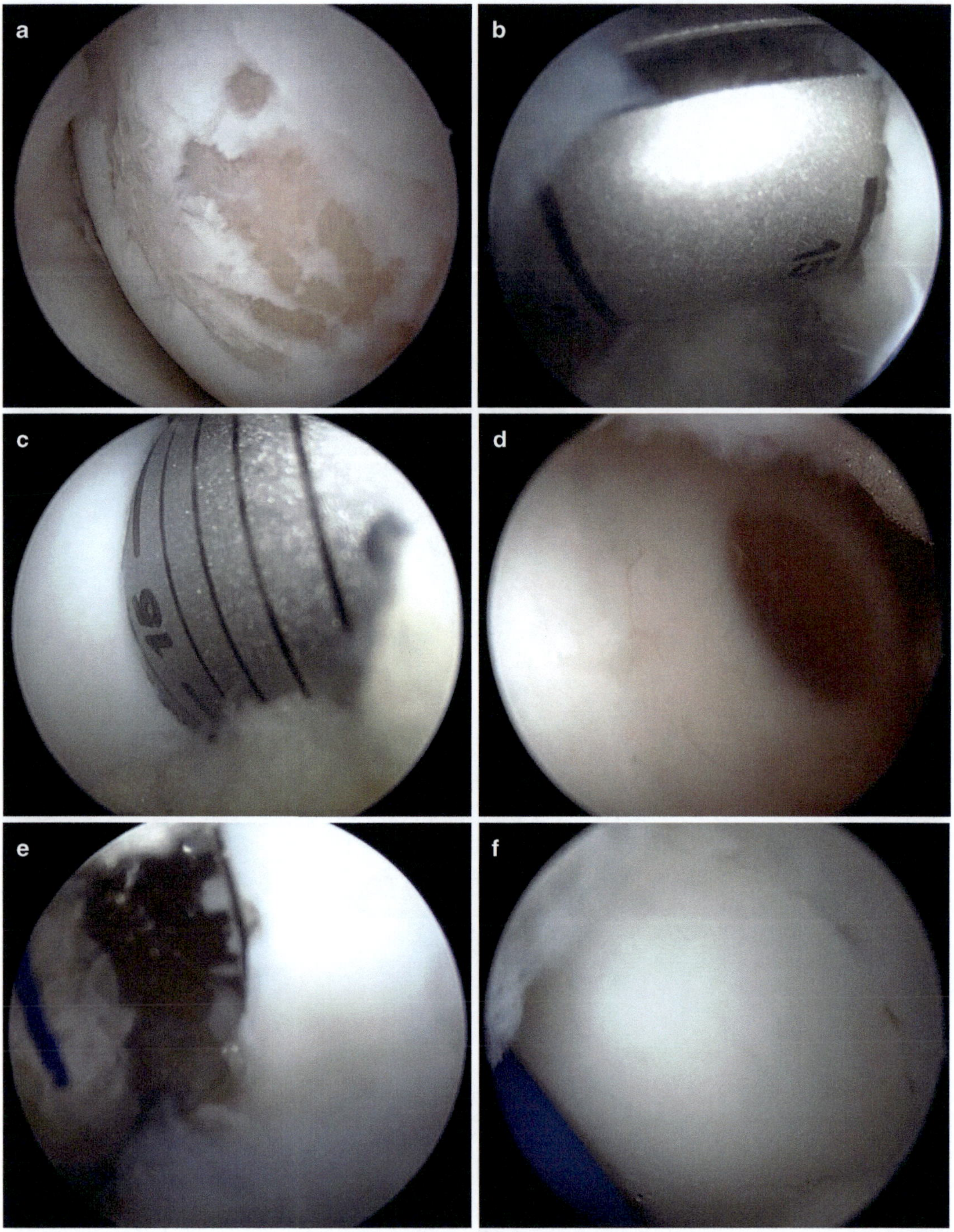

Fig. 15.4 (**a**–**f**) Step-by-step surgical technique of an arthroscopic OATS technique

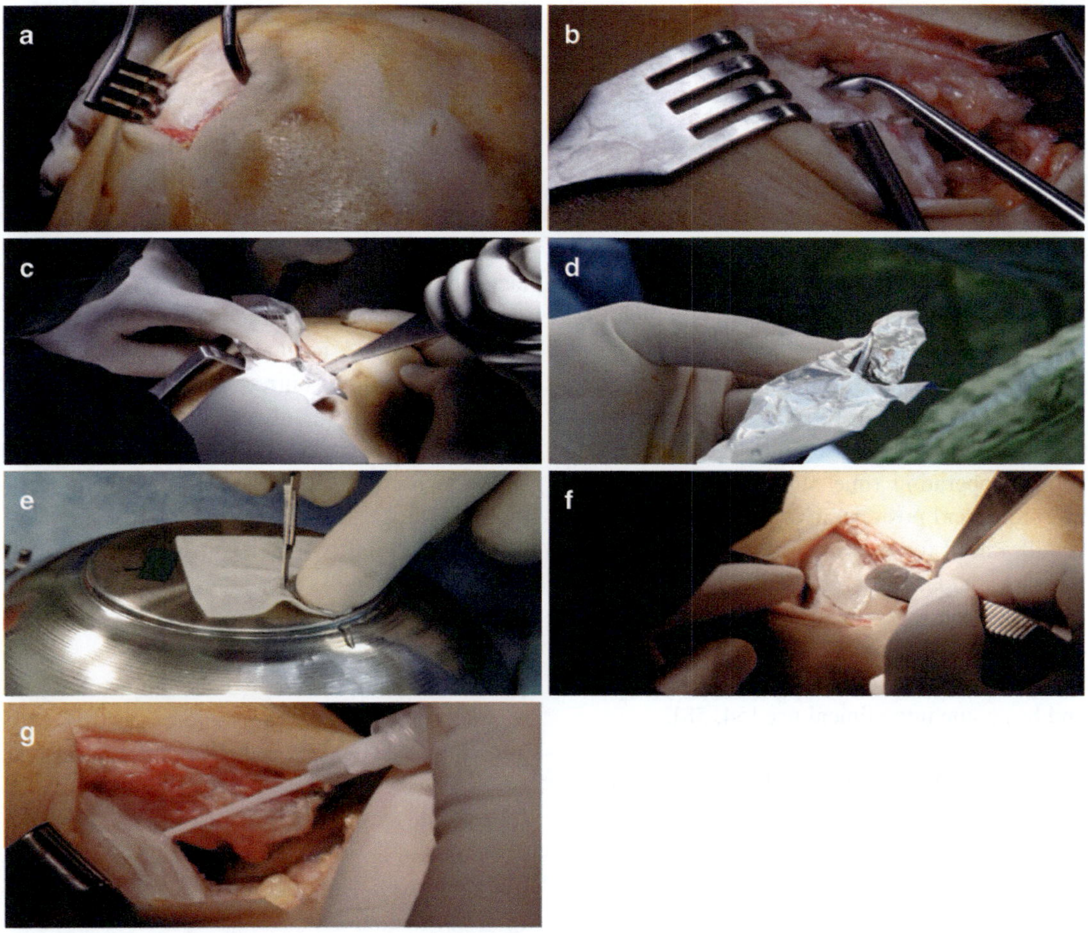

Fig. 15.5 (**a–g**) Step-by-step surgical technique of a matrix-induced autologous chondrocyte implantation technique

15.4.4.1 Autologous Chondrocyte Implantation (ACI)

The idea of culture and proliferation of autologous chondrocytes and re-implanting them to cartilage tissue beneath a periosteal membrane was first introduced by the Swedish doctor, Lars Peterson, in 1970. Studies of the method on a rabbit model were published by Grande et al. in 1987 [46]. The efficacy of the ACI technique in deep lesions was shown in the first patient series, comprising 23 patients by Brittberg, Peterson et al. in 1994 [19]. The surgical technique of ACI is composed of two stages.

During arthroscopic intervention, the localization and size of ICRS grade III–IV lesions should be recorded. For a patient candidate for ACI, a full-thickness cartilage biopsy is taken with a sharp curette from the non-weight-bearing medial or lateral edges of the trochlea in the affected joint or from around the intercondylar notch, including deep layers of cartilage, of 200–300 mg total weight, and is placed in transport solution [5, 7, 47, 48].

In the second session, after cell culture for 4–6 weeks, medial or lateral parapatellar mini-arthrotomy is applied depending on the defect localization. The defect area is debrided as far as the borders of healthy vertical cartilage. The periosteal cover taken from over the medial cortex of the proximal tibia at a size appropriate to the defect is sutured to the healthy cartilage with 6/0 vicryl sutures at 3 mm intervals. Solution containing mean 4.5 million cultured cells is injected into this area [49, 50].

The use of periosteal coverage in ACI, as a first-generation, conventional, cell-based therapy, has been reported to lead to various complications [19, 51, 52]. These include periosteal hypertrophy, loosening in the periosteal flap, ablation, and cell loss to the joint cavity [53, 54].

ACI-C, as a second-generation autologous chondrocyte implantation technique, which uses collagen membranes rather than periosteal graft, has obtained better results, and no periosteal hypertrophy has developed with this technique [55, 56]. Although the use of this technique shortens the operating time and reduces donor site morbidity and postoperative pain [55, 57], there is still a need for sutures, and the success of cell transport capacity is debatable. To overcome these problems, the matrix-induced autologous chondrocyte implantation technique (MACI), which uses cell-seeded scaffolds, was developed and has come into clinical use [54, 58].

15.4.4.2 Matrix-Induced Autologous Chondrocyte Implantation (MACI)

The MACI technique, which uses tissue engineering techniques, is the implantation to the debrided defect area of scaffolds formed of cultured autologous chondrocytes planted in a three-dimensional biodegradable matrix. There are several different forms of second-generation autologous chondrocyte implantation, produced in membrane and gel form. Both the different forms have their own advantages and disadvantages. The gel forms have been shown to provide better cell distribution, but the mechanical support is insufficient [59].

The tissue engineering techniques used in production are based on three basic factors: cells, matrix, and growth factors. Matrixes of cells forming supportive covers and which are kept within the defect form the basis of tissue engineering [60]. The first and the most commonly used matrixes are those formed of type I and type III collagen, which aim to provide equal distribution of cells [61]. The use of type II collagen rather than type I/III collagen has been shown to increase the incidence of arthritis and cartilage collapse [62–64]. There are two different surface structures of type I/III collagen membrane used for MACI. One is a more dense and smoother surface which is facing the joint during implantation, preventing leakage of the planted cells into the joint cavity. The other is a rough surface, richer in cells with wider pore intervals, which faces the subchondral side [65–67].

Indications include patient age of 18–55 years, defect >4 cm^2, normal BMI, focal lesions, impaired normal or corrected alignment, normal or corrected ligament stability, and healthy meniscus structure. Contra-indications for MACI are widespread cartilage defects, septic or rheumatoid arthritis, impaired uncorrected alignment, ligament instability, patella instability, and total meniscectomy [68].

Surgical Technique

The first stage of the surgical technique of MACI is arthroscopic evaluation and the taking of a biopsy, as for ACI. The biopsy taken in the manner described above is sent to the laboratory with appropriate transport procedures. With tissue engineering, approximately 12 million cells proliferate from 100,000 to 200,000 cells in a period of 6–8 weeks, and a three-dimensional cell-cover form is created by inoculating the formed scaffolds, so that there will be one million cells per square centimeter [57, 69].

In the second stage, a mini-arthrotomy is applied according to the location and size of the defect area. Debridement is performed with a closed curette until stable cartilage tissue is reached around the defect area and vertical angled edges are obtained. Following the debridement, the defect area is measured, and a template is made from aluminum foil or a similar material. The MACI implant is cut according to the template and is implanted with fibrin glue to the smooth cellular surface, so that it will face the subchondral bone [57, 66, 70] (Fig. 15.5a–g).

After a few minutes, when fixation is ensured, the stability is checked by moving the knee in flexion and extension [70–72]. If necessary, it can be supported with sutures. When ACI and MACI are compared in respect of surgical technique,

MACI can be said to be a less-invasive technique [70, 73]. As there is no periosteal or collagen cover in MACI, there is no need for microsuturation [74]. In a previous study that compared the operating times of ACI and MACI, the operating time of MACI was found to be mean 19.2 min shorter [45].

Although arthroscopic applications of MACI have been reported, which can be considered as early experience, advances are continuing [70, 75].

For the application of MACI, there should be no impairment of lower extremity alignment. The ACL and other ligamentous structures should be intact, >50% meniscus volume should be preserved, and there must be healthy subchondral bone. In cases of misalignment, the application can be made after correction with femoral or tibial osteotomies. ACL reconstruction should be performed before implantation, or in the same session, the meniscus should be sutured, or if the volume is <50%, integrity should be provided with allografts or synthetic meniscus implants.

The presence of healthy subchondral bone is extremely important for the application of MACI. When there is bone marrow edema before implantation, retrograde or antegrade drilling can be performed. In lesions deeper than 6–8 mm, the defect should be filled as far as the subchondral border with autograft taken from the iliac wing before implantation or in the same session [76]. In addition, the sandwich technique, which uses two layers of membrane, can be used in deep lesions. In this technique, the first membrane is placed on the rough, cell-enriched surface facing the joint, and the second membrane is placed on the rough surface facing the subchondral bone [48].

Clinical Results

Several studies showing the clinical results of the MACI technique have compared it with other treatment methods or presented the data of case series. These studies have shown that significant improvements in pain, activity, and function have been provided by MACI treatment through the use of instruments such as the Gillquist scale [5, 57, 75, 77], the Cincinnati scale, modified Cincinnati scale [44, 48, 57], Tegner–Lysholm scale [5, 57, 75, 77], VAS pain score [5, 47, 70], International Knee Documentation Classification (IKNC) scale [5, 75, 77], and the American Orthopedic Foot and Ankle Society clinical function scale [71].

As a result of arthroscopic examinations of patients applied with the treatment, the repair tissue formed has been shown to be a healthy cartilage structure, which is robust when examined with a probe, stable and integrated with the surrounding tissue. These studies have been supported by good results in MRI studies which have shown restoration of the joint surface with the tissue formed, filling of the defect area, and integration with surrounding tissue [47, 48, 77].

In addition, in histological examinations performed on the tissue at 6 months after MACI treatment, the tissue formed has been determined to be largely hyaline cartilage [58, 78]. In biopsies repeated at future times, the ratio of hyaline cartilage was seen to have increased, and no fibrous cartilage was determined [79].

Complications

Early stage complications have been reported as implant retention problems, superficial infection, deep vein thrombosis, and hemarthrosis. However, the rate of 26% hypertrophy in ACI has been reported to be significantly decreased in MACI [47]. Complications related to both the techniques have been significantly associated with arthrotomy [80].

References

1. Doral MN, Dönmez G, Atay OA, et al. Dejeneratif eklem hastalıkları. TOTBİD Dergisi. 2007;6:56–65.
2. Hijelle K, Solheim E, Strand T, Muri R, Brittberg M. Articular cartilage defects in 1000 knee arthroscopies. Arthroscopy. 2002;18:730–4.
3. Bhosale AM, Richardson JB. Articular cartilage: structure, injuries and review of management. Br Med Bull. 2008;87:77–95.
4. Hunter W. Of the structure and diseases of articular cartilages. 1743. Clin Orthop Relat Res. 1995;317:3–6.
5. D'Anchise R, Manta N, Prospero E, Bevilacqua C, Gigante A. Autologous implantation of chondrocytes on a solid collagen scaffold:clinical and histological

outcomes after two years of follow-up. J Orthop Traumatol. 2005;6:36–43.
6. Haddo O, Mahroof S, Higgs D, David L, Pringle J, Bayliss M, et al. The use of chondrogide membrane in autologous chondrocyte implantation. Knee. 2004;11(1):51–5.
7. Micheli LJ, Moseley JB, Anderson AF, Browne JE, Erggelet C, Arciero R, et al. Articular cartilage defects of the distal femur in children and adolescents: treatment with autologous chondrocyte implantation. J Pediatr Orthop. 2006;26(4):455–60.
8. Brittberg M. Evaluation of cartilage injuries and repair. J Bone Joint Surg Am. 2003;85-A(Suppl 2):58–69.
9. Rosenburg TD, Paulos LE, Parker RD, et al. The 45° posteroanterior flexion weight bearing radiograph of the knee. J Bone Joint Surg Am. 1988;70A:1479–83.
10. Oakley SP, Portek I, Szomor Z, Turnbull A, Murrell GA, Kirkham BW, Lassere MN. Accuracy and reliability of arthroscopic estimates of cartilage lesion size in a plastic knee simulation model. Arthroscopy. 2003;19(3):282–9.
11. Tiderius CJ, Svensson J, Leander P, Ola T, Dahlberg L. dGEMRIC (delayed gadolinium-enhanced MRI of cartilage) indicates adaptive capacity of human knee cartilage. Magn Reson Med. 2004;51(2):286–90.
12. Chung CB, Frank LR, Resnick D. Cartilage imaging techniques: current clinical applications and state of the art imaging. Clin Orthop. 2001;391S:S370–8.
13. Disler DG, McCauley TR, Kelman CG, et al. Fat-suppressed three-dimensional spoiled gradient-echo MR imaging of hyaline cartilage defects in the knee: comparison with standard MR imaging and arthroscopy. AJR Am J Roentgenol. 1996;167:127–32.
14. Recht MP, Piraino DW, Paletta GA, Schils JP, Belhobek GH. Accuracy of fat- three-dimensional spoiled gradient-echo FLASH MR imaging in the detection of patellofemoral articular cartilage abnormalities. Radiology. 1996;198:209–12.
15. Disler DG, McCauley TR, Wirth CR, Fuchs MD. Detection of knee hyaline cartilage defects using fat-suppressed three-dimensional spoiled gradient echo MR imaging: comparison with standard MR imaging and correlation with arthroscopy. AJR Am J Roentgenol. 1995;165:377–82.
16. Welsch GH, Mamisch TC, Zak L, Mauerer A, Apprich S, Stelzeneder D, et al. Morphological and biochemical T2 evaluation of cartilage repair tissue based on a hybrid double echo at steady state (DESS-T2d) approach. J Magn Reson Imaging. 2011;34(4):895–903.
17. Gold GE, Burstein D, Dardzinski B, Lang P, Boada F, Mosher T. MRI of articular cartilage in OA: novel pulse sequences and compositional/functional markers. Osteoarthr Cartil. 2006;14:76–86.
18. Potter HG. Magnetic resonance imaging of articular cartilage in the knee. An evaluation with use of fast-spin-echo imaging. J Bone Joint Surg Am. 1998;80(9):1276–84.
19. Brittberg M, Lindhal A, Nilsson A, Ohlsson C, Isaksson O, Peterson L. Treatment of deep cartilage defects in the knee with autologous chondrocyte transplantation of the knee. N Engl J Med. 1994;331:889–95.
20. Peterson L, Minas T, Brittberg M, Nilsson A, Sjogren-Jansson E, Lindhal A. Two- to 9-year outcome after autologous chondrocyte transplantation of the knee. Clin Orthop. 2000;374:212–34.
21. Hangody L, Kish G, Karpati Z, Udvarhelyi I, Szigeti I, Bely M. Mosaicplasty for the treatment of articular cartilage defects: application in clinical practice. Orthopedics. 1998;21:751–6.
22. Johnson L. Clinical methods of cartilage repair. Arthroscopic abrasion arthroplasty. A review. Clin Orthop. 2001;391S:306–17.
23. Rodrigo J, Steadman J, Silliman J. Improvement of full-thickness chondral defect healing in the human knee after debridement and microfracture using continuous passive motion. Am J Knee Surg. 1994;7:109–16.
24. Rae P, Noble J. Arthroscopic drilling of osteochondral lesions of the knee. J Bone Joint Surg Br. 1989;71B:534.
25. Bert J. Role of abrasion arthroplasty and debridement in the management of osteoarthritis of the knee. Rheum Dis Clin North Am. 1993;19:725–39.
26. Steadman J, Rodkey W, Rodrigo J. Microfracture: surgical technique and rehabilitation to treat chondral defects. Clin Orthop. 2001;391(suppl):S362–9.
27. Mithoefer K, Williams R, Warren R, et al. The microfracture technique for the treatment of articular cartilage lesions in the knee. A prospective cohort study. J Bone Joint Surg. 2005;87A:1911–20.
28. Meyers M, Akeson W, Convery F. Resurfacing of the knee with fresh osteochondral allograft. J Bone Joint Surg. 1989;71A:704–13.
29. Minas T, Chiu R. Autologous chondrocyte implantation. Am J Knee Surg. 2000;13:41–50.
30. Bugbee W, Convery F. Osteochondral allograft transplantation. Clin Sports Med. 1999;18:67–75.
31. Brittberg M, Tallheden T, Sjogren-Jansson B, Lindhal A, Peterson L. Autologous chondrocytes used for articular cartilage repair: an update. Clin Orthop. 2001;391(suppl):S337–48.
32. Pridie KH, Gordon G. A method of resurfacing osteoarthritic knee joints. J Bone Joint Surg. 1959;41(3):618–9.
33. Bert JM, Maschka K. The arthroscopic treatment of unicompartmental gonarthrosis: a five-year follow-up study of abrasion arthroplasty plus arthroscopic debridement and arthroscopic debridement alone. Arthroscopy. 1989;5(1):25–32.
34. Steadman JR, Briggs KK, Rodrigo JJ, Kocher MS, Gill TJ, Rodkey WG. Outcomes of microfracture for traumatic chondral defects of the knee: average 11-year follow-up. Arthroscopy. 2003;19(5):477–84.
35. Kreuz PC, Erggelet C, Steinwachs MR, et al. Is microfracture of chondral defects in the knee associated

with different results in patients aged 40 years or younger? Arthroscopy. 2006;22:1180–6.
36. Hangody L, Kish G, Kárpáti Z, Szerb I, Eberhardt R. Treatment of osteochondritis dissecans of the talus: use of the mosaicplasty technique-a preliminary report. Foot Ankle Int. 1997;18:628–34.
37. Iwasaki N, Kato H, Kamishima T, Suenaga N, Minami A. Donor site evaluation after autologous osteochondral mosaicplasty for cartilaginous lesions of the elbow joint. Am J Sports Med. 2007;35:2096–100.
38. Lightfoot A, Martin J, Amendola A. Fluorescent viability stains overestimate chondrocyte viability in osteoarticular allografts. Am J Sports Med. 2007;35:1817–23.
39. Abrams GD, Hussey KE, Harris JD, Cole BJ. Clinical results of combined meniscus and femoral osteochondral allograft transplantation: minimum 2-year follow-up. Arthroscopy. 2014;30(8):964–970.e1.
40. Briggs DT, Sadr KN, Pulido PA, Bugbee WD. The use of osteochondral allograft transplantation for primary treatment of cartilage lesions in the knee. Cartilage. 2015;6(4):203–7.
41. Gracitelli GC, Meric G, Briggs DT, et al. Fresh osteochondral allografts in the knee: comparison of primary transplantation versus transplantation after failure of previous subchondral marrow stimulation. Am J Sports Med. 2015;43(4):885–91.
42. Browne JE, Anderson AF, Arciero R, et al. Clinical outcome of autologous chondrocyte implantation at 5 years in US subjects. Clin Orthop Relat Res. 2005;436:237–45.
43. Farr J. Autologous chondrocyte implantation improves patellofemoral cartilage treatment outcomes. Clin Orthop Relat Res. 2007;463:187–94.
44. Amin AA, Bartlett W, Gooding CR, et al. The use of autologous chondrocyte implantation following and combined with anterior cruciate ligament reconstruction. Int Orthop. 2006;30:48–53.
45. Bartlett W, Krishnan SP, Skinner JA, Carrington RWJ, Briggs TWR, Bentley G. Collagen-covered versus matrix-induced autologous chondrocyte implantation for osteochondral defects of the knee: a comparison of tourniquet times. Eur J Orthop Surg Traumatol. 2006;16:315–7.
46. Grande DA, Singh IJ, Pugh J. Healing of experimentally produced lesions in articular cartilage following chondrocyte transplantation. Anat Rec. 1987;218(2):142–8.
47. Bartlett W, Skinner JA, Gooding CR, Carrington RW, Flanagan AM, Briggs TW, et al. Autologous chondrocyte implantation versus matrix-induced autologous chondrocyte implantation for osteochondral defects of the knee: a prospective, randomised study. J Bone Joint Surg Br. 2005;87(5):640–5.
48. Bartlett W, Gooding CR, Carrington RW, Skinner JA, Briggs TW, Bentley G. Autologous chondrocyte implantation at the knee using a bilayer collagen membrane with bone graft: a preliminary report. J Bone Joint Surg Br. 2005;87(3):330–2.
49. Ochi M, Uchio Y, Kawasaki K, Wakitani S, Iwasa J. Transplantation of cartilage-like tissue made by tissue engineering in the treatment of cartilage defects of the knee. J Bone Joint Surg Br. 2002;84:571–8.
50. Bentley G, Biant LC, Carrington RW, Akmal M, Goldberg A, Williams AM, et al. A prospective, randomised comparison of autologous chondrocyte implantation versus mosaicplasty for osteochondral defects in the knee. J Bone Joint Surg Br. 2003;85:223–30.
51. King PJ, Bryant T, Minas T. Autologous chondrocyte implantation for chondral defects of the knee: indications and technique. J Knee Surg. 2002;15:177–84.
52. Minas T, Nehrer S. Current concepts in the treatment of articular cartilage defects. Orthopedics. 1997;20:525–38.
53. Driesang IM, Hunziker EB. Delamination rates of tissue flaps in articular cartilage repair. J Orthop Res. 2000;18:909–11.
54. Marcacci M, Kon E, Zaffagnini S, Filardo G, Delcogliano M, Neri MP, et al. Arthroscopic second generation autologous chondrocyte implantation. Knee Surg Sports Traumatol Arthrosc. 2007;15(5):610–9.
55. Briggs TW, Mahroof S, David LA, Flannelly J, Pringle J, Bayliss M. Histological evaluation of chondral defects after autologous chondrocyte implantation of the knee. J Bone Joint Surg Br. 2003;85:1077–83.
56. Erggelet C, Kreuz PC, Mrosek EH, Schagemann JC, Lahm A, Ducommun PP, et al. Autologous chondrocyte implantation versus ACI using 3D- graft for the treatment of large full-thickness cartilage lesions of the knee. Arch Orthop Trauma Surg. 2009;130:957–64. [Epub ahead of print].
57. Cherubino P, Grassi FA, Bulgheroni P, Ronga M. Autologous chondrocyte implantation using a bilayer collagen membrane. J Orthop Surg. 2003;11:10–5.
58. Iwasa J, Engebretsen L, Shima Y, Ochi M. Clinical application of scaffolds for cartilage tissue engineering. Knee Surg Sports Traumatol Arthrosc. 2009;17(6):561–77.
59. Gavénis K, Schmidt-Rohlfing B, Mueller-Rath R, Andereya S, Schneider U. In vitro comparison of six different matrix systems for the cultivation of human chondrocytes. In Vitro Cell Dev Biol Anim. 2006;42:159–67.
60. Coutts RD, Healey RM, Ostrander R, Sah RL, Goomer R, Amiel D. Matrices for cartilage repair. Clin Orthop Relat Res. 2001;391(Suppl):S271–9.
61. Gikas PD, Bayliss L, Bentley G, Briggs TW. An overview of autologous chondrocyte implantation. J Bone Joint Surg Br. 2009;91:997–1006.
62. Courtenay JS, Dallman MJ, Dayan AD, Martin A, Mosedale B. Immunisation against heterologous type II collagen induces arthritis in mice. Nature. 1980;283:666–8.
63. Kim W-U, Cho M-L, Jung YO, et al. Type II collagen autoimmunity in rheumatoid arthritis. Am J Med Sci. 2004;327:202–11.

64. Trentham DE, Townes AS, Kang AH. Autoimmunity to type II collagen: an experimental model of arthritis. J Exp Med. 1977;146:857–68.
65. Ehlers E-M, Fuss M, Rohwedel J, Russlies M, Kuhnel W, Behrens P. Development of a biocomposite to fill out articular cartilage lesions: light, scanning and transmission electron microscopy of sheep chondrocytes cultured on a collagen I/III sponge. Ann Anat. 1999;181:513–8.
66. Gigante A, Bevilacqua C, Ricevuto A, Mattioli-Belmonte M, Greco F. Membrane-seeded autologous chondrocytes: cell viability and characterization at surgery. Knee Surg Sports Traumatol Arthrosc. 2007;15:88–92.
67. Russlies M, Behrens P, Wünsch L, Gille J, Ehlers E-M. A cell-seeded biocomposite for cartilage repair. Ann Anat. 2002;184:317–23.
68. Anders S, Schaumburger J, Schubert T, Grifka J, Behrens P. Matrix-associated autologous chondrocyte transplantation (MACT). Minimally invasive technique in the knee. Oper Orthop Traumatol. 2008;20:208–19.
69. Manfredini M, Zerbinati F, Gildone A, Faccini R. Autologous chondrocyte implantation: a comparison between an open periosteal-covered and an arthroscopic matrix-guided technique. Acta Orthop Belg. 2007;73:207–18.
70. Abelow SP, Guillen P, Ramos T. Arthroscopic technique for matrixinduced autologous chondrocyte implantation for the treatment of large chondral defects in the knee and ankle. Oper Tech Orthop. 2006;16:257–61.
71. Ronga M, Grassi FA, Montoli C, Bulgheroni P, Genovese E, Cherubino P. Treatment of deep cartilage defects of the ankle with matrix-induced autologous chondrocyte implantation (MACI). Foot Ankle Surg. 2005;11:29–33.
72. Marlovits S, Striessnig G, Kutscha-Lissberg F, et al. Early postoperative adherence of matrix-induced autologous chondrocyte implantation for the treatment of full-thickness cartilage defects of the femoral condyle. Knee Surg Sports Traumatol Arthrosc. 2005;13:451–7.
73. Bachmann G, Basad E, Lommel D, Steinmeyer J. MRI in the followup of matrix-guided autologous chondrocyte implantation (MACI and microfracture). Radiologe. 2004;44:773–82.
74. Willers C, Partsalis T, Zheng M-H. Articular cartilage repair: procedures versus products. Expert Rev Med Devices. 2007;4:373–92.
75. Ronga M, Grassi FA, Bulgheroni P. Arthroscopic autologous chondrocyte implantation for the treatment of a chondral defect in the tibial plateau of the knee. Arthroscopy. 2004;20:79–84.
76. Minas T, Peterson L. Advanced techniques in autologous chondrocyte transplantation. Clin Sports Med. 1999;18:13–44.
77. Behrens P, Bitter T, Kurz B, Russlies M. Matrix-associated autologous chondrocyte transplantation/implantation (MACT/MACI): 5-year follow-up. Knee. 2006;13:194–202.
78. Benya PD, Shaffer JD. Dedifferentiated chondrocytes re-express the differentiated collagen phenotype when cultured in agarose gels. Cell. 1982;30:215–24.
79. Brun P, Dickinson SC, Zavan B, Cortivo R, Hollander AP, Abatangelo G. Characteristics of repair tissue in second- look and third-look biopsies from patients treated with engineered cartilage: relationship to symptomatology and time after implantation. Arthritis Res Ther. 2008;10:R132.
80. Erggelet C, Sittinger M, Lahm A. The arthroscopic implantation of autologous chondrocytes for the treatment of full-thickness cartilage defects of the knee joint. Arthroscopy. 2003;19(1):108–10.

Posterior Knee Arthroscopy

16

Murat Bozkurt, Mustafa Akkaya, and Halil İbrahim Açar

16.1 Introduction

Knee arthroscopy is performed increasingly every day and constitutes an important place among the orthopedic surgical procedures. Significant improvement has been achieved in evaluating the compartments and treatment procedures with the increase in technical possibilities. However, posterior compartment of the knee, which is referred to as the "blind area," and treatments that will be applied to this area are not possible to be evaluated with standard anterior knee arthroscopy. Particularly, examination of the areas that are considered as the site of arthrosis onset in the knee joint such as posterior femoral condyle as well as the posterior roots of the meniscus and posterior cruciate ligament (PCL) is only possible with techniques such as posterior knee arthroscopy. In this chapter, we mention technical information and fields of application concerning posterior knee arthroscopy.

M. Bozkurt (✉) · M. Akkaya
Department of Orthopaedics and Traumatology, Faculty of Medicine, Ankara Yildirim Beyazit University, Ankara, Turkey

H. İ. Açar
Department of Anatomy, Faculty of Medicine, Ankara University, Ankara, Turkey

16.2 Background

Posterior aspect of the knee joint has always been considered mysterious and out of bounds due to the existing neurovascular structures. However, encouraging developments took place with arthroscopic imaging of the posteromedial compartment, which was initially performed by Gillquist et al. on 1232 patients in 1979 [1]. Then, Lewicky et al. described posterior knee arthroscopy in 1982 [1]. The technique they described was based on marking out the position of the portal that will be used in posterior knee arthroscopy after inserting a guide from the anteromedial portal (AMP). This method can be applied using standard arthroscopic hand instruments and imaging systems. It was stated that this method could be used for both posteromedial and posterolateral areas [1].

Today, the "transseptal" portal technique, which was described by Ahn et al. in 2000, is frequently used in imaging for posterior knee arthroscopy [2]. However, considering the fact that many specific pathologies of the knee joint are observed in the medial compartment, it is possible to say that isolated posteromedial knee arthroscopy will gain popularity day by day. In this chapter, we also provide information about the double posteromedial portal technique.

M. Bozkurt, H. İ. Açar (eds.), *Clinical Anatomy of the Knee*,
https://doi.org/10.1007/978-3-030-57578-6_16

16.3 Posterior Anatomy of the Knee

16.3.1 Osseous Structures

Evaluating the posterior aspect of the knee joint, bone structures of the femur, tibia, and fibula are important in terms of the basic anatomy of this area. Distal femur is comprised of lateral and medial femoral condyles (condylus lateralis et medialis). Both the condyles are separated in all planes by the intercondylar fossa (fossa intercondylaris). Lateral femoral condyle is wider in the posterior-anterior plane as compared to the medial femoral condyle and contains the attachment points of the lateral collateral ligament (ligamentum collaterale laterale) and popliteus tendon (Fig. 16.1). Except for the popliteus attachment, the lateral tibiofemoral compartment is covered with a synovial membrane and capsule.

Proximal tibia is comprised of lateral and medial condyles (condylus lateralis et medialis). Lateral and medial menisci (meniscus lateralis et medialis) sit on these condyles. Tibiofemoral joint is formed in the lateral and medial compartments with the femoral condyles. Proximal tibia has a curvature in the sagittal plane, and this curvature differs both in the lateral and medial condyle (Fig. 16.2). These anatomical features also provide the biomechanical details of knee joint movement.

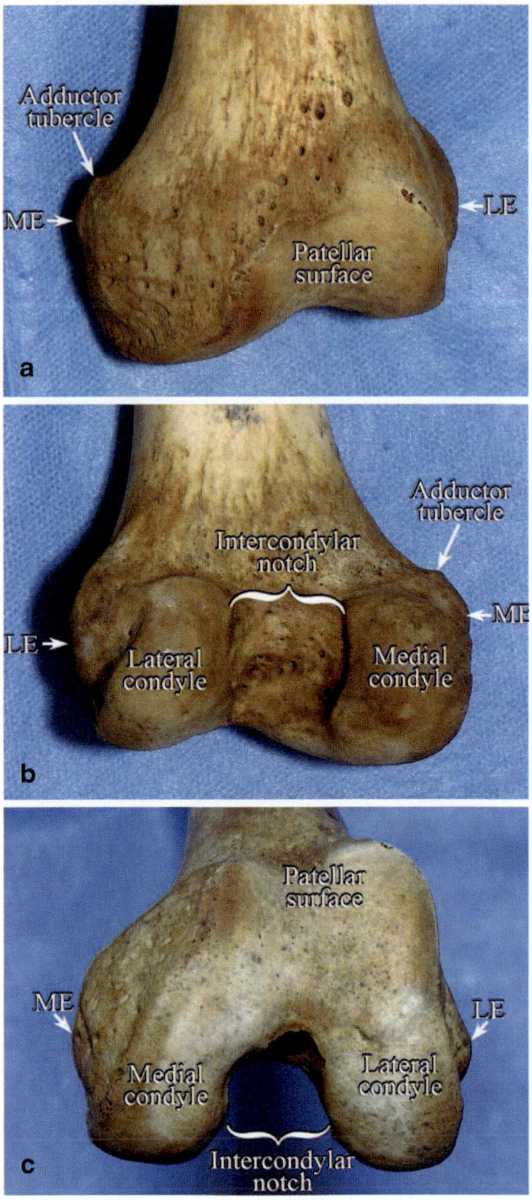

Fig. 16.1 Distal femoral condyle bony anatomy. (**a**) Anterior view of the left distal femur. (**b**) Posterior view of the left distal femur. (**c**) Inferior view of the left distal femur. *ME* medial epicondyle, *LE* lateral epicondyle

16.3.2 Extraosseous Structures

16.3.2.1 Synovia and Joint Capsule

In the knee joint, synovia exhibits a widespread and complex structure. While it joins the structure of bursae on the anterior aspect, it also shapes the joint space with plicae for the joints. The most important anatomical feature of the synovial membrane in the posterior aspect is its extension from the periphery of the popliteus tendon toward the proximal tibiofibular joint.

In the posterior aspect of the knee joint, joint capsule contains the vertical fibrils that emerge from the femoral condyle joint surface, proximal tibia, and intercondylar fossa and extend to the distal-medial aspect. Joint capsule thickened in the posteromedial aspect is known as oblique popliteal ligament (ligamentum popliteum obliquum).

16.3.2.2 Menisci

Medial meniscus (meniscus medialis) is thicker in the posterior section as compared to the

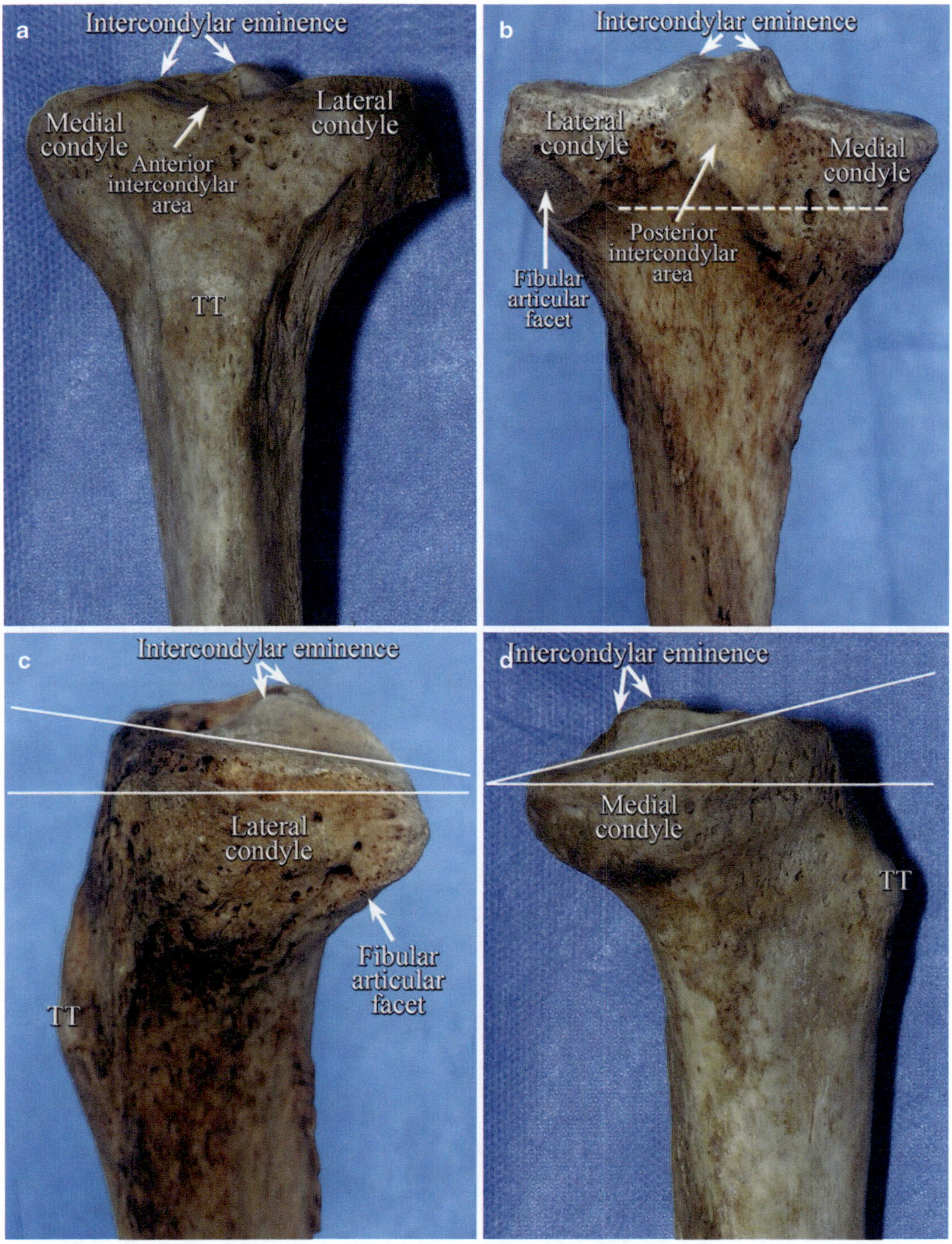

Fig. 16.2 Proximal tibial condyle bony anatomy. (**a**) Anterior view of the left proximal tibia. (**b**) Posterior view of the left proximal tibia. (**c**) Lateral view of the left proximal tibia. (**d**) Medial view of the left proximal tibia. TT tibial tuberosity. The dashed line shows the extension of the posterior criate ligament to the distal. The posterior slope of the tibial plateau is shown in lateral (**c**) and medial (**d**) views

anterior part. It connects to the anterior tibial process in the anterior aspect and to the posterior tibial intercondylar area in the posterior aspect in a manner to form the posterior root structure. The posterior area, in particular, is more firm and stable in comparison to the lateral meniscus.

Lateral meniscus (meniscus lateralis) is considerably different from the medial meniscus in terms of mobility, shape, and footprint. Anterior horn of the lateral meniscus is attached to the bone so as to form the anterior root structure right in the posterior and outer aspect of the anterior cruciate ligament. Posterior horn, on the other hand, is attached to the bone at a more anterior level of the posterior horn of the medial meniscus. Posterior part of the lateral meniscus is supported by anterior and posterior meniscofemoral ligaments (ligamentum Meniscofemorale anterius [ligament of Humphrey], ligamentum Meniscofemorale posterius [ligament of Wrisberg]) that extend from the meniscus to the medial femoral condyle right at the posterior aspect of the posterior cruciate ligament (Fig. 16.3).

16.3.2.3 Posterior Cruciate Ligament

Posterior cruciate ligament (ligamentum cruciatum posterius) is one of the intra-articular ligaments of the knee that rests between the medial femoral condyle and the posterior intercondylar area of the tibia (area intercondylaris posterior) in the posterior aspect of the knee joint. Its femoral insertion site is a kidney-shaped area with a mean longitudinal and transverse diameter of 24 mm and 12 mm, respectively. PCL attaches from this insertion site to the posterior intercondylar area that extends nearly 10 mm distal to the tibial plateau in the posterior aspect of the tibia. This insertion site is in close vicinity of the vascular-neural structures, wherein flexion and extension movements of the knee vary depending on this distance (Fig. 16.4).

16.3.2.4 Posteromedial Area

Anatomy of the posteromedial area of the knee has been described in two different ways in the literature. Warren et al. described the posteromedial area in three layers from the surface to the deeper section [3]. The first layer consists of deep fascia, the second layer consists of superficial medial collateral ligament (MCL), and the third layer consists of the joint capsule and deep MCL. This description system has drawbacks in terms of assessing the structures in the posterior aspect of the MCL. On the other hand, according to the system described by Robinson et al., the medial side of the knee—extending circumferentially from the medial edge of the patellar tendon

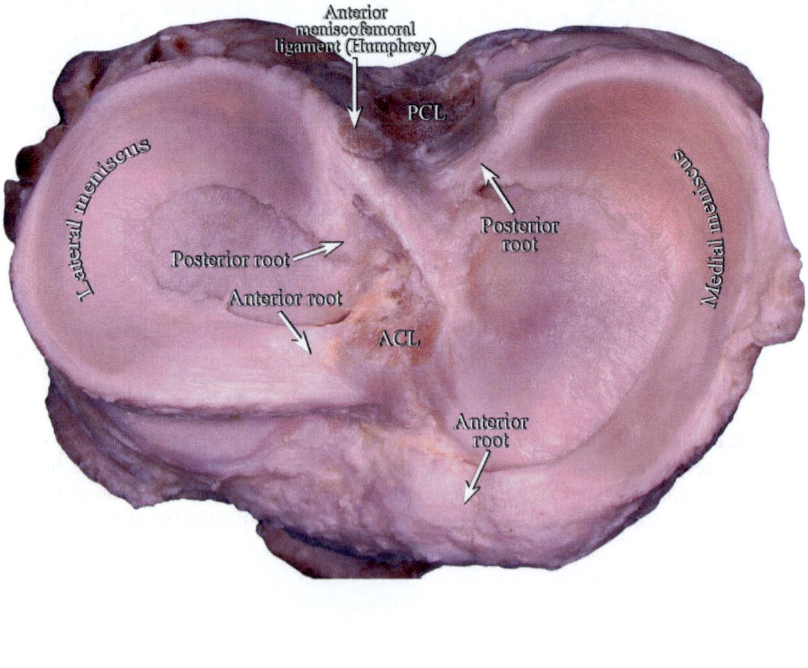

Fig. 16.3 Meniscus anatomy

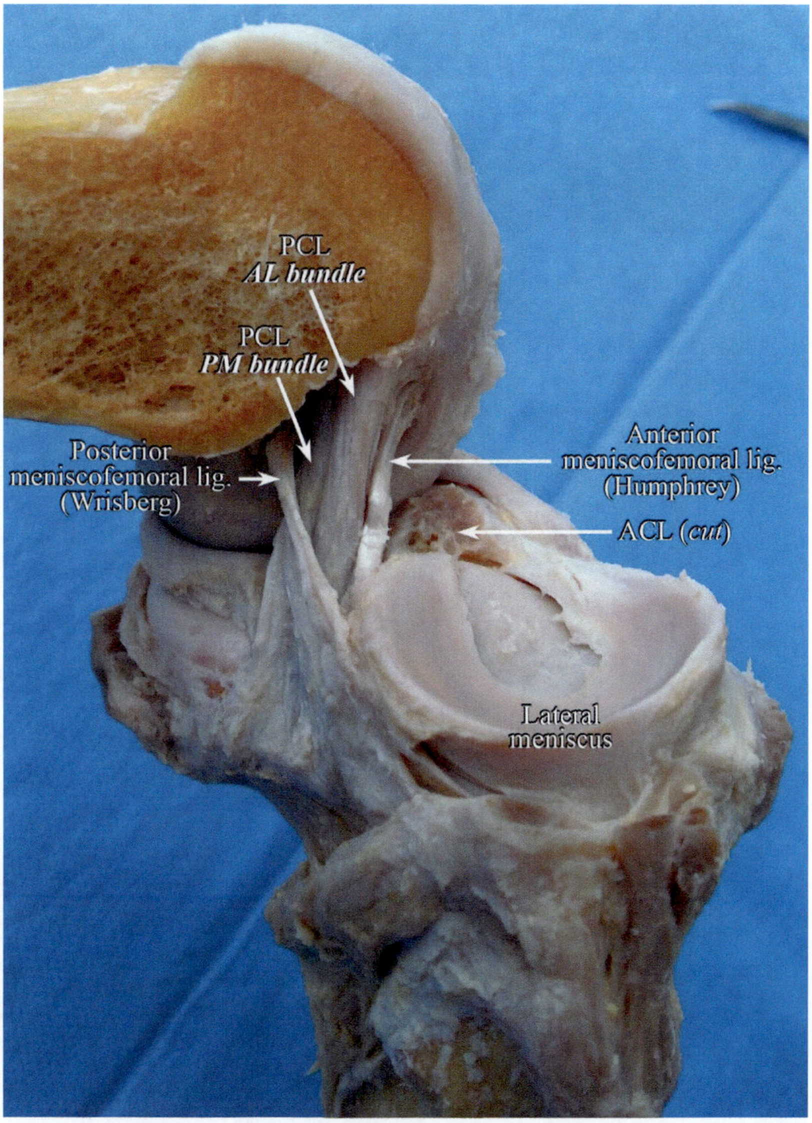

Fig. 16.4 Posterior cruciate ligament (PCL) anatomy

anteriorly to the most medial edge of the medial head of the gastrocnemius posteriorly—was divided into thirds, i.e., anterior, middle, and posterior. The anterior third lies between the medial margin of the patellar tendon and the anterior margin of the longitudinal fibers of the superficial MCL. The middle third consists of the longitudinal fibers of the MCL along the width. The posterior third, designated as the PMC of the knee, lies between the posterior margin of the longitudinal fibers of the superficial MCL and the medial edge of the medial head of the gastrocnemius muscle, which rests next to the medial margin of the PCL (Fig. 16.5).

16.3.2.5 Posterolateral Area

Lateral structures of the knee are also organized in layers similar to the medial structures. The first layer consists of the iliotibial band and the biceps femoris muscle. A major part of the iliotibial band ends at the "Gerdy's Tubercle," which is a bony protrusion located in the anterolateral side

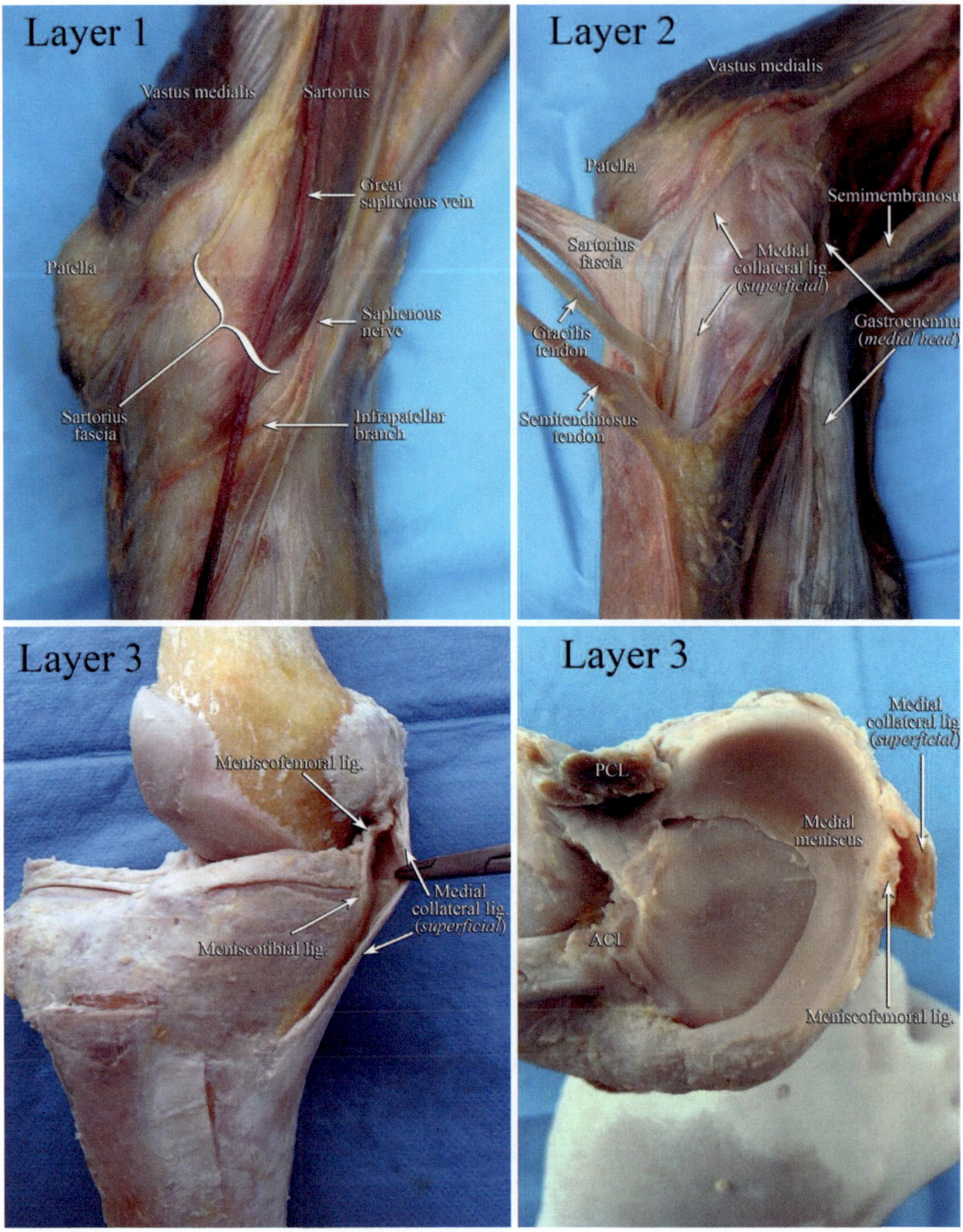

Fig. 16.5 Knee posteromedial area anatomy

of the proximal tibia. Biceps femoris muscle passes over the lateral collateral ligament (ligamentum collaterale fibulare) and ends at the lateral aspect of the head of fibula. The common peroneal nerve (n. fibularis [peroneus] communis) that runs parallel to the tendon of this muscle is also a significantly important structure in this layer and branches in the lateral aspect of the head of fibula. The second layer consists of the lateral patellar retinaculum (retinaculum patellae laterale) and lateral head of the gastrocnemius muscle (m. gastracnemius, caput laterale). The third layer is divided into superficial and deep sections. The most important structure in the superficial division is the lateral collateral ligament. It starts at the lateral epicondyle of the femur and ends at the lateral aspect of the head of fibula (caput fibulae). Its anatomical structure is considerably different from the medial collateral ligament. It has a round shape like a tendon. In addition, fabellofibular ligament is also located in this layer. The deep division consists of the popliteus tendon, popliteofibular ligament, and joint capsule (Fig. 16.6).

16.3.2.6 Blood Vessel Anatomy

Popliteal Artery (a. poplitea)

Popliteal artery is a continuation of the femoral artery extending toward the distal and lateral aspects, wherein it branches into the anterior and posterior tibial arteries at the lower border of the popliteal fossa. Particularly, the genicular arteries that originate from the popliteal artery around the knee are of utmost importance for blood flow in this area. Moreover, it provides blood flow to the hamstring (ischiocrural), gastrocnemius, soleus, and plantaris muscles in this area (Fig. 16.7).

Popliteal Vein

The popliteal vein is located lateral and superficial to the popliteal artery. It forms the posterior tibial vein distally (Fig. 16.7).

Lymph Nodes

Superficial popliteal lymph nodes accompany the terminal part of the small saphenous vein, and they receive lymphatic circulation associated with this vein. Deep popliteal lymph nodes, on the other hand, drain lymph that accompanies the popliteal veins and a major part of the lymph circulation in deep areas until the joint capsule.

16.3.2.7 Nerve Anatomy

The sciatic nerve (n. ischiadicus) divides into the tibial and fibular branches at the popliteal fossa. Tibial nerve (n. tibialis) is the larger branch located in the medial aspect. Although it is the most superficial structure at the popliteal fossa among vascular-neural structures, it is located deep in the popliteal fascia in a protected manner (Fig. 16.7). The tibial nerve gives off branches to the gastrocnemius muscle, plantaris muscle, soleus muscle, and popliteus muscle in the popliteal fossa. Superficial sensory branches and especially the medial sural cutaneous nerve (n. cutaneous surae medialis) are also important. This nerve joins the fibular nerve branch from the lateral aspect to form the sural nerve (n. suralis).

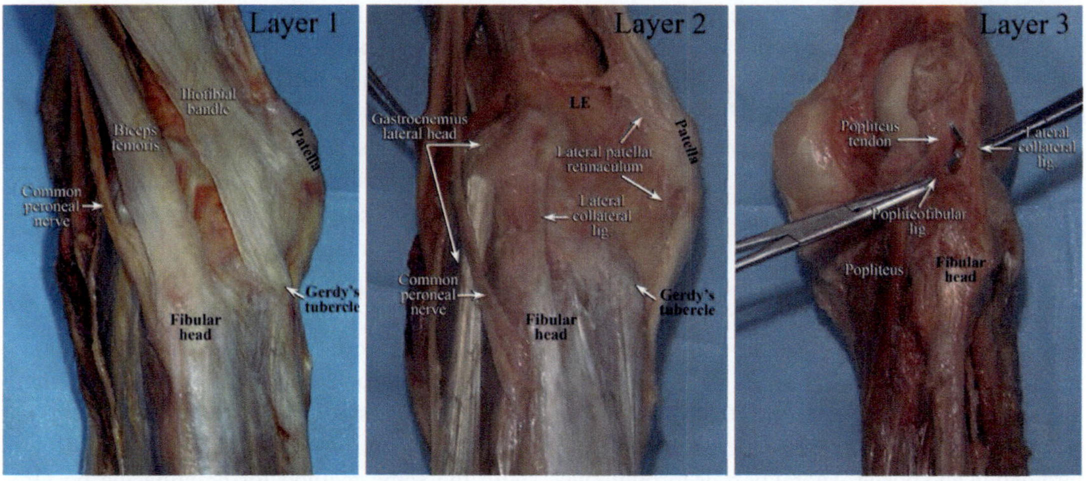

Fig. 16.6 Knee posteromedial area anatomy

Fig. 16.7 Knee posterior anatomy (upper part)

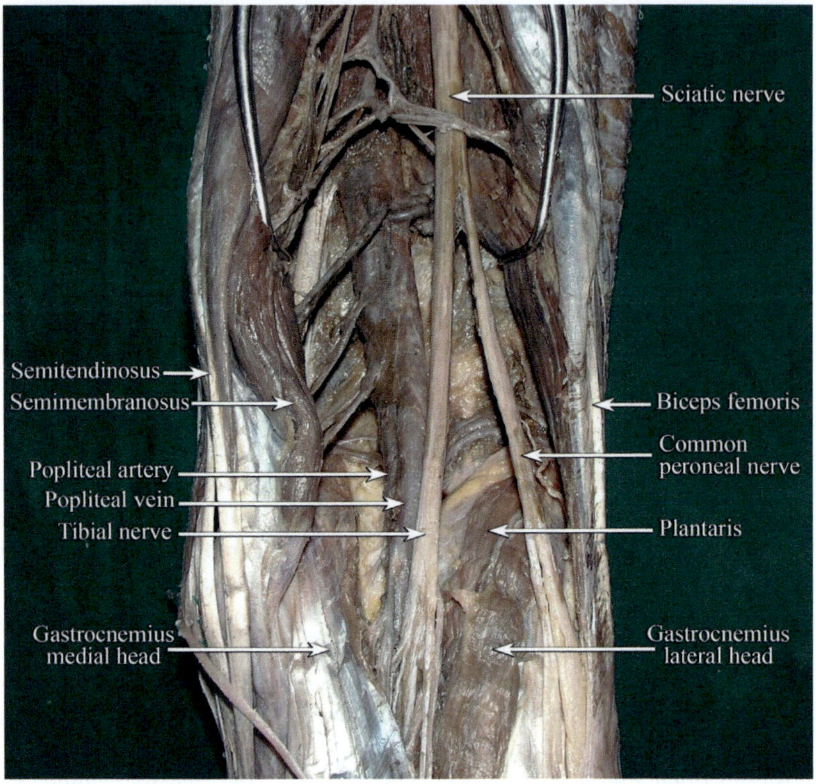

Common fibular nerve (n. fibularis [peroneus] communis), in other words the peroneal nerve, derives from the anterior branches of the L4-S2 roots of the sciatic nerve, starting from the upper border of the popliteal fossa and extending along the medial side of the biceps femoris muscle. Then, it leaves the popliteal fossa by extending along the lateral border of the gastrocnemius muscle. It gives sensory branches to the skin, joint capsule, and muscles around the neck of fibula in the proximal aspect of the fibula, and these branches are particularly important from the anatomical aspect (Fig. 16.8).

16.4 Technique

16.4.1 Trans-Septal Portal

According to the approach reported by Schreiber [4], the procedure is initiated by creating the posteromedial and posterolateral portals through two anterior portals. The only difference was that, instead of a 70° arthroscope, a 30° arthroscope was used in our method. The arthroscope is guided through the intercondylar notch from the anterolateral portal into the posteromedial compartment. Then, as described by Schreiber [4], guided by a cutaneous trans-illumination arthroscopic light, a 23-gauge spinal needle is inserted just behind the posterior medial condyle and 5 mm above the tibial articular surface (Fig. 16.9). It is possible to prevent damage to the saphenous nerve and adjacent vein by using a cutaneous trans-illumination arthroscopic guide [4]. The entry site, skin, and capsule are incised using a No. 11 knife anteriorly along the 23-gauge needle. Then, a hemostat is used to maintain the posteromedial portal opening. This is followed by placing an arthroscope in the posterolateral compartment with the use of intercondylar notch from the anteromedial portal. The posterolateral portal is then established using a 23-gauge spinal needle to localize the posterolateral portal site, with the same method for creating the posteromedial portal (Fig. 16.9). Injuries to the common

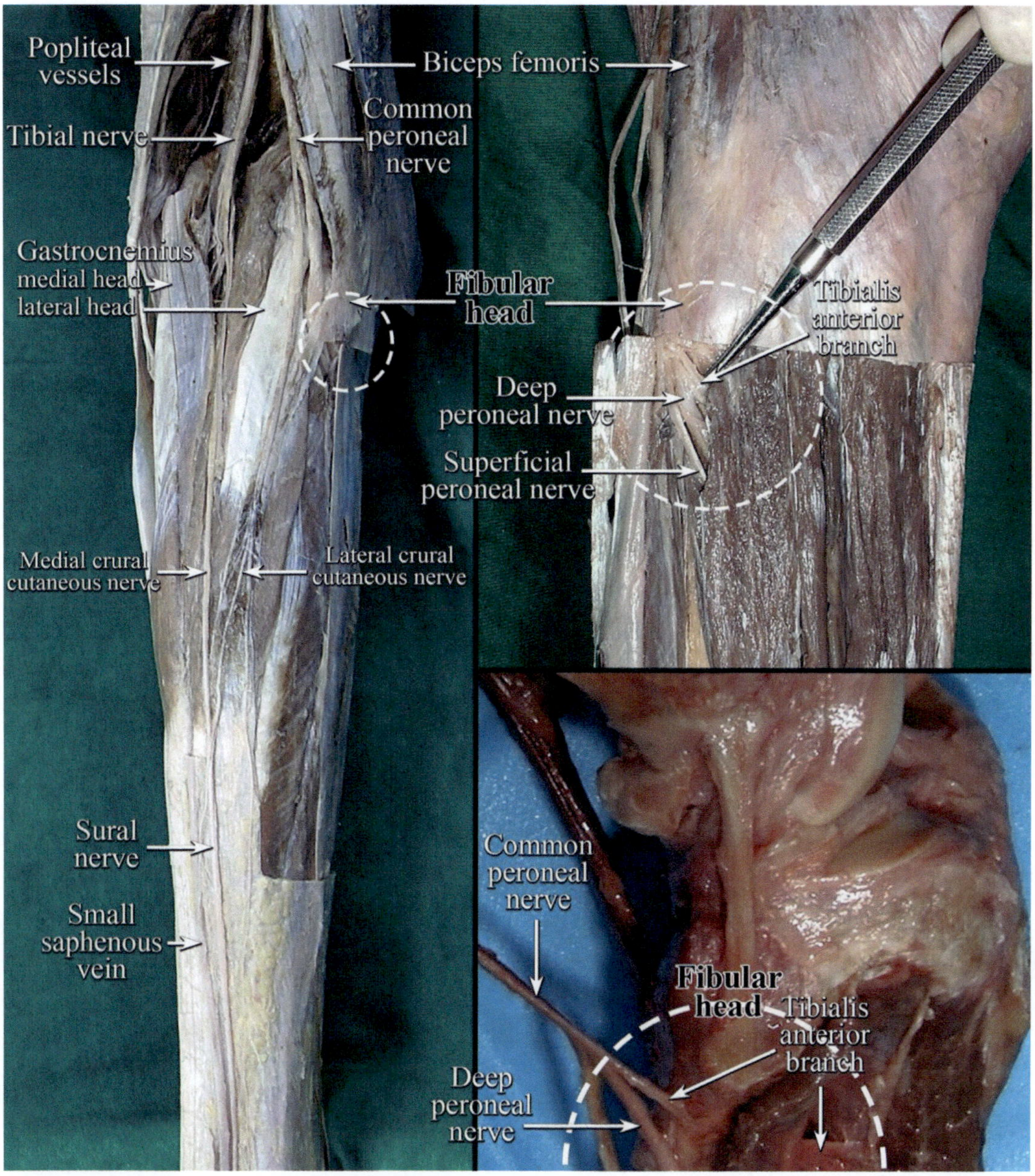

Fig. 16.8 Knee posterior anatomy (lower part)

peroneal nerve can be prevented by palpating the common peroneal nerve and the head of fibula. Afterwards, a switching rod with a sheath is inserted to the septum through the posterolateral portal. The camera head of the 30° arthroscope is turned to the septum in order to enable the operator to make sure that the tip of the rod is attached to the septum. Then, the rod is removed, while maintaining the sheath in place. The sheath is advanced into the septum, and the arthroscope is placed in the posteromedial portal. Kirschner wires 1.5 mm and 3.0 mm are respectively inserted to the septum through the sheath from the posterolateral portal, while in close contact with the posterior femoral condyle, by also maintaining a medial septum view with the arthro-

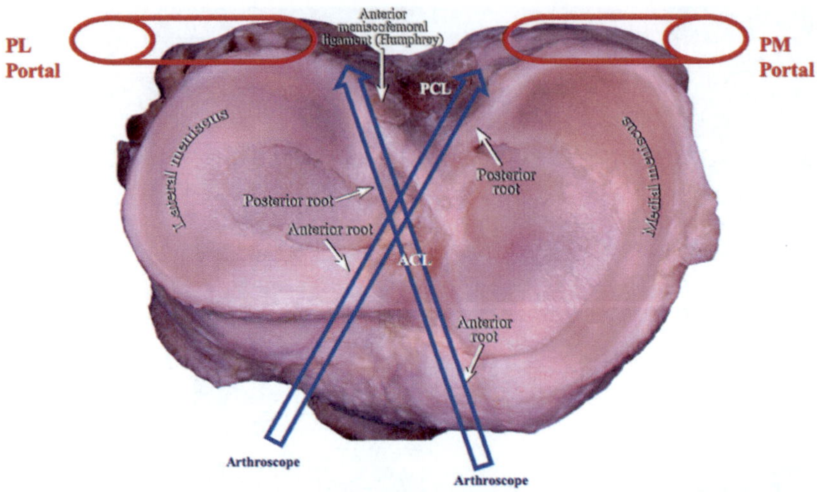

Fig. 16.9 Trans-septal portal technique

scope inserted through the posteromedial portal. This is followed by septum perforation. It is expected that the sheath will protect the posterolateral capsule. In order to enlarge the initial hole and enable easily passing the switching rod through the septum, Kirschner wires are pushed several times into the septum. Afterwards, the switching rod is inserted to the posteromedial portal from the posterolateral portal through the transseptal portal. After establishing the transseptal portal, one can easily alternate between the arthroscope and instruments through two posterior portals according to the posterior "back and forth" approach described by Louisia et al. [5].

16.4.2 Double Posteromedial Portal

An experienced orthopedic surgeon established a posteromedial portal (PM) using a trans-illumination technique with a 30° arthroscope (Tele Pack X LED, Karl Storz Endoskope, Tuttlingen, Germany) placed through the intercondylar notch (between the posterior cruciate ligament and the medial femoral condyle) via the anterolateral portal. This was followed by inserting a 21-gauge needle (Sterican®, BBraun, Melsungen, Hessen, Germany) (0.81-mm outer diameter, 40-mm length) to the posteromedial

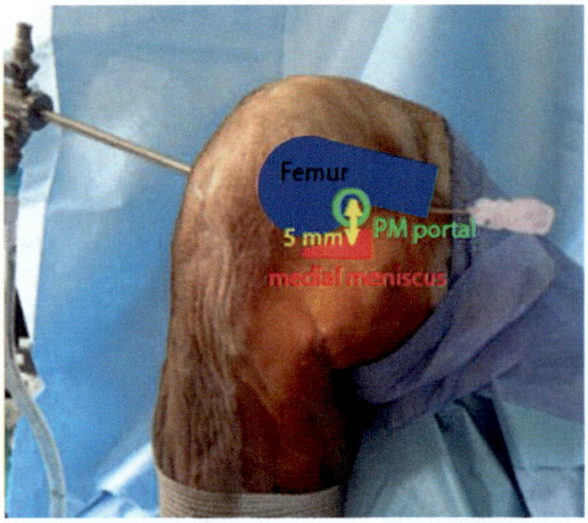

Fig. 16.10 Double posteromedial portal technique

compartment, positioned nearly 5 mm above the medial meniscus and just posterior to the medial femoral condyle (Fig. 16.10). Then, the high posteromedial portal (hPM) was established 3 cm above and 1 cm proximal to the PM (Fig. 16.11).

A 30° arthroscope was inserted through the hPM and placed in the posteromedial compartment, and instruments were advanced through a posteromedial portal, and the posterior capsule was resected in order to ensure a better view of the posterior aspect of the joint (Fig. 16.12).

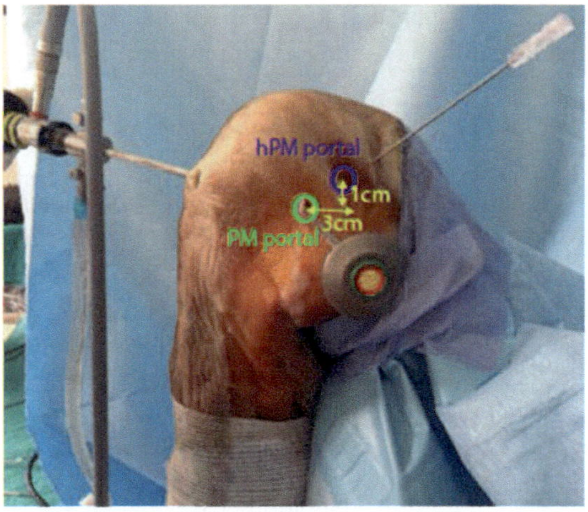

Fig. 16.11 Portals

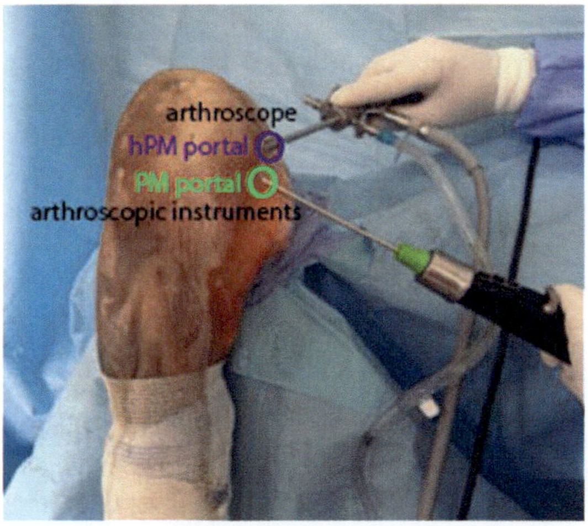

Fig. 16.12 Instruments and arthroscope position

Posterior horn and root of the medial meniscus in the posteromedial compartment, posterior cruciate ligament (PCL), posterior tibial bone margin, medial gastrocnemius insertion to the medial femoral condyle, and posterior capsule were viewed in the arthroscopic examination. Afterwards, the posterior soft tissue and bone structure relationship in the knee in deep flexion was observed by inserting a 30° arthroscope through the posteromedial and high posteromedial portals separately, during a dynamic arthroscopic knee examination (Fig. 16.13).

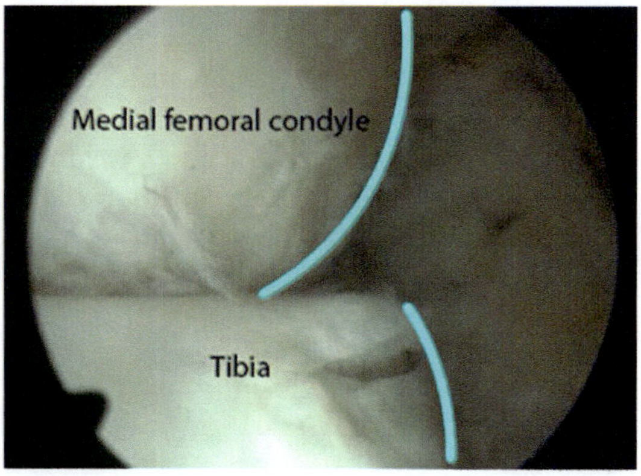

Fig. 16.13 Arthroscopic view of double posteromedial portal technique

16.5 Fields of Application and Advantages of Posterior Knee Arthroscopy

Posterior knee surgeries that can be performed with the posterior knee arthroscopy method include synovectomy, meniscectomy, PCL resection for ganglion cyst, meniscus and popliteal cyst decompression, PCL avulsion fractures, loose body excision, medial meniscus posterior root repairs, PCL reconstruction, etc. [6, 7]. In addition, cam-pincer lesions that start in the posterior femoral condyle-posterior tibial area and that are the initial manifestations of gonarthrosis can also be treated arthroscopically [8] (Fig. 16.14).

All the mentioned practices would lead to lower rate of morbidity and increased quality of life due to early stage treatment procedures.

16.6 Complications

Damage to the sartorial branch of the saphenous nerve is the most common complication of posteromedial portal placement. The saphenous nerve divides into the infrapatellar branch and sartorial branch at the point it exits the adductor canal. The sartorial branch descends inferiorly

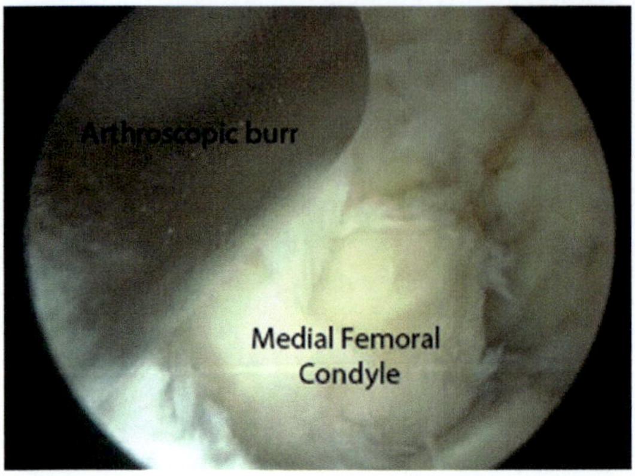

Fig. 16.14 Treatment of cam lesion with double posteromedial portal technique

just posterior to the long saphenous vein. Studies concerning anatomy in the literature have shown the distance from the sartorial branch of the saphenous nerve to the joint line [9, 10]. Other studies in the literature investigated the cases that developed nerve damage following the creation of the posteromedial portal. Gold et al. reported one saphenous nerve injury in their series consisting of 78 patients, whereas Ogilvie-Harris reported that they encountered three saphenous nerve injuries in a series that consisted of 179 cases [11, 12]. Injury can result in saphenous neuritis, which manifests with pain and dysesthesia in the distribution of the saphenous nerve. Desensitization therapy, padding, oral or topical analgesics, and tricyclic antidepressants are among the non-operative treatment methods for saphenous neuritis. There is no consensus or a proven study result concerning the effectiveness of these treatments in the literature.

Decompression of the adductor canal, neurolysis, neuroma excision, and neurectomy are among the surgical treatment options for saphenous neuritis. Studies in the literature have shown successful outcomes following decompression in the treatment of saphenous neuritis [13, 14].

Creation of a posterolateral accessory portal puts the common peroneal nerve at risk. This nerve courses along the medial margin of the biceps femoris tendon, lateral to the lateral head of the gastrocnemius, and around the fibular head and neck posterolaterally, where it divides into deep and superficial peroneal nerves. Studies in the literature have shown that the anatomy of the peroneal nerve is variable [15]. However, complications are associated with the creation of a posterolateral accessory portal.

Establishment of the transseptal portal apparently puts the most anterior structure of the popliteal neurovascular bundle, i.e., the popliteal artery, at risk. The highest risk of injury is encountered while resecting the inferior portion of the posterior septum. To the best of our knowledge, there are no reports of popliteal artery injury during the establishment of a transseptal portal until today.

16.7 Conclusion

Posterior knee arthroscopy has a challenging technique with a long learning curve. On the other hand, it is increasingly becoming more popular due to a high number of advantages as compared to the open surgical procedures. It is possible to apply all posterior surgical procedures for the knee joint after determining a safe and effective application procedure. Being aware of the possible complications and knowing how to manage these complications are the key points. A surgeon, who wants to start performing posterior knee arthroscopy, should first have a command of the anatomy of this area and then practice several times on a cadaver. The following applications on patients would in turn provide successful outcomes. It is probable that posterior knee arthroscopy will become a routine procedure just like anterior knee arthroscopy in the future.

References

1. Lewicky RT, Abeshaus MM. Simplified technique for posterior knee arthroscopy. Am J Sports Med. 1982;10(1):22–3.
2. Ahn JH, Ha CW. Posterior trans-septal portal for arthroscopic surgery of the knee joint. Arthroscopy. 2000;16(7):774–9.
3. Levy IM, Riederman R, Warren RF. An anteromedial approach to the posterior cruciate ligament. Clin Orthop Relat Res. 1984;190:174–81.

4. Schreiber SN. Arthroscopy update #9. Posterior compartment observation and instrumentation in the knee using anteromedial and anterolateral portals and an interchangeable cannula system. Orthop Rev. 1991;20(1):67–8, 73, 76–80.
5. Louisia S, Charrois O, Beaufils P. Posterior "back and forth" approach in arthroscopic surgery on the posterior knee compartments. Arthroscopy. 2003;19(3):321–5.
6. Kim SJ, Jung KA, Kwun JD, Kim JM. Arthroscopic synovectomy of the knee joint in rheumatoid arthritis: surgical steps for complete synovectomy. Arthroscopy. 2006;22(4):461–4.
7. Jang KM, Ahn JH, Wang JH. Arthroscopic partial meniscectomy of a posteriorly flipped superior leaflet in a horizontal medial meniscus tear using a posterior transseptal portal. Orthopedics. 2012;35(3):e430–3.
8. Bozkurt M, Akmese R, Cay N, Isik C, Bilgetekin YG, Kartal MG, Tecimel O. Cam impingement of the posterior femoral condyle in unicompartmental knee arthroplasty. Knee Surg Sports Traumatol Arthrosc. 2013;21(11):2495–500.
9. Arthornthurasook A, Gaew-Im K. The sartorial nerve: its relationship to the medial aspect of the knee. Am J Sports Med. 1990;18(1):41–2.
10. Dunaway DJ, Steensen RN, Wiand W, Dopirak RM. The sartorial branch of the saphenous nerve: its anatomy at the joint line of the knee. Arthroscopy. 2005;21(5):547–51.
11. Gold DL, Schaner PJ, Sapega AA. The posteromedial portal in knee arthroscopy: an analysis of diagnostic and surgical utility. Arthroscopy. 1995;11(2):139–45.
12. Ogilvie-Harris DJ, Biggs DJ, Mackay M, Weisleder L. Posterior portals for arthroscopic surgery of the knee. Arthroscopy. 1994;10(6):608–13.
13. Morganti CM, McFarland EG, Cosgarea AJ. Saphenous neuritis: a poorly understood cause of medial knee pain. J Am Acad Orthop Surg. 2002;10(2):130–7.
14. Dellon AL, Mont MA, Mullick T, Hungerford DS. Partial denervation for persistent neuroma pain around the knee. Clin Orthop Relat Res. 1996;329:216–22.
15. Deutsch A, Wyzykowski RJ, Victoroff BN. Evaluation of the anatomy of the common peroneal nerve. Defining nerve-at-risk in arthroscopically assisted lateral meniscus repair. Am J Sports Med. 1999;27(1):10–5.

17 Physiotherapy in Orthopedic Knee Injuries: Rehabilitation After Articular Cartilage Repair of the Knee

Mehmet Emin Şimşek and M. İ. Safa Kapıcıoğlu

17.1 Introduction

Cartilage defects in the knee joint cause increasing pain and functional disability in orthopedic and sports health. The avascular structure of articular cartilage provides a prolonged healing response, and therefore, because of insufficient healing, there is often a tendency to degeneration [1, 2]. Significant differences are seen in postoperative rehabilitation, and there should be personalization according to the lesion characteristics (size, depth, location, tissue, tissue quality), the patient (age, activity level, aims, tissue quality, lower extremity alignment, body mass index (BMI), general health and nutrition), and the surgical treatment applied (definitive procedure, tissue involvement, accompanying operations). In this section, the basic principles of rehabilitation are discussed following joint cartilage repair procedures, in addition to specific rehabilitation guidelines for debridement, chondroplasty, microfracture, osteochondral autograft transplantation (OAT), and autologous chondrocyte implantation (ACI) [3–5].

M. E. Şimşek (✉)
Department of Orthopaedics and Traumatology, Ankara Lokman Hekim University, Sincan Hospital, Ankara, Turkey

M. İ. S. Kapıcıoğlu
Department of Orthopaedics and Traumatology, Faculty of Medicine, Ankara Yıldırım Beyazıt University, Ankara, Turkey

17.2 Principles of Articular Cartilage Rehabilitation

One of the essential principles included in rehabilitation following joint cartilage repair procedures is the need for an individual approach to each patient. Several factors must be taken into consideration when developing an individual rehabilitation program for each patient [4]. The cartilage quality of each patient is affected by various factors such as age, BMI, general health status, nutrition, previous injuries, and genetic factors. Besides, when determining the rehabilitation approach, attention must be paid to the motivation of the patient and previous activity level to ensure that the targets of each patient have been taken into consideration [6]. The rehabilitation program should be personalized according to the specific demands of each patient for daily life, work, and sporting activities. The most crucial point is the exact location and size of the lesion. To avoid destructive impingement forces and prevent shear forces, lesions on the weight-bearing surface of the femoral condyle require a different rehabilitation approach than defects of the patella or trochlear surface.

Moreover, the size, depth, and involvement of each lesion must be considered. In broad, deep or degenerative joint cartilage lesions with inadequate coverage, it is necessary to progress with slower rehabilitation to provide sufficient time for healing of the repair tissue or graft. Lower

M. Bozkurt, H. İ. Açar (eds.), *Clinical Anatomy of the Knee*, https://doi.org/10.1007/978-3-030-57578-6_17

extremity alignment must also be carefully considered, and usually, treatment should be applied for this pathology [7].

A rehabilitation program in arthroscopic procedures such as chondroplasty or microfracture can progress at a different rate to those for surgical procedures such as OAT or ACI that require larger incisions. Also, concomitant procedures for alignment, stability, or meniscus function can change the rehabilitation program because of the need to protect the other healing tissues. The appropriate information related to the characteristics of each surgical procedure must be shared with the physiotherapy team to be able to provide the best program for each patient [8–10]. Pools and force platforms can be used in the early stages of rehabilitation for the performance of limited weight-bearing activities designed to facilitate a normal walking pattern and to develop resistance, proprioception, and balance. These techniques aim to start weight-bearing activities in the early protective phases of rehabilitation. Continuous passive movement (CPM) machines or manual passive range of movement (PROM) activities applied by a rehabilitation specialist are started immediately after surgery in a limited ROM to heal the cartilage in a joint with limited movement and prevent the formation of adhesions. The following rehabilitation principle includes the biomechanics of the tibiofemoral and patellofemoral joints during standard joint articulation.

Articulation between the femoral condyle and the tibial plateau is fixed throughout knee ROM. Articulation between the patella and the inferior edge of the trochlea starts at approximately 10°–30° knee flexion, depending on the size of the patella and the length of the patella tendon [11, 12]. As the knee moves into a higher degree of flexion, the contact area of the patellofemoral joint moves proximally along the patella. At 30°, the patellofemoral contact area (inferior facets) is approximately 2 cm^2, at 60° knee flexion, the mid-surface of the patella is in the trochlea, and at 90°, the patellofemoral joint contact area increases up to 6 cm^2 and are articulated with the superior patellar facet [10–12]. In contrast, lesions on the posterior condyle can prevent exercise in advanced knee flexion because of the rolling and slipping component of articulation during advanced knee flexion. In addition to the rehabilitation program for lesions on a non-weight-bearing surface, such as the trochlea, an adjustable angle knee support can be used to prevent excessive impingement in the repair region [13, 14]. Closed kinetic chain (CKC) exercises such as leg presses, quad dips, wall dips, and lateral step-ups are performed at 0°–30° initially, and then advanced up to 60°, when the tibiofemoral and patellofemoral joint reaction forces have reduced. As the repair region heals and the symptoms decrease, the ROM exercises applied are progressed to provide greater muscle strengthening in larger muscle movements. The exercises can be reorganized based on the symptoms of the patient and the clinical evaluation of swelling and crepitation [15–17].

17.3 Rehabilitation of Pain and Edema

A progressive reduction in quadriceps activity has been recorded when there is distension and increased pain in the knee. Therefore, reducing knee joint pain and swelling is very important to reduce reflective inhibition to a minimum and regain regular quadriceps activity. Also, any increase in the intra-articular temperature is a warning sign of proteolytic enzyme activity which harms joint cartilage. Of the treatment options to reduce swelling, cryotherapy, elevation, laser treatment, high-voltage stimulation, a knee sleeve, or compressive bandage can be used. PROM may also provide pain neuromodulation during acute or severe conditions. Pain can be passively reduced with the use of cryotherapy, transcutaneous electrical nerve stimulation, and analgesic medication. Immediately following injury or surgery, the use of a commercial cold wrap can be of great benefit [18, 19] (Fig. 17.1).

One of the most critical aspects of joint cartilage rehabilitation is the avoidance of arthrofibrosis. This is achieved through the restoration of passive knee extension, patellar movement, and soft tissue flexibility of the knee and the whole

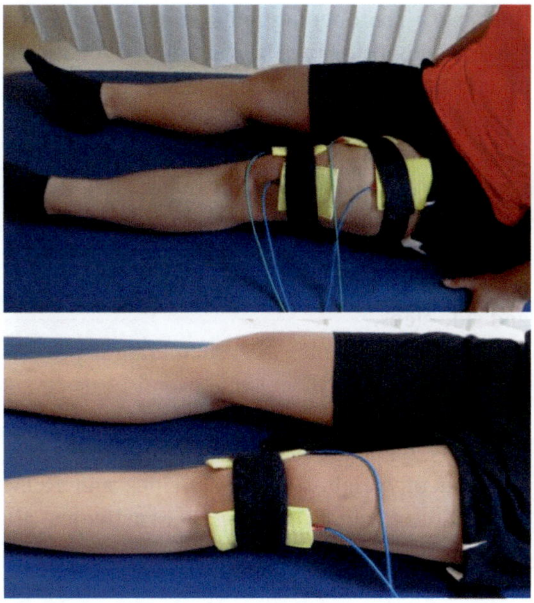

Fig. 17.1 Neuromuscular electrical stimulation for pain and edema

lower extremity. Incomplete lengthening of the knee results in abnormal joint arthrokinematics and an increase in patellofemoral and tibiofemoral joint contact pressure results in increased tension in the quadriceps and muscle fatigue. Therefore, a locking postoperative knee brace can be used to provide 0° extension during ambulation. The aim is to obtain 0° knee extension within the first few days after the operation. Loss of patellar movement following surgery can be due to various causes, primarily excessive scar tissue adhesions originating from the anterior incision in addition to adhesions occurring in the medial and lateral grooves. Loss of patellar movement can cause ROM complications and difficulties [20–22]. The next principle includes the regaining of muscle function. Electrical muscle stimulation (EMS) and biofeedback are often combined with therapeutic exercises to facilitate active contractions of the quadriceps. EMS and biofeedback facilitate the regaining of muscle activation in the quadriceps muscle system [23]. Clinically, while EMS is started immediately following the surgical intervention, quadriceps sets, straight-leg extension, and isometric and isotonic exercises, such as hip adduction and abduction, are performed. EMS is useful to attempt to gather the maximum amount of muscle fiber during active contraction and can be used throughout the rehabilitation period. When there is independent muscle activation, biofeedback can be included to facilitate more neuromuscular activation of the quadriceps. The patient should concentrate independently on neuromuscular control. Also, the importance of using trunk stability exercises cannot be ignored. It has been emphasized that training for proximal and distal placement of the trunk, hips, and ankle throughout the kinetic chain is helpful in producing and controlling the distribution of forces in the knee. The hip and ankle also help control the moments of abduction and adduction in the knee joint [24–26] (Fig. 17.1).

17.4 Rehabilitation Phases After Cartilage Repair of the Knee

Rehabilitation progression is designed based on the four biological phases of cartilage maturation: proliferation, transitional, remodeling, and maturation. The length of each phase varies according to the lesion, the age and general health of the patient, and the previously discussed operational characteristics. The following sections present a general view of the rehabilitation process for each of the four phases [1–3, 13, 17, 21, 27, 28].

17.4.1 Phase I: Early Protection Phase

The first phase of cartilage healing is the proliferation phase, which typically includes the first 6 weeks postoperatively. The healing process starts in this phase. Swelling must be reduced, passive ROM must be gradually increased, and control of the quadriceps must be increased in this rehabilitation period. A gradually weighted progression of controlled active ROM and PROM is an essential part of the rehabilitation process. PROM and controlled partial weight application help to support cartilage nutrition through the diffusion of synovial fluid. The patient can start partial weight-bearing activities with the use of crutches. Loading exercises that are applied in pools or on force platforms can be used gradually to increase the amount of weight applied to the weight-bearing surfaces of the joint. A pool or aquatic therapy can be useful for starting lower extremity exercises and gait training after proper healing of the incisions [29–33].

The use of CPM should typically start at 6–8 h postoperatively and should be used throughout at least 2–3 weeks. Also, there is usually a demand for active supported ROM throughout the day. To reduce the formation of scar tissue to a minimum and prevent the loss of movement, patella mobilization, soft tissue mobilization, and soft tissue flexibility exercises are applied. Low-intensity fixed cycling can be applied at this stage. With the simultaneous use of EMS, early strengthening exercises are applied to provide quadriceps control and neuromuscular control. The exercises performed at this stage are limited to the specific weight-bearing status of each patient and typically include early start proprioception exercises such as quadriceps dips, straight-leg raises, and weight shifting. These include strengthening exercises for the hip and basal corestabilization. At this stage, it is essential to strengthening the hip abductors, external rotators, and extensors to control the valgus collapse of the knee joint. In addition to the hip and ankle movement, manual therapy to correct lumbopelvic alignment is essential at this stage, and this is specific to each patient. Establishing this basis allows a more advanced rehabilitation approach after the removal of postoperative limitations in the later parts of the process [34, 36].

17.4.2 Phase II: Transition Phase

The second stage is the transition phase, typically in postoperative weeks 7–12. At this point, the repair tissue gains strength allowing progression of rehabilitation exercises. Full weight-bearing, full ROM, and soft tissue flexibility are gained at this stage. The ongoing maturation of the repair tissue is supported by a higher level of functional and ROM exercises. Patients can typically perform normal daily living activities at this stage. When the weight-bearing capability of the patient has returned to normal, the rehabilitation program continues with strengthening activities including CKC exercises and machine weights such as leg weights, anterior lungs, wall slides, and side steps.

At the same time, the rehabilitation process includes advanced therapeutic exercises designed to increase the knee function postoperatively gradually. As previously discussed, while the progression of weight-bearing activities and ROM restoration facilitate healing, the activities are advanced in stages to avoid the emergence of complications. This phase is based on the principle of gradual, staged loading. Common complications include limitations of movement and the formation of scar tissue. An exaggerated approach early in the rehabilitation program can cause an increase in pain, inflammation or effusion, and graft damage. This simple concept is applied as a progression through strengthening exercises, proprioception training, neuromuscular control drills, and functional drills [27, 37–41] (Fig. 17.2).

17.4.3 Phase III: Remodeling Phase

The third phase is the remodeling phase, which is generally at 13–26 weeks postoperatively. In this phase, there is ongoing remodeling to obtain a more regular structure of the tissue. The tissue strength and resistance continue to increase. When the tissue becomes tighter and more integrated, more functional training activities can be performed. In this stage, it is typically recorded that the symptoms have healed and the patient has normal ROM. For the patient to reach the highest level of strength and flexibility, they are encouraged to continue on their own, independent of the rehabilitation program. Activities with a low- to moderate-level effect can be gradually added, such as cycling, golf, and walking for pleasure [23, 27–30].

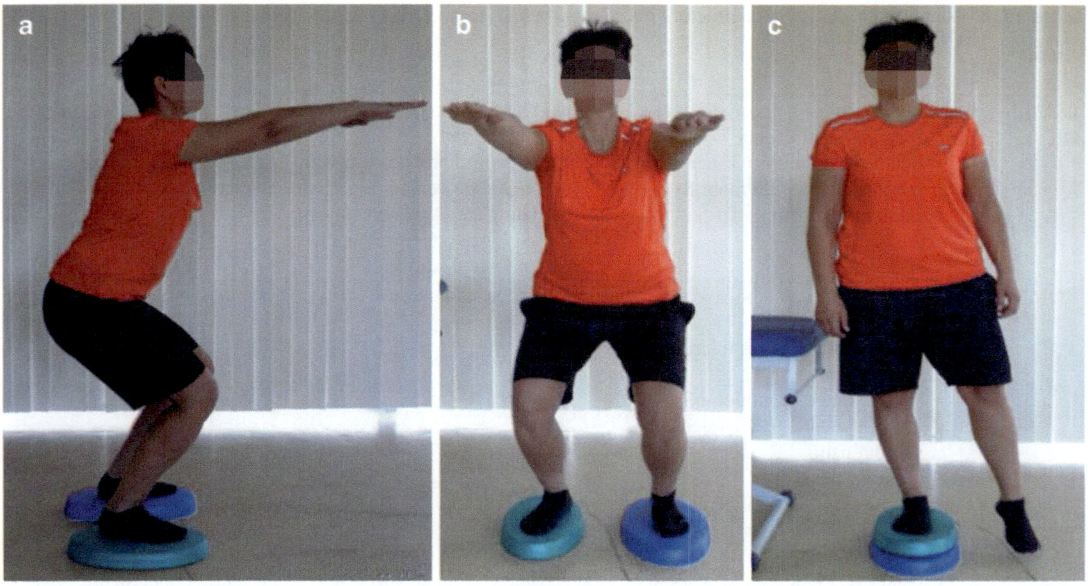

Fig. 17.2 Squat exercises. In the exercise, the arms are brought parallel to the floor in forward flexion (**a**, **b**). Side step-up exercise, the patient stands with the affected limb on the step and unaffected limb on the ground. Patient extends the unaffected knee, thereby affected limb off the ground (**c**)

17.4.4 Phase IV: Maturation Phase

The final phase typically starts at 13–26 weeks and can last up to 15–18 months, depending on the operation type, size, and localization of the lesion. The repair tissue reaches full maturation in this phase. The patient can gradually return to all premorbid activities as tolerated. Impact loading activities are applied gradually. Although procedures such as OAT and ACI have been designed to restore function rather than facilitate a return to high impact athletic activities, the return to sporting activities is determined based on each patient's specific presentation [32–36].

17.4.4.1 Rehabilitation After Debridement and Chondroplasty

Rehabilitation following arthroscopic debridement or chondroplasty is extremely simple because of the nature and aim of the procedure is to facilitate tissue healing rather than create repair tissue. The first weight-bearing is limited to the use of underarm crutches in the first 3–5 days but can be increased according to the tolerance to weight-bearing. PROM exercises are immediately tolerated without postoperative restrictions of movement. Full passive ROM is generally obtained within 2–3 weeks [27, 28, 32, 41]. Initially, OKC exercises are performed, and at the end of the first week, CKC strengthening exercises and cycling are typically included. After the 4th week, the patient can return to full functional activities and can start moderate level activities such as light running and sports. If there are evident degenerative changes in the knee, or if pain or effusion symptoms continue, the progression of weight-bearing can be delayed. The rates of progression are incredibly variable and should be based on the degree of the procedure and the individual status of each patient [36–39] (Table 17.1).

17.4.4.2 Rehabilitation After Microfracture

The surgical procedure referred to as microfracture of the knee joint is the most commonly applied procedure for the treatment of focal articular cartilage lesions of the knee. Rehabilitation after microfracture progresses more cautiously than after debridement or chondroplasty. The program is based on size, local-

Table 17.1 Rehabilitation after microfracture and chondroplasty procedures

	Phase I (0–6 weeks)	Phase II (7–12 weeks)	Phase III (13–26 weeks)
GOALS	• Protection of the post-surgical knee pain and edema • Restore normal knee range of motion and patellar mobility • Eliminate effusion	Single leg stand control • Normalize gait • Good control and no pain with functional movements, including step-up/down, squat, partial lunge (staying less than 60° of knee flexion)	Good control and no pain with sport- and work-specific movements, including impact
ROM	Knee extension on a bolster • Prone hangs • Supine wall slides as tolerated without pain • Passive range of motion off the end of the table as tolerated without pain • CPM machine • Biking—Use contralateral leg to create ipsilateral passive range of motion	• Full range of motion	• Full range of motion
WB	• Weeks 0–2 non-weight-bearing • Weeks 3–4 touch down weight-bearing • Weeks 5–6 weight-bearing as tolerated	• Normal gait on all surfaces	• Normal gait on all surfaces
Precautions	• Avoid post-activity swelling • Avoid loading knee deep flexion angles	• No impact activities until 12 weeks post-op	
BRACE	• Use axillary crutches, to follow the weight-bearing guidelines below This is essential for proper healing. For special situations and in winter months, a brace may also be used For patellofemoral lesions, use axillary crutches, in locked knee brace for 6 weeks	• No brace	• No brace
Therapeutic exercise, treatment recommendations, and return to sport and work	• Straight-leg raises • Four-way leg lifts in standing with brace on for balance and hip strength • Patellar mobilizations • Begin pool activity at the start of week 5. Exercises may include gait drills (forward walk, march walk, skate step, step and balance) with depth of water at the level of the axilla. Deep water running, vertical kicking or biking can also be included Upper body circuit training or upper body ergometer	• Non-impact balance and proprioceptive drills • Stationary bike • Gait drills • Hip and core strengthening • Stretching for patient-specific muscle imbalances • Quad strengthening—Closed chain exercises short of 60° knee flexion • Continue pool program—Alternating days with land program • Non-impact endurance training; stationary bike, Nordic track, swimming, deep water run, cross-trainer	Impact control exercises beginning 2 ft to 2 ft, progressing from 1 foot to other and then 1 foot to same foot • Movement control exercises beginning with low velocity, single plane activities and progressing to higher velocity, multi-plane activities • Sport/work-specific balance and proprioceptive drills • Hip and core strengthening • Stretching for patient-specific muscle imbalances • Replicate sport- or work-specific energy demands

ization, number of treated areas and concomitant procedures. The early protective phase starts immediately after the operation and lasts for up to 4 weeks. In this period, defects in the membrane start to fill with fibrin clots, although there is no fibrocartilage. For most lesions, a 2- to 6-week period of non-weight-bearing is applied postoperatively. In cases with small focal lesions, it seems to be possible to start early controlled weight-bearing without applying harmful forces to the repair area. In well-controlled patients, it is recommended to start weight-bearing first with controlled touch-down weight for localized lesions<2.0 cm² [27, 39–42]. For patients with patellofemoral lesions, a locked knee brace is applied during weight-bearing to prevent pure harmful forces in the healing repair area. PROM is applied because of the arthroscopic nature of the procedure. Full PROM of at least 0°–120° is generally obtained with little difficulty within 3–4 weeks. The transition phase starts in the 4th week and continues to the 8th week. In this phase, the patient may eventually be full weight-bearing and can progress to more functional CKC exercises [43–47]. In the 6th postoperative week, a fine tissue layer covers the lesion base. Even if there is still not full repair, it has been determined that there is fibrocartilaginous tissue and some tissues have hyaline-like characteristics in 8 weeks. At up to 12 weeks, the defect is filled, and there is a significant improvement in the quality of the cartilaginous tissue. Therefore, gentle CKC exercises can be started at up to 8 weeks, but they should not be significantly increased before the 12th week. During the maturation phase at 4–6 months, the patient can start to return to previous activities gradually. However, in cases in cases with more extensive lesions, it may be necessary to delay progression to high-impact activities until 8 months [41, 45, 46] (Figs. 17.1 and 17.2) (Table 17.1).

17.4.4.3 Rehabilitation After Osteochondral Autograft Transplantation (OAT) Procedure

Rehabilitation after OAT procedures requires new harmful forces to be avoided, so that the healed transplanted bone plugs are not damaged. Thus, the rate of the rehabilitation program after OAT procedures is based not only on the size of the lesion but also on the number and location of the transplanted bone plugs. When many plugs have been used, the program should be progressed with caution because of the less compatible surface potential. The early protective phase lasts up to the eighth postoperative week. Partial weight-bearing generally starts at 2–4 weeks postoperatively, depending on the lesion size and the number of transplanted bone plugs. Even if the original hyaline cartilage is healthy and can be applied, the strength of the bone plugs is a limiting factor when designing the postoperative rehabilitation program [47–49]. Subchondral integration with spongious bone plug starts to occur up to 4 weeks. Despite integration, a decrease of 63% is seen in graft hardness in the 6th week. This period increases repair gradually in the membrane with weight-bearing. In the 8th week postoperatively, the fibrocartilaginous surface emerges, the donor and recipient areas form hyaline cartilage, and full weight-bearing can be given. For cases with patellofemoral lesions, a locking knee brace is applied, and weight-bearing can be started immediately, and full mobilization can be obtained at approximately 6–8 weeks postoperatively without the brace. During the early protective phase, ROM is gradually increased to prevent adhesion forming and loss of movement. Due to the large incision and invasive nature of the procedure, the movement is progressed gradually to reduce effusion to a minimum [50, 51]. During the transition phase, full ROM and weight-bearing typically occur at 8–10 weeks, but for more extensive lesions, it may be necessary to delay progression to full weight-bearing to 12–14 weeks. At this point, a strengthening program is progressed to include weight-bearing CKC and machine exercises. In this phase, patients can return to low-impact functional activities. During the remodeling and maturation phases, strength, proprioception, and neuromuscular control during impact are developed. Despite later progression in rehabilitation, patients can return to various sports activities. Generally, low-impact sports such as golf, swimming, cycling, and walking for exercise are permitted within 4–5 months postoperatively. Moderate-level sports such as tennis and walking are permitted at 6–8 months, depending on the size and location of the lesion. At 8–10 months, running and aerobics are permitted and sports such as tennis, basketball, and baseball at 12–18 months [44, 45, 49–52] (Table 17.2).

Table 17.2 Rehabilitation after osteochondral autograft transplantation procedure

	Phase I (0–6 weeks)	Phase II (7–12 weeks)	Phase III (13–26 weeks)	Phase IV (26 weeks and after)
Goals	• Protection of knee after surgery • Restore normal knee ROM (range of motion) and patellar mobility • Restore full control over leg	• Single leg stand control • Normalize gait • Good control and no pain with functional movements, including step-up/down, squat, partial lunge (staying less than 60° of knee flexion and avoiding excessive weight-bearing at the position of the lesion)	• Good control and no pain with sport- and work-specific movements, including impact	• Good control and no pain with sport- and work-specific movements, including impact
ROM	• Continuous passive motion (CPM) machine: 6–8 h per day, 6–8 weeks • Set CPM to 1 cycle per min, starting at 40° flexion • Advance 10° per day until flexion is achieved (goal: 100° by week 6) • PROM/AAROM and stretching under guidance of PT	• Advance to full/painless ROM (should obtain 130° of flexion)	• Full/painless ROM	• Full/painless ROM
WB	• Week 1–3 = non-weight-bearing • Week 4–6 = touch down to 25% weight-bearing • 0–6 weeks = locked extension lock splint brace	• Begin progressive weight-bearing as tolerated with axillary crutches and no brace	• Begin progressive weight-bearing as tolerated with no brace	• Begin progressive weight-bearing as tolerated with no brace
Precautions		• Avoid post-activity swelling • Avoid loading knee a deep flexion angles • No impact activities until 12 weeks after surgery	• Post-activity soreness should resolve within 24 h • Avoid post-activity swelling • Avoid knee pain with strengthening	• Post-activity soreness should resolve within 24 h • Avoid post-activity swelling • Avoid knee pain with impact

Table 17.2 (continued)

	Phase I (0–6 weeks)	Phase II (7–12 weeks)	Phase III (13–26 weeks)	Phase IV (26 weeks and after)
Brace	• Week 1: Hinged knee brace locked in extension; removable for CPM and rehabilitation • Weeks 2–6: Gradually open brace in 20° increments as quad strengthens • D/C brace when able to perform straight-leg raise without extension lag	• No brace	• No brace	• No brace
Therapeutic exercise, treatment recommendations	• Patellar mobilization • Quad/hamstring/adductor/gluteal sets: Straight-leg raise, ankle pumps • Stationary bike for ROM • Upper body circuit training or upper body ergometer	• Closed chain exercises: Wall sits, shuttle, mini-squats, toe raises. • Gait training • Patellar mobilization • Begin unilateral stance activities • Non-impact endurance training, swimming (stiff knee flutter kick), deep water run, upper body circuits	• Advanced closed chain strengthening exercises, proprioception activities. • Sport-specific rehabilitation • Maintenance program for strength and endurance • Replicate sport- or work-specific energy demands	• Impact control exercises beginning 2 ft to 2 ft, progressing from 1 foot to other and then 1 foot to same foot • Movement control exercise beginning with low velocity, single plane activities and progressing to higher velocity, multi-plane activities • Sport/work-specific balance and proprioceptive drills • Hip and core strengthening • Stretching for patient-specific muscle imbalances
Return to sport and work			• Dynamic neuromuscular control with multi-plane activities, without pain or swelling	• 8 months post-surgery; and good dynamic neuromuscular control with multi-plane activities, without pain or swelling • 10 months post-surgery; and good dynamic neuromuscular control with multi-plane activities, without pain or swelling

17.4.4.4 Rehabilitation After Autologous Chondrocyte Implantation (ACI)

The rehabilitation program after ACI and more recent matrix-origin ACI procedures are of critical importance for the success and long-term results of patients. Early controlled ROM and weight-bearing are necessary to stimulate chondrocyte development, but excessive activity should be avoided as it can cause cell damage or graft delamination. Knowledge of the biological healing response is vital for the development of appropriate rehabilitation guidelines [49–53]. Rehabilitation can start early with CPM, during which chondrocytes are aligned and bind to the underlying surface. First, the position of the patient with a patellar defect in the prone position is a contra-indication for CPM because it will create an abnormal gravity effect. Second, the use of CPM immediately postoperatively is seen to be necessary for "equal distribution of the implanted cells within the defect" [54, 55]. There are currently no in vivo or clinical studies that have been published that support this hypothesis. Chondrocyte proliferation emerges in the first 6 weeks after implantation. Within the first 24 h of cell implantation, the cells from the base of the lesion, and proliferate several times to produce a matrix that will be filled with soft repair tissue to the level of the defect periosteal cover. At this time, PROM and partial weight-bearing will be useful to support cellular nutrition through synovial fluid diffusion and will provide appropriate stimulation for the production of specific matrix markers. In this first stage, controlled PROM and regular weight-bearing are the two most essential components of the rehabilitation process. In knees with smaller lesions, toe-touch weight-bearing with 20% of body weight is applied at 2–4 weeks, 50% of body weight at 5–6 weeks, and full weight-bearing in the 8th week [31, 44, 52]. For lesions within the patella or trochlea, weight-bearing, as tolerated, is permitted immediately postoperatively with a locking knee brace in full extension (Fig. 17.3).

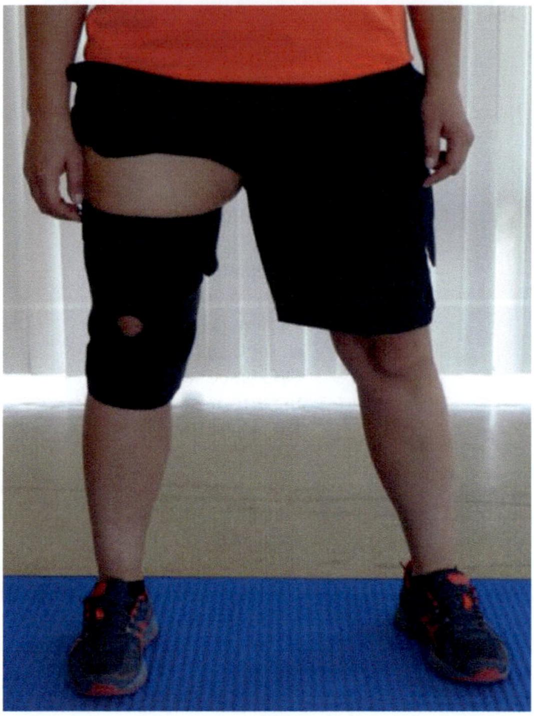

Fig. 17.3 Knee immobilization brace after surgery

Similar to the procedure following OAT, to prevent swelling, caution must be applied in the progression of ROM, with at least 90° flexion in the 1st week, 105° at 2–3 weeks, 115° in the 4th week, and 120° in the 6th week. Earlystrengthan dproprioceptiveexercisesareappliedwithinthewei ght-bearingstatus of the patient. The ongoing maturation of the repair tissue is supported with higher level functional and movement exercises. In patients with patellofemoral lesions, attention must be paid to exercises which produce hard forces. The remodeling phase is from 12 weeks up to 32 weeks postoperatively. In this phase, there is continuous matrix production with remodeling to a more organized structural tissue. When the tissue becomes tighter and integrated, more functional training activities can be applied in addition to elliptical bicycle and a graduated walking program. The final maturation phase can last until 15–18 months depending on the size and location of the lesion. The stiffness of the cartilage is similar to that of the surrounding tissue [56–63] (Table 17.3).

Table 17.3 Rehabilitation after autologous chondrocyte implantation (ACI)

	Phase I (0–6 weeks)	Phase II (7–12 weeks)	Phase III (13–26 weeks)	Phase IV (26 weeks and after)
Goals	0–7 days • Avoid shear forces generated across the articular surface • Maintain lower-limb joint mobility, muscle tone, and circulation • Reduce postoperative pain and edema 2–3 weeks Pain-free and full passive knee extension • Proficient heel-toe gait at 30% BW with 2 crutches • Regain active quadriceps control • Reduce postoperative pain and edema • Proficiency in undertaking home exercise program 4–6 weeks • Pain-free knee flexion to 125° • Proficiency in performing all home-based exercises, including an SLR • Pain-free gait using 1 crutch and 60% BW through operated limb	7–12 weeks • Pain-free and full active knee ROM within anatomical limits • Pain-free 6-min walk test without the use of walking aids • Pain-free upright cycling without a protective knee brace • Proficiency in performing home- and clinic-based exercises for the independent continuation of post-discharge rehabilitation	13–24 weeks • Normal gait pattern without pain, walking aids, or a knee brace • Ability to negotiate stairs and mild gradients without pain • Return to work (dependent on occupational demands) • Proficiency in performing all full • WB strengthening and proprioception activities 24–26 weeks • Hamstring and calf strength within 90% of the contralateral limb • Ability to tolerate pain-free walking distances of more than 5 to 10 km • Ability to effectively negotiate uneven terrain and soft sand • Return to preoperative low-impact recreational activities	• Quadriceps strength within 90%of the contralateral limb • Ability to perform all activities of daily living pain free • Ability to commence a progressive running program at 9 to 26 weeks and after • Resumption of dynamic recreational and sporting activities; however, sports that generate high compressive, shear, and torsional loads should be avoided until 12 to 18 months post-surgery
ROM	0–7 days • Passive and active assisted knee ROM from 0–30 degree 2–3 weeks • Progress active knee ROM from 30 to 90 degree 4–6 weeks • Progress active knee ROM from 110° to 120°	7–12 weeks • Progress to full active knee ROM	13–24 weeks Full and pain-free active knee ROM 24–26 weeks Full and pain-free active knee ROM	Full and pain-free active knee ROM

(continued)

Table 17.3 (continued)

	Phase I (0–6 weeks)	Phase II (7–12 weeks)	Phase III (13–26 weeks)	Phase IV (26 weeks and after)
WB	0–7 days • ≤20% of BW through the operated limb 2–3 weeks • Progress from ≤20% BW (weeks 1 and 2) to 30% BW 4–5 weeks • Progress from 40%BW (week 4) to 60% BW	7–12 weeks • Progress from 80%BW (week 7) to full WB (week 8), pending clearance from orthopedic specialist	13–24 weeks Full WB, no crutches	Full WB, no crutches
Precautions	0–7 days • Avoid shear forces generated across the articular surface • Reduce postoperative pain and edema 2–3 weeks •Pain-free and full passive knee extension 44–6 weeks • Pain-free knee flexion to 125° • Proficiency in performing all home-based exercises, including an SLR • Pain-free gait using 1 crutch and 60% BW through operated limb	7–12 weeks • Pain-free and full active knee ROM within anatomical limits • Pain-free 6-min walk test without the use of walking aids • Pain-free upright cycling without a protective knee brace • Proficiency in performing home- and clinic-based exercises for the independent continuation of post-discharge rehabilitation	13–26 weeks • Proficiency in performing home- and clinic-based exercises for the independent continuation of post-discharge rehabilitation	
Brace	0–7 days • Protective brace set at 0° to 30°, worn 24 h per day 2–3 weeks • Knee brace 0° to 30° (week 1) to 0° to45° (week 3)	No brace	No brace	No brace full WB, no crutches

Therapeutic exercise, treatment recommendations, and return to sport and work	0–7 days • Passive and active assisted knee flexion and extension ROM exercises (0°-30°) • CPM (0°–30°) 12 to 24 h post-surgery for a minimum of 1 h daily • Active ankle dorsi flexion and plantar flexion • Isometric contraction of the quadriceps, hamstrings, adductors, and gluteal musculature • Instruction and practice in proficient toe-touch WB ambulation (≤20% BW) using 2 crutches 2–3 weeks • Continue phase 1 exercises • Patellar mobilization in all directions • CPM employed during clinic visits only • Isometric quadriceps and quadriceps-hamstring co-contraction activities with and without the use of NMES • Introduce SLR activities (hip flexion, abduction, adduction, extension) • Introduce hydrotherapy: Deep-water walking (forward, backward, sideways), heel raises, squats, straight-leg hip flexion, extension, abduction, and circumduction 4–6 weeks • Continue phase 1 and 2 exercises • Increase demand of plinth-based exercises, including SLR with external hip rotation, side lying gluteal exercises in knee flexion • Introduce hamstring and calf stretches • Introduce seated or standing machine-based exercises, including standing calf raises (dependent on WB status), weighted hip abduction and adduction, trunk strengthening (supine sit-ups, weight-supported trunk flexion) • Introduce recumbent cycling (week 5) • Hydrotherapy: Continue phase 2 exercises, plus active knee flexion (with resistive devices), shallow-water walking (waist depth), heel raises, step-ups/step-downs, lunges, cycling, scissor kicks, lower-limb flexibility and proprioception exercises	7–12 weeks • Continue phase 2 and 3 exercises • Further knee flexion through intensive quadriceps Stretching and passive rowing ergometry • Introduce bridging exercises and weighted knee flexion (week 8) • Introduce proprioceptive WB exercises • Introduce upright stationary cycling (weeks 9–12) • Continue phase 2 and 3 hydrotherapy exercises, plus "patter kicking"	13–24 weeks • Continue phase 3 and 4 exercises • intRoduce modified non-WB exercises (e.g., terminal leg extension), dictated by graft location and size • Introduce modified WB exercises (e.g., single-leg heel raises, leg press, squats, lunges, steps) • Increase duration and intensity of stationary and outdoor road cycling • Introduce rowing ergometry and elliptical trainers • Hydrotherapy generally not required; pool may be used for ongoing cardiovascular exercise 24–26 weeks • Continuation of phase 3 to 5 exercises • Ongoing progression of WB and non-WB activities with respect to duration, intensity, proprioceptive component, and overall complexity • Exercises employed should begin to replicate what is required for the patient's individual activity goals	• Continuation of phase 3 to 6 exercises pertinent to the patient's individual activity goals • Ongoing progression of WB activities with respect to duration, intensity, proprioceptive Component, and overall complexity • Introduction of agility and plyometric drills relevant to the patient's individual activity goals

17.5 Conclusion

The rehabilitation process following articular cartilage repair is of critical importance in the long-term success and functional results of the patient. The rehabilitation programs discussed are based on the natural healing response of the joint cartilage observed after articular cartilage repair procedures. Rehabilitation is based on several fundamental principles designed to facilitate the repair process by creating a healing environment avoiding destructive forces which can overload healing tissue. These principles must be included in the rehabilitation process together with the knowledge of the basic science and maturation process of each specific repair procedure. The surgical team must take the long-term results of patients into consideration and the 1-, 2-, and 5-year functional results after surgery. There is a need for long-term studies to determine the efficacy of each surgical technique and the postoperative rehabilitation programs discussed in this section.

References

1. Bhosale AM, Richardson JB. Articular cartilage: structure, injuries and review of management. Br Med Bull. 2008;87:77–95.
2. Micheli LJ, Moseley JB, Anderson AF, Browne JE, Erggelet C, Arciero R, et al. Articular cartilage defects of the distal femur in children and adolescents: treatment with autologous chondrocyte implantation. J Pediatr Orthop. 2006;26(4):455–60.
3. Disler DG, McCauley TR, Wirth CR, Fuchs MD. Detection of knee hyaline cartilage defects using fat-suppressed three-dimensional spoiled gradient echo MR imaging: comparison with standard MR imaging and correlation with arthroscopy. AJR Am J Roentgenol. 1995;165:377–82.
4. Rodrigo J, Steadman J, Silliman J. Improvement of full-thickness chondral defect healing in the human knee after debridement and microfracture using continuous passive motion. Am J Knee Surg. 1994;7:109–16.
5. Meyers M, Akeson W, Convery F. Resurfacing of the knee with fresh osteochondral allograft. J Bone Joint Surg. 1989;71A:704–13.
6. Steadman J, Rodkey W, Rodrigo J. Microfracture: surgical technique and rehabilitation to treat chondral defects. Clin Orthop. 2001;391(suppl)
7. Hijelle K, Solheim E, Strand T, Muri R, Brittberg M. Articular cartilage defects in 1000 knee arthroscopies. Arthroscopy. 2002;18:730–4.
8. Benthien JP, Behrens P. The treatment of chondral and osteochondral defects of the knee with autologous matrix-induced chondrogenesis (AMIC): method description and recent developments. Knee Surg Sports Traumatol Arthrosc. 2011;19(8):1316–9.
9. Aston JE, Bentley G. Repair of articular surfaces by allografts of articular and growth-plate cartilage. J Bone Joint Surg. 1986;68B:29–34.
10. Amiel D, Coutts RD, Abel M, Stewart W, Harwood F, Akeson WH. Rib perichondrial grafts for the repair of full thickness articular cartilage defects. A morphological and biochemical study in the rabbits. J Bone Joint Surg. 1985;67A:911–20.
11. Mitthoefer K, Hambly K, Logerstedt D, Ricci M, Silvers H, Della VS. Current concepts for rehabilitation and return to sport after knee articular cartilage repair in the athlete. J Orthop Sports Phys Ther. 2012;42(3):254–73.
12. Wilk KE, Macrina LC, Reinold MM. Rehabilitation following microfracture of the knee. Cartilage. 2010;1(2):96–107.
13. Brittberg M, Faxén E, Peterson L. Carbon fiber scaffolds in the treatment of early knee osteoarthritis. A prospective 4-year follow-up of 37 patients. Clin Orthop Relat Res. 1994;307:155–64.
14. Gross AE, Kim W, Las Heras F, Backstein D, Safi r O, Pritzker KP. Fresh osteochondral allografts for posttraumatic knee defects: long-term follow-up. Clin Orthop Relat Res. 2008;466(8):1863–70.
15. Hangody L, Kárpáti Z. New possibilities in the management of severe circumscribed cartilage damage in the knee. Magy Traumatol Ortop Kezseb Plasztikai Seb. 1994;37(3):237–43.
16. Windt TS, Concaro S, Lindahl A, Saris DB, Brittberg M. Strategies for patient profi ling in articular cartilage repair of the knee: a prospective cohort of patients treated by one experienced cartilage surgeon. Knee Surg Sports Traumatol Arthrosc. 2012;20(11):2225–32.
17. Buckwalter JA, Mow VC. Cartilage repair in osteoarthritis. In: Moskowitz RW, Howell DS, Goldberg VM, Mankin HJ, editors. Osteoarthritis. Diagnosis and medical/surgical management. 2nd ed. Philadelphia: WB Saunders; 1994. p. 71–108.
18. Bert JM. Role of abrasion arthroplasty and debridement in the management of osteoarthritis of the knee. Rheum Dis Clin North Am. 1993;19:725–39.
19. Knutsen G, Drogset JO, Engebretsen L, Grøntvedt T, Isaksen V, Ludvigsen TC, Roberts S, Solheim E, Strand T, Johansen O. A randomized trial comparing autologous chondrocyte implantation with microfracture. Findings at five years. J Bone Joint Surg Am. 2007;89(10):2105–12.
20. Kreuz PC, Erggelet C, Steinwachs MR, Krause SJ, Lahm A, Niemeyer P, Ghanem N, Uhl M, Südkamp N. Is microfracture of chondral defects in the knee

associated with different results in patients aged 40 years or younger? Arthroscopy. 2006;22(11):1180–6.
21. Minas T. Autologous chondrocyte implantation for focal chondral defects of the knee. Clin Orthop Relat Res. 2001;391(Suppl):S349–61.
22. O'Driscoll SW, Keeley FW, Salter RB. The chondrogeneic potential of free autogenous periosteal grafts for biological resurfacing of major full thickness defects in joint surfaces under the influence of continuous passive motion. J Bone Joint Surg. 1986;68A:1017–35.
23. Mahomed MN, Beaver RJ, Gross AE. The long-term success of fresh, small-fragment osteochondral allografts used for intra-articular posttraumatic defects in the knee joint. Orthopaedics. 1992;15:1191–9.
24. Minas T, Peterson L. Autologous chondrocyte transplantation. Oper Tech Sports Med. 2000;8:144–57.
25. Brittberg M, Lindahl A, Nilsson A, Ohlsson C, Isaksson O, Peterson L. Treatment of deep cartilage defects in the knee with autologous chondrocyte transplantation. N Engl J Med. 1994;331:889–95.
26. Briggs TW, Mahroof S, David LA, Flannelly J, Pringle J, Bayliss M. Histological evaluation of chondral defects after autologous chondrocyte implantation of the knee. J Bone Joint Surg Br Vol. 2003;85:1077–83.
27. Irrgang JJ, Pezzullo D. Rehabilitation following surgical procedures to address articular cartilage lesions in the knee. J Orthop Sports Phys Ther. 1998;28:232–40.
28. O'Driscoll SW, Keeley FW, Salter RB. Durability of regenerated articular cartilage produced by free autogenous periosteal grafts in major full-thickness defects in joint surfaces under the influence of continuous passive motion. A follow-up report at one year. J Bone Joint Surg Am Vol. 1988;70:595–606.
29. Suh J, Aroen A, Muzzonigro T, Disilvestro M, Fu F. Injury and repair articular cartilage; related scientific issues. Oper Tech Orthopead. 1997;7:270–8.
30. Gillogly SD, Voight M, Blackburn T. Treatment of articular cartilage defects of the knee with autologous chondrocyte implantation. [review] [64 refs]. J Orthopaed Sports Phys Ther. 1998;28:241–51.
31. Gomoll A, Probst C, Farr J, Cole B, Minas T. Use of a type I/III bilayer collagen membrane decreases reoperation rates for symptomatic hypertrophy after autologous chondrocyte implantation. Am J Sports Med. 2009;37:20S–3S.
32. Buckwalter JA. Effects of early motion on healing of musculoskeletal tissues. Hand Clin. 1996;12:13–24.
33. Buschmann MD, Gluzband YA, Grodzinsky AJ, Hunziker EB. Mechanical compression modulates matrix biosynthesis in chondrocyte/agarose culture. J Cell Sci. 1995;108(pt 4):1497–508.
34. Ebert JR, Fallon M, Robertson WB, et al. Radiological assessment of accelerated versus traditional approaches to postoperative rehabilitation following matrix-induced autologous chondrocyte implantation. Cartilage. 2011;2:60–72. https://doi.org/10.1177/1947603510380902.
35. Curl WW, Krome J, Gordon ES, Rushing J, Smith BP, Poehling GG. Cartilage injuries: a review of 31,516 knee arthroscopies. Arthroscopy. 1997;13:456–60.
36. Carey-Smith R, Ebert JR, Davies H, Garrett S, Wood DJ, Janes GC. Arthroscopic matrix-induced autologous chondrocyte implantation: a simple surgical technique. Tech Knee Surg. 2010;9:170–5.
37. Dennis DA, Komistek RD, Nadaud MC, Mahfouz M. Evaluation of off-loading braces for treatment of unicompartmental knee arthrosis. J Arthroplasty. 2006;21:2–8. https://doi.org/10.1016/j.arth.2006.02.099.
38. Basad E, Ishaque B, Bachmann G, Sturz H, Steinmeyer J. Matrix-induced autologous chondrocyte implantation versus microfracture in the treatment of cartilage defects of the knee: a 2-year randomised study. Knee Surg Sports Traumatol Arthrosc. 2010;18:519–27. https://doi.org/10.1007/s00167-009-1028-1.
39. Arokoski JP, Jurvelin JS, Väätäinen U, Helminen HJ. Normal and pathological adaptations of articular cartilage to joint loading. Scand J Med Sci Sports. 2000;10:186–98. https://doi.org/10.1034/j.1600-0838.2000.010004186.x.
40. Kowal MA. Review of physiological effects of cryotherapy. J Orthop Sports Phys Ther. 1983;5:66–73. https://doi.org/10.2519/jospt.1983.5.2.66.
41. Kannus P. Immobilization or early mobilization after an acute soft-tissue injury? Phys Sportsmed. 2000;28:55–63. https://doi.org/10.3810/psm.2000.03.775.
42. Manfredini M, Zerbinati F, Gildone A, Faccini R. Autologous chondrocyte implantation: a comparison between an open periosteal-covered and an arthroscopic matrix-guided technique. Acta Orthop Belg. 2007;73:207–18.
43. Hirschmüller A, Baur H, Braun S, Kreuz PC, Südkamp NP, Niemeyer P. Rehabilitation after autologous chondrocyte implantation for isolated cartilage defects of the knee. Am J Sports Med. 2011;39:2686–96. https://doi.org/10.1177/0363546511404204.
44. Laskowski ER, Newcomer-Aney K, Smith J. Refining rehabilitation with proprioception training: expediting return to play. Phys Sportsmed. 1997;25:89–104. https://doi.org/10.3810/psm.1997.10.1476.
45. Lautamies R, Harilainen A, Kettunen J, Sandelin J, Kujala UM. Isokinetic quadriceps and hamstring muscle strength and knee function 5 years after anterior cruciate ligament reconstruction: comparison between bone-patellar tendon-bone and hamstring tendon autografts. Knee Surg Sports Traumatol Arthrosc. 2008;16:1009–16. https://doi.org/10.1007/s00167-008-0598-7.
46. Lee GM, Poole CA, Kelley SS, Chang J, Caterson B. Isolated chondrons: a viable alternative for studies of chondrocyte metabolism in vitro. Osteoarthr Cartil. 1997;5:261–74.
47. Lephart SM, Pincivero DM, Rozzi SL. Proprioception of the ankle and knee. Sports Med. 1998;25:149–55.

48. Mithoefer K, Hambly K, Logerstedt D, Ricci M, Silvers H, Della Villa S. Current concepts for rehabilitation and return to sport after knee articular cartilage repair in the athlete. J Orthop Sports Phys Ther. 2012;42:254–73. https://doi.org/10.2519/jospt.2012.3665.
49. Quinn TM, Grodzinsky AJ, Buschmann MD, Kim YJ, Hunziker EB. Mechanical compression alters proteoglycan deposition and matrix deformation around individual cells in cartilage explants. J Cell Sci. 1998;111(pt 5):573–83.
50. Roberts D, Fridén T, Stomberg A, Lindstrand A, Moritz U. Bilateral proprioceptive defects in patients with a unilateral anterior cruciate ligament reconstruction: a comparison between patients and healthy individuals. J Orthop Res. 2000;18:565–71. https://doi.org/10.1002/jor.1100180408.
51. Sturnieks DL, Besier TF, Hamer PW, et al. Knee strength and knee adduction moments following arthroscopic partial meniscectomy. Med Sci Sports Exerc. 2008;40:991–7. https://doi.org/10.1249/MSS.0b013e318167812a.
52. Van Assche D, Staes F, Van Caspel D, et al. Autologous chondrocyte implantation versus microfracture for knee cartilage injury: a prospective randomized trial, with 2-year follow-up. Knee Surg Sports Traumatol Arthrosc. 2010;18:486–95. https://doi.org/10.1007/s00167-009-0955-1.
53. Saris DB, Vanlauwe J, Victor J, et al. Treatment of symptomatic cartilage defects of the knee: characterized chondrocyte implantation results in better clinical outcome at 36 months in a randomized trial compared to microfracture. Am J Sports Med. 2009;37(suppl 1):10S–9S. https://doi.org/10.1177/0363546509350694.
54. Wallis JA, Taylor NF. Pre-operative interventions (non-surgical and non-pharmacological) for patients with hip or knee osteoarthritis awaiting joint replacement surgery—a systematic review and meta-analysis. Osteoarthr Cartil. 2011;19:1381–95. https://doi.org/10.1016/j.joca.2011.09.001.
55. Wondrasch B, Zak L, Welsch GH, Marlovits S. Effect of accelerated weight bearing after matrix-associated autologous chondrocyte implantation on the femoral condyle on radiographic and clinical outcome after 2 years: a prospective, randomized controlled pilot study. Am J Sports Med. 2009;37(suppl 1):88S–96S. https://doi.org/10.1177/0363546509351272.
56. Reider B, Arcand MA, Diehl LH, et al. Proprioception of the knee before and after anterior cruciate ligament reconstruction. Arthroscopy. 2003;19:2–12. https://doi.org/10.1053/jars.2003.50006.
57. Petersen W, Zelle S, Zantop T. Arthroscopic implantation of a three dimensional scaffold for autologous chondrocyte transplantation. Arch Orthop Trauma Surg. 2008;128:505–8. https://doi.org/10.1007/s00402-007-0348-1.
58. Reiser RF 2nd, Broker JP, Peterson ML. Knee loads in the standard and recumbent cycling positions. Biomed Sci Instrum. 2004;40:36–42.
59. Stevens JE, Mizner RL, Snyder-Mackler L. Neuromuscular electrical stimulation for quadriceps muscle strengthening after bilateral total knee arthroplasty: a case series. J Orthop Sports Phys Ther. 2004;34:21–9. https://doi.org/10.2519/jospt.2004.34.1.21.
60. Ronga M, Grassi FA, Bulgheroni P. Arthroscopic autologous chondrocyte implantation for the treatment of a chondral defect in the tibial plateau of the knee. Arthroscopy. 2004;20:79–84. https://doi.org/10.1016/j.arthro.2003.11.012.
61. Kean CO, Birmingham TB, Garland SJ, Bryant DM, Giffin JR. Preoperative strength training for patients undergoing high tibial osteotomy: a prospective cohort study with historical controls. J Orthop Sports Phys Ther. 2011;41:52–9. https://doi.org/10.2519/jospt.2011.3490.
62. Foley A, Halbert J, Hewitt T, Crotty M. Does hydrotherapy improve strength and physical function in patients with osteoarthritis—a randomised controlled trial comparing a gym based and a hydrotherapy based strengthening programme. Ann Rheum Dis. 2003;62:1162–7.
63. Løken S, Ludvigsen TC, Høysveen T, Holm I, Engebretsen L, Reinholt FP. Autologous chondrocyte implantation to repair knee cartilage injury: ultrastructural evaluation at 2 years and long-term follow-up including muscle strength measurements. Knee Surg Sports Traumatol Arthrosc. 2009;17:1278–88. https://doi.org/10.1007/s00167-009-0854-5.

18 Physiotherapy in Orthopedic Knee Injuries: Rehabilitation Program Following Treatment of Meniscus Repair

Mehmet Emin Şimşek and M. İ. Safa Kapıcıoğlu

18.1 Introduction

Meniscus rehabilitation protocols show differences according to meniscus repair type, size, and location of the lesion and whether or not simultaneous procedures were applied (e.g., ligament reconstructions) [1, 2]. The surgeon is responsible for informing the physical therapy team of the details related to the type of repair made. Meniscus repairs made with all internal fixators have low holding strength, and a few sutures are often used [3–6]. These repairs require more protection throughout the first 6 weeks postoperatively to provide healing. Internal–external meniscus repair techniques include many vertical divergent sutures and have superior holding strength. Peripheral red-red (R/R) zone heal rapidly, and mixed repairs extending to the central red-white (R/W) zone tend to heal more slowly and require greater attention. Also, when there is a significant joint cartilage damage during the arthroscopic procedure, changes can be made to the postoperative exercise program [1–7]. Patients should be given instructions preoperatively related to the postoperative protocol so that they fully understand what is expected postoperatively. Patients are warned about the risk of a repeated meniscus tear or a tear requiring a transplant in the early period as a result of the impact, jogging, or strenuous activity such as deep knee flexion or pivot. These characteristics are valid for the first 4–6 months postoperatively. Preoperative considerations must include any deep vein thrombosis (DVT) risk factors for surgery and an inherited clotting disorder in the patient or family history. A supervised rehabilitation program is supported by daily home exercises [2, 5–8]. The therapist applies the appropriate protocol and examines the patient regularly in the clinic for progression. For successful rehabilitation, the necessary treatment procedures and methods are used [4, 8, 9]. On average, 11–16 physical therapy sessions are required in a 9- to 12-month period for the desired results to be obtained. Lateral and anterior plain radiographs are taken 1 week postoperatively to confirm the position of the osseous component of the transplants, and at 6–8 weeks to confirm healing and attachment of the bone section within the transplant slot or tunnels. The onset of pain or clicking in the tibiofemoral joint line can show that meniscus repair or transplantation has not been successful and the surgeon must be informed immediately for re-evaluation [2, 5].

M. E. Şimşek (✉)
Department of Orthopaedics and Traumatology, Ankara Lokman Hekim University, Sincan Hospital, Ankara, Turkey

M. İ. S. Kapıcıoğlu
Department of Orthopaedics and Traumatology, Faculty of Medicine, Ankara Yıldırım Beyazıt University, Ankara, Turkey

M. Bozkurt, H. İ. Açar (eds.), *Clinical Anatomy of the Knee*,
https://doi.org/10.1007/978-3-030-57578-6_18

18.2 Early Postoperative Management

Significant early postoperative symptoms for follow-up by the therapist include effusion, pain, gait, knee flexion and extension, patellar mobility, lower extremity strength and control, lower extremity flexibility and tibiofemoral symptoms showing meniscus tear. Careful follow-up is made of the DVT prophylaxis protocol including aspirin (300 mg/day), compression dressings, thromboembolic stockings, early ambulation, and active ankle pumps (hourly) [10–14]. On a postoperative day 1, physical therapy is applied with postoperative dressings and bilateral underarm crutches. In addition to compression, cryotherapy is very important throughout this period [9–13]. The patient receives a cold unit that is used 6–8 times a day. The use of various cryotherapy machines in the clinic provides compression at the same time as the cooling program. In the first week, lower extremity elevation should be provided as often as possible. A portable neuromuscular electric stimulator may be useful for quadriceps re-education and pain management. This device is used for 15 min, 4–6 times per day until the patient shows first voluntary quadriceps contraction. The first response to surgery and progression in the first 2 weeks determine the initial stages of the rehabilitation program. Commonly seen postoperative complications include excessive pain or swelling, quadriceps shutdown or loss of voluntary isometric contraction, limited ROM and saphenous nerve irritations for medial repairs. The patient must be observed for complaints such as posteromedial or infrapatellar burning, posteromedial sensitivity along the distal pes anserine tendons, the sensitivity of the Hunter canal along the medial thigh, excessive sensitivity to mild pressure, and excessive sensitivity to changes in temperature [4, 6, 7, 12, 13] (Tables 18.1, 18.2 and 18.3).

18.3 Brace and Crutch Support

A long-leg postoperative brace is applied to all patients immediately postoperatively to prevent hyperflexion or knee rotation that could lead to repair or transplantation damage. The brace is angled from 0° to 90°, but for the first 2 weeks, it is locked at 0° at nights. Then, in patients who cannot obtain 0° extension, the brace is routinely not locked. The brace is used for 4 weeks. Partial weight-bearing crutches are recommended for all the patients for the first 4 weeks. Patients with radial meniscus tear repair are not permitted any weight-bearing for 4 weeks to protect the repair region [15–19] (Figs. 17.3 and 18.1).

Table 18.1 Rehabilitation program after meniscectomy

	Phase I (0–2 weeks)	Phase II (3–8 weeks)	Phase III (8 weeks and after)
Goals	• Diminish pain, edema • Restore knee range of motion (goal 0–115, minimum of 0 degrees extension to 90 degrees of flexion to 120 degrees. • Reestablish quadriceps muscle activity/re-education • Educate the patient regarding weight-bearing as tolerated, use of crutches, icing, elevation, and the rehabilitation process	• Restore and improve muscular strength and endurance • Reestablish full pain-free ROM • Gradual return to functional activities • Restore normal gait without an assistive device • Improve balance and proprioception	• Enhance muscular strength and endurance • Maintain full ROM • Return to sport/functional activities/work tasks

Table 18.1 (continued)

	Phase I (0–2 weeks)	Phase II (3–8 weeks)	Phase III (8 weeks and after)
ROM	Passive range of motion: • Prone hang or heel prop (with or without ice) Active range of motion: • Ankle pumps • Quadriceps sets • Straight-leg raises • Hip abduction, adduction, extension • Gluteal sets • Heel slides • Stationary bike (no resistance, for range of motion only) Strengthening: • Partial squats (complete with the supervision of physical therapist or certified athletic trainer) Stretching: • Active assistive range of motion stretching (quadriceps, gastroc, soleus, and hamstring)	Passive range of motion: • Continue all from phase I Active range of motion: • Continue all from phase I Strengthening: • Toe raises and calf raises • Hamstring curls • Lunges—Front and lateral • Leg press (not past 90 degrees) • Step-up, step down, lateral step-up • Four-way hip: Standing • Knee extension exercise: 90 to 40 degrees (do not use machine) • TKE • Closed kinetic chain Stretching: • Continue all from phase I	• Begin plyometrics/ vertical jumps • Initiate running program • Agility drills progression • Forward/ backwards running, cutting, figure eight and carioca • Sports-specific drills/activities (begin with non-contact)
WB	• Weight-bearing as tolerated Use two crutches initially progressing to weaning crutches as swelling and quadriceps status dictates	• Patients may progress to full weight-bearing as tolerated without analgia • Patients may require one crutch or cane to normalize gait before ambulating without assistive device	
Precautions			
Brace	• Knee immobilizer • TED hose to be worn until ambulation has returned to normal pre-surgery level or 2 weeks • Crutches: Weight-bearing as tolerated	• None	• None
Therapeutic exercise, treatment recommendations, and return to sport and work	• Quadriceps sets • SLR • Hip adduction, abduction, and extension • Ankle pumps • Gluteal sets • Heel slides • Active-assisted ROM stretching, emphasizing full knee extension (flexion to tolerance) • Hamstring and gastroc/soleus and quadriceps stretches • Use of compression wrap or brace • Bicycle for ROM when patient has sufficient knee ROM • Begin partial revolutions to recover motion if the patient does not have sufficient knee flexion	• Continue all exercises as needed from phase I • Toe raises and calf raises • Hamstring curls • Continue bike for motion and endurance • Cardio equipment—Stair master, elliptical trainer, treadmill and bike as above • Lunges—Lateral and front • Leg press • Lateral step-ups, step-downs, and front step-ups • Knee extension 90 to 40 degrees • Closed kinetic chain exercise terminal knee extension • Four-way hip exercise in standing • Proprioceptive and balance training • Stretching exercises—As above, may need to add ITB and/or hip flex or stretches	• Continue to emphasize closed-kinetic chain exercises • May begin plyometrics/ vertical jumping • Begin running program and agility drills (walk-jog) progression, forward and backward running, cutting, figure of eight and carioca program • Sport-specific drills

18.4 Range of Knee Motion

On a postoperative day 1, passive knee flexion and passive and active/active-supported knee extension exercises are started. Active knee flexion is limited to avoid stretching the hamstrings. The ROM exercises are initially applied from 0° to 90° in a seated position. Flexion is gradually increased to 120° in the third–fourth week and to 135° in the sixth week. ROM may have to be restricted to 0° to 90° in the first 2 weeks in patients with comprehensive repairs. The knee movement exercises are applied 3–4 times per day until normal movement is obtained. Hyperextension must be avoided in patients with anterior horn meniscus repair [13, 17, 19–22]. When the knee movements at 0° to 90° cannot be easily achieved at the end of the first week, the patient may be at risk of knee movement complications. To elevate the hamstrings and

Table 18.2 Rehabilitation program after meniscus repair

	Phase I (0–6 weeks)	Phase II (7–12 weeks)	Phase III (13–26 weeks)
Goals	• Protection of the post-surgical knee • Restore normal knee extension • Eliminate effusion (swelling) • Restore leg control • Pain-free gait without crutches • No effusion (swelling)	• Single leg stand control • Normalize gait • Good control and no pain with functional movements, including step-up/down, squat, partial lunge (between 0° and 60° of knee flexion)	• Good control and no pain with sport- and work-specific movements
ROM	ROM 0–120° • Knee extension on a bolster • Prone hangs • Supine wall slides • Heel slides • Knee flexion off the edge of the table	• Passive, 0–135° (full) • Gastroc/soleus stretch • Hamstring/quad/ITB stretch • Prone hang to reach goal as needed • Patellar mobilization	• Continue all stretching activities • Continue all exercises from previous phases
WB	• Pain-free gait without crutches • 2 weeks after surgery toe-touch weight-bearing • ROM with crutches • 4 weeks after surgery PWB to FWB with crutches as tolerated	• Normal gait on all surfaces • Ability to carry out functional movements without unloading affected	
Precautions	• The patient may gradually wean from two crutches to one crutch to no crutches as long as the knee is in the locked knee brace, and there is no increase in pain or swelling for 4 weeks	• No forced flexion with passive range of motion with knee flexion or weight-bearing activities that push the knee past 60° of knee flexion • Avoid post-activity swelling • No impact activities	• Post-activity soreness should resolve within 24 h • Avoid post-activity swelling • Avoid posterior knee pain with end range knee flexion
Brace	• Knee brace locked for all weight-bearing activities for 4 weeks • Do not flex the knee past 90°	As needed	

Table 18.2 (continued)

	Phase I (0–6 weeks)	Phase II (7–12 weeks)	Phase III (13–26 weeks)
Therapeutic exercise, treatment recommendations	• Quadriceps sets • Straight-leg raises • 4-way leg lifts in standing with brace on for balance and hip strength • Abdominal isometrics • Upper body circuit training or upper body ergometer	• Non-impact balance and proprioceptive drills • Stationary bike • Gait drills • Hip and core strengthening • Stretching for patient-specific muscle imbalances • Quadriceps strengthening, making sure that closed chain exercises occur between 0° and 60° of knee flexion • Non-impact endurance training: Stationary bike, Nordic track, swimming, deep water running or cross-trainer	• Impact control exercises beginning 2 ft to 2 ft, progressing from 1 foot to the other and then 1 foot to the same foot • Movement control exercises beginning with low velocity, single-plane activities and progressing to higher velocity, multi-plane activities • Strength and control drills related to sport-specific movements • Sport/work-specific balance and proprioceptive drills • Hip and core strengthening • Stretching for patient-specific muscle imbalances • Water walking • Swimming • Backward run • Lateral shuffle • Initiate light plyometric program box hops, level, double-leg, and sport-specific drills

gastrocnemius that permit the knee to move to full extension, a towel is placed over the foot and ankle. To stretch the posterior capsule, weights of 10–15 pounds can be added to the distal thigh and knees. This program is applied for 10 min, 6–8 times per day [19–22].

Flexion exercises are applied in the sitting position using the contralateral extremity to provide excess pressure. Chair dips, wall dips, passive quadriceps stretching and ROM devices are helpful in regaining full knee flexion. Squatting exercises should not be performed for at least 5 months as these exercises form great tension forces on posterior meniscus repair and transplants. Patellar mobilization accompanying ROM exercises is of the most important to achieve full knee movement [23]. Flexibility exercises starting with the hamstring and gastrocnemius-soleus, are applied three times a week starting on postoperative day 1. Quadriceps and iliotibial band flexibility exercises are added at 7–8 weeks postoperatively. Patients who have undergone combined procedures may require close observation and additional exercises to correct natural knee movements successfully. There is no difference between medial and lateral meniscus repair or transplants in respect to knee movement complications [6, 11, 17, 23, 24] (Fig. 18.2).

18.5 Balance, Proprioception, and Neuromuscular Training

The restoration of normal neuromuscular function following meniscus repair and transplantation is essential for a successful outcome. Knee

Table 18.3 Rehabilitation program after meniscus allograft transfer

	Phase I (0–8 weeks)	Phase II (9–12 weeks)	Phase III (13–20 weeks)	Phase IV (20 weeks and after)
Goals	• ROM 0–90° • Control pain, inflammation, and effusion	• Partial WB to full WB • ROM 0–135° • Control pain, inflammation, and effusion • Increase lower extremity strength • Enhance proprioception, balance, and coordination	• Maintain full range of motion • Increase lower extremity strength and endurance • Initiate functional activity • Initiate sport-specific activity	• Enhance lower extremity strength and endurance • Return to previous activity level • Return to sport-specific functional level
ROM	• 0 to 2 weeks: 0 to 60 degrees with CPM • 2 to 4 weeks: 0 to 90 degrees with CPM • 4 to 8 weeks: Full motion	• Full active range of motion	• Full active range of motion	• Full active range of motion
WB	• Foot flat for balance for the first 10 days • Progress to weight-bearing as tolerated after suture removal	• Progression to full weight-bearing normalized gait pattern; no limping	Full weight-bearing	Full weight-bearing
Precautions				
Brace	• Locked in full extension for sleeping and all activity off for exercises and hygiene • 2–8 weeks: Locked 0–90° off at night • Discontinue brace at 8 weeks	• No brace	• No brace	• No brace

	Phase I (0–8 weeks)	Phase II (9–12 weeks)	Phase III (13–20 weeks)	Phase IV (20 weeks and after)
Therapeutic exercise, treatment recommendations, and return to sport and work	• Patellar mobilization • No weight-bearing with flexion >90° • Addition of heel raises • Activities with brace until 6 weeks; then without brace as tolerated • No weight-bearing with flexion >90° 0 to 2 weeks • Prone hangs, heel props, quad sets, straight-leg raises (SLR), hamstring isometrics; complete exercises in brace if quad control is inadequate • Core proximal program • Normalize gait • Functional electrical stimulation (FES) biofeedback as needed 4 to 6 weeks • Addition of heel raises, total gym adduction/abduction exercises and incorporate use of stationary bike (high seat, low resistance) • No weight-bearing with flexion more than 60 degrees during phase I	• Progress closed chain activities • Begin hamstring work, lunges/leg press 0–90°, proprioception exercises, balance/core/hip/glutes • Begin stationary bike • Begin closed chain activities: Mini-squats 0 to 45 degrees progressing to step-ups, leg press • 0 to 60 degrees, closed chain terminal knee extensions, toe raises, balance activities, hamstring curls, increase to moderate resistance on bike	• Progress phase III exercises and functional activities: Single leg balance, core, glutes, hamstrings, elliptical, and bike • Advance bilateral and unilateral closed chain exercises with emphasis on concentric/eccentric control • Continue with biking, elliptical and walking on treadmill • Progress balance activities	• Advance to sport-specific drill sand running/jumping once cleared by MD

joint proprioception and balance are essential components of neuromuscular function. Therefore, balance and proprioception exercises are typically started as soon as there is partial weight-bearing in the first week after surgery. Crutches are used for support during these exercises until full weight-bearing is permitted [3, 24–27]. Tandem balance is started during the partial weight-bearing phase to help a sense of place and balance. A mini-trampoline is used to make this exercise more difficult after mastering it on a hard surface. There are several devices to help with balance and gait, primarily, styrofoam hemispheres and cylinders, and the Biomechanic Ankle Platform System (BAPS). Patients cannot walk (without assistance) on styrofoam half rolls to develop a balance center, central quadriceps control, and postural location [19–22, 24, 25]. These devices provide visual feedback for assistance in various balance activities. More advanced exercises are needed to support the neuromuscular function. These include forward steps, sidesteps, single-leg balance drills, advanced perturbation training with a therapist, and movement exercises in various directions using a resistance band on an unstable surface. Correct posture training is essential for the patient when doing these exercises [4, 11, 21, 28]. This includes preventing varusor valgus alignment of the lower extremity, maintaining knee flexion to prevent knee hyperextension, preventing hip adduction and internal rotation, maintaining balance and control throughout the exercise and a slight lowering to reduce the ground reaction forces. The progression of exercises to running, pivot/cutting, plyometric and a

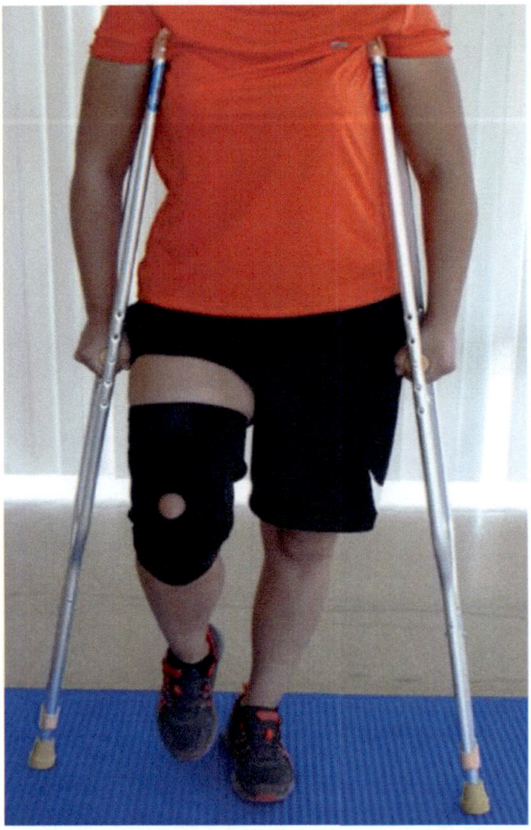

Fig. 18.1 Using two crutches

Fig. 18.2 Balance and strengthening training exercises with different surface stability types

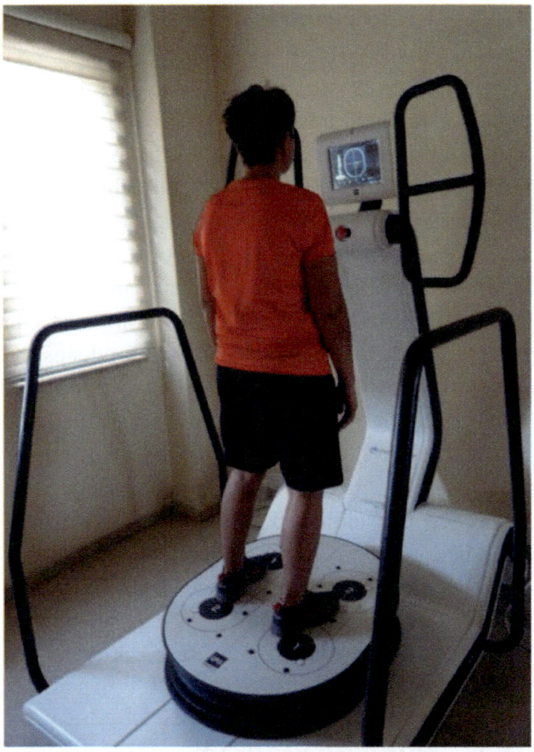

Fig. 18.3 Neuromuscular balance training device

return to full activities is based on specific criteria explained in later sections. Catheter prevention exercises are especially important in young athletic patients with ACL reconstruction [4, 19, 28–30] (Figs. 18.2 and 18.3) (Tables 18.1, 18.2 and 18.3).

18.6 Strengthening

The strengthening program starts on postoperative day 1. initially, straight-leg raises—elevation exercises are performed in the flexion plane only. Before adding straight-leg raises in the other three planes (abduction, adduction, and extension), the patient must achieve sufficient quadriceps contraction to eliminate extensor delay [9, 14, 22, 26].

These exercises are applied as 3–10 repetitions in 3–5 sets, and this set/repetition rule allows the systematic progression of ankle weights as tolerated. For gastrocnemius-soleus strengthening, foot raises, wall-sitting and mini-squats for quadriceps strengthening are started when there is 50% weight-bearing. Sitting leaning against the wall exercises are started at 5–6 weeks after meniscus transplantation surgery. To protect the posterior horn of the meniscus, these activities must be limited to 0°-60° flexion [12, 17, 19]. By changing the exercise technique, the wall sitting isometry can be made more difficult. First, when the patient reaches a maximum knee flexion angle of 30–45°, the quadriceps muscle tone can be voluntarily adjusted. This contraction and knee flexion position is held until the muscles tire. In a second modification designed to support a stronger vastus medialis obliquus contraction, the patient contracts the hip in adduction while squeezing a ball between the distal thighs [3, 19, 24, 30]. In a third variation, the patient holds dumbbells to increase body weight, and this encourages stronger quadriceps contraction. To simulate a single contraction or to sit against the wall on one leg, the patient shifts the body weight to the proper side. A TheraBand or surgical pipe is used as a resistance mechanism. The depth of the squat is controlled to protect the meniscus repair, transplant, and the patellofemoral joint. Open kinetic chain, non-weight-bearing exercises are started at 5–6 weeks postoperatively. Knee extension resistance exercises to protect the patellofemoral joint start at 90°-30°. In this protective ROM, minimal force is placed along the peripheral and mid-substance repair regions by the quadriceps exercises. There is a progression from ankle weights to machines when the number of weight increases in the patient exercise program. Quadriceps control is of critical importance for the progression of the program. Attention must be paid to avoid hyperextension. This exercise is postponed until at least 7–8 weeks after a complex meniscus repair and until 9–12 weeks after meniscus transplantation. Isolated resistant hamstring curls are restricted in complex repairs and transplants because of the medial hamstring insertion along the posteromedial joint capsule. This restriction is designed to reduce the potential shear forces applied to the repair area. Lateral lying straight-leg raises are added to the rehabilitation program in the early stage. Then, when patients have access to equipment, an abduction-adduction machine, and a multi-column machine or cable column for hip flexion, extension, abduction, and adduction are included in the exercise program. These activities are applied at 5–6 weeks after surgery [2–4, 21, 22, 25, 26, 28–30] (Figs. 18.4 and 18.5).

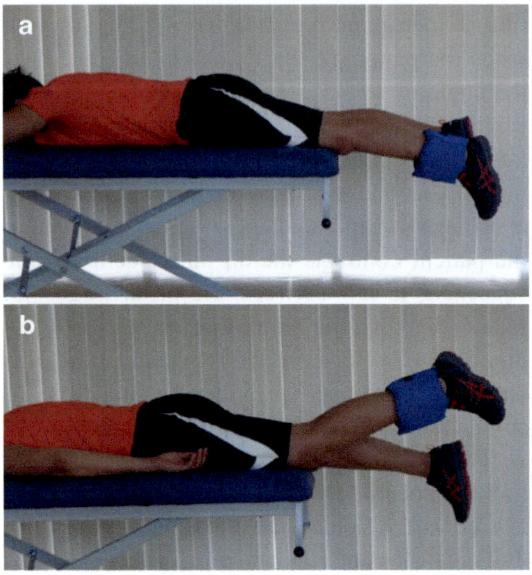

Fig. 18.4 Prone knee hangs to restore full passive knee extension (**a**), active 30°knee flexion to restore hamstring strength (**b**)

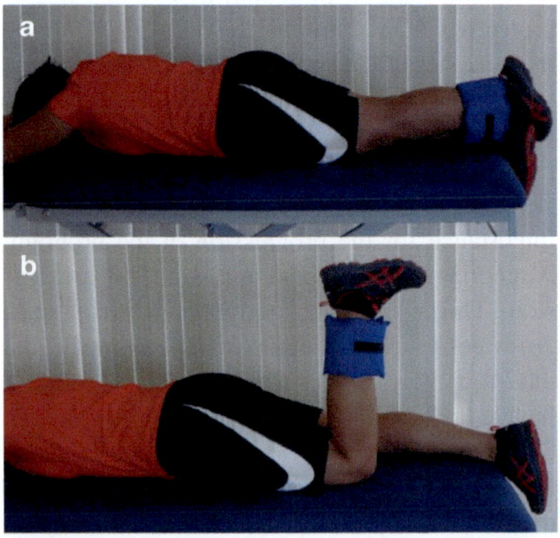

Fig. 18.5 Outer-range resisted isometric hamstring curl, 20° knee flexion (**a**), single-leg dynamic resisted hamstring curl (**b**)

18.7 Conditioning

A cardiovascular program can start at 2–4 weeks postoperatively. Fixed cycling can be started at 7–8 weeks with the saddle adjusted to the highest level according to the patient's height, and low resistance is used. A reclining bicycle can be used for patients where there is anterior knee pain or where patellofemoral joint articular cartilage may be damaged. In this period, walking in water can be applied, as water up to waist level reduces the impact on the knee by 50%. In patients with symptoms or joint cartilage damage, it is necessary to protect the patellofemoral joint against high stress. If a stair-climbing machine can be tolerated, a short step is established with low resistance levels. The cardiovascular program should be applied for 30–40 min at least three times a week, and the exercises should be performed at least 75%–90% of maximum heart rate [2, 22, 26, 27, 30].

18.8 Running Program

Isometric tests made on a Biodex dynamometer for patients with peripheral meniscus repair, which provide mean 80% healing at peak torque for the quadriceps and hamstrings, can be started as a running program at approximately 18–20 weeks postoperatively. This program can be postponed to 30 weeks or up to 1 year for patients with complicated meniscus repair or meniscus transplantation. Initially, patients work at 30%–60% of average speeds. When patients can run straight at full speed, lateral and transverse maneuvers are added. To work on speed and agility, short distances are used. Side-to-side running on cups can be used to facilitate agility and proprioception [2, 4, 16, 23, 29–32] (Tables 18.1, 18.2 and 18.3).

18.9 Plyometric Training

This training program is generally started 6 months postoperatively for patients with a large peripheral or complex repair. For patients with radial meniscus repair, it can be postponed

Fig. 18.6 Single-leg dynamic resisted knee extension

until 9 months because of impairment occurring with the stress on the meniscus ring. The majority of patients with meniscus transplantation have deteriorated articular joint cartilage and are not suitable candidates for strenuous plyometric training. Long-term training starts with individual sessions in the same way as recurrent training. Initially, the length of the exercise period includes a rest period 2–3 times longer, and this is gradually reduced to the exercise duration. Stress meter is applied 2–3 times a week added to the strength and cardiovascular resistance exercises. Plyometrics should be applied on a hard but forgiving surface such as a wooden sports hall floor and hard surfaces such as concrete must be avoided. Cross-training or running shoes should be worn, which provide sufficient stability for the foot as well as shock-absorbance. For patients with combined meniscus repair and ACL reconstruction, advanced-level plyometric training is recommended before resuming high-risk sporting activities such as football or basketball [25–29, 31, 32] (Figs. 18.4, 18.5 and 18.6).

18.10 Return to Sports Activities

Isokinetic testing is generally repeated once a month depending on the athletic targets of the patient, progressing throughout the first 6 months from isometric tests to isokinetic tests at rates of 180 and 300 degrees per second. This test not only gives feedback about performance but also assists the clinician in advancing the program. The test targets should be to meet at least 70% of bilateral torque initially and to be 90% and approximately 60% agonist/antagonist ratio for full activity. Torque/body weight ratios are based on the parameters of age, gender, and body weight. Return to sports activities are based on successful completion of the running and functional training program [4, 18, 23, 24, 26, 28–30]. Attempting a function is encouraged, and throughout this period, the patient is observed for excessive symptoms.

Due to impaired articular joint cartilage in the majority of patients who have undergone meniscus transplantation, they are not suitable candidates for strenuous plyometric training or sports activities. A return to low-impact activities is recommended for these patients [2, 5, 11, 15, 24, 29, 33] (Tables 18.1, 18.2 and 18.3).

References

1. Ulrich GS, Aronczyk SP. The basic science of meniscus repair. Tech Ortho. 1993;8(2):56–62.
2. Arnoczky SP, Warren RF. Microvasculature of the human meniscus. Am J Sports Med. 1982;10:90–5.
3. Morgan CD, Wojtys EM, Casscells CD, Casscells SW. Arthroscopic meniscus repair evaluated by second-look arthroscopy. Am J Sports Med. 1991;19:632–7.
4. Albrecht-Olsen P, Lind T, Kristensen G, Falkenberg B. Failure strength of a new meniscus arrow repair technique: biomechanical comparison with horizontal suture. Arthroscopy. 1997;13:183–7.
5. Bryant D, Dill J, Litchfield R, et al. Effectiveness of bioabsorbable arrows compared with inside-outsuturing for vertical, reparable meniscal lesions: a randomized clinic altrial. Am J Sports Med. 2007;35:889–96.
6. Bullough PG, Munuera L, Murphy J, Weinstein AM. The strength of the menisci of the knee as it relates to their fine structure. J Bone Joint Surg Br. 1970;52:564–70.
7. Buma P, Ramrattan NN, van Tienen TG, Veth RP. Tissue engineering of the meniscus. Biomaterials. 2004;25:1523–32.
8. Dugdale TW, Noyes FR, Styer D. Preoperative planning for high tibial osteotomy: the effect of lateral tibio femoral separation and tibio femoral length. Clin Orthop Relat Res. 1992;274:248–64.

9. Meniscal repair: a clinical and magnetic resonance imaging evaluation. Arthroscopy. 1996;12(6):680–6.
10. Barber FA, Harding NR. Meniscal repair rehabilitation. In: AAOS Instructional Course Lectures, vol. 49; 2000. p. 207–9.
11. Buseck MS, Noyes FR. Arthroscopic evaluation of meniscal repairs after anterior cruciate ligament reconstruction and immediate motion. Am J Sports Med. 1991;19(50):489–94.
12. DeHaven KE. Basic science, indications for repair, and open repair. J Bone Jt Surg. 1994;76A(1):140–52.
13. DeHaven KE. Meniscus repair. Am J Sports Med. 1999;27:242–50.
14. Davies GJ, Zillmer DA. Functional progression of exercise during rehabilitation in knee ligament rehabilitation: Ellenbecker; 2000. p. 345–60.
15. McClure PW, Blackburn LG, Dusold C. The use of splints in the treatment of joint stiffness: biological rational and algorithm for making clinical decisions. Phys Ther. 1994;74:1101–7.
16. Dowdy PA, Miniaci A, Arnoczky SP, Fowler PJ, Boughner DR. The effect of cast immobilizationon meniscal healing. An experimental study in the dog. Am J Sports Med. 1995;23(6):721–8.
17. Eggli S, Wegmuller H, Kosina J, Huckell C, Jakob RP. Long-term results of arthroscopic meniscal repair. An analysis of isolated tears. Am J Sports Med. 1995;23(6):715–20.
18. Johnson MJ, Lucas GL, Dusek JK, Henning CE. Isolated arthroscopic meniscal repair: a long-term outcome study (more than 10 years). Am J Sports Med. 1999;27(1):44–9.
19. Nyland J, Chang H, Kocabey Y, et al. A cyclic testing comparison of FasT-Fix and RapidLoc devices in human cadaveric meniscus. Arch Orthop Trauma Surg. 2008;128:489–94.
20. Klein L, Player JS, Heiple KG. Isotopic evidence for resorption of soft tissues and bone in immobilized dogs. J Bone Jt Surg. 1982;64:225–30.
21. Mariani PP, Santori N, Adriani E, Mastantuono M. Accelerated rehabilitation after arthroscopic.
22. Thompson WO, Thaete FL, Fu FH, Dye SF. Tibial meniscal dynamics using three-dimensional reconstruction of magnetic resonance images. Am J Sports Med. 1991;19:210–6.
23. McCarty EC, Marx G, Dehaven KE. Meniscus repair: considerations in treatment and update of clinical results. Clin Orthop Relat Res. 2002;1(402):122–34.
24. Mintzer CM, Richmond JC, Taylor J. Meniscal repair in the young athlete. Am J Sports Med. 1998;26:630–3.
25. Sapega AA, Quedenfeld TC. Biophysical factors in range of motion exercises. Phys Sports Med. 1981;9:57–65.
26. Siebold R, Dehler C, Boes L, Ellermann A. Arthroscopic all inside repair using the meniscus arrow: long-term clinical follow up of 113 patients. Arthroscopy. 2007;23:394–9.
27. Lee GP, Diduch DR. Deteriorating outcomes after meniscal repair using the meniscus arrow in knees undergoing concurrent anterior cruciate ligament reconstruction: increased failure rate with long-term follow-up. Am J Sports Med. 2005;33:1138–41.
28. Shelbourne KD, Patel DV, Adsit WS, Porter DA. Rehabilitation after mensical repair. Clin Sports Med. 1996;15(3):595–612.
29. Albrecht-Olsen PM, Bak K. Arthroscopic repair of the bucket handle meniscus. 10 failures in 27 stable knees followed for 3 years. Acta Orthop Scand. 1993;64:446–8.
30. Andersson-Molina H, Karlsson H, Rockborn P. Arthroscopic partial and total meniscectomy: a long-term follow-up study with matched controls. Arthroscopy. 2002;18:183–9.
31. Arnoczky SP, Warren RF. The microvasculature of the meniscus and its response to injury. An experimental study in the dog. Am J Sports Med. 1983;11:131–41.
32. Dugdale TW, Noyes FR, Styer D. The effect of lateral tibio femoral separation and tibio femoral length. Clin Orthop Relat Res. 1992;274:248–64.
33. Ellermann A, Siebold R, Buelow JU, Sobau C. Clinical evaluation of meniscus repair with a bioabsorbable arrow: a 2- to 3-year follow-up study. Knee Surg Sports Traumatol Arthrosc. 2002;10:289–93.

19 Physiotherapy in Orthopedic Knee Injuries: Rehabilitation Program Following Treatment of Posterior Cruciate Ligament Rupture

Mehmet Emin Şimşek and M. İ. Safa Kapıcıoğlu

19.1 Introduction

Physical therapy starts on the first postoperative day with a posterior icepack on the postoperative dressing and a long-leg knee brace locked in full extension with bilateral axillary crutches. The postoperative bandage and dressing are changed to allow the application of thigh-high compression stockings and a Jones bandage [1–4]. Early control of postoperative effusion is important for pain management and early quadriceps re-education. In addition to compression, cryotherapy is important in this period. Throughout the first week, the patient is instructed to hold the lower extremity as tight as possible. The response to surgery and progression in the first 2 weeks determines the initial stages of the rehabilitation program. Commonly seen post-operative complications include excessive pain or swelling, quadriceps shutdown, and limited joint range of motion (ROM). Early identification and treatment of these problems are critical for a successful result. In patients with quadriceps tendon autograft, pain may increase during knee flexion and quadriceps contraction. To be sure that the rehabilitation targets are reached, care must be taken in the first four postoperative weeks [2, 3]. When the patient is doing ROM exercises, an electrode is placed in the center of the hamstring muscle. Cryotherapy is started postoperatively in the recovery room. There are several different modalities for use both in the clinic and at home. For most patients, a bag of ice or a commercial cold pack is applied, and patients can have a commercial cold unit to be used at home, 4–8 times per day. Of the cold therapy units which can be purchased, there are mobile cold units that provide fixed hot and cold water circulation along a pad providing excellent pain control [1, 2, 4–6]. Gravity-flow units also provide effective pain management, although it is harder to maintain a fixed temperature with these devices. The temperature can be controlled by releasing the water and using reverse flow gravity, and when necessary the sleeve can be filled with fresh iced water. Standard treatment is applied for 30 min five times a day depending on the degree of pain and swelling. In some cases, the treatment time can be extended depending on the thickness of the buffer used between the skin and the device. Cryotherapy is generally used when necessary for pain and swelling control or after exercise and is retained as a part of the whole postoperative rehabilitation protocol [3–5].

M. E. Şimşek (✉)
Department of Orthopaedics and Traumatology, Ankara Lokman Hekim University, Sincan Hospital, Ankara, Turkey

M. İ. S. Kapıcıoğlu
Department of Orthopaedics and Traumatology, Faculty of Medicine, Ankara Yıldırım Beyazıt University, Ankara, Turkey

M. Bozkurt, H. İ. Açar (eds.), *Clinical Anatomy of the Knee*,
https://doi.org/10.1007/978-3-030-57578-6_19

19.2 Brace Support

In patients who have undergone PCL reconstruction, a long-leg hinged brace with a posterior calf pad is applied throughout the first postoperative 6 weeks. To prevent sudden knee flexion movements, the brace is worn for 24 h a day, including sleeping time. For the first 3–4 weeks, the brace is fully extended, and then it is angled to 0–90°. For patients evaluated with physiological joint laxity or poor lower extremity muscle control, the brace is used for up to 12 weeks postoperatively [5–7]. At this stage, a patient with indications for a higher level of occupational or sports activity can be measured for a functional PCL brace. For patients returning to lower level activities or those who develop patellofemoral symptoms, a patellofemoral knee sleeve may be used for long periods of standing or walking. Standard soft hinged support is not accepted as sufficient for the protection of complex reconstructions including both the PCL and posterolateral structures. A sufficient degree of healing should be obtained in the posterolateral ligamentous reconstructive procedure in the first four postoperative weeks [4, 5, 7] (Fig. 18.1).

19.3 The Range of Knee Motion

Passive knee movements of 0–90° are permitted immediately postoperatively. A return to 135° full knee flexion causes high shear forces in the PCL reconstruction, and the time that this degree of flexion is safe is not known. Accordingly, in the first 8 weeks, the rehabilitation program aims to regain 120° and flexion is obtained gradually in 10–12 weeks. When there are no sharp or occupational indications, knee flexion of more than 135° is not encouraged. To limit excessive wear on the PCL graft in the tibial and femoral tunnels, knee movement sets are restricted to 60 per day for the first 4 weeks (20 sets—three times per day). To prevent excessive scarring in the intercondylar notch or posterior capsule contracture, it is essential to regain knee extension in as short a time as possible [1, 3, 6, 7].

On the 7th day postoperatively, patients who cannot reach this extension target have extra pressure included in the program [4, 8]. The foot and ankle are raised onto a towel or another device that elevates the hamstrings and gastrocnemius, allowing full extension of the knee. This exercise is maintained for 10 min and is repeated approximately six times a day. To stretch the posterior capsule, 5 kg weights can be added to the distal thigh and knees. To prevent posterior shift regressions, care must be taken to keep the suspended weight away from the proximal tibia. The knee is taken into 0° extension from postoperative week 3 until week 6. To protect the healing PCL graft, hyperextension must be avoided. Passive knee flexion exercises are applied in a sitting position using anterior manual pressure to provide proximal tibial support to prevent falling from the posterior. During passive knee flexion, a 5 kg anterior pull is held on the proximal tibia, because PCL forces significantly increase after 70° flexion [9–12]. Care must be taken not to activate the hamstrings. Other extra pressure techniques that can be applied to assist flexion include an open chair, wall shift using the other extremity, commercial devices and passive quadriceps stretching exercises [8, 11].

19.4 Weight-Bearing

Weight-bearing of 25% of the body weight is permitted throughout postoperative weeks 1 and 2. Gradual advances in weight-bearing and the use of crutches generally terminate in week 6. After the support is unlocked, the weight sequence progresses by limiting the locked knee position and encouraging normal flexion throughout the gait cycle using a usual walking technique [13, 14]. This allows normal regulation of toe-to-heel ambulation, quadriceps contraction in the mid-gait cycle and hip and knee flexion. When patients progress to full-weight-bearing, they should be warned to avoid squatting, hills or ramps or sudden slower movements that will place high forces on the PCL graft. These precautions should be maintained through at least 6 months postoperatively [7, 11, 12].

19.5 Patellar Mobilization

Patellar mobilization is essential to support the full sequence of knee movements. Loss of patellar movement is generally related to the development of patella infera or knee complications. Patellar shifts occur in all four planes (superior, inferior, medial, lateral) with continuous pressure applied to the patellar edge for at least 10 s. This exercise is applied 5 min before the completion of the ROM exercises. Care must be taken if extensor lag is determined, because this could be associated with poor superior migration of the patella, and shows that extra emphasis must be placed on this exercise. Patellar mobilization is applied for approximately 8 weeks postoperatively [12, 13].

19.6 Flexibility

Hamstring and gastrocnemius-soleus flexibility exercises are started on postoperative day 1. A continuous static stretch is held for 30 s and repeated five times. The most common hamstring exercise is the modified obstructed stretching exercise, and the most common gastrocnemius-soleus stretching exercise is the towel pull. These exercises assist in controlling pain that occurs with the reflex response. The towel pull exercise also assists in reducing discomfort in the calf, Achilles tendon, and ankle. These exercises constitute a critical component of the ROM program as the ability of these muscle groups to loosen is necessary for the success of full passive knee extension [6, 7, 12–14]. The patient must be instructed not to make dynamic stretching which could result in the activation of the hamstring muscles. The PCL reconstruction is always protected by protecting the knee extension with a soft posterior calf pad. In the ninth postoperative week, iliotibial band stretching exercises are started. These exercises help to control lateral hip and thigh stricture and provide full knee flexion. When designing a flexibility program, the therapist must consider the position or physical requirements of the activity in addition to the patient's desire to return to specific sports or activities. The flexibility program should be continued after the official discharge of the patient [12, 14–16] (Figs. 19.1 and 19.2).

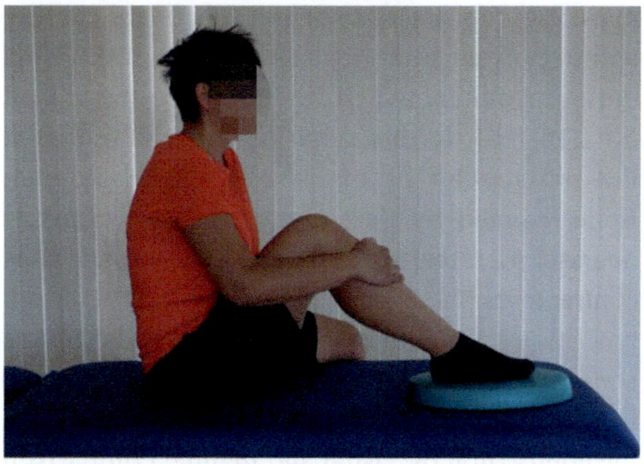

Fig. 19.1 Passive knee flexion on exercise table

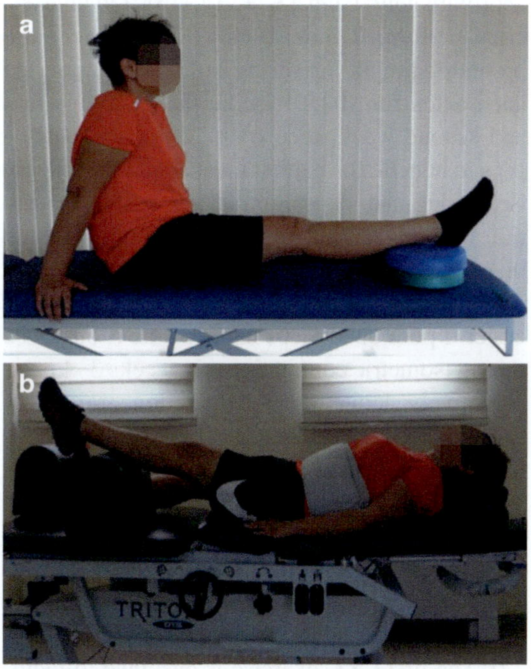

Fig. 19.2 Supine static passive knee hyperextension mobilization (**a**).The heel of the foot is placed in a raised position while the patient lies supine which allows hamstring muscle to relax. The patient pushes a controlled handle to apply force that will put the knee into hyper extension (**b**)

19.7 Strengthening

The strengthening program should be started on the first postoperative visit. It is of critical importance for early rehabilitation of the quadriceps muscle group, a safe return to functional activity and to prevent posterior subluxation of the tibia during activities which flex the knee more than 50° [15, 16]. In this stage of the rehabilitation, the progression of the strengthening program is determined by the start of proper voluntary quadriceps contraction. Isometric quadriceps contractions should be applied for 10 s with ten repetitions, ten times a day. Evaluation of the contractions by both the patient and the therapist is critical. The patient can observe the quality of the contraction by visual and manual comparison with the contralateral side. During contraction, the patella superior-inferior movement should be approximately 1 cm [3, 7, 14, 15]. Other exercises performed immediately after surgery are prone to straight-leg raises and active-supported knee extension (0–70° in the postoperative first and second week, then 0–90°). For the quadriceps to benefit, the patient must achieve sufficient isometric quadriceps contraction with leg elevation. Initially, 2 kg ankle weights are used, and then up to 5 kg ankle weights [1–8, 14].

Active-supported extension exercises are made easier if poor quadriceps muscle tone is observed during isometric contractions. In weeks 3–4 postoperatively, adduction and abduction straight-leg raises are added. Abduction foot raises can be postponed to weeks 7–8 for patients with knees that have undergone simultaneous posterolateral procedures. In week 9, prone extension leg raises are started and continued until at least week 12. When partial weight-bearing starts, closed kinetic chain (CKC) exercises are started [13, 17]. The first CKC exercise is designed to facilitate sufficient quadriceps control when walking to prevent knee hyperextension occurring. When there is 50%–75% weight-bearing, toe raises to strengthen the gastrocnemius-soleus, wall-sitting isometry for quadriceps control and mini-squats for quadriceps strengthening, are started. Wall-sitting aims to strengthen the quadriceps until muscle fatigue [12, 15]. This exercise can be modified to reduce patellar pain or to place additional stress on the quadriceps muscle. Additional stress can be placed on the quadriceps during wall-sitting with various methods [16, 17]. First, when the patient reaches maximum knee flexion of 30–45°, the quadriceps muscle can be voluntarily adjusted. This contraction and knee flexion position is held until the muscle tires, and the exercise is repeated 2–3 times, eight times a day. In the second modification, hip adduction contraction is applied by squeezing a ball between the distal thighs [18–22]. In a third modification, the patient holds dumbbells to increase body weight, and this supports a stronger quadriceps contraction. Finally, to stimulate single-leg contraction, the body weight is shifted to the proper side. These exercises are beneficial and each time are typically held until maximum quadriceps muscle fatigue which cannot be obtained with other exercises and this results in exposure to a real muscle burn. Ideally, the sets should be performed twice and repeated four times a day. In the final CKC exercise, the patient uses his body weight as resistance, and a TheraBand or surgical pipe is used as the resistance mechanism [3, 12, 17, 18, 21]. The depth of the squat is controlled to protect the patellofemoral joint. A fast, smooth squat is repeated at a high-set/high-repetition tempo to increase muscle fatigue. It is essential to observe the hip position for emphasis on the quadriceps. Increased body flexion facilitates increased hamstring contractions, and therefore, this must be carefully monitored for at least 3–6 months to prevent strong hamstring contractions [19–23]. Open kinetic chain (OKC) exercises are included in the rehabilitation program because of the advantage of muscle group isolation provided by weights machines [15, 19] (Figs. 19.3 and 19.4).

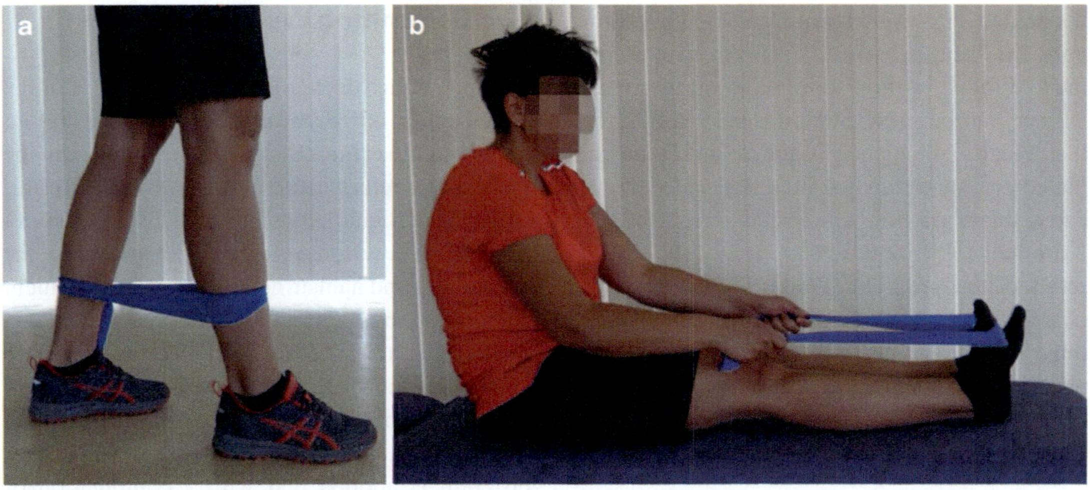

Fig. 19.3 Rhythmic stabilization with band kicking (**a**). Band stretches, while sitting, the patient pushes down on the leg, with the hang just proximal to the patella. The patient pushes down on the leg while pulling on the towel, bringing the knee into hyper extension (**b**)

Fig. 19.4 Core stabilization program. Bridges (**a**), rose-wall slides, (**b**) and side-lying bridges (**c**)

19.8 Balance, Proprioceptive, and Perturbation Training

Balance and proprioceptive training are started at approximately 4–6 weeks postoperatively when the patient is partial weight-bearing. The first exercise includes changing weight from side to side and from front to back. This activity helps the patient to start to understand there liability of the ability of the weight-bearing leg to resist pressure and the stimulating knee joint position [23, 24]. The second exercise is cup-walking which is designed to support walking and to develop symmetry between the operated and non-operated extremities. Cup-walking control ship and knee flexion, quadriceps control in the mid-phase, the transition from pelvic control in the stomach, sufficient gastrocnemius-soleus control during pushing and the hip when walking. When full weight-bearing is achieved, another useful activity for balance control is the single-leg balance exercise [25, 26]. The standing position is vital for the benefit of this exercise. The patient is instructed to push the foot straight-forward, bend the knee to approximately 20–30°, extend the arms outwards horizontally, and with the trunk over the hips, keep the hips vertical over the ankles. This exercise aims to hold the position until the balance is lost. To make this exercise more difficult, a mini-trampoline or an unstable surface can be used. Standing on a soft surface that creates imbalance requires more dynamic leg control than standing on a flat surface. In the early stages of full, unassisted weight-bearing, foam half-rolls are used as a part of the walking re-education program. Walking on half-rolls is helpful in developing the balance and dynamic muscle control necessary to preserve the upright position of the patient and to walk from one end of the roll to the other [21, 24–29] (Figs. 18.2 and 18.3).

19.9 Conditioning

In approximately the third or fourth week postoperatively, a cardiovascular program can be started. To reduce swelling in the leg to a minimum, the surgeon should elevate the limb. This exercise is for tolerance. At postoperative 5–6 weeks, fixed cycling can be started, with the saddle adjusted to the highest level for the patient's height and low resistance should be used during the exercise. To reduce hamstring involvement, foot clips should be avoided. Gradually between the 9th and 12th weeks, cross-skiing, elliptical cross-trainer and stair-climbing machines are included. Patients with symptoms against high stress in the patellofemoral joint or impaired joint cartilage must be well protected [23, 25, 26, 28]. The aim of early conditioning exercises is full ROM, gait retraining and cardiovascular renewal. It is thought that generally greater cardiovascular activity and resistance is provided by a high percentage of maximum heart rate. A full cardiovascular exercise program is an essential component of the stages following rehabilitation. In addition to the previously explained exercises, a hydrotherapy program is started including freestyle or breaststroke swimming, walking in water, water aerobics, and running in deep water. Which cardiovascular exercises are most suitable depends on each patient [16, 19].

19.10 Running Program

Current studies do not allow the prediction of the return of the strength of PCL grafts. Therefore conservative estimates should be made on the subject of a return to strenuous activities. To start a running program, a deficit of >25% should not be seen in low torque in the isometric test for quadriceps and hamstrings, anterior displacement should not be >3 mm in the arthrometry test, and it should be at least 6 months postoperatively. No clear rules have been established about when to start more strenuous activities for patients with allografts [4, 8, 9, 16]. An initial-level running program is applied first with a combination of straightforward running/exercises. Running distances for both forwards and backward should be 20, 40, and 60 m. The initial running speed should be a quarter of the average speed of the patient, and this progresses gradually to one third and then full pace. A rest-train-rest approach is applied with the rest phase 2–3 times longer than the training phase. When the patient can run at full speed straight, the program can be advanced to include side running and passing maneuvers [29, 30]. Short distances, such as 20 m, are used to train for speed and strength. To aid proprioception, walking sideways on glasses can be used. The fourth stage of the running program is identified by cutting designs. These patterns at 45 and 90°, include changes of direction allowing the patient to advance from mild to sharp turns [22–27, 30].

19.11 Plyometric Training

To minimize bilateral changes in neuromuscular function and proprioception, continuous training is started on successful completion of the running program. This training starts at 9 months postoperatively for patients who want to return to active sports. When PCL allografts have been used, this training should be delayed for 12 months on an empirical basis [17, 28]. The patient should not show >20% deficit in the isometric tests for quadriceps and hamstrings. When performing plyometric exercises, the important parameters are the surface, the shoes, and the temperature. Cross-training or running shoes should be worn, which are shock absorbent and provide sufficient stability to the foot. The shape of wear on the shoe sole should be checked to prevent injuries of over-use. The patient must be given detailed instructions to prevent knee hyperextension and a general valgus lower extremity position when jumping and landing with the knees bent and at shoulder width and keeping the body weight on the balls of the feet. The first exercise is the level surface box jump using both the legs. A grid of four equal boxes is created on the floor with tape. The patient is instructed to jump from box 1 to box 3 (from front to back) than from 1 to 2 (side to side). At the second level, this exercise includes jumping on 1 leg in both directions and left and right [2–6, 15–18, 25].

Level 3 advances to cross-shuffles, and level 4 includes pivot jumps at 90° and 180°. After double-jumping in four directions, the exercise is repeated on one leg. At the next stage, the exercise is vertical jumping onto a box [1, 5, 23–27]. The importance of the intensity and stress of the continuous exercise in the program should be underlined. Individual sessions can be applied in a similar way to interval training. Initially, the rest period is 2–3 times longer than the exercise period and is slowly reduced to 2 times longer, then of equal duration. Improvement is measured by the number of jumps made in a specified period. The duration of the first exercise is 15 s. The patient is instructed to make as many jumps as possible between the squares in 15 s. The program progresses when the number of jumps increases together with patient confidence [4, 6, 21, 30, 31] (Figs. 19.5, 19.6 and 19.7) (Table 19.1).

Fig. 19.5 Double-leg bodyweight squat (**a**, **b**)

Fig. 19.6 Single-leg dumbbell squat

Fig. 19.7 Clam exercise

Table 19.1 Rehabilitation program after posterior cruciate ligament reconstruction

	Phase I (0–2 weeks)	Phase II (3–6 weeks)	Phase III (7–12 weeks)	Phase IV (13 weeks and after)
Goals	• Control joint pain, swelling, hemarthrosis • ROM 0–30° • Independent in HEP • Adequate quad/VMO control	• Regain normal knee range of motion • Regain a normal gait pattern • Regain normal lower extremity strength	• Regain normal proprioception, balance, and coordination	• Maximize lower extremity strength and endurance • Return to previous activity level • Return to specific functional level
ROM	• Passive, 0–30° • Patella mobilization • Ankle pumps • Gastroc/soleus stretching	• Passive, 0–90° (wk 3–4) 0–110° (wk 5–6) • Patella mobilization • Ankle pumps • Initiate light hamstring stretch	• Passive, 0–135° • Patella mobilization • Hamstring/ITB stretch • Gastroc/soleus stretch	• Continue with all stretching activities • Full ROM
WB	• 2 Crutches and brace	• Full WB with/without crutches as strength allows	• Full WB with/without crutches as strength allows	• Full WB without crutches
Precautions	• WBAT with crutches and brace	• Enhance proprioception, balance, and coordination	• Control pain, inflammation, and effusion	
Brace	• Limited from 0 to 30°	• Continue with brace, unlocked to 90°	• Discharge post-op brace at week 6 • Functional brace to be fitted	No brace
Therapeutic exercise, treatment recommendations, and return to sport & work	• E-stim/biofeed back as needed • Ice 15–20 min	• E-stim/biofeed back as needed • Ice 15–20 min • Weight shift (side-to-side, forward/backward) • Single-leg balance work • Hesitation/cup walking • Steam boats balance work	• Single-leg balance training • Wobble board balance activities • Foam roller balance activities • Balance/agility work	• Continue with advanced balance/agility training • Single-leg work on advanced surfaces • Initiate running on mini tramp and progress to treadmill as tolerated • Backward walking on treadmill • Walking program • Swimming program (kicking) • Bike for strength and endurance • EFX for strength and endurance • High-speed training • Initiate sport-specific training

References

1. Bergfeld JA, McAllister DR, Parker RD, et al. A biomechanical comparison of posterior cruciate ligament reconstruction techniques. Am J Sports Med. 2001;29:129–36.
2. Trent PS, Walker PS, Wolf B. Ligament length patterns, strength, and rotational axes of the knee joint. Clin Orthop Relat Res. 1976;117:263–70.
3. Robinson JR, Bull AM, Thomas RR, Amis AA. The role of the medial collateral ligament and posteromedial capsule in controlling knee laxity. Am J Sports Med. 2006;34:1815–23.
4. Markolf KL, Feeley BT, Jackson SR, McAllister DR. Biomechanical studies of double-bundle posterior cruciate ligament reconstructions. J Bone Joint Surg Am. 2006;88:1788–94.
5. Haimes JL, Wroble RR, Grood ES, Noyes FR. Role of the medial structures in the intact and anterior cruciate ligament–deficient knee. Limits of motion in the human knee. Am J Sports Med. 1994;22:402–9.
6. Gill TJ, DeFrate LE, Wang C, et al. The effect of posterior cruciate ligament reconstruction on patellofemoral contact pressures in the knee joint under simulated muscle loads. Am J Sports Med. 2004;32:109–15.
7. Blickenstaff KR, Grana WA, Egle D. Analysis of a semi tendinosus autograft in a rabbit model. Am J Sports Med. 1997;25:554–9.
8. Gupte CM, Bull AM, Atkinson HD, et al. Arthroscopic appearances of the menisco femoral ligaments: introducing the "meniscal tug test". Knee Surg Sports Traumatol Arthrosc. 2006;14:1259–65.
9. Makris CA, Georgoulis AD, Papageorgiou CD, et al. Posterior cruciate ligament architecture: evaluation under microsurgical dissection. Arthroscopy. 2000;16:627–32.
10. Pasque C, Noyes FR, Gibbons M, et al. The role of the popliteo fibular ligament and the tendon of popliteus in providing stability in the human knee. J Bone Joint Surg Br. 2003;85:292–8.
11. Covey DC, Sapega AA, Marshall RC. The effects of varied joint motion and loading conditions on posterior cruciate ligament fiber length behavior. Am J Sports Med. 2004;32:1866–72.
12. Shearn JT, Grood ES, Noyes FR, Levy MS. One- and two strand posterior cruciate ligament reconstructions: cyclic fatigue testing. J Orthop Res. 2005;23:958–63.
13. Deehan DJ, Salmon LJ, Russell VJ, Pinczewski LA. Endoscopic single-bundle posterior cruciate ligament reconstruction: results at minimum 2-year follow-up. Arthroscopy. 2003;19:955–62.
14. LaPrade RF, Tso A, Wentorf FA. Force measurements on the fibular collateral ligament, popliteo fibular ligament, and popliteus tendon to applied loads. Am J Sports Med. 2004;32:1695–701.
15. Markolf KL, Graves BR, Sigward SM, et al. Effects of postero lateral reconstructions on external tibial rotation and forces in a posterior cruciate ligament graft. J Bone Joint Surg Am. 2007;89:2351–8.
16. Simonian PT, Sussmann PS, van Trommel M, et al. Popliteo meniscal fasciculi and lateral meniscal stability. Am J Sports Med. 1997;25:849–53.
17. McAllister DR, Markolf KL, Oakes DA, et al. A biomechanical comparison of tibial inlay and tibial tunnel posterior cruciate ligament reconstruction techniques. Graft pretension and knee laxity. Am J Sports Med. 2002;30:312–7.
18. Dandy DJ, Pusey RJ. The long-term results of unrepaired tears of the posterior cruciate ligament. J Bone Joint Surg Br. 1982;64:92–4.
19. Noyes FR, Barber-Westin SD. Posterior cruciate ligament revision reconstruction, part 1: causes of surgical failure in 52 consecutive operations. Am J Sports Med. 2005;33:646–54.
20. Noyes FR, Barber-Westin SD. Posterolateral knee reconstruction with an anatomical bone–patellar tendon–bone reconstruction of the fibular collateral ligament. Am J Sports Med. 2007;35:259–73.
21. Petersen WJ, Loerch S, Schanz S, et al. The role of the posterior oblique ligament in controlling posterior tibial translation in the posterior cruciate ligament–deficient knee. Am J Sports Med. 2008;36:495–501.
22. Sekiya JK, Haemmerle MJ, Stabile KJ, et al. Biomechanical analysis of a combined double-bundle posterior cruciate ligament and posterolateral corner reconstruction. Am J Sports Med. 2005;33:360–9.
23. Weiler A, Peine R, Pashmineh-Azar A, et al. Tendon healing in a bone tunnel. Part I: biomechanical results after biodegradable interference fit fixation in a model of anterior cruciate ligament reconstruction in sheep. Arthroscopy. 2002;18:113–23.
24. Wiley WB, Askew MJ, Melby A 3rd, Noe DA. Kinematics of the posterior cruciate ligament/posterolateral corner–injured knee after reconstruction by single- and double-bundle intra articular grafts. Am J Sports Med. 2006;34:741–8.
25. Bergfeld JA, Graham SM, Parker RD, et al. A biomechanical comparison of posterior cruciate ligament reconstructions using single- and double-bundle tibial inlay techniques. Am J Sports Med. 2005;33:976–81.
26. Kennedy JC, Alexander IJ, Hayes KC. Nerve supply of the human knee and its functional importance. Am J Sports Med. 1982;10:329–35.
27. Camara-Arrigunaga F, Camposeco-Longo P, Nieves-Silva J, D'Apuzzo M. Fixation with anchors for posterior cruciate ligament avulsion by arthroscopy: a new technique. Arthroscopy. 2005;21:1146.
28. Fanelli GC, Edson CJ. Arthroscopically assisted combined anterior and posterior cruciate ligament reconstruction in the multiple ligament injured knee: 2- to 10-year follow-up. Arthroscopy. 2002;18:703–14.

29. Chen CH, Chuang TY, Wang KC, et al. Arthroscopic posterior cruciate ligament reconstruction with hamstring tendon autograft: results with a minimum 4-year follow-up. Knee Surg Sports Traumatol Arthrosc. 2006;14:1045–54.
30. Markolf KL, O'Neill G, Jackson SR, McAllister DR. Effects of applied quadriceps and hamstrings muscle loads on forces in the anterior and posterior cruciate ligaments. Am J Sports Med. 2004;32:1144–9.
31. Peterson CA II, Warren RF. Management of acute and chronic posterior cruciate ligament injuries. Am J Knee Surg. 1996;9:172–84.

20 Physiotherapy in Orthopedic Knee Injuries: Rehabilitation Program Following Primary and Revision Anterior Cruciate Ligament Reconstruction

Mehmet Emin Şimşek and M. İ. Safa Kapıcıoğlu

20.1 Introduction

Protocols are based on evaluations; in other words, progress through the program is restructured using the principles of anatomy, physiology, biomechanics, and surgery, and continuous evaluation is made with the understanding of the general aim of rehabilitation:

- Regain standard knee stability: <3 mm anteroposterior displacement in the knee arthrometry test and negative pivot shift.
- Control of joint pain, swelling, and hemarthrosis.
- Regain normal knee range of motion (ROM).
- Regain a normal gait and neuromuscular control.
- Preserve the normal muscle strength of the lower extremity.
- Regain normal proprioception, balance, coordination, and neuromuscular control for desired activities.
- Obtain the best orthopedic and functional result.

The appropriate program for each patient takes into consideration the sporting and occupational goals of the patient, the status of the joint surface, meniscus and other knee ligaments, surgical procedures applied at the same time as ACL reconstruction, the graft-type used, postoperative healing and response to surgery, and the biological principles of graft healing and regeneration [1–9]. Protocols are separated into seven phases according to the postoperative periods. At each stage, four main categories explain the exercises to be performed by the patient and the factors evaluated by the therapist:

- General observation of the patient status.
- Evaluation and measurement of specific variables for each parameter with identified targets.
- Duration of treatment and exercise program according to the frequency.
- Attainment of the rehabilitation goals necessary to advance to the next stage.

The first protocol has been designed for patients who have undergone primary ACL bone-patellar tendon-bone (B-PT-B) autogenous reconstruction and wish to return to strenuous sports or work activities as soon as possible postoperatively [3, 6, 7]. Patients who develop postoperative problems such as knee movement complications, chronic effusion, patellofemoral pain, or patellar tendinitis are recommended to

M. E. Şimşek (✉)
Department of Orthopaedics and Traumatology, Ankara Lokman Hekim University, Sincan Hospital, Ankara, Turkey

M. İ. S. Kapıcıoğlu
Department of Orthopaedics and Traumatology, Faculty of Medicine, Ankara Yıldırım Beyazıt University, Ankara, Turkey

M. Bozkurt, H. İ. Açar (eds.), *Clinical Anatomy of the Knee*,
https://doi.org/10.1007/978-3-030-57578-6_20

slow down the rate of progression until the problems are resolved [4, 7, 10]. To determine whether or not a patient is ready to move from one stage to the next, specific criteria are evaluated throughout both rehabilitation programs. Both protocols include an estimated number of official physical therapy visits plus a home self-managed program. For patients to return to strenuous activity, it may be necessary to have more supervised sessions between 6 and 12 months postoperatively for patients following an advanced training program. A specific neuromuscular re-education program is recommended for all patients returning to high-risk activities. Signs such as postoperative joint swelling, pain, gait pattern, knee movements, patellar mobility, muscle strength, flexibility, and AP migration are continuously monitored in all patients. Any patient developing a complication or experiencing difficulty in progressing through the protocol will need extra supervision in the clinic [8–13]. The first postoperative week is a critical time for all patients in respect of control of knee joint pain and swelling, sufficient quadriceps muscle contraction, the time of starting knee movements, and providing sufficient limb elevation. Compression dressings are used for 48 h after which additional compression stockings can be used if necessary. Patients are encouraged to have above-the-heart elevation for the first 5–7 days and only under take exercises and personal hygiene activity. Deep vein thrombosis prophylaxis consists of one aspirin per day for 10 days, short periods of ambulation with crutches 6–8 times a day, ankle pumps once an hour every waking hour, and close observation of the lower extremity. Knee joint hemarthroses require aspiration. Non-steroid anti-inflammatory drugs are used for at least 5 days postoperatively. Appropriate pain relief is prescribed to provide comfort and allow the application of the next exercise protocol described below [4, 9, 11] (Table 20.1).

Table 20.1 Rehabilitation program following primary and revision anterior cruciate ligament reconstruction

	Phase I (0–2 weeks)	Phase II (2–6 weeks)	Phase III (7–12 weeks)	Phase IV (13 weeks and after)
Goals	• Full passive extension • Flexion to 90 degrees • Decrease pain and swelling • Increase range of motion and restore full extension	• ROM 0–110 degrees • No effusion • No extensor lag	• Full ROM • Normal gait	• Full ROM • Normal gait
ROM	• Full passive extension • Flexion to 90 degrees • Active ROM 0–90 degrees (passive extension, active flexion, heel slides)	• ROM 0–110 degrees • ROM through full range as tolerated • SLR × 4 on mat • ROM of mini-squats as tolerated	• Full ROM	• Full ROM
WB	• Weight-bearing as tolerated Use two crutches initially progressing to weaning crutches as swelling and quadriceps status dictates	• Patients may progress to full weight-bearing as tolerated without analgia • Patients may require one crutch or cane to normalize gait before ambulating without assistive device	• Full WB	• Full WB

Table 20.1 (continued)

	Phase I (0–2 weeks)	Phase II (2–6 weeks)	Phase III (7–12 weeks)	Phase IV (13 weeks and after)
Precautions	• WB with crutches and brace	• Enhance proprioception, balance, and coordination	• Control pain, inflammation, and effusion	
Brace	• Limited from 0 to 120 degree	• No brace	• No brace	• No brace
Therapeutic exercise, treatment recommendations, and return to sport & work	• Patellar mobilization (teach patient) • Calf pumping • Passive extension with heel on bolster or prone hangs • Electrical stimulation in full extension with quad sets and SLR • Quad sets, co-contraction squads/hamstring • Straight-leg raise × 4 on mat	• Continue appropriate previous exercises and following without brace • Scar massage when incision healed • Leg press 0–60 degrees with light resistance (up to ¼ body weight) • Hamstring curls—Carpet drags or rolling stool (closed chain)	• Continue appropriate previous exercises • Standing SLR ° × 4 with light TheraBand bilaterally • Wall squats 0–45 degrees, progress to single leg • Leg press 0–60 degrees with resistance as tolerated • Hamstring curls with resistance as tolerated • Forward, lateral, and retro step-downs in parallel • No knee flexion past 45 degrees (small step) • Single-leg heel raises • Proprioceptive training • Single-leg standing in parallel bars • Double-leg BAPS for weight shift	• Isokinetic testing at 180 and 300 degrees/s—Must have 80% of opposite leg to clear for straight line running • Continue appropriate previous exercises • Home/gym program • Agility drills/ plyometrics • Sit-up progression • Treadmill—Running progression program if cleared • Transition to home/gym program

20.2 Rehabilitation Protocol

Knee effusion must be controlled in the early postoperative period (1–3 days) to prevent the quadriceps inhibition phenomenon. Electro galvanic stimulation or high-voltage electrical muscle stimulation (EMS), ice to control swelling and a compression and elevation program for strengthening can be used. Effusion and swelling have a negative electrical load, so negative electrodes on the knee and positive (dispersive) electrodes below the waist or on the opposite thigh are helpful in removing fluid from the joint which could be reabsorbed [3, 6, 8]. The treatment lasts approximately 30 min, the intensity is set according to patient tolerance, and the frequency is 3–6 times a day. After the joint effusion has been brought under control, functional EMS is started to regain muscle strength and facilitate quadriceps contractions. The use of EMS to develop and facilitate quadriceps contractions is based on the evaluation of quadriceps and vastus medialis obliquus (VMO) muscle tone. One electrode is placed on the VMO and a second electrode on the central and lateral side of the upper third of the belly of the quadriceps. Treatment lasts for 15–20 min [12, 13]. EMS is continued until muscle grade is evaluated as good. Bio feedback is useful to increase hamstring loosening when there is difficulty in reaching full knee extension secondary to muscle spasm or knee pain. For the surface electrode to provide positive feedback to the patient and clinician about the quality of selective or voluntary quadriceps contractions, it can be placed over the selected muscle junction [7, 8, 13] (Fig. 17.1).

Cryotherapy is the most widely used modality after ACL reconstruction and starts in the recovery room after the operation. The cost of various cryotherapy options and patient compliance are two critical factors in the successful control of postoperative pain and swelling [5, 9]. The standard cold treatment method is for a bag of ice or a commercial cold pack to be held in place as necessary. Empirically patients prefer a mobile cold unit. These units provide excellent pain control with the circulation of iced water at a fixed temperature in a pad. Gravity flow units are also useful, although temperature maintenance with these devices is more difficult than with mobile cold units [7, 13].

The temperature can be controlled by releasing the water and using reverse flow gravity, and when necessary, the sleeve can be filled with fresh iced water. Cryotherapy lasts for 20 min, 3–4 times a day depending on the degree of pain and swelling. In some cases, the treatment time can be extended depending on the thickness of the buffer used between the skin and the device. Motorized units include a thermostat which is useful when cold therapy is used for an extended period. Vasopneumatic devices present another option for cold therapy. The Game Ready device allows the clinician to adjust the temperature according to patient tolerance, and at the same time, there are four different compression levels [2, 4, 7–11]. Cryotherapy is generally applied when necessary for pain and swelling control or after exercise and can be used to complete the postoperative rehabilitation protocol [12, 14] (Table 20.1).

20.3 Postoperative Bracing

The use of postoperative braces after ACL reconstruction is controversial. Screening patients for personality type, pain tolerance, and program compliance may provide insight into which patients require postoperative brace protection. The brace should be rigid and hold the knee at 0° initially. The brace is gradually opened according to the protocol, allowing normal knee flexion during ambulation. Periodic evaluation of the body must be made to check the position on the leg and ensure that maximum benefit is provided [6–9, 13] (Fig. 17.3).

20.4 Range of Knee Motion (ROM)

ROM of 0–90° is obtained in the first week postoperatively. Passive and active ROM exercises are performed in 10-min sessions, 4–6 times a day. Hamstrings and gastrocnemius exercises are done with the foot and ankle elevated with the support of a towel or another device allowing full extension of the knee. This position is held for 10 min and is repeated 4–6 times a day. To provide high pressure for stretching of the posterior capsule, 10 lb. (pound imperial weight) weights can be added to the distal thigh and knees. Full knee extension should be made in the second postoperative week. The aim is to obtain 0–3° hyperextension within the normal knee movement limit. Inpatients with physiological bilateral knee hyperextension, it is recommended that the reconstructed knee is returned to 3–5° hyperextension to resemble the more considerable amount of hyperextension in the contralateral knee. Re-gaining >5° hyperextension is also recommended because of potentially harmful forces on the healing graft. Knee flexion reaches 120° in the 4th–5th week postoperatively and 135° by the end of the 5th week. Passive knee flexion exercises are traditionally applied in a sitting position using the contralateral extremity to provide extra pressure. Other methods that are useful to obtain flexion of >90° include chair exercises, wall shift, knee flexion devices, and passive quadriceps stretching exercises. Patients who have difficulty reaching 0–90° flexion by the 5th week may require additional treatment interventions to relieve pain with joint movement such as bringing the knee gently into flexion under local nerve blocks and possibly anesthesia with appropriate drugs [5, 6, 11, 13, 14] (Table 20.1).

20.5 Patellar Mobilization

To regain normal knee ROM, normal patellar mobility is critical. Loss of patellar movement is often related to arthrofibrosis and in severe cases, the development of patella infera. Patellar shifts start in the first plane of all four posts (superior, inferior, medial, lateral) with continuous pressure applied to the patellar edge for at least 10 s. This exercise is applied 5 min before the ROM exercises. Care must be taken if extensor lag is determined because this could be associated with poor superior migration of the patella [12–14].

20.6 Weight Bearing

In the period when pain and swelling are brought under control, and voluntary quadriceps contraction is seen, partial weight-bearing is permitted. Initially, two underarm crutches are used, and 50% of the body weight is placed on the affected leg. The amount of weight permitted on the affected limb is adjusted to full weight-bearing in the third and fourth weeks postoperatively. It is essential that a regular walking pattern is maintained encouraging normal knee flexion throughout the gait cycle and preventing a locked knee position. This technique provides a typical pattern of heel-to-toe ambulation, quadriceps contraction in the mid-gait cycle, and hip and knee flexion during the gait cycle. Thus a locked knee position associated with the potential development of a deviant quadriceps walking pattern is avoided [15–18].

20.7 Flexibility

Hamstring and gastrocnemius-soleus flexibility exercises are started on the first day after ACL reconstruction. A continuous static stretch is held for 30 s and repeated five times. The most common hamstring exercise is the modified obstructed stretching exercise, and the most common gastrocnemius-soleus stretching exercise is the towel pull. These exercises are performed in knee flexion and help pain control because of the reflex response occurring in the hamstrings [13, 14, 16]. The towel pull exercise also assists in reducing discomfort in the calf, Achilles tendon, and ankle. These exercises constitute a critical component of the ROM program as the ability of these muscle groups to loosen is necessary for the success of full passive knee extension. Quadriceps and iliotibial band flexibility exercises are applied to help to control lateral hip and thigh tension and provide full knee flexion. The full evaluation of the lower extremity will reveal flexible areas in the deficit areas requiring correction. When designing a flexibility program, the therapist must consider the position or physical requirements of the activity in addition to the patient's desire to return to specific sports or activities. Flexibility is included in the program to be applied after discharge of the patient [12, 13, 15] (Figs. 18.4, 18.5, 18.6, 19.2, 19.3, 19.4, 19.5, 19.6, 20.1, 20.2 and 20.3) (Table 20.1).

20.8 Strengthening

Lower extremity muscle atrophy and weakness is a challenging and unresolved problem after ACL reconstruction. Therefore, the strengthening program should be started on the first postoperative visit. Isometric quadriceps contractions are applied for 10 s with ten repetitions, ten

Fig. 20.1 Stretching technique of rectus femoris

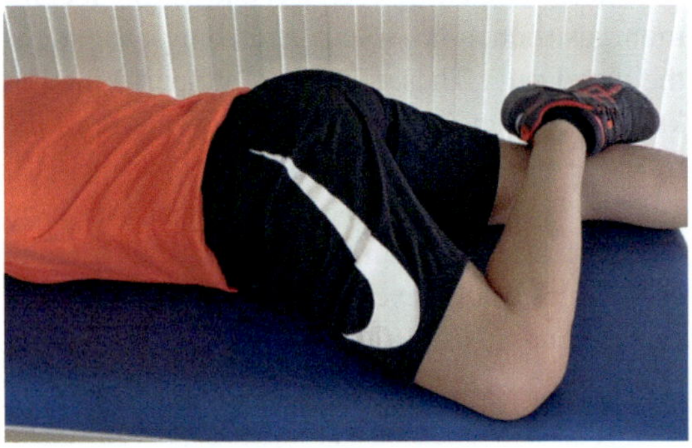

Fig. 20.2 Stretching technique of adductor longus

Fig. 20.3 Stretching technique of gastrocnemius and soleus muscle

times a day. Evaluation of the of contractions by both the patient, and the therapist is essential. The patient can observe the quality of the contraction by visual and manual comparison with the contralateral side. During contraction, the patient can evaluate the superior patella migration, which should be approximately 1 cm, and during the first loosening of the contraction, there should be inferior migration of the patella [6, 7, 13]. During isometric contractions, the patient should not allow the knee to go into hyperextension but should hold the knee in a flexion position throughout the exercise. Biofeedback can also be used if necessary to strengthen a proper quadriceps contraction. Straight-leg raises are started on postoperative day 1 in the hip movement on all fours. Adduction straight-leg raises have a beneficial effect on the vastus medialis obliques (VMO). <<Supine straight-leg should include a sufficient isometric quadriceps contraction to be of benefit to the quadriceps [14, 16]. When these exercises become more comfortable, ankle weights are added for advanced muscle strengthening. Initially, weights of 1 and2 kg are used and up to 5 kg can be added, not to exceed 15% of the patient's body weight. If a poor tone is observed during isometric contractions, active-supported ROM can also be used to facilitate quadriceps contraction. These exercises are used throughout the first 8 weeks postoperatively until pain and swelling are controlled, full ROM is regained, early quadriceps control and proximal stabilization are provided, and a normal walking pattern is maintained. In postoperative week 1, closed kinetic chain (CKC) exercises are started with mini-squats of 0–45° as tolerated by the patient [2, 14, 17] (Table 20.1).

Initially, the body weight of the patient is used as resistance, and then a TheraB and or surgical tubes are gradually used as resistance mechanisms. A fast, smooth, rhythmic squat is repeated at a high-set/high-repetition tempo to increase muscle fatigue. It is essential to observe the hip position for support of the quadriceps contraction [4, 8, 11].

Foot-raises to strengthen the gastrocnemius-soleus, and wall-sitting isometry for quadriceps control is started in the second week. Wall-sitting aims to strengthen the quadriceps until muscle fatigue. If anterior knee pain is experienced, the sitting knee flexion angle can be reduced. The exercises can be modified to force the quadriceps more. Typically when the patient reaches maximum knee flexion of 30–45°, the quadriceps can be voluntarily adjusted. This contraction and knee flexion position is held until muscle fatigue occurs and the exercise is repeated 3–5 times. Finally, the patient shifts the body weight to the affected side compatible with single leg contraction [7, 8, 13]. This exercise has been suggested to have excellent properties for use at home 4–6 times a day in respect of providing quadriceps fatigue in safe knee flexion not inducing anterior tibial translation. When the patient has reached the capacity for full weight-bearing, side steps are started. These are gradually increased according to the height of the step and patient tolerance. Due to muscle isolation obtained when the knee joint is stable, weight machines are beneficial. The patient uses these on both legs and a single leg. If the lowest weight of the machine is as heavy as can be lifted by the affected extremity alone when weight lifted by both legs is applied there may be eccentric contraction which is reduced on the affected side [16, 17]. Eccentric contractions can be used in the advanced stages of strength training. As this muscle system plays a role in the dynamic stabilization of the knee joint, hamstring strength is of critical importance for the general success of the rehabilitation program. Weight training is used through put the advanced program and continues in return to activities and care stages of rehabilitation. Open kinetic chain (OKC) extension exercises are included within several weeks for further development of quadriceps muscle strength. These exercises must be applied with care because of potential problems that could be created for the healing graft and the patellofemoral joint [3–8, 18] (Figs. 18.4, 18.5, 18.6, 19.2, 19.3, 19.4, 19.5, 19.6, 20.1, 20.2 and 20.3) (Table 20.1).

20.9 Balance, Proprioceptive, and Perturbation Training

Regaining normal neuro muscular function after ACL reconstruction is essential for the application of sports-specific exercises and a return to all activities. Knee joint proprioception is an essential component of neuro muscular function, and necessary for healing and a successful outcome. Balance and proprioceptive training are started in the first week postoperatively [19, 20]. Initially, the patient only shifts weight from one side to the other and from front to back. This activity increases the confidence in the ability of the leg to resist the pressure of weight-bearing and signals the knee joint position. When full weight-bearing is started, cup-walking is started to support the symmetry between the operated and non-operated limbs. This exercise helps to prevent knee hyper extension by developing quadriceps control and hip and knee flexion during walking. Cup-walking also controls excessive hip and pelvic movement, and gastrocnemius-soleus activity during the push phase of walking [15, 18, 19].

These components of gait control are of critical importance in the early stages of rehabilitation to reduce forces in the healing graft. Balance exercises on both legs and one leg in the standing position are useful in the early postoperative period. In the single-leg exercise, the foot is moved straightforward, the knee is bent at 20–30°, the arms are extended outwards horizontally, and the trunk is held vertical with the hips over the ankles and the shoulder over the hips. The aim is to hold the position until the balance is lost. To make this exercise more difficult, a mini-trampoline or an unstable surface can be used, because these types of surfaces require more dynamic leg control than is required to stand on a flat surface [21, 22].

The primary thought behind a conditioning program during running rehabilitation is to apply stress to the cardiovascular system without compromising the knee joint. A cardiovascular program is started depending on the attainment of a vertical position that can be sufficiently maintained with upper extremity ergometry. To minimize lower extremity swelling, the surgeon should elevate the limb. If this exercise is tolerated, fixed cycling is started in the third postoperative week. When the surgical wound has healed, walking in water is started [13, 19, 21]. The early targets of these programs are full ROM, gait reduction, and cardiovascular renewal. To develop cardiovascular resistance, this program should be applied for 20–30 min at least three times a week, and the exercises should be performed at least 60%–85% of maximum heart rate. Greater cardiovascular efficiency and resistance is thought to be provided at a high percentage of maximum heart rate. The use of cross-skiing and stair-climbing machines can be added in the fifth to sixth weeks postoperatively [22, 23].

The patellofemoral joint is well protected against high stress. During fixed cycling, the saddle is adjusted to the highest level according to the patient's height, and low resistance is used at the beginning. The stair-climbing machine is adjusted to produce a short step with low resistance [18, 19, 23]. An effective cardiovascular exercise program is an important component of the stages after rehabilitation. In addition to the previously explained exercises, a hydrotherapy program is encouraged including freestyle or breaststroke swimming, walking in water, water aerobics, and running in deep water. Which cardiovascular exercises are most suitable depends on the preference of each patient and the availability of equipment [12, 21–26] (Figs. 17.2, 17.2 and 17.3).

20.10 Return to Sports Activities

A return to sports activities is based on the successful completion of the occupational and functional training and plyometric programs and the following criteria [4, 12, 15, 16, 19, 21, 24, 25]:

1. Knee examination.
   - ROM: normal oral most normal.
   - Lachman test: Normal.
   - Pivot shift test: Normal.
   - Patellar pain: none.
   - Effusion: none.
2. KT-2000 (Medometrik).
   - Reconstruction of <3 mm compared to 134-N total AP displacement of the contralateral knee.
3. Quadriceps and hamstrings muscle strength and resistance tests: <10% deficit compared with the contralateral side according to the available equipment:
   - Isokinetic 180°/s and 300°/s.
   - Isometric portable fixed or manual dynamometer: quadriceps 60°flexion, hamstring 60° or 90° flexion, each repeated three times and the average is taken.
   - If isokinetic or isometric equipment is not available, but there is an equipped weights room, and there is a person experienced in this test, and sufficient time, maximum press and leg pressure is recommended.
4. Single-leg jumping tests: ≤15% deficit in lower extremity symmetry in the single-jump and triple-jump tests.

After a successful return to activity, the patient is encouraged to continue with a care program. Two exercise programs a week are recommended in the sports season, and before or out of season, the program should be applied three times a week to maximize the benefits of flexibility, strength, and cardiovascular resistance [23–28] (Fig. 20.4 and Table 20.1).

20.11 Rehabilitation Protocol with Delayed Parameters for Revision ACL Reconstruction, Allografts, and Complex Knees

In this protocol, the return to full weight-bearing and knee flexion is delayed. The start of specific strengthening, conditioning, running, and agility movements includes the return to restricted activities for knees which have undergone ACL revision, allograft reconstruction, major concomitant operative procedures (complex meniscus repairs or transplants, other ligament reconstructions, joint cartilage restorative procedures, patella femoral corrective procedures, or osteotomies) or knees with significant articular cartilage dam-

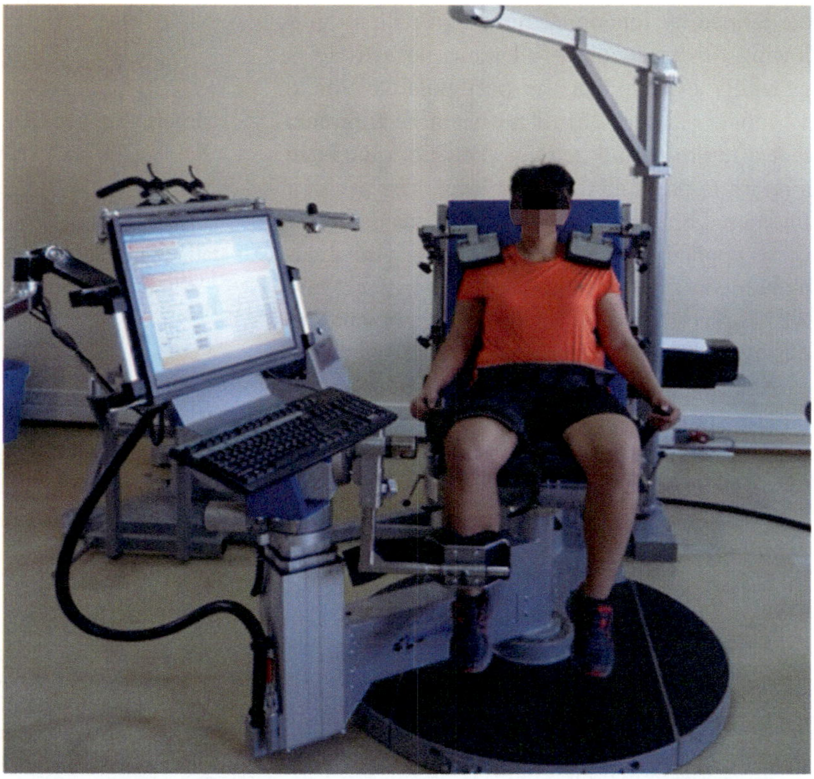

Fig. 20.4 Neuromuscular training device

age. Weight-bearing only in the form of toe-touch is permitted in the second week postoperatively [21, 22, 25]. At the end of this period, it is permitted according to the patient's weight, depending on the surgical procedure applied, postoperative pain and swelling, and evaluation of quadriceps muscle control and ROM. Most patients should be able to manage without the support of crutches by weeks 6–8. The transition of knee flexion to at least 135° can be postponed according to the concomitant procedures applied [21, 24, 26].

For knees which have undergone a posterolateral reconstructive procedure, a 50 bivalent long-leg brace is applied for the first 4 weeks. For the ROM exercises to be performed several times a day, the clip is removed, and 0° full extension is reached, but the patient is instructed to avoid hyperextension. Patients applied with concomitant proximal patellar realignment are permitted 0–75° flexion in the first 2 postoperative weeks. Flexion is slowly increased to 135° by the eighth week. In knees that have undergone PCL reconstruction and complex meniscus repair, knee flexion is initially limited [17, 27]. In patients with abnormal hyperextension (≥10°) showing physiological loosening, knee extension is limited to 0–5° for approximately 3 weeks to allow sufficient healing before pushing the stress to 0°. A postoperative long-leg, a hinged knee brace is used throughout the first 8 weeks, except for those applied with a posterolateral procedure. The brace provides support and protects the healing tissues, providing patient comfort in this period. Changes in strengthening, conditioning, and resistance training depend on concomitant procedures. A return to activity is postponed for at least 6 months to allow the healing of all the repaired and reconstructed tissues and the full return of joint and muscle functions [27–31]. In comparison with autografts, the maturation of allografts is slower, and there is currently empirical evidence of the postoperative time limitations in respect of the return to full activity. Evaluation of the symptom evaluation and knee movement examination is a critical component allowing a functional program to be started including muscle strength and ligament stability. The total evaluation is not a single parameter but should be defined by functional return. In patients following this protocol, a full return to activity is generally expected in the postoperative 9th to 12 month [23, 26, 30]. It must not be forgotten that a return to full activity does not guarantee a return to pre-injury activity level. The use of a functional brace or a derotation is considered for those applied with ACL revision or multi-ligament reconstruction or patients who have >3 mm AP displacement postoperatively compared to the contralateral extremity.

Also, patients who are concerned about returning to strenuous activities or are subjectively uncertain are candidates for functional support [28, 30–39] (Table 20.1).

References

1. Aglietti P, Buzzi R, Zaccherotti G, De Biase P. Patellar tendon versus doubled semitendinosus and gracilis tendons for anterior cruciate ligament reconstruction. Am J Sports Med. 1994;22(2):211–7; discussion 217–218. Retrieved from http://www.ncbi.nlm.nih.gov/pubmed/8198189
2. Shelbourne KD, Patel DV. Timing of surgery in anterior cruciate ligament-injured knees. Knee Surg Sports Traumatol. 1995;3:148–56.
3. Oeffinger DJ, Shapiro R, Nyland J, Pienkowski D, Caborn DNM. Delayed gastrocnemius muscle response to sudden perturbation in rehabilitated patients with anterior cruciate ligament reconstruction. Knee Surg Sports Traumatol Arthrosc. 2001;9(1):19–27.
4. Wright RW, Haas AK, Anderson J, Calabrese G, Cavanaugh J, Hewett TE, et al. Anterior cruciate ligament reconstruction rehabilitation: MOON guidelines. Sports Health. 2015;7(3):239–43. https://doi.org/10.1177/1941738113517855.
5. Shani RH, Umpierez E, Nasert M, Hiza EA, Xerogeanes J. Biomechanical comparison of quadriceps and patellar tendon grafts in anterior cruciate ligament reconstruction. Arthroscopy. 2016;32(1):71–5. https://doi.org/10.1016/j.arthro.2015.06.051.
6. Meyers MC, Sterling JC, Marley RR. Efficacy of stairclimber versus cycle ergometry in postoperative anterior cruciate ligament rehabilitation. Clin J Sport Med. 2002;12(2):85–94.
7. Risberg MA, Mork M, Jenssen HK, Holm I. Design and implementation of a neuromuscular training program following anterior cruciate ligament reconstruction. J Orthop Sports Phys Ther. 2001;31(11):620–31. http://www.ncbi.nlm.nih.gov/pubmed/11720295.
8. van Grinsven S, van Cingel RE, Holla CJ, van Loon CJ. Evidence-based rehabilitation following anterior cruciate ligament reconstruction. KneeSurg Sports Traumatol Arthrosc. 2010;18(8):1128–44. https://doi.org/10.1007/s00167-009-1027-2.
9. Beynnon BD, Johnson RJ, Abate JA, Fleming BC, Nichols CE. Treatment of anterior cruciate ligament injuries, part 2. Am J Sports Med. 2005a;33(11):1751–67. https://doi.org/10.1177/0363546505279922.
10. McHugh MP, Tyler TF, Nicholas SJ, Browne MG, Gleim GW. Electromyographic analysis of quadriceps fatigue after anterior cruciate ligament reconstruction. J Orthopaed Sports Phys Ther. 2001;31(1):25–32.
11. Beynnon BD, Johnson RJ, Naud S, Fleming BC, Abate JA, Brattbakk B, Nichols CE. Accelerated versus nonaccelerated rehabilitation after anterior cruciate ligament reconstruction. Am J Sports Med. 2011;39(12):2536–48. https://doi.org/10.1177/0363546511422349.
12. Shrier I. Stretching before exercise does not reduce the risk of local muscle injury: a critical review of the clinical and basic science literature. Clin J Sport Med. 1999;9:221–7.
13. Beynnon BD, Uh BS, Johnson RJ, Abate JA, Nichols CE, Fleming BC, Roos H, et al. Rehabilitation after anterior cruciate ligament reconstruction: a prospective, randomized, double-blind comparison of programs administered over 2 different time intervals. Am J Sports Med. 2005;33(3):347–59. http://www.ncbi.nlm.nih.gov/pubmed/15716250.
14. Birmingham TB, Kramer JF, Kirkley A, Inglis JT, Spaulding SJ, Vandervoort AA. Knee bracing after ACL reconstruction: effects on postural control and proprioception. Med Sci Sports Exerc. 2001;33(8):1253–8. http://www.ncbi.nlm.nih.gov/pubmed/11474323.
15. Brewer BW, Cornelius AE, Van Raalte JL, Tennen H, Armeli S. Predictors of adherence to home rehabilitation exercises following anterior cruciate ligament reconstruction. Rehabil Psychol. 2013;58(1):64–72. https://doi.org/10.1037/a0031297.
16. Bynum EB, Barrack RL, Alexander AH. Open versus closed chain kinetic exercises after anterior cruciate ligament reconstruction. A prospective randomized study. Am J Sports Med. 1995;23(4):401–6. http://www.ncbi.nlm.nih.gov/pubmed/7573647.
17. Duthon VB, Barea C, Abrassart S, Fasel JH, Fritschy D, Menetrey J. Anatomy of the anterior cruciate ligament. KneeSurg Sports Traumatol Arthrosc. 2006;14(3):204–13. https://doi.org/10.1007/s00167-005-0679-9.
18. Fleming BC, Oksendahl H, Beynnon BD. Open- or closed-kinetic chain exercises after anterior cruciate ligament reconstruction? Exerc Sport Sci Rev. 2005;33(3):134–40. http://journals.lww.com/acsm-essr/Fulltext/2005/07000/Open__or_Closed_Kinetic_Chain_Exercises_After.6.aspx.
19. Goldblatt JP, Fitzsimmons SE, Balk E, Richmond JC. Reconstruction of the anterior cruciate ligament: meta-analysis of patellar tendon versus hamstring tendon autograft. Arthroscopy. 2005;21(7):791–803. https://doi.org/10.1016/j.arthro.2005.04.107.

20. Ochi M, Iwasa J, Uchio Y, Adachi N, Sumen Y. The regeneration of sensory neurons in the reconstruction of the anterior cruciate ligament. J Bone Jt Surg (Br). 1999;81:902–6.
21. Grant JA. Updating recommendations for rehabilitation after ACL reconstruction: a review. Clin J Sport Med. 2013;23(6):501–2. https://doi.org/10.1097/JSM.0000000000000044.
22. Ha TP, Li KC, Beaulieu CF, Bergman G, Ch'en IY, Eller DJ, Herfkens RJ, et al. Anterior cruciate ligament injury: fast spin-echo MR imaging with arthroscopic correlation in 217 examinations. Am J Roentgenol. 1998;170(5):1215–9. http://www.ajronline.org/content/170/5/1215.abstract.
23. Wilk KE. Anterior cruciate ligament injury prevention and rehabilitation: let's get it right. J Orthop Sports Phys Ther. 2015;45(10):729–30. https://doi.org/10.2519/jospt.2015.0109.
24. Shiraishi M, Mizuta H, Kubota K, Otsuka Y, Nagamoto N, Takagi K. Stabilometric assessment in the anterior cruciate ligament-reconstructed knee. Clin J Sport Med. 1996;6(1):32–9.
25. Keays SL, Newcombe PA, Bullock-Saxton JE, Bullock MI, Keays AC. Factors involved in the development of osteoarthritis after anterior cruciate ligament surgery. Am J Sports Med. 2010;38(3):455–63. https://doi.org/10.1177/0363546509350914.
26. Lindstrom M, Wredmark T, Wretling ML, Henriksson M, Fellander-Tsai L. Post-operative bracing after ACL reconstruction has no effect on knee joint effusion. A prospective, randomized study. Knee. 2015;22(6):559–64. https://doi.org/10.1016/j.knee.2015.04.015.
27. Perry MC, Morrissey MC, King JB, Morrissey D, Earnshaw P. Effects of closed versus open kinetic chain knee extensor resistance training on knee laxity and leg function in patients during the 8- to 14-week post-operative period after anterior cruciate ligament reconstruction. Knee Surg Sports Traumatol Arthrosc. 2005;13(5):357–69. https://doi.org/10.1007/s00167-004-0568-7.
28. Raynor MC, Pietrobon R, Guller U, Higgins LD. Cryotherapy after ACL reconstruction: a meta-analysis. J Knee Surg. 2005;18(2):123–9. http://www.ncbi.nlm.nih.gov/pubmed/15915833.
29. Beynnon BD, Johnson RJ, Abate JA, Fleming BC, Nichols CE. Treatment of anterior cruciate ligament injuries, part I. Am J Sports Med. 2005b;33(10):1579–602. https://doi.org/10.1177/0363546505279913.
30. Rodriguez-Merchan EC. Evidence-based ACL reconstruction. Arch Bone Jt Surg. 2015;3(1):9–12. http://www.ncbi.nlm.nih.gov/pubmed/25692162.
31. Ross MD, Denegar CR, Winzenried JA. Implementation of open and closed kinetic chain quadriceps strengthening exercises after anterior cruciate ligament reconstruction. J Strength Cond Res. 2001;15(4):466–73. http://www.ncbi.nlm.nih.gov/pubmed/11726258.
32. Seto JL, Orofino AS, Morrissey MC, Medeiros JM, Mason WJ. Assessment of quadriceps/hamstring strength, knee ligament stability, functional and sports activity levels five years after anterior cruciate ligament reconstruction. Am J Sports Med. 1988;16(2):170–80. http://www.ncbi.nlm.nih.gov/pubmed/3377102.
33. Shino K, Inoue M, Horibe S, Nagano J, Ono K. Maturation of allograft tendons transplanted into the knee. An arthroscopic and histological study. J Bone Jt Surg Br. 1988;70(4):556–60. http://www.ncbi.nlm.nih.gov/pubmed/3403597.
34. Vaishya R, Agarwal AK, Ingole S, Vijay V. Current trends in anterior cruciate ligament reconstruction: a review. Cureus. 2015;7(11):e378. https://doi.org/10.7759/cureus.378.
35. Wilk KE, Macrina LC, Cain EL, Dugas JR, Andrews JR. Recent advances in the rehabilitation of anterior cruciate ligament injuries. J Orthop Sports Phys Ther. 2012;42(3):153–71. https://doi.org/10.2519/jospt.2012.3741.
36. Xie X, Liu X, Chen Z, Yu Y, Peng S, Li Q. A meta-analysis of bone-patellar tendon-bone autograft versus four-strand hamstring tendon autograft for anterior cruciate ligament reconstruction. Knee. 2015;22(2):100–10. https://doi.org/10.1016/j.knee.2014.11.014.
37. Arna Risberg M, Lewek M, Snyder-Mackler L. A systematic review of evidence for anterior cruciate ligament rehabilitation: how much and what type? Phys Ther Sport. 2004;5(3):125–45. http://linkinghub.elsevier.com/retrieve/pii/S1466853X0400029X?showall=true.
38. Mikkelsen C, Werner S, Eriksson E. Closed kinetic chain alone to combined open and closed kinetic chain exercises for quadriceps strengthening after ACL reconstruction with respect to return to sports: a prospective matched follow-up study. Knee Surg Sports Traumatol Arthrosc. 2000;8(6):337–42.
39. O'Connor JJ. Can muscle co-contraction protect knee ligaments after injury or repair? J Bone Jt Surg (Br). 1993;75:41–8.

21 Physiotherapy in Orthopedic Knee Injuries: Rehabilitation Program Following Tibial and Femoral Osteotomies

Mehmet Emin Şimşek and M. İ. Safa Kapıcıoğlu

21.1 Introduction

The protocol described in this section has been designed for open-wedge high tibial osteotomy (HTO) and distal femoral osteotomy (DFO). As there is higher torque in the femoral osteo to my region, iliac crest autograft is preferred in an open-wedge DFO to accelerate healing. Full weight-bearing is not permitted until recommended by the surgeon based on the evidence of radiographic healing [1–3]. Also, the physical therapy specialist must be informed about whether or not locking plate and screw fixation has been used to provide a more rigid fixation. If smaller non-locking plate and screws have been used, weight-bearing must be avoided until the healing of the osteotomy has progressed. In rare cases where the wedge is entrapped in the lateral tibial cortex or the medial femoral cortex, no weight-bearing is permitted until full healing of the osteotomy is confirmed. The patient is instructed preoperatively about the postoperative protocol, so that they fully understand what to expect from the procedure [3, 4].

M. E. Şimşek (✉)
Department of Orthopaedics and Traumatology, Ankara Lokman Hekim University, Sincan Hospital, Ankara, Turkey

M. İ. S. Kapıcıoğlu
Department of Orthopaedics and Traumatology, Faculty of Medicine, Ankara Yıldırım Beyazıt University, Ankara, Turkey

Daily exercises at home support a supervised rehabilitation program. The therapist communicates the postoperative protocol safely and effectively to the patient in the clinic and uses what is necessary to safely and effectively provide therapeutic procedures and modalities [5]. The general aims of osteotomy and rehabilitation are to control pain, swelling and hemarthrosis, to regain normal knee flexion and extension, to protect the osteotomy to prevent displacement, to maintain neuromuscular stability and a regular walking pattern for ambulation, to regain lower extremity muscle strength, proprioception, balance and coordination for the activities desired, and to obtain the optimal functional outcome based on the patient targets [2, 6, 7].

Immediately after the operation, the lower extremity is wrapped in cotton wool with extra padding on the posterior, and then a double-compression cotton bandage is applied and hinged, bilateral ankle-foot compression boots. A commercial ice distribution system is used with the bladder placed over the first cotton bandage to be several layers from the wound. The neurovascular status is checked immediately in the operating room and is closely monitored in the early postoperative period. To encourage venous blood flow, a call for foot suppression system is used [8–11]. Aspirin is prescribed, and occasionally for high-risk patients, low molecular weight heparin (LMWH) or warfarin sodium. Throughout the first week postoperatively, the patients are on their

M. Bozkurt, H. İ. Açar (eds.), *Clinical Anatomy of the Knee*,
https://doi.org/10.1007/978-3-030-57578-6_21

feet for short periods, the leg is elevated, and they are instructed to stay indoors and not resume normal activities. This program helps in the healing of soft tissue edema that can occur after the operation (4,5. 9) (Table 21.1).

Deep vein thrombosis (DVT) prophylaxis includes intermittent calf compression or boots on both extremities, immediate knee movement exercises, anti-embolism stockings, and hourly ankle pump sand aspirin (300 mg/day for 10 days).

If a patient has abnormal calf sensitivity, a positive Homans sign or increased edema, Doppler ultrasound should be applied.

Important postoperative signs which must be monitored [3, 5, 6, 9–11]:

- Loss of correction, reformation of varus or valgus incompatibility.
- Swelling of the knee joint or soft tissue.
- Abnormal pain response increased the pain on weight-bearing.
- Abnormal gait.
- Insufficient flexion or extension, limited patellar movement.
- The weakness of the lower extremity (strength/control).
- Insufficient flexibility of the lower extremity.
- Peroneal nerve paralysis.
- DVT (calf pain, Homans test (+), tibial edema).

Table 21.1 High tibial osteotomy and distal femoral osteotomy rehabilitation protocol

	Phase I (0–2 weeks)	Phase II (3–6 weeks)	Phase III (7–12 weeks)	Phase IV (13 weeks and after)
Goals	• Control joint pain, swelling, hemarthrosis • ROM 0–30° • Independent in HEP • Adequate quad/VMO control	• Regain normal knee range of motion • Regain a normal gait pattern • Regain normal lower extremity strength	• Regain normal proprioception, balance, and coordination	• Maximize lower extremity strength and endurance • Return to previous activity level • Return to specific functional level
ROM	• 0–90°of flexion out of brace	• CPM if knee flexion is at least 90°	• Full ROM	• Full ROM
WB	• Partial WB	• Full weight-bearing with brace locked in extension	• As tolerated with crutches begin to advance to normalized gait pattern without crutches	• Full WB, without use of crutches and with a normalized gait pattern
Brace	• Locked in full extension	• Unlocked for ambulation	• Discontinue use	• No brace
Therapeutic exercise, treatment recommendations, and return to sport and work	• Calf pumps, quad sets • SLR in brace, modalities	• Progress non-weight-bearing flexibility, modalities • Begin floor-based core and glutes exercises • Advance quad sets, patellar mobs, and SLR	• Advance closed chain quads, progress balance, core/pelvic, and stability work • Begin stationary bike at 6 weeks • Advance SLR, floor-based exercise; hip/core	• Progress flexibility/strengthening, progression of functional balance, core, glutes program • Advance bike, add elliptical at • 12 weeks as tolerated swimming at 12 weeks

21.2 Postoperative Rehabilitation Protocol

To prevent quadriceps muscle inhibition or shutdown in the postoperative period, knee pain and effusion must be controlled. In addition to ice, elevation, and compression, electrogalvanic stimulation or high-voltage electrical muscle stimulation (EMS) can be used to control swelling [8, 12]. The treatment lasts approximately 30 min, and the intensity is adjusted according to patient tolerance. After the control of joint effusion, functional EMS is started to re-educate the muscles and facilitate adequate quadriceps contraction. One electrode is placed on the VMO, and the second electrode on the central and lateral side of the upper third of the belly of the quadriceps. The patient works the quadriceps muscle at the same time as machine stimulation. The treatment lasts for 20 min. A portable EMS machine for home use may be necessary for patients with poor muscle strength. EMS is continued until muscle grade is evaluated as good [11–13]. Biofeedback is useful in facilitating sufficient quadriceps contraction in the early stages of postoperative treatment. For the surface electrode to provide feedback to the patient and clinician about the quality of selective or voluntary quadriceps contractions, it can be placed over the selected muscle junction. Biofeedback is useful to increase hamstring loosening when there is difficulty in reaching full knee extension secondary to muscle spasm or knee pain. When the patient is doing ROM exercises, an electrode is placed in the center of the hamstring muscle [14, 15] (Figs. 17.3 and 18.1).

Cryotherapy is started immediately postoperatively. The standard method is the application of a bag of ice or a commercial cold pack. However, patients prefer mobile cold units with fixed hot and cold water circulation providing excellent pain control. Gravity-flow units are also effective pain management, but it is harder to maintain a fixed temperature with these devices compared to the mobile cold units. Cryotherapy is applied for 20 min three times a day depending on the degree of pain and swelling. In some cases, the treatment time can be extended depending on the thickness of the buffer used between the skin and the device. Cryotherapy is generally used when necessary for pain and swelling control or after exercise and is retained as a part of the whole postoperative rehabilitation protocol [16, 17].

21.3 Postoperative Bracing and Weight-Bearing

In the first 7 weeks postoperatively, a long-leg brace is applied. Then, the use of a brace is terminated, but if pain persists, Visco (Bauerfeind) heel pads are applied to drain the affected area. While a lateral wedge is used for patients applied with HTO, a medial wedge is used for those with DFO. If pain persists, drainage support is recommended, initially used in a neutral position and then it can be adjusted according to the patient response [18].

Only touch-based weight-bearing is permitted initially. In this period, patients are encouraged to touch the toes to the floor while standing as often as possible. In the fourth week postoperatively, bilateral standing, full-length anteroposterior radiographs are taken to determine the healing of the osteotomy. If the radiographs show a sufficient degree of healing of the osteotomy position and implant fixation, weight-bearing of 25% of the body weight is permitted. Throughout the walking cycle, a normal walking pattern is maintained supporting normal knee flexion. This technique of normal correction of heel-to-toe ambulation allows hip and knee flexion during the gait cycle and quadriceps contraction midcycle. Healing of the fracture region occurs at mean 8–10 weeks postoperatively. In patients where allograft has been used to fill the osteotomy defect, full weight-bearing can be delayed until 12 weeks to allow bone union [14, 17].

21.4 Range of Knee Motion

In the first 2 weeks, ROM flexion is encouraged from 0° to 90°. Passive and active ROM exercises are performed 3–4 times a day in 10-min sessions. Patients experiencing difficulty in returning to 0° in the second week can be applied with extra pressure using the propped foot/ankle position. This position is held for 10 min and repeated 4–6 times a day. To provide high pressure to stretch the posterior capsule, a 5 kg weight can be added to the distal thigh and knee. If this method is not successful in obtaining the full extension, an extension board or additional weights of 7–10 kg can be used up to 6 times a day. Knee flexion is obtained of 110° in 3–4 weeks, 130° in the fifth week, and 135° in the seventh to eighth week [14, 15, 17] Figs. 18.4, 18.5, 18.6, 19.2, 19.3, 19.4, 19.5, 19.6, 20.1, 20.2 and 20.3).

21.5 Patellar Mobilization

To regain normal knee ROM, normal patellar mobility is critical. Loss of patellar movement is often related to arthrofibrosis and in severe cases, the development of patella infera. The patellar shift occurs with continuous pressure applied to the patellar edge for at least 10 s starting in all four planes (superior, inferior, medial, lateral). This exercise is applied 5 min before the ROM exercises. Care must be taken if extensor lag is determined because this could be associated with poor superior migration of the patella, and this shows that extra emphasis must be placed on this exercise. Patellar mobilization is applied throughout approximately 8 weeks postoperatively [7–14].

21.6 Flexibility

Hamstring and gastrocnemius-soleus flexibility exercises are started on postoperative day 1. A continuous static stretch is held for 30 s and repeated five times. To stretch these muscle groups, the modified obstructed stretching exercise and the towel pull are widely used methods. These exercises are helpful in controlling pain created by the reflex response in the hamstrings. The towel pull exercise also assists in reducing discomfort in the calf, Achilles tendon, and ankle. In the ninth postoperative week, quadriceps and iliotibial band stretching exercises are started [15, 16].

21.7 Strengthening

The strengthening program is started on the first postoperative visit. Isometric quadriceps contractions are applied for 10 s with ten repetitions, ten times a day. The patient is taught to monitor the quality of the contraction by visual and manual comparison with the contralateral side. During contraction, the patient can evaluate the patella superior migration, which should be approximately 1 cm, and during the first loosening of the contraction, there should be inferior migration of the patella. During isometric contractions, the knee is held in mild flexion. Biofeedback can also be used if necessary to strengthen a good quadriceps contraction. OKC extension exercises are added for a few weeks to develop quadriceps muscle strength. These exercises must be applied with caution because of potential problems they can create for the patellofemoral joint. Resistant knee extension is started with Velcro ankle weights at 30–90°. The extension terminal phase is avoided because of the high forces placed on the patellofemoral joint [1, 5, 8, 11, 18].

The patellofemoral joint must be monitored in respect of changes in pain, swelling, and crepitus to prevent patella reversion which develops with joint cartilage damage of painful patellofemoral crepitus. A point which requires attention in DFO cases is that the patient must be warned against turning and bending movements that could damage the osteotomy region before healing. Partial weight-bearing only is included in ambulation, and other activities are avoided in the early postoperative period (6–8 weeks). Straight-leg raises are started on a postoperative day 1 in the hip flexion plane (supine position). These exercises progress throughout 3 weeks to include the hip extension. Straight-leg raises in adduction/abduction planes are started at 4–6 weeks depending on the radiographic findings of healing for HTO patients and at 6–8 weeks for DFO patients. When the exercises become more comfortable for the patient to perform, ankle weights are added for advanced muscle strengthening. Initially, 1–2 kg weights are added up to 5 kg, not exceeding 10% of the patient's body weight. The weight is initially placed over the patella to control the control forces. To provide increased resistance, a rubber tube can provide benefit in the eighth week. Another widely used exercise that is useful in quadriceps re-education is to increase the time of straight-leg raises. The quadriceps are contracted, and the leg is raised approximately 6 in. (15 cm) from the table or chair, held for 15 s, then lowered and rested for 45 s. To increase the difficulty of the exercise, extra ankle weights can be added. This can be started in the third to fourth week according to tolerance. CKC exercises can be started in postoperative weeks 5–6 [7, 13, 14, 19, 20] (Figs. 18.4, 18.5, 18.6, 19.2, 19.3, 19.4, 19.5, 19.6, 20.1, 20.2 and 20.3).

References

1. Agneskirchner JD, Hurschler C, Stukenborg-Colsman C, et al. Effect of high tibial flexion osteotomy on cartilage pressure and joint kinematics: a biomechanical study in human cadaveric knees. Winner of the AGA-DonJoy Award 2004. Arch Orthop Trauma Surg. 2004;124:575–84.
2. Barrett SL, Dellon AL, Rosson GD, Walters L. Superficial peroneal nerve (superficial fibularis nerve): the clinical implications of anatomic variability. J Foot Ankle Surg. 2006;45:174–6.
3. Noyes FR, Barber-Westin SD, Albright JC. An analysis of the causes of failure in 57 consecutive posterolateral operative procedures. Am J Sports Med. 2006;34:1419–30.
4. Backstein D, Morag G, Hanna S, et al. Long-term follow-up of distal femoral varus osteotomy of the knee. J Arthroplasty. 2007;22(4 suppl 1):2–6.
5. Song EK, Seon JK, Park SJ. How to avoid unintended increase of posterior slope in navigation-assisted open-wedge high tibial osteotomy. Orthopedics. 2007;30(10 suppl):S127–31.
6. Koshino T, Yoshida T, Ara Y, et al. Fifteen to twenty-eight years' follow-up results of high tibial valgus osteotomy for osteoarthritic knee. Knee. 2004;11:439–44.
7. Marti CB, Gautier E, Wachtl SW, Jakob RP. Accuracy of frontal and sagittal plane correction in open-wedge high tibial osteotomy. Arthroscopy. 2004;20:366–72.
8. Guettier JH, Glisson RR, Stubbs AJ, et al. The triad of varus malalignment, meniscectomy, and chondral damage: a biomechanical explanation for joint degeneration based on pressure and force distribution within the medial knee compartment. In: Meeting of the American Orthopaedic Society of Sports Medicine. Edited, Orlando, FL, 2002.
9. Eckhoff DG, Bach JM, Spitzer VM, et al. Threedimensional mechanics, kinematics, and morphology of the knee viewed in virtual reality. J Bone Joint Surg Am. 2005;87:71–80.
10. Naudie DD, Amendola A, Fowler PJ. Opening wedge high tibial osteotomy for symptomatic hyperextension-varus thrust. Am J Sports Med. 2004;32:60–70.
11. Noyes FR, Mayfield W, Barber-Westin SD, et al. Opening wedge high tibial osteotomy: an operative technique and rehabilitation program to decrease complications and promote early union and function. Am J Sports Med. 2006;34:1262–73.
12. Giffin JR, Vogrin TM, Zantop T, et al. Effects of increasing tibial slope on the biomechanics of the knee. Am J Sports Med. 2004;32:376–82.
13. Magyar G, Toksvig-Larsen S, Lindstrand A. Hemicallotasis open-wedge osteotomy for osteoarthritis of the knee. Complications 308 operations. J Bone Joint Surg Br. 1999;81:449–51.
14. Noyes FR, Barber-Westin SD. Anterior cruciate ligament revision reconstruction: results using a quadriceps tendon-patellar bone autograft. Am J Sports Med. 2006;34:553–64.
15. Stoffel K, Stachowiak G, Kuster M. Open wedge high tibial osteotomy: biomechanical investigation of the modified Arthrex Osteotomy Plate (Puddu Plate) and the TomoFix Plate. Clin Biomech (Bristol, Avon). 2004;19:944–50.
16. Warden SJ, Morris HG, Crossley KM, et al. Delayed- and non-union following opening wedge high tibial osteotomy: surgeons' results from 182 completed cases. Knee Surg Sports Traumatol Arthrosc. 2005;13:34–7.
17. Noyes FR, Wojtys EM, Marshall MT. The early diagnosis and treatment of developmental patella infera syndrome. Clin Orthop. 1991;265:241–52.
18. Noyes FR, Barber SD, Simon R. High tibial osteotomy and ligament reconstruction in varus angulated, anterior cruciate ligament deficient knees. A two to seven year follow-up study. Am J Sports Med. 1993;21:2–12.
19. Bonin N, Ait Si Selmi T, Donell ST, et al. Anterior cruciate reconstruction combined with valgus upper tibial osteotomy: 12 years follow-up. Knee. 2004;11:431–7.
20. Huang TL, Tseng KF, Chen WM, et al. Preoperative tibio femoral angle predicts survival of proximal tibia osteotomy. Clin Orthop Relat Res. 2005;432:188–95.

Morphometric Analysis of the Knee: A Comprehensive Evaluation of Knee Morphology in Designing Arthroplasties of Knee

22

Mohamed Elfekky and Samih Tarabichi

Salient Features

- A conceptual knowledge of knee architecture is important in designing various implants in patients with severe osteoarthritis and other bone deformities.
- Two types of methodologies have been followed for knee morphology assessments: (1) clinical intraoperative measurements and (2) computational analysis directly on medical images of knee.
- The metrics used for width are the mediolateral (ML) width of the proximal tibia, distal femur, and patellar, whereas for length determination is the anterolateral (AP) length of the proximal tibia and distal femur, and the proximal distal (PD) length of the patella.
- The angle between inclination of the tibial plateau and the long axis of the tibia shaft and the posterior slope of the tibial component influences various aspects of the knee kinematics and therefore plays an important role in implant fixation and wear of polyethylene insert.
- The commonly accepted threshold for implant overhang is 3 mm. Excessive overhang causes pain and worsens outcomes of knee arthroplasties. However, they can be tackled intraoperatively by reducing the size of implant.
- Novel metrics considered in tibial architecture are areas for the entire resection surface and each of the medial and lateral plateaus, areas of the bounding boxes for the entire resection and each of the medial and lateral plateaus, and radii of the tibial anterior periphery on the medial and lateral plateaus.
- Patellar stock thickness should be at least 12 mm, and bone stock should be maintained after resection to provide sufficient biomechanical strength of the composite.
- Computational methods such as shape models provide an analytic tool for the study of anatomy such as individual bone types in the knee or even the entire knee joint complex.
- The intrinsic differences in knee morphology across ethnics and genders that may in part explain the variable clinical outcomes.
- The disease progression or deformity in the knee may lead to alternation of its bony structure. Therefore, clinicians commonly believe that resections with a deviation within ±3° in alignment are acceptable, while the account of bone resected may be less under control depending on the reference point for resection depth, device thickness to match, and sometimes the quality of the bone.
- The anatomical designs exhibited improved latero-posterior (LP) coverage than the symmetrical standard designs in the ML and MP dimensions.

M. Elfekky (✉)
Hatta Hospital, Dubai Health Authority, Dubai, UAE

S. Tarabichi
Consultant Orthopaedic Surgery, Dubai, UAE

M. Bozkurt, H. İ. Açar (eds.), *Clinical Anatomy of the Knee*,
https://doi.org/10.1007/978-3-030-57578-6_22

22.1 Introduction

The primary objective of knee arthroplasty is to alleviate chronic pain and reinstate functionality of the impaired knee [1, 2]. Over the history of its application, the design of the prothesis has been improved tremendously in order to satisfy elevated demand and expectation from the patients. The modern knee prothesis are no longer designed for only providing preservation of the knee joint, pain relief, and (maybe) partial restoration of some functionalities for daily living. Nowadays, the implants are refined into optimal features in the design size and shape, advanced materials, surgical instruments and approach, and advanced postoperative recovering management and rehabilitation regime [3]. For example, in TKA surgery, the current implants are expected to have more than 90% 10-year survivorship with many implantations survived much longer.

It has been reported by meta-analysis that 82% of the total knee replacements lasted over 25 years after the surgery [4]. The leading etiologies for TKA failures have been identified as microbial infections, instability, malfunctioning of the prosthetic, aseptic loosening, implant fracture, gangrene in the patella, and improper axial positioning [5, 6]. However, even with a well-implanted knee confirmed by clinical evidences such as radiographs and in-office functional tests, there is still a group of approximately 20% of the TKA patients who still complain about the outcomes of the surgery [7].

One important contributing factor for implant loosening and satisfaction may be the anatomic fit of the implants. Excessive implant overhang over the bony resection boundary, especially in the mediolateral direction, has been shown to associate with soft tissue impingement and inflammation and cause significantly worse knee outcomes and pain [8–10]. A commonly accepted threshold for implant overhang is 3 mm as demonstrated in several studies to cause significant impact in the knee biomechanical and clinical outcomes [8–10]. Intraoperatively, a surgeon can avoid excessive overhang by reducing the size of the implant. However, the downsizing of the implant may lead to another issue, which is the reduced bony coverage by the implant. This inevitably results in the implant sits on the "softer" (cancellous) area of the resection surface and therefore may compromise the fixation and optimal load transfer between the implant and the bone with the reduced cortical support.

With the TKA implantation on the tibia, internally rotating the tibial implant may avoid the downsizing of the implant, but the internal rotation can be detrimental to the proper rotational alignment of the implantation, leading to heightened risks of patellofemoral complications, pain, and implant failure [11–14]. The resolution to the struggle between minimizing implant overhang, improving implant bony coverage, and maintaining proper rotation is the anatomical appropriateness in the implant design. To this point, it is critical to study knee morphology and its variation globally for a better understanding of current knee designs and the development of the next generation that properly fits the target patient population.

22.2 Morphological Analysis in Knee Arthroplasty

Arthroplasties in the knee should ideally consider the bony anatomy of the specific patient and operative site. This requires a good understanding of the knee morphology concerning the target treatment population. Studies of the knee morphology with respect to knee arthroplasty have been mainly using two types of methodologies: (1) clinical intraoperative measurements [15, 16] and (2) computational analysis directly on medical images of the knee or the virtual surface segmented from the medial images [17–21]. Although intraoperative measurements have the strength of providing real-world information on the clinically observed morphology from the actual targeted patient population, there are many setbacks in these studies. First, surgical variability is involved in the preparation of the bony resections; especially the reported studies were based on conventional surgery without the availability of intraoperative guidance (robots or surgical navigation) to minimize the surgical error.

Second, the manual measurements on the bony surface can be impacted by human error, joint environment (soft tissue and cartilage), and the inconsistency of measurement locations across samples. Last, the number of available samples is often small and imbalanced in terms of patients' ethnicity and gender due to the difficulty of recruiting patients into a large-scale study under clinical setting.

In contrast, computational studies pose the strength regarding removing variability in the data by the inclusion of multi-ethnic bone database with the potential of constant expansion of the study size by adding more bones and consistent and precise execution of the measurement following predefined algorithms. In addition, the ability to fully control the computational morphological studies provides possibility to compare the results from multiple studies given they are performed under the same computational protocol.

Irrespective to the methodology and bone type (femur, tibia, and patella), the morphometric measurements in the knee area focused on a common set of size and shape metrics. These measurements are mainly driven by the focus of proper implant fit during arthroplasty, as insufficient bony coverage may be detrimental to the longevity of biomechanical fixation of the implant, and the presence of excessive implant overhang has been shown to associate with soft tissue impingement and inflammation and cause significantly worse knee outcomes and pain [8–10]. This set of metrics are generally used to measure the dimensions of the articulating area in the bone, namely (1) width: the mediolateral (ML) width of the proximal tibia, distal femur, and patellar; and (2) length: the anterolateral (AP) length of the proximal tibia and distal femur, and the proximal distal (PD) length of the patella. Furthermore, the ratio between medial and lateral sides of the bone in these dimensional measurements (aspect ratio) is commonly used to depict the asymmetry between the two compartments, which is a representation of the shape of the bone. These metrics have been used to quantify knee morphology based on either the native bone or specific arthroplasty resection scenarios (Table 22.1).

In a 2011 study, morphometric measurements were reported by Yue et al. [19] on the size of the native intact proximal tibia by the AP lengths on each of the medial and lateral plateaus and the ML width. Similarly, for the native intact distal femur, the overall AP and ML dimensions were measured. From the measured size metrics, aspect ratios were calculated as the ratio between the measured ML and AP dimensions for the femur and tibia, respectively. Although the studies on the native intact knee provided insights on

Table 22.1 Illustration on AP, AM measurements on the femur and tibia

	Morphometrics	
Bone	Size	Shape
Native tibia	AP dimension (overall, medial, lateral) ML dimension	Overall aspect ratio
Resected tibia (TKA as example)	AP dimension (overall, medial, lateral) ML dimension Resection area (overall, medial, lateral) Boxiness (overall, medial, lateral) Anterior radii (medial, lateral)	Aspect ratio (overall, medial, lateral) Asymmetry between medial and lateral plateaus (area, boxiness, anterior radii)
Femur	AP dimension (overall, medial, lateral) ML dimension (multiple locations)	Aspect ratio Posterior condylar offset Trochlear orientation Trochlear sulcus angle
Native patella	ML dimension PD dimension ML location of the ridge Thickness	Aspect ratio

the general joint morphology, most studies were interested in the assessment under a surgical scenario as it provided a further understanding of knee morphology specifically related to a particular arthroplasty application. Yang et al. [15] studied in tibial implants that the AP and ML dimensions were measured on the resected proximal tibia from actual TKA patients intraoperatively [15].

A 2008 study shed a light on the difference between the medial and lateral plateaus under the application of UKA [22]. The data cumulated from numerous resection-specific measurements not only provide the knowledge on the average size and shape of the bone restated to arthroplasty and their associated variability (standard deviation), but also serve as the basis for the identification of common characteristics in the bone morphology to guide development of new arthroplasty implants for the application across multiple patient populations.

One well-established, universally applicable knowledge is that based on either native or resected tibia, a positive correlation between the size metrics, such as ML and AP dimensions, bone length, and patient's height is revealed. There is also a well-accepted consensus that the medial and lateral compartments of each bone type in the knee demonstrated an asymmetry (ML/AP aspect ratio) that revealed the medial compartment to be larger. The specific correlation formula and aspect ratio calculated may be used as a meaningful indication for the design and use of associated implants for the treatment population.

In addition to the common list of dimensional measurements and aspect ratios mentioned above, other metrics were used to quantify knee morphology specifically to individual bone types and arthroplasty applications. Quantified as the angle between inclination of the tibial plateau and the long axis of the tibia shaft (multiple definitions exist by using mechanical axis, anatomical axis, or anterior cortex of the proximal tibial) [23], the posterior slope of the tibial component influences various aspects of the knee kinematics and therefore plays an important role in implant fixation and wear of polyethylene insert [24, 25]. An increased posterior slope has been shown to result in greater anterior translation of the tibia during weight-bearing activities and increased strain on the ACL with more substantial compressive loads [26–31]. Additionally, some authors have recently discussed modification of the TS through tibial deflexion osteotomy as an important surgical consideration in patients with ACL injuries with excessive TS. However, data currently remain limited to support this approach [32–36].

Novel metrics have been observed in an Indiana study [37] for the morphology of tibial resection during TKA, including: (1) areas for the entire resection surface and each of the medial and lateral plateaus; (2) areas of the bounding boxes for the entire resection and each of the medial and lateral plateaus; and (3) radii of the tibial anterior periphery on the medial and lateral plateaus. Using the newly introduced size metrics, several definitions of asymmetry were developed to descript the shape of the resection morphology as the ratio between the medial and lateral plateaus. These additional morphometric provided further understanding of tibial morphology in relation to TKA application for the evaluation and development of modern anatomic implant designs. A number of additional femur-specific metrics were also introduced, such as posterior offset of the femoral condyles; depth, width, and angle of the femoral sulcus; and orientation of the trochlear groove. This expanded list of measurements on the femur is needed as the metrics correspond to great clinical relevance to specific arthroplasty considerations. Specifically, the restoration of the original femoral posterior condylar offset (PCO), measured as the maximum thickness of the posterior condyle projecting posteriorly to the tangent of the posterior cortex of the femoral shaft, is important to maximize postoperative range of motion (ROM), avoid impingement, improve knee kinematics, and minimize flexion instability after TKA [38–41]. The measurements on the sulcus morphology, as well as the orientation of the trochlear groove, all have profound indications in both disease progression and outcomes of the patella-femoral joint [42, 43].

In the patella, the thickness of the bone stock is a critical morphometric to be considered under resurfacing surgery, as at least 12 mm bone stock should be maintained after resection to provide sufficient biomechanical strength of the composite [44] and the original thickness should be restored as close as possible after the resurfacing for the preservation of the extensor efficiency. The position of the medial ridge on the patellar articular surface serves as morphological reference for patellar alignment. It has been shown that proper positioning of the patellar implant with the medial ridge reduces the Q angle and helps in restoring kinematics post-surgery [45]. Beyond the measurements purely based on the patellar bone, there are additional focuses on the relative position of the patellar relative to the femoral trochlea, including patellar displacement and patellar tilt measured in the sunrise view, and patellar height assessed with a slight knee flexion in the sagittal view. These additional measurements serve as important indications for assessing extensor mechanism and patellar tracking.

Morphological considerations related to arthroplasty in the knee are not limited to just the morphology of proximal tibial and distal femur. Research efforts also expanded to the understanding of the anatomical shape of the tibial and femoral shaft and its impact on alignment and clinical outcomes related to the use of stemmed implants and intramedullary nailing. In the coronal plane, a normal "straight" femur shows no bowing of its shaft and a 3° valgus angle in the condylar surface with respect to the femoral mechanical axis, whereas the proximal tibial surface is at a corresponding 3° of varus with reference to the tibial mechanical axis [46–48]. However, a fair amount of bowing exists in the population, especially prevalent in the Asian ethnicities [49, 50]. With the use of intramedullary guide during surgery, the presence of lateral bowing can lead to varus alignment of the femoral component, and varus inclination of the tibia surface had clinical implications in gap balancing requiring increased medial release or femoral implant external rotation. In the sagittal plane, studies have shown that overlooking sagittal femoral bowing can cause improper femoral implant sizing during conventional TKA and notching in navigated mechanically aligned TKA [51]. These surgical mistakes heighten the risk of excessive flexion of the femoral implant, limited knee extension, compromised fixation of stemmed femoral implant, postoperative supracondylar femoral fracture, and polyethylene post wear caused by cam-post impingement in posterior stabilized (PS) TKA [51–53]. Thereupon, care needs to be taken during intramedullary guided procedure as the bowing morphology of the femur and tibia may lead to alignment outliers that are detrimental to the clinical outcomes and longevity of the surgery [54, 55].

With the advances of computational methods, the investigation of knee morphology progressed into the era of population-based analysis and is no longer restricted to limited number of discrete measurement metrics. One powerful tool for such analysis is statistical shape modeling. Shape models provide an analytic tool for the study of anatomy such as individual bone types in the knee or even the entire knee joint complex. By disseminating a complex bony anatomy into a mathematical formula using a set of principal components, the variability in morphology across can be understood with the identification of the primary driving mode of variations. Morphological data from the application of this advanced tool has been applied successfully in detecting variability in native distal femoral and proximal tibial morphology [56] and TKA tibial resection surface [57] across populations as inputs to drive anatomical designs in the knee, including tibial plates for internal fracture fixation, fibular plates, and TKA tibial base plate [58, 59]. Several studies characterized the entire joint anatomy by looking at individual bones coupled with their relative position to each other. Using one shape model, Fitzpatrick et al. quantified the combined variability of the resected profiles of the patella, femur, and tibia during TKA [60]. Some studies advanced the analysis to combine the morphology of the knee joint structure and limb alignment in the shape modeling, quantify variability due to morphology and relative alignment [61].

22.3 Gender Variations in Knee Morphology

With the expansion of knee arthroplasty from its western origin to around the world and increased application volume in genders, recurrent gender- and ethnic-based inferior outcomes were discovered which led to an extensive research. It has been realized that there are intrinsic differences in knee morphology across ethnics and genders that may in part explain the variable clinical outcomes.

Numerous studies have documented the differences between male and female knee morphology. Consistent trends have been reported that within any specific ethnic population, male knees are on average larger in size than the female knees in all dimensional measurements, while the gender differences found in knee shape are less prominent [37, 62–65]. In a 2012 study by Yan et al., the male knees showed significantly greater coronal dimensions of the trochlea than the female knees [63]. The authors suggested these dimensional discrepancies contributed to the higher prevalence of prosthetic overhang in women with some standard implants. Koh et al. revealed that although the posterior condylar offset was larger in the male knee compared to the female knee, the same trend in gender did not stand in the ratio between the sagittal AP shape of the knee. Female knees exhibited higher posterior condylar offset proportional to the total AP size of the distal femur than the male knees [62].

In a study carried out by Asseln et al., an extensive list containing 33 features of the femur and 21 features of the tibia were used to investigate gender differences [64]. The results demonstrated significant larger values in all linear dimensional measurements (size) but only selective angular measurements (shape). A systematic review of the PubMed database was performed on published studies on more than 9000 knees from four ethnic groups [65]. The key dimensions in the knee (ML and AP) were all shown to have higher values in males compared to females, while the differences in the aspect ratios were more subtle and variable. Several studies showed that when the dimensional measurements are normalized, gender-specific differences dissipate. Fehring et al. reported no significant gender difference in the height of femoral lateral condyle and nominal differences regarding the medial condyles [66].

Voleti et al. found gender-specific differences in femoral medial and lateral posterior condylar offsets. However, they disappeared after normalization by the condylar height [67]. Dai et al. used a comprehensive list of morphometric to quantify the size and shape of the tibial TKA resection surface [37]. Although it was revealed that male knees were bigger in all size metrics in each of the three ethnic groups investigated (Caucasian, Indian, and Japanese), the correlation between the ML and AP dimensions shared very similar slopes between the two genders, suggesting limited difference between the two genders in aspect ratio of the resection plateau [37]. Further statistical shape analysis in the same study revealed that the driving factor for the variability over served between male and female resided in the general size difference between the two genders, while in general shape remains constant.

As the knee size in females increases, the aspect ratio of femur decreases. However, in males, the aspect ratio remains constant irrespective of the knee sizes [68–70]. Therefore, it is necessary to implicate gender-specific implants to minimize the femoral overhang issue post implant. To put this into practice, orthopedic surgeons of Korea demonstrated that in cases where overhang was difficult to overcome while inserting traditional implants, the incidence of femoral component overhang was reduced by 34.6% by using the gender-specific implants [71].

22.4 Ethnic Variations in Knee Morphology

In the history of the application of knee arthroplasty, nearly all knee prostheses were originally designed based on the morphological features of western knees from primarily white Caucasian patients. Several studies have passed judgment that the tailor-made arthroplasty prostheses for Caucasian patients are not appropriate in cases with other ethnic background. Investigations that have detailed anthropometric differences according to ethnicity have primarily been focused between Caucasian and Asian populations [19,

56, 72], as well as numerous individual reports on ethnic-specific knee morphology [20, 21, 45, 73–76]. Caucasian knees have been shown to be generally larger than Asian knees [37, 56, 65, 73], which may give rise to the risk of implant oversize when used in Asian patients. In addition, for a given AP dimension, Caucasian knees have been shown to have a higher aspect ratio compared to Asian knees [19] (Fig. 22.1).

In a statistically defined shape analysis of the knee architecture, Mahfouz and his colleagues, identified differences in shape between the distal femur and proximal tibia with respect to the African American, Asian, and Caucasian [56]. Both the locations and magnitudes of morphological deviations of the distal femur and proximal tibia were identified during the paired comparison between the ethnicities investigated. Studies also showed considerable variations in normal alignment between ethnicities [49, 50, 77–79]. Asian population is reported to have more deviation in the angular alignment than Caucasian. In a 2008 study by Harvey et al., Asian knees were found to have a substantially more valgus anatomic axis, valgus condylar angles, and valgus condylar-plateau angles compared to the Caucasian knees [72]. Femurs in the Asian population also are substantially more tibial slope and externally rotated than Western patients [80, 81]. Thereafter, many Asian studies have found severe varus inclination in cases of advanced osteoarthritis with femoral lateral bowing and obliquity of the proximal tibial joint surface in knees [49, 50, 77–79].

The accumulated data by studies across geographic regions, along with the availability of advanced population-based analyses, provided a vast amount of knowledge to reveal gender and ethnic impact on knee morphology. Table 22.2 summarizes a collection of reported data on common measurements across gender and ethnic populations.

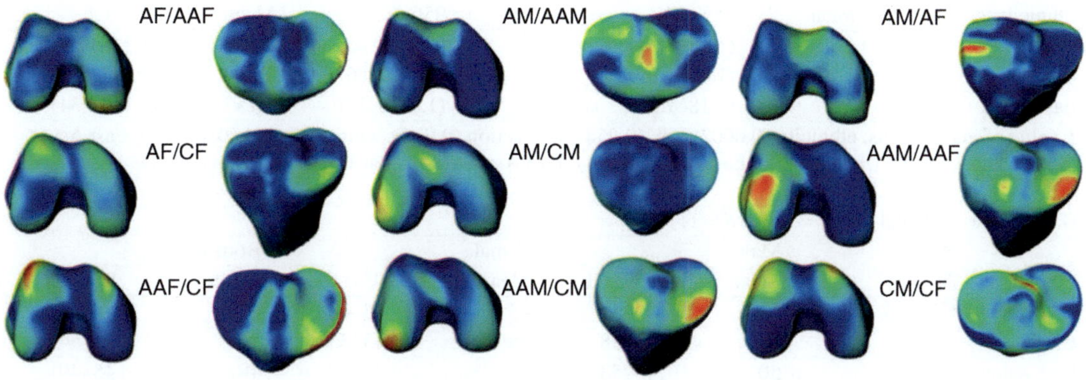

Fig. 22.1 The differences between high (red) and low (blue) global shape variations among gender and ethnic background portrayed by the second to ninth principal components are shown in this figure. *AF* East Asian Female, *CM* Caucasian male, *AAM* African American male, *AAF* African American Female [18]

Table 22.2 Summary of a list of common gender and ethnic specific morphological measurements in ML, AP, aspect ratio, etc. (*N* = No. of knees, *S* = No. of studies) [45, 65, 75, 76] Measures of Femoral AP (*N* = 360; *S* = 13)

	Males		Females		Both sexes	
Ethnicity	Mean	95% CI	Mean	95% CI	Mean	95% CI
White	64	60–69	59	54–64	62	57–66
Black	66	61–70	61	55–67	63	58–68
East Asian	61	57–66	56	52–60	59	54–63
Indian	61	45–77	55	39–70	59	42–73

Measurements in mm; *p* values of main effects: ethnicity (<0.001); sex (<0.001); interaction (0.954); white versus black (0.639), East Asian (<0.001), Indian (0.957); black versus white (0.639), East Asian (0.012), Indian (0.900); East Asian versus black (0.012), white (<0.001), Indian (0.999); Indian versus black (0.900), white (0.957), East Asian (0.999)

(continued)

Table 22.2 (continued)

A. Measures of Femoral mediolateral aspect ($N = 1884$; $S = 15$)

	Male		Female		Both sexes	
Ethnicity	Mean	95% CI	Mean	95% CI	Mean	95% CI
White	79	75–83	69	65–72	74	70–77
Black	71	65–77	67	60–75	69	64–74
East Asian	76	73–79	67	64–70	71	69–74
Indian	70	59–80	61	49–73	65	55–76

Measurements in mm; *p* values of main effects: ethnicity (0.167); sex (<0.001); interaction (0.564); black versus while (0.254), East Asian (0.560), Indian (0.458); black versus white (0.254), East Asian (0.738), Indian (0.911); East Asian versus black (0.738), while (0.560), Indian (0.670); Indian versus black (0.911), white (0.458), East Asian (0.670)

B. Measures of Femoral medial AP ($N = 2183$; $S = 8$)

	Male		Female		Both sexes	
Ethnicity	Mean	95% CI	Mean	95% CI	Mean	95% CI
White	65	61–68	59	55–62	62	58–65
Black	65	61–70	63	56–70	M	59–69
East Asian	60	57–64	56	52–59	58	54–62

Measurements in mm; *p* values of main effects: ethnicity (0.009); sex (0.004); interaction (0.156); while versus black (0.338), East Asian (0.012); black versus white (0.338), East Asian (0.022); East Asian versus black (0.022), white (0.012)

C. Determination of Femoral aspect ratio ($N = 4825$; $S = 14$)

	Male		Female		Both sexes	
Ethnicity	Mean	95% CI	Mean	95% CI	Mean	95% CI
White	1.22	(1.13–1.31)	1.17	(1.08–1.26)	1.20	(1.11–1.29)
Black	1.19	(1.09–1.29)	1.19	(1.08–1.26)	1.19	(1.02–1.37)
East Asian		(1.18–1.35)	1.23	(1.15–1.32)	1.25	(1.16–1.34)

P value of main effects: ethnicity (0.0002); sex (0.558); interaction (0.915); white versus black (0.996), East Asian (0.001); black versus white (0.996), East Asian (0.694); East Asian versus black (0.694), white (0.001)

D. Measures of Tibial AP ($N = 3553$; $S = 11$)

	Male		Female		Both sexes	
Ethnicity	Mean	95% CI	Mean	95% CI	Mean	95% CI
White	52	49–54	45	43–48	48	46–51
Black	53	48–58	48	43–53	50	46–54
East Asian	50	48–53	45	43–47	48	45–49
Indian	48	40–56	44	36–52	46	38–54

Measurements in mm: *p* values of main effects: ethnicity (0.401); sex (<0.001); interaction (0.662); white versus black (0.664), East Asian (0.646), Indian (0.904); black versus white (0.664), East Asian (0.409), Indian (0.722); East Asian versus black (0.409), white (0.646), Indian (0.969); Indian versus black (0.722), white (0.904), East Asian (0.969)

E. Measures of Tibial mediolateral aspect ($N = 4194$; $S = 14$

	Male		Female		Both sexes	
Ethnicity	Mean	95% CI	Mean	95% CI	Mean	95% CI
While	79	78–81	69	68–71	74	73–76
Black	80	76–83	67	63–70	73	71–76
East Asian	77	76–78	69	68–70	73	72–74
Indian	77	74–79	69	66–71	73	71–75

Measurements in mm; *p* values of main effects: ethnicity (0.039); sex (<0.001); interaction (0.013); while versus black (0.771), East Asian (0.036), Indian (0.361); black versus white (0.771), East Asian (0.984), Indian (0.990); East Asian versus black (0.984), while (0.036), Indian (1.000); Indian versus black (0.990), while (0.361), East Asian (1.000)

Table 22.2 (continued)

F. Measures of Tibial medial AP ($N = 3541$; $S = 12$)

	Male		Female		Both sexes	
Ethnicity	Mean	95% CI	Mean	95% CI	Mean	95% CI
White	53	51–55	47	45–49	50	48–52
Indian	51	48–53	45	42–53	48	45–50
East Asian	52	50–53	46	45–48	49	48–51

Measurements in mm; p values of main effects: ethnicity (0.096); sex (<0.001); interaction (0.466); white versus East Asian (0.598), Indian (0.079); East Asian versus white (0.598), Indian (0.287); Indian versus white (0.079), East Asian (0.287)

G. Determination of Tibial aspect ratio ($N = 1653$; $S = 5$)

	Male		Female		Both sexes	
Ethnicity	Mean	95% CI	Mean	95% CI	Mean	95% CI
White	1.57	1.42–1.73	1.54	1.38–1.69	1.55	1.40–1.71
Black	1.54	1.38–1.70	1.43	1.27–1.59	1.49	1.33–1.64
East Asian	1.53	1.38–1.69	1.54	0.39–1.70	1.54	1.39–1.69

P values of main effects: ethnicity (0.006); sex (0.003); interaction (0.005); while versus black (0.005), East Asian (0.382); black versus white (0.005), East Asian (0.057); East Asian versus black (0.057), white (0.382)

22.5 Additional Considerations Regarding Knee Morphology

Although disease progression or deformity in the knee may lead to alternation of its bony structure, most of the large-scale morphological studies still focused on healthy knees, possibility due to the challenges in obtaining sufficient number of specimens with controlled pathological conditions and other confounding factors of the patients. A number of studies made the effort to report on OA knee morphology. Mullaji et al. performed a radiographic analysis on Asian varus osteoarthritic (OA) knees with a healthy cohort as control [50]. The study discovered that compared to healthy knees, varus OA knees exhibited significantly lower condyle–mechanical axis angle and a higher deviation between femoral mechanical axis and the axis of the distal intramedullary canal. The evaluation of anatomic variations and their outcomes were studied on the operative techniques practiced in total knee arthroplasty (TKA). Nagamine [49] assessed anatomic variations specific to OA patients in six morphological parameters based on their preoperative AP radiographs and identified the signature of the medial OA knees as bowing of the femoral shaft and proximal tibia vara with lateral offset of the tibial shaft concerning the tibial plateau's center. Leveraging statistical shape modeling, Fitzpatrick et al. explored morphological variations in TKA resection geometries from OA patients, revealing the variation of size and shape with the tested population and highlighted the domination of size on resection morphology [60]. These disease-specific morphological measurements provided valuable insights regarding special considerations that worth attention in surgical treatments of the affected knee under clinical setting.

The most important and notable thing is that almost all analyses performed to aid the understanding of knee morphology under bony resection situation relative to a specific arthroplasty application were based on a single resection scenario. It is hard to ignore that as the nature of human manual work, the variability in surgical resections, especially from conventional instrumented cases, should be well expected under clinical setting. For example, during TKA resection, variation can exist both during visual and manual identification of the landmarks for the establishment of critical anatomical references and surgeon's preference in using a slight variable definition of reference. Clinicians commonly believe that resections with a deviation within ±3° in alignment are acceptable, while the account of bone resected may be less under control depending on the reference point for resection depth, device thickness to match, and sometimes the quality of the bone. It still remains largely unclear how to properly interpolate and what difference should be expected when transfer the published morphological knowledge based on single resection scenario into clinical setting. Limited investigations have attempted to shed light on this topic.

In a computational study by Dai et al. [82], the influence of variabilities at each step of proximal tibial resections, aiming to quantify the influence of variability in landmark detection on resection parameters on TKA resections. One important finding was that landmark variability influenced key dimensions of the resected plateau by several millimeters, significant enough to impact clinical decisions based on morphology. The morphology of the proximal tibia at different levels of resection was studied by Nakamura et al. with the depth ranging from primary to revision TKA (10–25 mm) [83]. Deeper resection depth led to substantial internal rotation of the resection surface relative to the tibial shaft up to 23° in the range investigated, with significant changes in the aspect ratio. The authors cautioned surgeons to pay attention to morphological changes depending on the specific amount of bone taken from the patients. The existing limited reports suggested that currently knowledge on single resection scenario may not be entirely conclusive. The investigations need to be advanced further to fully understand the scope of impact caused by the involvement of surgical variability.

Many studies have been suggested that anatomical designs exhibited improved latero-posterior (LP) coverage than the symmetrical standard designs in the ML and MP dimensions [84]. For example, in TKA designs, whether there is a clear clinical advantage provided by asymmetric designs over the symmetric designs is still debated over. Dai et al. reported that tibial designs based on the anatomical structure are more durable with high alignment precision than tibial structures designed in accordance with symmetric and asymmetric framework. Wernecke and his associates also concluded that in asymmetrical implants, there is more LP coverage than symmetrical implants in rotational controlled MRI study [85, 86]. Yang et al. [15] reflects that asymmetric tibial components are more fitting than other designs because the medial and lateral tibial surfaces are asymmetrical. Therefore, other prosthetic components would not completely overlay the tibial surface leading to extensive lateral overhang or undersized medial component. The functionality and efficacy of both the designs can be fully unraveled only after methodical clinical studies.

22.6 Conclusion

As application of knee arthroplasty is on the rise, innumerable explorative research has been conducted to measure the anatomical features in the knee and identify associated gender and ethnic differences. Variations are detected by computational statistical methods for morphologic analysis. Considering these variations, the use of gender-specific and ethnic-specific implants may lead to positive outcomes after knee arthroplasties. These results may help surgeons and manufacturers to better understand their patient population and improve the fit of the designed implants. Further studies need to be carried out to acquire more evidences on the benefits offered by designs driven by morphological inputs and expand the knowledge of knee morphology and its implication in surgical technique and variabilities to address good long-term outcomes and patient satisfaction across the population.

References

1. Kim K, Snir N, Schwarzkopf R. Modern techniques in knee arthrodesis. Int J Orthopaed. 2016;3(1):487–96.
2. Buckland-Wright JC, Macfarlane DG, Lynch JA, Jasani MK, Bradshaw CR. Joint space width measures cartilage thickness in osteoarthritis of the knee: high resolution plain film and double contrast

macro radiographic investigation. Ann Rheum Dis. 1995;54(4):263–8.
3. Curtin B, Malkani A, Lau E, Kurtz S, Ong K. Revision after total knee arthroplasty and unicompartmental knee arthroplasty in the Medicare population. J Arthroplasty. 2012;27(8):1480–6.
4. Evans JT, Walker RW, Evans JP, Blom AW, Sayers A, Whitehouse MR. How long does a knee replacement last? A systematic review and meta-analysis of case series and national registry reports with more than 15 years of follow-up. Lancet. 2019;393(10172):655–63.
5. Lum ZC, Natsuhara KM, Shelton TJ, Giordani M, Pereira GC, Meehan JP. Mortality during total knee periprosthetic joint infection. J Arthroplasty. 2018;33(12):3783–8.
6. Rand JA, Bryan RS. Revision after total knee arthroplasty. Orthop Clin North Am. 1982;13(1):201–12.
7. Scott CE, Howie CR, MacDonald D, Biant LC. Predicting dissatisfaction following total knee replacement: a prospective study of 1217 patients. J Bone Jt Surg Br Vol. 2010;92(9):1253–8.
8. Mahoney OM, Kinsey T. Overhang of the femoral component in total knee arthroplasty: risk factors and clinical consequences. JBJS. 2010;92(5):1115–21.
9. Chau R, Gulati A, Pandit H, Beard DJ, Price AJ, Dodd CA, Gill HS, Murray DW. Tibial component overhang following unicompartmental knee replacement—does it matter? Knee. 2009;16(5):310–3.
10. Gudena R, Pilambaraei MA, Werle J, Shrive NG, Frank CB. A safe overhang limit for unicompartmental knee arthroplasties based on medial collateral ligament strains: an in vitro study. J Arthroplasty. 2013;28(2):227–33.
11. Bédard M, Vince KG, Redfern J, Collen SR. Internal rotation of the tibial component is frequent in stiff total knee arthroplasty. Clin Orthop Relat Res. 2011;469(8):2346–55.
12. Berger RA, Rubash HE. Rotational instability and malrotation after total knee arthroplasty. Orthoped Clin. 2001;32(4):639–47.
13. Thompson JA, Hast MW, Granger JF, Piazza SJ, Siston RA. Biomechanical effects of total knee arthroplasty component malrotation: a computational simulation. J Orthop Res. 2011;29(7):969–75.
14. Nicoll D, Rowley DI. Internal rotational error of the tibial component is a major cause of pain after total knee replacement. J Bone Jt Surg Br Vol. 2010;92(9):1238–44.
15. Yang B, Song CH, Yu JK, Yang YQ, Gong X, Chen LX, Wang YJ, Wang J. Intraoperative anthropometric measurements of tibial morphology: comparisons with the dimensions of current tibial implants. Knee Surg Sports Traumatol Arthrosc. 2014;22(12):2924–30.
16. Westrich GH, Haas SB, Insall JN, Frachie A. Resection specimen analysis of proximal tibial anatomy based on 100 total knee arthroplasty specimens. J Arthroplasty. 1995;10(1):47–51.
17. Hartel MJ, Loosli Y, Gralla J, Kohl S, Hoppe S, Röder C, Eggli S. The mean anatomical shape of the tibial plateau at the knee arthroplasty resection level: an investigation using MRI. Knee. 2009;16(6):452–7.
18. Mahfouz MR, Merkl BC, Fatah EE, Booth R Jr, Argenson JN. Automatic methods for characterization of sexual dimorphism of adult femora: distal femur. Comput Methods Biomech Biomed Eng. 2007;10(6):447–56.
19. Yue B, Varadarajan KM, Ai S, Tang T, Rubash HE, Li G. Differences of knee anthropometry between Chinese and white men and women. J Arthroplasty. 2011;26(1):124–30.
20. Uehara K, Kadoya Y, Kobayashi A, Ohashi H, Yamano Y. Anthropometry of the proximal tibia to design a total knee prosthesis for the Japanese population. J Arthroplasty. 2002;17(8):1028–32.
21. Cheng FB, Ji XF, Lai Y, Feng JC, Zheng WX, Sun YF, Fu YW, Li YQ. Three dimensional morphometry of the knee to design the total knee arthroplasty for Chinese population. Knee. 2009;16(5):341–7.
22. Servien E, Saffarini M, Lustig S, Chomel S, Neyret P. Lateral versus medial tibial plateau: morphometric analysis and adaptability with current tibial component design. Knee Surg Sports Traumatol Arthrosc. 2008;16(12):1141–5.
23. Luo CF. Reference axes for reconstruction of the knee. Knee. 2004;11(4):251–7.
24. Hofmann AA, Bachus KN, Wyatt RW. Effect of the tibial cut on subsidence following total knee arthroplasty. Clin Orthop Relat Res. 1991;269:63–9.
25. Whiteside LA, Amador DD. The effect of posterior tibial slope on knee stability after Ortholoc total knee arthroplasty. J Arthroplasty. 1988;3:S51–7.
26. Fening SD, Kovacic J, Kambic H, McLean S, Scott J, Miniaci A. The effects of modified posterior tibial slope on anterior cruciate ligament strain and knee kinematics–a human cadaveric study. J Knee Surg. 2008;21(03):205–11.
27. Feucht MJ, Mauro CS, Brucker PU, Imhoff AB, Hinterwimmer S. The role of the tibial slope in sustaining and treating anterior cruciate ligament injuries. Knee Surg Sports Traumatol Arthrosc. 2013;21(1):134–45.
28. Shelburne KB, Kim HJ, Sterett WI, Pandy MG. Effect of posterior tibial slope on knee biomechanics during functional activity. J Orthop Res. 2011;29(2):223–31.
29. Akamatsu Y, Mitsugi N, Mochida Y, Taki N, Kobayashi H, Takeuchi R, Saito T. Navigated opening wedge high tibial osteotomy improves intraoperative correction angle compared with conventional method. Knee Surg Sports Traumatol Arthrosc. 2012;20(3):586–93.
30. Allen MR, Newman CL, Smith E, Brown DM, Organ JM. Variability of in vivo reference point indentation in skeletally mature inbred rats. J Biomech. 2014;47(10):2504–7.
31. Baier C, Maderbacher G, Springorum HR, Zeman F, Fitz W, Schaumburger J, Grifka J, Beckmann J. No difference in accuracy between pinless and conventional computer-assisted surgery in total knee arthroplasty. Knee Surg Sports Traumatol Arthrosc. 2014;22(8):1819–26.

32. Weinberg DS, Williamson DF, Gebhart JJ, Knapik DM, Voos JE. Differences in medial and lateral posterior tibial slope: an osteological review of 1090 tibiae comparing age, sex, and race. Am J Sports Med. 2017;45(1):106–13.
33. Lee YS, Moon GH. Comparative analysis of osteotomy accuracy between the conventional and devised technique using a protective cutting system in medial open-wedge high tibial osteotomy. J Orthop Sci. 2015;20(1):129–36.
34. Marriott K, Birmingham TB, Kean CO, Hui C, Jenkyn TR, Giffin JR. Five-year changes in gait biomechanics after concomitant high tibial osteotomy and ACL reconstruction in patients with medial knee osteoarthritis. Am J Sports Med. 2015;43(9):2277–85.
35. Westermann RW, DeBerardino T, Amendola A. Minimizing alteration of posterior tibial slope during opening wedge high tibial osteotomy: a protocol with experimental validation in paired cadaveric knees. Iowa Orthop J. 2014;34:16.
36. Zeng C, Yang T, Wu S, Gao SG, Li H, Deng ZH, Zhang Y, Lei GH. Is posterior tibial slope associated with noncontact anterior cruciate ligament injury? Knee Surg Sports Traumatol Arthrosc. 2016;24(3):830–7.
37. Dai Y, Bischoff JE. Comprehensive assessment of tibial plateau morphology in total knee arthroplasty: influence of shape and size on anthropometric variability. J Orthop Res. 2013;31(10):1643–52.
38. Bellemans J, Banks S, Victor J, Vandenneucker H, Moemans A. Fluoroscopic analysis of the kinematics of deep flexion in total knee arthroplasty: influence of posterior condylar offset. J Bone Jt Surg Br Vol. 2002;84(1):50–3.
39. Kang KT, Koh YG, Son J, Kwon OR, Lee JS, Kwon SK. A computational simulation study to determine the biomechanical influence of posterior condylar offset and tibial slope in cruciate retaining total knee arthroplasty. Bone Joint Res. 2018;7(1):69–78.
40. Kapoor A, Mishra SK, Dewangan SK, Mody BS. Range of movements of lower limb joints in cross-legged sitting posture. J Arthroplasty. 2008;23(3):451–3.
41. Onodera T, Majima T, Nishiike O, Kasahara Y, Takahashi D. Posterior femoral condylar offset after total knee replacement in the risk of knee flexion contracture. J Arthroplasty. 2013;28(7):1112–6.
42. Iriuchishima T, Ryu K, Murakami T, Yorifuji H. The correlation between femoral sulcus morphology and osteoarthritic changes in the patellofemoral joint. Knee Surg Sports Traumatol Arthrosc. 2017;25(9):2715–20.
43. Kulkarni SK, Freeman MA, Poal-Manresa JC, Asencio JI, Rodriguez JJ. The patellofemoral joint in total knee arthroplasty: is the design of the trochlea the critical factor? J Arthroplasty. 2000;15(4):424–9.
44. Seo JG, Moon YW, Park SH, Lee JH, Kang HM, Kim SM. A case-control study of spontaneous patellar fractures following primary total knee replacement. J Bone Jt Surg Br Vol. 2012;94(7):908–13.
45. Kim TK, Chung BJ, Kang YG, Chang CB, Seong SC. Clinical implications of anthropometric patellar dimensions for TKA in Asians. Clin Orthop Relat Res. 2009;467(4):1007–14.
46. Hsu RW, Himeno S, Coventry MB, Chao EY. Normal axial alignment of the lower extremity and load-bearing distribution at the knee. Clin Orthop Relat Res. 1990;255:215–27.
47. Lombardi AV Jr, Nett MP, Scott WN, Clarke HD, Berend KR, O'Connor MI. Primary total knee arthroplasty. JBJS. 2009;91(Supplement_5):52–5.
48. Moreland JR, Bassett LW, Hanker GJ. Radiographic analysis of the axial alignment of the lower extremity. J Bone Jt Surg. 1987;69(5):745–9.
49. Nagamine R, Miura H, Bravo CV, Urabe K, Matsuda S, Miyanishi K, Hirata G, Iwamoto Y. Anatomic variations should be considered in total knee arthroplasty. J Orthop Sci. 2000;5(3):232–7.
50. Mullaji AB, Marawar SV, Mittal V. A comparison of coronal plane axial femoral relationships in Asian patients with varus osteoarthritic knees and healthy knees. J Arthroplasty. 2009;24(6):861–7.
51. Ko JH, Han CD, Shin KH, Nguku L, Yang IH, Lee WS, Kim KI, Park KK. Femur bowing could be a risk factor for implant flexion in conventional total knee arthroplasty and notching in navigated total knee arthroplasty. Knee Surg Sports Traumatol Arthrosc. 2016;24(8):2476–82.
52. Yehyawi TM, Callaghan JJ, Pedersen DR, O'Rourke MR, Liu SS. Variances in sagittal femoral shaft bowing in patients undergoing TKA. Clin Orthop Relat Res. 2007;464:99–104.
53. Lee JH, Wang SI. Risk of anterior femoral notching in navigated total knee arthroplasty. Clin Orthop Surg. 2015;7(2):217–24.
54. Fang DM, Ritter MA, Davis KE. Coronal alignment in total knee arthroplasty: just how important is it? J Arthroplasty. 2009;24(6):39–43.
55. Lotke PA, Ecker ML. Influence of positioning of prosthesis in total knee replacement. J Bone Jt Surg. 1977;59(1):77–9.
56. Mahfouz M, Abdel Fatah EE, Bowers LS, Scuderi G. Three-dimensional morphology of the knee reveals ethnic differences. Clin Orthop Relat Res. 2012;470(1):172–85.
57. Dai Y, Scuderi GR, Bischoff JE, Bertin K, Tarabichi S, Rajgopal A. Anatomic tibial component design can increase tibial coverage and rotational alignment accuracy: a comparison of six contemporary designs. Knee Surg Sports Traumatol Arthrosc. 2014;22(12):2911–23.
58. Kozic N, Weber S, Büchler P, Lutz C, Reimers N, Ballester MÁ, Reyes M. Optimisation of orthopaedic implant design using statistical shape space analysis based on level sets. Med Image Anal. 2010;14(3):265–75.
59. Schulz AP, Reimers N, Wipf F, Vallotton M, Bonaretti S, Kozic N, Reyes M, Kienast BJ. Evidence based

development of a novel lateral fibula plate (VariAx fibula) using a real CT bone data based optimization process during device development. Open Orthop J. 2012;6:1.
60. Fitzpatrick CK, FitzPatrick DP, Auger DD. Size and shape of the resection surface geometry of the osteoarthritic knee in relation to total knee replacement design. Proc Inst Mech Eng H J Eng Med. 2008;222(6):923–32.
61. Rao C, Fitzpatrick CK, Rullkoetter PJ, Maletsky LP, Kim RH, Laz PJ. A statistical finite element model of the knee accounting for shape and alignment variability. Med Eng Phys. 2013;35(10):1450–6.
62. Koh YG, Nam JH, Chung HS, Kim HJ, Chun HJ, Kang KT. Gender differences in morphology exist in posterior condylar offsets of the knee in Korean population. Knee Surg Sports Traumatol Arthrosc. 2019;27(5):1628–34.
63. Yan M, Wang J, Wang Y, Zhang J, Yue B, Zeng Y. Gender-based differences in the dimensions of the femoral trochlea and condyles in the Chinese population: correlation to the risk of femoral component overhang. Knee. 2014;21(1):252–6.
64. Asseln M, Hänisch C, Schick F, Radermacher K. Gender differences in knee morphology and the prospects for implant design in total knee replacement. Knee. 2018;25(4):545–58.
65. Kim TK, Phillips M, Bhandari M, Watson J, Malhotra R. What differences in morphologic features of the knee exist among patients of various races? A systematic review. Clin Orthop Relat Res. 2017;475(1):170–82.
66. Fehring TK, Odum SM, Hughes J, Springer BD, Beaver WB Jr. Differences between the sexes in the anatomy of the anterior condyle of the knee. JBJS. 2009;91(10):2335–41.
67. Voleti PB, Stephenson JW, Lotke PA, Lee GC. No sex differences exist in posterior condylar offsets of the knee. Clin Orthop Relat Res. 2015;473(4):1425–31.
68. Conley S, Rosenberg A, Crowninshield R. The female knee: anatomic variations. JAAOS. 2007;15:S31–6.
69. Merchant AC, Arendt EA, Dye SF, Fredericson M, Grelsamer RP, Leadbetter WB, Post WR, Teitge RA. The female knee: anatomic variations and the female-specific total knee design. Clin Orthop Relat Res. 2008;466(12):3059–65.
70. Greene KA. Gender-specific design in total knee arthroplasty. J Arthroplasty. 2007;22(7):27–31.
71. Kim JM, Kim SB, Kim JM, Lee DH, Lee BS, Bin SI. Results of gender-specific total knee arthroplasty: comparative study with traditional implant in female patients. Knee Surg Relat Res. 2015;27(1):17.
72. Harvey WF, Niu J, Zhang Y, McCree PI, Felson DT, Nevitt M, Xu L, Aliabadi P, Hunter DJ. Knee alignment differences between Chinese and Caucasian subjects without osteoarthritis. Ann Rheum Dis. 2008;67(11):1524–8.
73. Choi KN, Gopinathan P, Han SH, Han CW. Morphometry of the proximal tibia to design the tibial component of total knee arthroplasty for the Korean population. Knee. 2007;14:295–300.
74. Moghtadaei M, Moghimi J, Shahhoseini G. Distal femur morphology of Iranian population and correlation with current prostheses. Iran Red Cresc Med J. 2016;18(2).
75. Baldwin JL, House CK. Anatomic dimensions of the patella measured during total knee arthroplasty. J Arthroplasty. 2005;20(2):250–7.
76. Huang AB, Luo X, Song CH, Zhang JY, Yang YQ, Yu JK. Comprehensive assessment of patellar morphology using computed tomography-based three-dimensional computer models. Knee. 2015;22(6):475–80.
77. Khattak MJ, Umer M, Davis ET, Habib M, Ahmed M. Lower-limb alignment and posterior tibial slope in Pakistanis: a radiographic study. J Orthop Surg. 2010;18(1):22–5.
78. Tang WM, Zhu YH, Chiu KY. Axial alignment of the lower extremity in Chinese adults. JBJS. 2000;82(11):1603.
79. Yau WP, Chiu KY, Tang WM, Ng TP. Coronal bowing of the femur and tibia in Chinese: its incidence and effects on total knee arthroplasty planning. J Orthop Surg. 2007;15(1):32–6.
80. Yip DK, Zhu YH, Chiu KY, Ng TP. Distal rotational alignment of the Chinese femur and its relevance in total knee arthroplasty. J Arthroplasty. 2004;19(5):613–9.
81. Chiu KY, Zhang SD, Zhang GH. Posterior slope of tibial plateau in Chinese. J Arthroplasty. 2000;15(2):224–7.
82. Dai Y, Seebeck J, Henderson AD, Bischoff JE. Influence of landmark and surgical variability on virtual assessment of total knee arthroplasty. Comput Methods Biomech Biomed Eng. 2014;17(10):1157–64.
83. Nakamura S, Morita Y, Ito H, Kuriyama S, Furu M, Matsuda S. Morphology of the proximal tibia at different levels of bone resection in Japanese knees. J Arthroplasty. 2015;30(12):2323–7.
84. Jin C, Song EK, Prakash J, Kim SK, Chan CK, Seon JK. How much does the anatomical tibial component improve the bony coverage in total knee arthroplasty? J Arthroplasty. 2017;32(6):1829–33.
85. Wernecke GC, Harris IA, Houang MT, Seeto BG, Chen DB, MacDessi SJ. Comparison of tibial bone coverage of 6 knee prostheses: a magnetic resonance imaging study with controlled rotation. J Orthop Surg. 2012;20(2):143–7.
86. Ho WP, Cheng CK, Liau JJ. Morphometrical measurements of resected surface of femurs in Chinese knees: correlation to the sizing of current femoral implants. Knee. 2006;13(1):12–4.

The Biomechanics of the Knee Joint

23

Peter Theobald, Samih Tarabichi, and Mohamed Elfekky

The field of biomechanics serves to link the seemingly disparate specialisms of biology and mechanical engineering. Application of biomechanics to the musculoskeletal system is documented from the early twentieth century, as innovators and pioneers began investigating synovial joint behavior with a view to understanding the link between its structure and function. These early attempts, to quantify mechanical parameters of the mammalian synovial joint, provided the building blocks for the current researchers to explore equivalent human parameters, in vivo. Biomechanical engineering now extends to describe a breadth of mechanical engineering applied to the human body, extending from the micro-scale cellular biomechanics, to the gross-scale sports biomechanics. The field is now overseen by international bodies including the American Society of Biomechanical Engineering, the European Society of Biomechanics, and the International Society of Biomechanics in Sport.

Comprising the patellofemoral, tibiofemoral, and, to a lesser extent, the tibiofibular joint, highly complex biomechanics underpin knee joint behavior and its ability to withstand intense mechanical loading for typically 60–70 years. Musculoskeletal biomechanics allows quantification of these multifactorial interactions, enabling an enhanced understanding of the relationship between the tissue structure and its function, for example, quantifying the friction when the fibrous/fluidic articular cartilage surfaces compress against and then translate past, one another. First, however, appreciating the fundamental biomechanics will assist in understanding these more complex behaviors.

P. Theobald
School of Engineering, Cardiff University, The Parade, Cardiff, UK

S. Tarabichi
Orthopaedic Surgery, Dubai, UAE

M. Elfekky (✉)
Hatta Hospital, Dubai Health Authority, Dubai, UAE

23.1 Musculoskeletal Mechanics

Musculoskeletal soft tissues are typically collagenous-based structures, bathed in a fluid of water and salts [1]. The interaction of these solid and fluid components produces a complex behavior when the tissues are compressed, extended, or subject to a shear loading [1]. This is caused by the fluid resisting the deformation of the loaded, fibrous network. Fluidic resistance increases when the rate of deformation is increased, meaning the tissue behavior depends on the loading speed. This is called viscoelastic behavior and, in soft tissues, allows greater deformation at lower strain rates. Key traits of any viscoelastic material are its creep and stress relaxation behavior. The former describes the extension of the soft tissue when exposed to a constant load, while the latter

M. Bozkurt, H. İ. Açar (eds.), *Clinical Anatomy of the Knee*,
https://doi.org/10.1007/978-3-030-57578-6_23

describes the reduction in load when maintaining a constant length. The precise stress relaxation and creep behavior of a tissue are governed by its structural and fluidic compositions [2, 3].

Articular cartilage is perhaps the single most important structure in maintaining synovial joint longevity, given that it is the ease of the femoral surface translating against the tibial surface that is common with the healthy joint. Wet articular cartilage comprises 20% collagen and 5% proteoglycans, which form strong structural networks that resist mechanical loading [3]. These networks, however, are unevenly distributed through the cartilage thickness, with the collagen fibers in the upper 10% aligned parallel to the cartilage surface. Collagen fiber bundles in the deepest 30% are radially orientated, while the middle 60% has the lowest fiber density. This tissue network is then bathed in water and salts, which comprises the remaining wet weight. It is the flow and frictional drag of this fluid that govern the articular cartilage's viscoelastic behavior [2].

The tendons and ligaments of the synovial joint serve to transmit motion and provide stability, respectively. In the knee, the tendons overlay the anterior and posterior aspects of the joint, transferring contractile force from the muscle groups of the upper leg and so generating motion. Ligaments constrain tibiofemoral translation and rotation, so providing the requisite skeletal rigidity (Fig. 23.1). Wet tendons and ligaments comprise approximately 30% collagen and 70% water. Collagen forms approximately 80% dry weight, with the vast majority comprising collagen type I [4]. Collagen fibrils are inherently wavy, or "crimped," providing the unloaded tendon with an innate flexibility. Several fibrils are grouped, embedded within an amorphous matrix and then wrapped in connective tissue, to form a collagen fiber. An accumulation of fibers forms a bundle, with the matrix enabling force transfer force along its length. Multiple bundles are then wrapped to produce a fascicle and multiple fascicles gathered to form the final tissue, which itself is wrapped in a loose, connective tissue to facilitate gliding against neighboring structures. While a tendon or ligament can be broadly considered analogous to a multi-strand rope, being stiff in tension though offering negligible resistance in compression, their behaviors are subtly different as the collagen fibers in the former tend to be better aligned, as they carry greater uniaxial

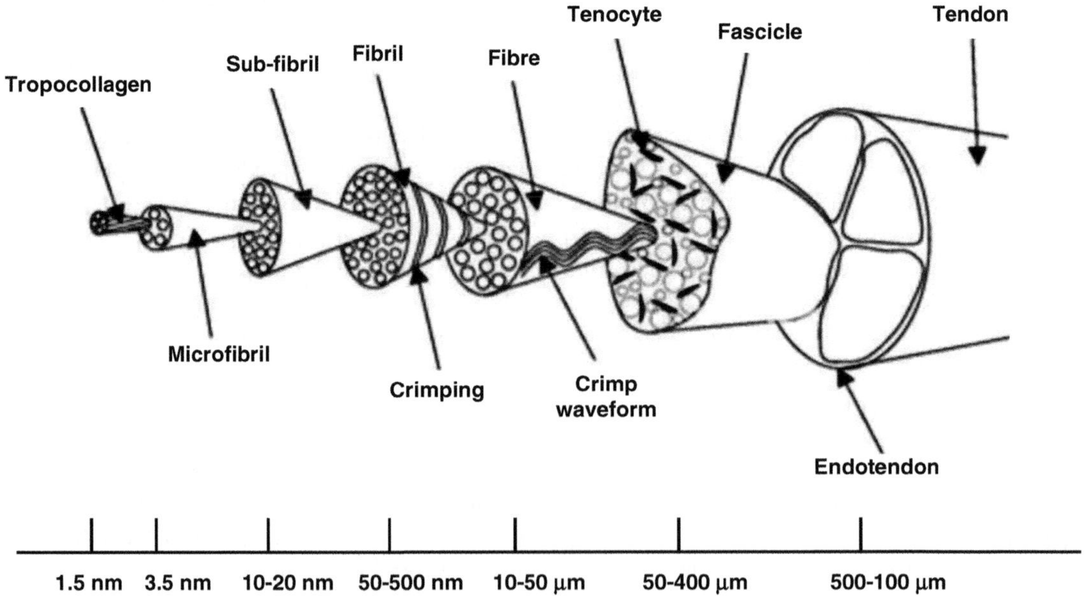

Fig. 23.1 Demonstrating how the collagen fibers are bundled together to form a tendon [4]

load [5]. Ligaments experience greater off-axis loads, meaning its fibers are less aligned.

23.2 Knee Kinematics

Knee functionality focusses on achieving and maintaining joint mobility and joint stability. Synovial joint movement is described by quantifying the extent of translation and rotation, or the range of motion, in the three orthogonal planes. Tibiofemoral rotation is the main source of knee joint mobility, achieving approximately 160° flexion extension. During the initial 30° flexion, internal tibial rotation also occurs, guided by the medial and lateral collateral ligaments. This represents the reversal of the "screw-home mechanism," with external rotation during the last 30° extension providing greater stability when the knee is fully extended [6]. This is achieved as the medial condyle is larger, generating rotation as it translates 5–10 mm across the tibial plateau. Lateral condylar translation is less pronounced. Full flexion achieves a posteriorly located tibio femoral contact region. In the frontal plane, 6–8° varus-valgus is evident in extension and is accompanied by 1–2 mm medio-lateral translation.

The need to constrain motion is critical to achieving an efficient gait. Excessive motion in the lower limb joints may cause greater medio-lateral or antero-posterior displacement in the overall center of mass, necessitating the generation of greater muscular force to retain balance. Knee motion is constrained by the two cruciate ligaments forming a "modified four-bar linkage mechanism," where the other two bars are represented by the bone that links the femoral insertion sites and the tibial insertion sites. During joint motion, the angles between the bars change, though stability is maintained as the four bars remain a broadly constant length [7]. Conventional linkages have four stiff bars. Marginal ligament extension enables the internal-external rotation, associated with the screw-home mechanism.

By balancing the conflicting demands of mobility and stability, the biomechanical properties of ligaments and tendons underpin the functionality of the natural joint. This behavior is typically described via a load–extension graph (Fig. 23.2). To plot these data independent of cross-sectional area and thereby enable comparison across tissues and individuals, the stress (the load divided by the area) and strain (the extension achieved, relative to the original length) are plotted to establish a correlation. Plots typically exhibit an initial toe region, where the crimped collagen fibrils achieve relatively great strain for a given stress. Additional extension then requires a greater load, due to the taut fibers providing greater resistance. During this phase, the stress–

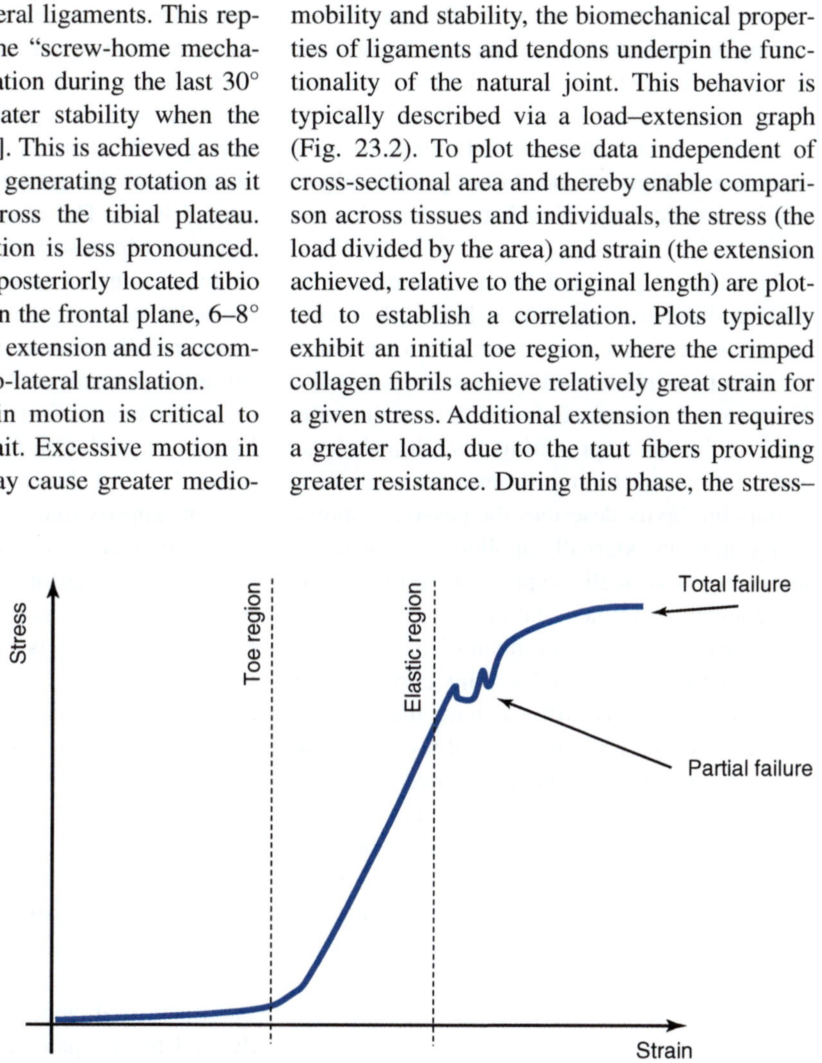

Fig. 23.2 The stress–strain relationship that is commonly observed when tensioning a tendon (author's drawing)

strain correlation will remain consistent, until such time as individual fibrils begin to fail. This failing will be represented by a reducing stress, caused by a relative reduction in the load required for extension. The stress will then continue to rise as greater load is required for ultimate failure, though distributed across fewer fibrils, until the tissue finally ruptures [8].

23.3 Knee Stability

The physical behavior and clinical observations of the healthy and pathological knee are often described by simplified terminology and subjective measurements, yet motion at the knee joint is a complex mechanical phenomenon. Terms such as instability and laxity are used interchangeably to describe a pathologic deviation from normal knee behavior; however, the definitions of these terms are ambiguous and do not provide objective information regarding knee kinematics and function.

Stability is provided by a combination of static and dynamic structures that work in concert to prevent excessive movement, or instability, which is inherent in various knee injuries. The anterior cruciate ligament is a main stabilizer of the knee, providing both translational and rotatory constraints. From a biomechanical standpoint, laxity describes the passive response of a joint to an externally applied force, whereas instability is typically expressed by the patient and constitutes a functional measure. Laxity and instability can be better quantified by understanding the stiffness of a joint complex [9]. Tissues with greater stiffness have the effect of limiting overall joint motion, although abnormally high tissue stiffness can over-constrain joint motion, as is possible in procedures such as extra-articular tenodesis. An increased laxity is not tantamount to knee instability nor an inferior outcome, as patients compensate differently and experience differing levels of disability.

23.4 Tibiofemoral Joint Kinematics and Forces

Static analysis can be used to estimate the compressive forces acting on the tibiofemoral and patellofemoral joints. This principle freezes a dynamic event, enabling analysis of a single time-point by considering only the predominant forces. Static analysis is typically used to investigate relative joint reaction forces (JRF), in this instance typically the tibiofemoral force, by calculating the resultant vector of the ground reaction force (GRF) and muscular force. This "inverse" method enables comparison of different every-day activities, or to track rehabilitation after joint replacement surgery. The GRF is measured using force plates, which may be embedded in the floor in a laboratory environment or used as a portable device for in situ testing. GRF represents the equal and opposite reaction of the ground when loaded by the human body. The muscular force can then be calculated by resolving the moments about the center of joint rotation (i.e., where the JRF acts through), enabling the JRF to be calculated by summing all forces to zero. When the knee is in full extension, there is no resultant moment nor muscular contraction, meaning that in double-leg stance, each tibiofemoral joint is exposed to one-half total body weight (the lower leg weight, approximately 5% BW, is ignored for simplicity). A static analysis of more complex scenarios, for example, walking, demonstrate that the JRF is greatest at the greatest knee flexion angle, which is approximately 20° when walking on the flat. Different peak flexion angles when ascending (~60° flexion) and descending stairs (~85° flexion) are examples of producing different JRFs during tasks that may seem similar.

23.5 Patellofemoral Joint Kinematics and Forces

Patellofemoral range of motion is predominantly defined by the patella's posterior anatomy and that of the underlying femoral condyles. The patella articulates with the lateral femoral ridge as it approaches full extension, ultimately fitting between the medial and lateral condyles. It glides into the intercondylar notch at full flexion. The patellofemoral JRF increases during flexion due to the increasing patella tendon force, approaching 4 × BW during stair climbing. The increased patella tendon force also increases the patellofemoral contact area, which serves to better dissipate the greater JRF and so moderate joint stress [10].

23.6 Joint Tribology

While mobility and stability are the key traits of the knee and wider synovial joint family, maintaining these functionalities over a lifetime is highly demanding. The longevity of the synovial joints is remarkable given the amount of translation and rotation, which risks degrading the surfaces as they move past one another. Further still, the musculoskeletal system is frequently exposed to loads multiple-times body weight, acting on very small contact areas between the articulating surfaces in the lower limbs. This can generate extreme contact pressures. The very slow cartilage repair process means that the synovial joint needs to be hard-wearing and resilient, as structural damage is likely to encourage further degradation.

The ability of the healthy articular cartilage to withstand this demanding mechanical environment has long been investigated, with studies from the early twentieth century, reporting the oozing of synovial fluid from the tissue surface when exposed to a compressive load [11]. It was, and is still, believed that this process is of critical importance to protecting the interacting surfaces as the femoral condyles glide over the tibia, minimizing articular cartilage wear. The challenge of explaining how this process works, however, remains unsolved, limiting the ability of biomechanical engineers to design replacement components. Biomechanical engineers have long been seeking solutions from their mechanical-based colleagues, who can mathematically predict the extent of lubrication between synthetic bearing surfaces, enabling design changes that leverage enhanced performance.

The synovial joint has a small volume of synovial fluid, produced via the synovium and retained within the joint by the synovial capsule. When there is no motion between the articulating surfaces, it is typical for the two articulating surfaces to be devoid of fluid, being either soaked into the articular cartilage or pooled at the lowest point in the joint, by gravity. Larger proteins are, however, adhered to the surface by chemical attraction, providing a constant source of "boundary" lubrication [12]. The extracellular matrix and proteoglycans that form the fibrous basis of the tissue retain synovial fluid near the tissue surface during periods of boundary lubrication such that, on joint loading, it is squeezed into the intra-articular space. This fluid is then drawn in between the two cartilage surfaces, seemingly enabling the femoral condyles to "aqua-plane" over the tibial plateau. This is currently the main theory as to how the synovial joint achieves such little wear and remains healthy for so long; however, this concept cannot be proven due to an inability to visualize synovial fluid via dynamic imaging.

Quite how fluid is drawn between the two opposing surfaces with the intra-articular space remains unknown, defying existing mechanical engineering theories. Mathematical and experimental analyses demonstrate that generation of a "fluid film" is dependent on bearing surface stiffness, smoothness, relative rotational velocity, and contact load. These rules mean that mechanical applications demanding low wear typically use metallic or ceramic bearing materials. When these concepts are replicated in prostheses, longevity rarely extends beyond 20 years and so highlights the natural tissue's remarkable efficacy.

Biomechanical engineers continue to try and understand how the natural joint is so well lubricated. Explants of articular cartilage have been slid against an array of standardized surfaces, enabling cross-comparison of articular cartilage frictional characteristics versus known materials. This has led to the adoption and modification of established mathematical equations to analytically describe cartilage behavior; however, no models are able to fully explain how a synovial joint can generate highly effective lubrication at very low translation speeds and relatively high compressive loads. Engineers have proposed methods of fluid film generation including the "squeeze film," "boosted," and "weeping" systems, to try and explain how optimal lubrication is achieved in sub-optimal conditions. One of the most recent ideas, of a micro-elastohydrodynamic lubrication system, proposes that the cartilage surface is flattened under loading, making it smoother and so more likely to generate a fluid film. Engineers have now created a mathematical solution that can be considered in the design of replacement joints [13].

23.7 Biomechanical Causes of Knee Degeneration

Osteoarthritis (OA) is the predominant outcome from knee degeneration, with lifestyle and genetics both underlying risk factors. OA is now recognized as a joint disease and is a common cause of disability in people over 65 years. It can lead to the development of bony spurs and changes to the tensile behavior of the ligaments and tendons, changing the articular cartilage's biomechanical environment. An atypical biomechanical environment will cause unequal distribution of loading through the knee, with increased unicompartmental loading thought to be one cause of accelerated joint degeneration [14].

The external knee adduction moment distributes 60–80% of total intrinsic knee compressive load to the medial tibiofemoral compartment, with the lower limb mechanical axis accounting for 50% variation, emphasizing the need for dynamic evaluation. Walking with larger knee adduction moments can increase medial compartment pressure and reduce medial joint space, although this does not appear to cause reduced cartilage thickness, potentially because of deformation in soft structures such as menisci.

While biomechanical factors are also likely to directly contribute to the articular cartilage pathogenesis, the effect of such variables on joint morphology remains unknown. Until recently, biomechanical studies typically considered cross-sectional, rather than longitudinal, knee OA, making differentiation of factors that cause, or result from the disease, difficult. Increased knee loading is known to increase bone mineral density, though little is understood about cartilage response to repetitive, altered load. There is, however, emerging evidence that cartilage volume will become a useful measure in OA pathogenesis, with studies needing to capture the response of human tissues such as hyaline cartilage, to controlled biomechanical variables. The relation between muscle weakness and knee OA is also becoming better understood. Longitudinal studies anecdotally report quadriceps weakness as a feature common to knee OA presentation and degeneration. Baseline knee extensor strength has been shown to be lower in women without radiographic knee OA at the initial examination who later developed OA changes, compared with unaffected women. It may be that weak quadriceps reduce the net extensor moment, which may help to counteract the lateral knee joint opening and medial compression that would occur if the knee adduction moment acted as an unopposed force. Value may be gained by investigating the association between quadriceps strength and the knee adduction moment during gait.

Knee joint laxity, defined as displacement or rotation of the tibia with respect to the femur, is another biomechanical variable argued to contribute to OA pathogenesis. Varus-valgus laxity is reportedly greater in the unaffected knees of patients with unilateral OA than in healthy control subjects, suggesting that knee joint laxity may predispose to disease. It has also been shown that varus and valgus alignment of the lower limb is associated with the progression of medial and lateral compartment knee OA, as determined by joint space narrowing and deterioration of physical function. Moreover, changes resulting from the relation between alignment and disease progression can be detected after only 18 months of observation. This suggests that over a relatively short intervention time frame, the correction of biomechanical variables in people with established knee OA may delay the progression of disease. Earlier results, however, suggested that women with reduced quadriceps strength have a greater risk of developing knee OA, meaning greater quadriceps baseline strength may be associated with increased OA progression in malaligned and lax knees. Although these results infer that strong quadriceps reduce the risk of developing knee OA, they also suggest that strong quadriceps are a risk factor for the progression of disease in people with malaligned and lax arthritic knees.

Longitudinal studies in normal subjects are still required to determine whether biomechanical variables, such as the knee adduction moment, predate the onset of OA or occur after presentation. Other studies in subjects with OA will be required to clarify the role of biomechanical variables in disease progression, to identify potentially modifiable factors to alter the course of the disease.

23.8 Biomechanics of Implant Design

Designing total joint replacement systems that exhibit both compliance with clinical need and longevity remains a technical challenge. Sir John Charnley was innovating initial joint replacement systems in the 1950s, selecting metallic surfaces to replicate the nuanced attributes of the articular cartilage. Seventy years thence, designers have trialed various systems—including those with stiffnesses more comparable to the soft tissues, though have reverted to adopting traditional engineering materials. Key to the design process is an ability to replicate the generation of the fluid film lubrication layer that separates the two articulating surfaces on translation, minimizing abrasion and so maximizing longevity. While fluid film generation in the healthy joint remains poorly understood, achieving this between two engineering materials requires a high sliding speed, although this typically necessitates an increased sliding distance and so wear. Materials are also required to be hard and smooth, minimizing the layer thickness required to achieve complete separation. Positive conditions are also achieved by having a relatively low contact load, though there is little that can be designed into new components to markedly change these parameters (i.e., this is predominantly a factor of body mass).

The main opportunity for improvement is to focus on achieving high-quality components, as only a small number of materials are available to designers, due to the need for biocompatibility. Efforts to increase the resilience of materials have been successful, with cross-linked ultra-high molecular weight polyethylene producing fewer wear particles than its predecessor. Wear particles are known to accelerate further wear, becoming sandwiched between the two surfaces, like grit. Other improvements have focused on enhancing the implant surface smoothness, to reduce film thickness requirements.

Emerging techniques relate to patient-specific systems, enabling engineers to best match the implant design to an individual's anatomy and so providing opportunity to minimize peak load; however, such systems are more expensive than those 'off the shelf', meaning that they are not yet commonplace. Systems that enable the use of commercially available arthroplasty, though fitted using patient-specific instrumentation, are becoming more popular, potentially offering a more favorable balance between increased longevity and increased cost.

References

1. Mow VC, Holmes MH, Lai WM. Fluid transport and mechanical properties of articular cartilage: a review. J Biomech. 1984;17(5):377–94.
2. Mow VC, et al. Biphasic indentation of articular cartilage—II. A numerical algorithm and an experimental study. J Biomech. 1989;22(8–9):853–61.
3. Mow VC, Ratcliffe A. Structure and function of articular cartilage and meniscus. In: Mow VC,

Hayes WC, editors. Basic orthopaedic biomechanics. Philadelphia: Lippincott-Raven; 1997.
4. Kastelic J, Galeski A, Baer E. The multicomposite structure of tendon. Connect Tissue Res. 1978;6(1):11–23.
5. Amiel D, et al. Tendons and ligaments: a morphological and biochemical comparison. J Orthop Res. 1984;1(3):257–65.
6. Matsumoto H, et al. Axis location of tibial rotation and its change with flexion angle. Clin Orthop Relat Res. 2000;371:178–82.
7. O'Connor JJ, et al. The geometry of the knee in the sagittal plane. Proc Inst Mech Eng H. 1989;203(4):223–33.
8. Butler DL, et al. Biomechanics of ligaments and tendons. Exerc Sport Sci Rev. 1978;6:125–81.
9. Fu FH, et al. Biomechanics of knee ligaments: basic concepts and clinical application. Instr Course Lect. 1994;43:137–48.
10. Reilly DT, Martens M. Experimental analysis of the quadriceps muscle force and patello-femoral joint reaction force for various activities. Acta Orthop Scand. 1972;43(2):126–37.
11. Jones ES. Joint lubrication. Lancet. 1934;223(5783):1426–7.
12. Swann DA, Mintz G. The isolation and properties of a second glycoprotein (LGP-II) from the articular lubricating fraction from bovine synovial fluid. Biochem J. 1979;179(3):465–71.
13. Jin ZM, Dowson D, Fisher J. Analysis of fluid film lubrication in artificial hip joint replacements with surfaces of high elastic modulus. Proc Inst Mech Eng H. 1997;211(3):247–56.
14. Setton LA, Elliott DM, Mow VC. Altered mechanics of cartilage with osteoarthritis: human osteoarthritis and an experimental model of joint degeneration. Osteoarthr Cartil. 1999;7(1):2–14.

GPSR Compliance

The European Union's (EU) General Product Safety Regulation (GPSR) is a set of rules that requires consumer products to be safe and our obligations to ensure this.

If you have any concerns about our products, you can contact us on ProductSafety@springernature.com

In case Publisher is established outside the EU, the EU authorized representative is:

Springer Nature Customer Service Center GmbH
Europaplatz 3
69115 Heidelberg, Germany

Batch number: 10731383

Printed by Printforce, the Netherlands